Environmental Mutagenesis

The HUMAN MOLECULAR GENETICS series

Series Advisors

D.N. Cooper, *Institute of Medical Genetics, University of Wales College of Medicine, Cardiff, UK*

S.E. Humphries, *Division of Cardiovascular Genetics, University College London Medical School, London, UK*

T. Strachan, *Department of Human Genetics, University of Newcastle-upon-Tyne, Newcastle-upon-Tyne, UK*

Human Gene Mutation
From Genotype to Phenotype
Functional Analysis of the Human Genome
Molecular Genetics of Cancer
Environmental Mutagenesis

Forthcoming titles

Human Genome Evolution
HLA and MHC: genes, molecules and function

SEVEN DAY LOAN

This book is to be returned on
or before the date stamped below

UNIVERSITY OF PLYMOUTH

PLYMOUTH LIBRARY

Tel: (01752) 232323
This book is subject to recall if required by another reader
Books may be renewed by phone
CHARGES WILL BE MADE FOR OVERDUE BOOKS

Environmental Mutagenesis

David H. Phillips and Stanley Venitt
Section of Molecular Carcinogenesis, The Haddow Laboratories, Institute of Cancer Research, Sutton, UK

© **BIOS Scientific Publishers Limited, 1995**

First published 1995

A CIP catalogue record for this book is available from the British Library.

ISBN 1 872748 19 8

BIOS Scientific Publishers Ltd
9 Newtec Place, Magdalen Road, Oxford OX4 1RE, UK.
Tel. +44 (0)1865 726286. Fax +44 (0)1865 246823

DISTRIBUTORS

Australia and New Zealand
    DA Information Services
    648 Whitehorse Road, Mitcham
    Victoria 3132

India
    Viva Books Private Limited
    4346/4C Ansari Road
    Daryaganj
    New Delhi 110002

Singapore and South East Asia
    Toppan Company (S) PTE Ltd
    38 Liu Fang Road, Jurong
    Singapore 2262

USA and Canada
    Books International Inc.
    PO Box 605, Herndon,
    VA 22070

Typeset by Touchpaper, Abingdon, UK.
Printed by Information Press Ltd, Eynsham, UK.

Contents

Contributors

Albertini, R.J. Genetics Laboratory, College of Medicine, University of Vermont, 32 North Prospect Street, Burlington, VT 05401, USA

Barrett, J.C. Environmental Carcinogenesis Program, Laboratory of Molecular Carcinogenesis, National Institutes of Environmental Health Sciences, MD C2-15, PO Box 12233, Research Triangle Park, NC 27709, USA

Bishop, J.B. Department of Health and Human Services, National Institutes of Health, National Institute of Environmental Health Sciences, PO Box 12233, Research Triangle Park, NC 27709, USA

de Boer, J. Centre for Environmental Health and the Department of Biology, University of Victoria, Victoria, British Columbia, V8W 2Y2, Canada

Clive, D. Glaxo Wellcome, 3030 Cornwallis Road, Research Triangle Park, NC 27709-2700, USA

Crespi, C.L. GENTEST Corporation, 6 Henshaw Street, Woburn, MA 01801, USA

Eastmond, D.A. Environmental Toxicology Graduate Program, Department of Entomology, University of California, Riverside, CA 92521, USA

Fahrig, R. Fraunhofer-Institut für Toxikologie und Aerosolforschung, Nikolai-Fuchs-Strasse 1, D-30625 Hannover, Germany

Farmer, P.B. MRC Toxicology Unit, Hodgkin Building, University of Leicester, PO Box 138, Lancaster Road, Leicester LE1 9HN, UK

Forster, R. Transgenic Research, C.I.T., Miserey, PO Box 563, 27005 Evreux Cedex, France

Glickman, B.W. Centre for Environmental Health and the Department of Biology, University of Victoria, Victoria, British Columbia, V8W 2Y2, Canada

Harnden, D.G. Paterson Institute for Cancer Research, Christie Hospital NHS Trust, Wilmslow Road, Manchester M20 9BX, UK

Heddle, J.A. Department of Biology, York University, Toronto, Ontario M3J 1P3, Canada

Kotturi, G. Centre for Environmental Health and the Department of Biology, University of Victoria, Victoria, British Columbia, V8W 2Y2, Canada

Kusser, W. Centre for Environmental Health and the Department of Biology, University of Victoria, Victoria, British Columbia, V8W 2Y2, Canada

Mirsalis, J.C. Toxicology Laboratory, SRI International, Menlo Park, CA 94025-3493, USA

O'Neill, J.P. Genetics Laboratory, College of Medicine, University of Vermont, 32 North Prospect Street, Burlington, VT 05401, USA

Parry, E.M. School of Biological Sciences, University of Wales at Swansea, Singleton Park, Swansea SA2 8PP, UK

Parry, J.M. School of Biological Sciences, University of Wales at Swansea, Singleton Park, Swansea SA2 8PP, UK

Phillips, D.H. Section of Molecular Carcinogenesis, The Haddow Laboratories, Institute of Cancer Research, Cotswold Road, Sutton, Surrey SM2 5NG, UK

Rupa, D.S. Environmental Toxicology Graduate Program, Department of Entomology, University of California, Riverside, CA 92521, USA

Smith, C.A.D. Department of Pathology, University of Edinburgh Medical School, Teviot Place, Edinburgh EH8 9AG, UK

Smith, G. Biomedical Research Centre, Ninewells Hospital and Medical School, Dundee DD1 9SY, UK

Tice, R.R. Integrated Laboratory Systems, PO Box 13501, Research Triangle Park, NC 27709, USA

Venitt, S. Section of Molecular Carcinogenesis, The Haddow Laboratories, Institute of Cancer Research, Cotswold Road, Sutton, Surrey SM2 5NG, UK

Witt, K.L. Oak Ridge Institute for Science and Education, Oak Ridge, TN 37831, USA

Wolf, C.R. Biomedical Research Centre, Ninewells Hospital and Medical School, Dundee DD1 9SY, UK

Zeiger, E. Environmental Toxicology Program, National Institute of Environmental Health Sciences, PO Box 12233, Research Triangle Park, NC 27709, USA

Abbreviations

AAF	acetylaminofluorene
AFB	aflatoxin B_1
AGT	alkyltransferase
Ah	aryl hydrocarbon
ARE	antioxidant response element
Arnt	aryl hydrocarbon nuclear translocator
ASTM	American Society of Testing and Materials
A-T	ataxia telangiectasia
BD	1, 3-butadiene
BHA	butylated hydroxy/anisole
BHT	butylated hydroxy/toluene
BP	benzo[a]pyrene
BPDE	benzo[a]pyrene diol-epoxide
BrdU	5-bromo-deoxyuridine
BS	Bloom's sydrome
BVDU	(E)-5-(2-bromovinyl)2'-deoxyuridine
CB	cytochalasin B
CDA	4-cyanodimethylaniline
CDCE	constant denaturing capillary electrophoresis
CE	cloning efficiency
CHO	Chinese hamster ovary
CI	chemical ionization
CIS-DDP	cis-diammine dichloroplatinum
CPD	cyclobutane pyrimidine dimer
CREST	calcinosis, Raynaud's phenomenon, oesophageal dismobility, telangiectasia, schlerodactyly
CS	Cockayne's syndrome
CYP	cytochromes P450
DAB	dimethylaminoazobenzene
DAP	2, 6-diaminopurine
DAPI	4, 6-diamidino-2-phenylindole
DAT	diaminotoluene
DELFIA	dissociation-enhanced lanthanide fluoroimmunoassay
DGGE	denaturing gradient gel electrophoresis
DMBA	7, 12-dimethylbenz[a]anthracene
DMN	dimethylnitrosamine
DNT	dinitrotoluene
EBV	Epstein–Barr virus
EDTA	ethylene diamine tetraacetic acid
EI	electron impact
ELISA	enzyme-linked immunosorbent assay
EM	extensive metabolizer
EMS	ethyl methanesulphonate

EndoIII	endonuclease III
ENU	N-ethyl-N-nitrosourea
EO	ethylene oxide
ES	embryonic stem (cells)
ESI	electrospray ionization
FA	Fanconi's anaemia
Fapy-G	2, 6-diamino-4-hydroxy-5-formamidopyridine
FDA	Food and Drug Administration (US)
FISH	fluorescence in situ hybridization
FLNS	fluorescence line-narrowing spectroscopy
GC	gas chromatography
GPA	glycophorin A
GST	glutathione S-transferase
Hb	haemoglobin
HGPRT	hypoxanthine-guanine phosphoribosyl tyransferase
HLA	human leukocyte antigen
HNPCC	hereditary non-polyposis colorectal cancer
HPB	4-hydroxyl-l-(3-pyridyl)-1-butanone
HPLC	high performance liquid chromatography
HPRT	hypoxanthine phosphoribosyltransferase
IAC	immunoaffinity chromatography
IARC	International Agency for Research on Cancer
IPCS	International Program on Chemical Safety (WHO)
IQ	2-amino-3-methylimidazo (4,5-f)quinoline
ISH	in situ hybridization
lcr	locus control region
LOH	loss of heterozygosity
LTR	long terminal repeat
MALDI	matrix-assisted laser desorption ionization
MDA	4,4'-methylenedianiline
MDR	multidrug resistance
MelQx	2-amino-3, 8-dimethylimidazo[4,5-f]quinoline
Mf	mutant frequency
MGMT	methylguanine DNA methyltransferase
MLA	mouse lymphoma assay
MLV	murine leukaemia virus
MMS	methyl methanesulphonate
MN	micronuclei
MNNG	N-methyl-N'-nitro-N-nitrosoguanidine
MNU	N-methyl-N-nitrosourea
MOCA	4-4'-methylene-bis (2-chloroaniline)
MPTP	1-methyl-4-phenyl-1,2,3,6-tetrahydropyridine
MS	mutator strain, mass spectroscopy
NAT	N-acetyl transferase
NCI	National Cancer Institute (US)
NDEA	N-nitrosodiethylamine
NDMA	N-nitrosodimethylamine
NER	nucleotide excision repair
NMRI	nuclear magnetic resonance imaging

NNK	4-(methylnitrosamino)-1-(3-pyridyl)-butanone
NNN	N'-nitrosonornicotine
NTP	National Toxicology Program (US)
NTS	non-transcribed strand
OR	oxidoreductase
PAH	polycyclic aromatic hydrocarbon
PCC	premature chromosome condensation
PCR	polymerase chain reaction
PD	Parkinson's disease
PEPCK	phosphoenolpyruvate carboxykinase
pfu	plaque-forming unit
P-gp	P-glycoprotein
PHA	phytohaemagglutinin
PhIP	2-amino-1-methyl-6-phenylimidazo[4,6-b]pyridine
PM	poor metabolizer
PYR	pyrene
RBC	red blood cell
RFLP	restriction fragment length polymorphism
RIA	radioimmunoassay
RIT	radioimmunotherapy
RSS	recombination signal sequences
RT	reverse transcriptase
SC	synaptonemal complex
SCE	sister chromatid exchange
SCG	single cell gel (assay)
SFS	synchronous fluorescence spectroscopy
SHE	Syrian hamster embryo
SIM	selective ion monitoring
SOD	superoxide dismutase
SS	stable strain
SSCP	single-stranded conformational polymorphism
SV	Simian virus
T-ALL	T-cell acute lymphoblastic leukaemia
TCDD	2,3,6,8-tetrachlorodibenzo-p-dioxin
TCR	T-cell receptor
TFT	trifluorothymidine
TG	6-thioguanine
TK	thymidine kinase
TLC	thin-layer chromatography
TPA	12-O-tetradecanoyl-phorbol-13-acetate
TS	transcribed strand
UDS	unscheduled DNA synthesis
Ung	uracil-N-glycosylase
Vf	variant frequency
WME	Williams Medium E
XP	xeroderma pigmentosum
XRE	xenobiotic response element
YAC	yeast artificial chromosome
YOYO-1	benzoxazolium-4-quinolinum, oxazole yellow dimer

Preface

Research in the last 50 years has established beyond doubt that the environment contains both natural and anthropogenic mutagens. The title of this book – *Environmental Mutagenesis* – carries the suggestion that there is a connection between exposure to environmental agents and an increase in the burden of mutations in the populations of organisms which receive that exposure. An increase in mutation also implies an increase in somatic mutation, leading to cancer in those organisms susceptible to this set of diseases. While we acknowledge that it is possible that the entire biosphere could be at risk of mutation by man-made chemicals, this book is determinedly anthropocentric and is devoted almost entirely to those methods that are designed to protect the human population from germ-line mutation and cancer.

The book aims to present a dispassionate view of the current state of knowledge of the origins, mechanism of action and biological significance of environmental mutagens. The contributions to this volume divide into three sections. In chapters 1–5, the current view of our understanding concerning environmental mutagens and carcinogens and their mechanisms of action are reviewed, together with the influence of host factors, determining the genetic diversity of the human population, on these processes. The middle section of the book, chapters 6–12, provides a critical survey of established procedures for determining the genotoxicity of an agent or chemical. The remaining chapters, 13–18, review the current state of development of new methods for screening chemicals, for elucidating the mechanism of mutagenesis in humans and for assessing the impact of environmental mutagens on the human genome.

David H. Phillips (*Sutton, UK*)
Stanley Venitt (*Sutton, UK*)

The importance of environmental mutagens in human carcinogenesis and germline mutation

Stanley Venitt and David H. Phillips

1.1 The history of environmental mutagenesis

The publication by Muller (1928) in which he showed that X-rays could induce heritable gene mutations in *Drosophila melanogaster* could be said to mark the beginning of the era of the study of environmental mutagenesis, since it gave investigators the opportunity to produce mutant organisms and study mutation under experimental conditions with an agent that did not require metabolism to exert its effects, and whose dosimetry could be strictly controlled. Eighteen years elapsed before the announcement by Auerbach and Robson (1946) of the first chemical mutagen. This was mustard gas [bis-(2-dichloroethyl) sulphide, also known as sulphur mustard], a difunctional alkylating agent. This horrifying vesicant (blistering agent), which is a human carcinogen, was used extensively in the Great War of 1914–18 and in Ethiopia in 1936, and was manufactured and stockpiled, but not used, by various combatants during World War II (IARC, 1975). Mustard gas is the archetypal environmental chemical mutagen. It still poses a threat to the human population, both military and civilian. It was used extensively in the Iran–Iraq war between 1980 and 1988, and artillery shells containing mustard gas, both in the waters where they were dumped and on former battlefields, present a problem in peace-time, especially for those who collect wartime memorabilia and those, including children, who stumble across such shells by accident (Ruhl *et al.*, 1994). Blistering, hair loss, immunosuppression, induction of chromosomal abnormalities, characteristic changes in the bone marrow, and delayed lethality due to leukopenia, are all typical of heavy exposure to X-rays and it was this 'radiomimetic' effect of mustard gas that stimulated interest in its possible mutagenicity. It also led to the recognition, by analogy with X-rays, that the related difunctional nitrogen mustards might be useful anti-cancer agents, and ultimately to the development, at the Chester Beatty Research Institute (now the Institute of Cancer Research), of clinically useful drugs such as melphalan and chlorambucil (Lawley, 1994b; Ross, 1953).

The detonation of the atomic bombs over Hiroshima and Nagasaki in 1945 was a defining moment for environmental mutageneticists. It was a spectacular demonstration of the ability of the human race to gamble with its germline (and germlines in general) on an unprecedented scale and was followed by a succession of atmospheric nuclear test explosions carried out in the knowledge that each explosion released large amounts of radioactivity and radionuclides into the environment. Robert Oppenheimer, a key player in the development of the atomic bomb, has been quoted as saying "We have sown the wind, and we shall reap the whirlwind" (a slight adaptation from the Bible, Hosea, chapter 8, verse 7). The whirlwind in the current context could be interpreted as an increase in the burden of genetic disease in the offspring of those exposed to the radioactive fallout at Hiroshima and Nagasaki. However, as described below, the whirlwind does yet not appear to have followed the wind, in that there has been no measurable increase in mutation in the children of the survivors of those atomic bombs.

The rise of modern genetics, the history of research into the nature of mutation and the role of mutation in human disease have been reviewed in detail by Cooper and Krawczak (1993). An account of the types, frequencies and mutation rates of genetic diseases in the human population is given in COM (1989). Lawley (1994a, b) has reviewed the history of how mutagenesis became linked to carcinogenesis.

1.2 The creation of a new chemical environment

A second stimulus to the study of and concern with environmental mutagenesis has been the growing awareness that a new chemical environment has been created in the last 200 years. Large-scale release of chemicals into the environment started with the industrial revolution in the mid-18th century, when steam power superseded other less convenient means of energy, such as people, draft animals, windmills and waterwheels. Since then, boiling water has been the primary source of power for modern industrial civilization. Carbon-based fuels such as peat, coal, oil and natural gas are burnt, releasing large quantities of complex chemical mixtures into the environment (air, soil and water). The by-products of burning and pyrolysing carbon-based fuels, and the fuels themselves, feed the petrochemical industry which provides feedstocks for production of plastics, paints and a host of industrial products, and for pharmaceuticals and cosmetics. Each activity releases waste chemicals into the environment. The products themselves are made into other products, or are consumed as medicines, food and feed additives, and once used are excreted, dumped, destroyed, burnt or recycled. Each of these processes results in a net environmental gain of chemicals, many of which are unknown in nature. Internal combustion engines and gas turbines also devour enormous quantities of carbon-based fuels and inevitably add to the burden of man-made environmental chemicals. Concomitant with and essential to these developments was the increased production and use of metals, ceramics and other inorganic materials such as asbestos. Smelting and refining metals consumes large amounts of energy (resulting in more carbon-based pollution) and produces both organic and inorganic wastes. These too find their way into the environment.

These developments have led to a rapid growth in the human population, extension of the life span and complex urbanized societies that have come to expect constant supplies of cheap food in great variety. This demand has been met by the industrialization of agriculture which depends on the use of high inputs of fertilizers and pesticides for intensive production of staple crops and heavy use of potent pharmaceutical products to support an otherwise unsustainably intense form of animal husbandry. Thus modern agriculture places another chemical burden on the environment and the human population in the form of runoff of chemicals to water and to soil, and pesticide and drug residues in the food.

Release of radioactive materials into the environment is another product of modern civilization. The major releases from tests of nuclear bombs are, it is hoped, a thing of the past. However, many countries have come to depend, to varying extents, on generation of electricity by nuclear fission (another way of boiling water). Under normal circumstances, this results, it is said, in the release of small quantities of radioactivity to the environment. However, large quantities of radioactive waste are produced. How to store or dispose of this waste, and how to deal with decommissioned nuclear power stations in the long term have not been worked out.

Universal adult literacy is a product of industrial societies, and brings with it (it is said) enlightenment and personal growth. But it produces a strong thirst for printing inks, dyes, paper and their means of distribution, and latterly, an increased demand for all the chemicals and energy necessary to manufacture computers and their media (for example, floppy discs, CD-ROMS). Recorded music is now an indispensible accompaniment to many people's lives and it too creates demands for products of the chemical industry.

The personal environment ('lifestyle') of large sections of the population has also changed as a consequence of industrialization and mass production. The single most important development in terms of personal environmental pollution was the invention of machines for the mass production of cigarettes. This took place in the mid-1850s and by the early part of the 20th century cigarette smoking was firmly established in northern Europe and the USA. Although cigarette consumption is now static or declining in these areas, growth in eastern Europe and Asia more than makes up for this contraction. For every cigarette less smoked in the developed world in 1985–92, three more cigarettes were smoked in China alone. The world smoked 5170 billion factory-made cigarettes in 1993. This figure is forecast to grow by 4.2% by 2000, with Asian volumes growing by 54.3% (Oram and Tomkins, 1995).

The now almost universal use of over-the-counter medicines and cosmetics is another route by which man-made chemicals could pose a threat to the personal environment, as is the zealous pursuit of personal hygiene which requires the use of enormous quantities of soaps, detergents and deodorants, which also threaten the general environment.

The extraordinary speed at which industrialization occurred in the 'first' and 'second' worlds (accelerated by two world wars) and the rapid growth of this process in the highly populated areas that constitute the 'third world' has left little time for quiet and reasoned contemplation of the deleterious effects that this chemical assault might have on the human population, the biosphere and indeed on the planet itself. Fortunately, such contemplation is now happening and, indeed, has become fashionable, to the extent that no political party worth its salt dares to present a manifesto that does not contain a reference to the 'environment'.

Of the possible long-term harmful consequences of this new chemical environment, mutation and cancer are of particular concern. In this context, 'mutation' is seen as deleterious germline mutations leading to impairment and disease in future generations, and 'cancer' is seen as somatic mutation in the present population and, of course, in succeeding generations suffering similar exposures. The 'environmental mutagenesis' movement is driven by the idea that we, the present generation, must take responsibility for protecting the existing gene pool, since it determines the form and function of all succeeding generations. Key elements in this enterprise are to identify those agents that could pose a threat to the gene pool by testing them for mutagenicity, and to determine whether exposure to those agents that are found to possess this property increases the risk of mutation or cancer. This is by no means an unambitious undertaking, since only a small fraction of the very large numbers of chemicals already known to be present in the environment have been tested, and some 500 new chemicals enter commerce each year (Maltoni and Selikoff, 1988). The arguments we have presented to defend the environmental mutagenesis industry apply not only to the human population but also to the biosphere in general. This aspect of environmental mutagenesis has not been addressed in detail in this book. However, a review of methods for monitoring environmental gentoxicants is given in MacGregor et al. (1994).

1.3 Environmental mutagenesis: fact or fantasy?

It could be argued that fears that man-made mutagens pose a serious threat to the germline of the human population and to the biosphere in general are unfounded. After all, the study of mutation in children of survivors of the atomic bombs dropped on Hiroshima and Nagasaki ("the single most expensive and extensive genetic undertaking on record") found no excess over matched controls for eight different indicators of genetic damage (Neel et al.,

1990). Moreover, there is no convincing evidence that smoking, a habit known to bathe the internal organs of smokers in a rich broth of mutagens and to cause cancer in most major organs, induces germ-cell mutations (Little and Vainio, 1994). Many regimes of cancer chemotherapy deliver high parenteral doses of mutagens to humans. Nevertheless, so far there is no evidence for a genetic effect in the offspring of patients treated with such agents (Mulvihill, 1990). It might be that human germ cells are more resistant to mutagenesis than are those of mice (the only other species subject to intensive and extensive study) or that such studies lack the power to detect small increases in mutation which would, nonetheless, be important to public health. Indeed, Neel *et al.* (1990) state that "we can scarcely doubt that some genetic damage occurred in consequence of that [atomic bomb] exposure" and "it may be confidently assumed that some mutations were induced".

In the same vein, Little and Vainio (1994) state that

"Thus, although no instance of genetic disease resulting from 'mutagenic lifestyles' has been unequivocally demonstrated in humans so far, the induction of heritable mutations has been demonstrated repeatedly in experimental mammals and a wide range of organisms. Current knowledge of genetics, mutagenesis, and genetically based diseases leaves no room for doubt that genetic disease can result from exposure to mutagens."

In the absence of data for induced mutagenesis *per se*, other manifestations of toxicity might inform the debate as to whether environmental agents can (or have) affected the human germline. There are persistent claims that the sperm count has declined over the last 50 years (Giwercman *et al.*, 1993), although these claims have been disputed (Bromwich *et al.*, 1994; Olsen *et al.*, 1995). If this decline is real there are as yet no clues as to whether the putative causative agent or agents are environmental or mutagenic. However, the increasing incidence, in Western countries, of congenital abnormalities of the male genital tract, such as cryptorchidism and hypospadias, and of testicular cancer (Giwercman *et al.*, 1993; Higginson *et al.*, 1992) may not be unrelated.

Nevertheless, as things stand, there is no evidence that three proven means of delivering mutagens to the human body (γ–radiation + a small component of neutrons; tobacco smoke, which contains representatives of most classes of chemical mutagens; and cancer chemotherapy) have actually caused an increase in human germline mutations. Moreover, there is no convincing evidence that *any* exogenous agent has increased the burden of human germline mutations over the existing 'spontaneous' or background level. In the light of this, why is there continuing concern about environmental mutagenesis?

That there is concern is shown in several ways.

(i) Mandatory screening of new chemicals for mutagenicity is now enshrined in legislation in most industrialized countries (see Sections 1.7 and 1.8).

(ii) There is a flourishing and growing literature on all aspects of mutation research, given renewed vigour by the advances in molecular biology and mapping the fine structure of genes.

(iii) The link between mutation and cancer is beyond doubt, and cancer is seen as a genetic disease of cells and of the cell cycle; in short, mutagens are likely to be carcinogens.

(iv) We are led to believe that grant-giving agencies do respond to expressions of public concern and they are still happy to provide resources for conducting research in environmental mutagenesis.

(v) There is a general feeling in the wider world that ionizing radiation, chemical pollution, vehicle exhausts, pesticides, food additives and all the other less attractive accompaniments of life in the late 20th century are dangerous to health. To what extent fear of germ-line mutation contributes to this feeling is difficult to determine, but there is little doubt that cancer is a major preoccupation.

It is often argued by those who accept that environmental pollution is a necessary but manageable by-product of affluence that such fears are inflamed by the general population's ignorance of the issues and fanned by the activities of 'environmentalists', some of whom are driven by politics rather than by science, and by cynical manipulation by the media. Efron's account of "how environmental politics controls what we know about cancer" (Efron, 1984), though pre-dating the explosion of knowledge that confirms cancer as a somatic genetic disorder, provides a stimulating portrayal of the hectic late 1970s when, in the USA, ill-founded scientific absolutism confronted the vested interests of the chemical industry. The debate continues, fuelled by recent revelations in the USA and in the UK of experiments with ionizing radiation on children and adults (Masood, 1995), and by the controversy surrounding the 'Gardner hypothesis' which claimed that the excess of childhood leukaemia around the nuclear re-processing plant in Sellafield, UK, was due to germline mutation in fathers occupationally exposed to radiation within the plant (see Chapter 4). Proctor (1995) provides a more up-to-date account of the argument about the relative contributions of man-made and natural carcinogens to the burden of human cancer. Pollution of the environment and the workplace with mutagens and carcinogens is a matter of public record, as are past attempts by Industry and Government to suppress information on such episodes and to mislead the public (Ashford, 1994). These activities do little to inspire public confidence in the ability or desire of governments to protect populations and the environment against genetic and carcinogenic hazards. Obsessive official secrecy is another obstacle to trust; it engenders conspiracy theories and prevents open scientific debate.

Despite the fact that we cannot answer with any confidence the question as to whether environmental mutagenesis is a fact or fantasy, either for the human population or for other species, the question cannot be ignored, if for no other reason than that public opinion, informed or otherwise, will not let the matter rest. However, there are defensible scientific arguments for maintaining and perhaps increasing the effort put into the study of environmental mutagenesis:

(i) There is clear evidence of exposure to mutagens, and exposure on a wide scale (*Table 1.1*).

(ii) There is the clear structural and physiological similarity between human germ cells and those of other species – most notably rodents, the only mammals for which extensive data are available. The same is broadly true of the processes that precede mutagenesis – absorption, distribution, metabolism and excretion – and the processes that can ameliorate or (in some circumstances) exacerbate mutagenesis – DNA replication, proof-reading and repair. It is prudent, therefore, to assume that human germ cells are mutable by chemical and physical agents.

(iii) There is a large body of evidence from studies of rodents that somatic-cell mutagens tend also to be germ-cell mutagens. Thus, an agent capable of inducing somatic mutations in humans should be regarded as a potential human germ-cell mutagen.

(iv) There is the question of how to regard the lack of evidence for human mutagenesis in populations known to have suffered substantial exposure to ionizing radiation, tobacco smoke or cancer chemotherapeutic drugs. None of these studies has taken full advantage of the newer techniques of molecular biology, such as the polymerase chain reaction, and in particular the ability to identify human genes and detect mutations in them. Unless and until the full might of molecular biology has been unleashed on the problem, it could be argued that the current, comforting view that the human genome is resistant to induced chemical or physical mutagenesis is wrong, and that the studies carried out to date are simply inadequate to answer the question.

Table 1.1. Sources of potential human exposure to mutagens

Type	Examples
Endogenous	Nitric oxide
	Oxygen free radicals
	Endogenously formed nitrosamines
Occupational	Petrochemical manufacture
	Iron and steel production
	Nuclear power production
Diet	Mutagens naturally present in food
	Mutagens generated during cooking
	Mutagens generated by food preservation
	Mutagens generated during food spoilage
Lifestyle	Tobacco and other 'recreational' drugs
	Exposure to sunlight
Medical	Cancer chemotherapy
	Psoralens + UV therapy for psoriasis
Radiation	'Natural' radiation – radon
	Medical exposure – diagnostic X-rays and radiotherapy
	Fallout from nuclear weapons testing
	Emissions from nuclear power stations (e.g. Chernobyl)
	Exposure to nuclear waste
Pollution	Industrial effluents
	By-products of water chlorination
	Motor-vehicle emissions
	Pesticides used in agriculture
	Burning carbon sources for fuel
	Waste incineration
Biological	Generation of mutagens as a consequence of chronic infection with viruses, bacteria or parasites

The deficiencies of current methods for detecting the induction of human germline mutation by ionizing radiation and by chemicals are discussed in detail by Sankaranarayanan (1994). He examines the impact of advances in knowledge on the molecular biology of human Mendelian diseases on the estimation of genetic risks of exposure to ionizing radiation and to chemical mutagens. In particular he considers:

"whether and to what extent naturally occurring Mendelian diseases can be used as a baseline for efforts in this area. Data on the molecular nature and mechanisms of origin of spontaneous mutations underlying naturally occurring Mendelian diseases and on radiation-induced mutations in experimental systems suggest that for ionizing radiation, naturally occurring Mendelian diseases may not constitute an entirely adequate frame of reference and that current risk estimates for this class of diseases are conservative; these estimates however provide a margin of safety in formulating radiation protection guidelines. Currently available data on mechanisms and specificities of action of chemical mutagens, molecular dosimetry, repair of chemically induced adducts in the DNA, adduct–mutation relationships, etc., permit the tentative conclusion that naturally occurring Mendelian diseases may provide a better baseline for genetic risk estimation for chemical mutagens than for ionizing radiation. With both ionizing radiation and chemical mutagens, the question of which Mendelian diseases are potentially inducible will become answerable in the near future when more molecular data on human genetic diseases become available. It is therefore essential that risk estimators keep abreast of advances in human genetics and integrate these into their conceptual framework. However, induced Mendelian diseases (especially the dominant ones which are of more immediate concern) are likely to represent a very small fraction of the adverse genetic effects of induced mutations. More attention therefore needs to be devoted to studies on the heterozygous effects of induced mutations."

Finally there is the fact that the present generation is the guardian of the genome – not only the human genome, but also the genomes of any species that happens to get in the way of the collective anthropogenic effluvium that has become an inevitable accompaniment to human existence. We owe it to posterity to protect these genomes from mutation by preventing mutagens entering the environment.

1.4 Human exposure to mutagens

Preoccupation with the possibility of global catastrophe brought about by man-made chemicals has fostered the notion that 'natural' things are good, harmless, wholesome and kind to the planet, whereas things deemed to be 'unnatural' – in other words, chemical, or artificial or synthetic – are characterized as being harmful and likely to lead, in the end, to disaster. In the world of environmental mutagenesis and carcinogenesis, such a judgement is easily refuted. Because there are no examples of human mutagens *per se*, our knowledge of exposure to potential mutagens is based on evaluation of those agents considered to be carcinogenic to humans. Most of these are mutagenic and man-made (*Table 1.2*). However, this may simply reflect the past bias in selecting agents to test and to evaluate. Doll and Peto (1981), in reviewing the causes and prevention of cancer, have argued persuasively that the proportion of cancers caused by what we have called the 'new chemical environment' (see Section 1.2) is likely to be low compared with the proportion of cancers attributable to tobacco, diet, infections and natural carcinogens; this view has been endorsed and reinforced more recently by Ames *et al.* (1995). It might be wondered, therefore, why there should be such a worry over the new chemical environment which some consider to be a relatively minor source of mutagens. The justification for such concern is that additional exposure (exogenous over endogenous) represents a net increase in the mutagenic burden and that the putative excess risk needs to be evaluated and, if possible, minimized. Furthermore, earlier concern exclusively for human health has given way in the last quarter of the 20th century to a less anthropocentric view of the hazards, whereby a mutagenic consequence for virtually any species is deemed to be detrimental to the ecology of the planet. However, from a primarily human perspective, the potential sources of exposure to mutagens are summarized in *Table 1.1.*

1.4.1 Food

Potent mutagens and carcinogens are present as natural products in fungi and green plants and some enter the human food chain as constituents of the food itself, or as contaminants or are formed during cooking. Diet is an important determinant of cancer risk, but precisely why is still a matter for debate (Rogers *et al.*, 1993).

Mutagens/carcinogens as food constituents. To what extent mutagens that occur as natural products in widely consumed foodstuffs contribute to those cancers linked with diet is impossible to determine in the present state of knowledge. For example, the flavonoid quercetin is present in a wide variety of fruits and vegetables. It is genotoxic to bacteria and mammalian cells *in vitro*. At doses very much higher than those in the normal human diet it induces tumours of the urinary tract in rats (Dunnick and Hailey, 1992). Clearly, quercetin has been identified as a genotoxic hazard, but what risk, if any, does it pose to humans when ingested as a normal component of a mixed diet, bearing in mind the consistent epidemiological evidence that a diet rich in fruit and vegetables protects against several human cancers such as those of lung, stomach and oesophagus (Block *et al.*, 1992)? Fruit and vegetables that contain quercetin also contain a variety of anti-mutagens and anti-carcinogens whose activity has been detected in *in vitro* and *in vivo* assays. These substances include anti-oxidants such as vitamins C and E and a plethora of non-nutrient constituents that inhibit mutagenesis and/or carcinogenesis by several different mechanisms (Wattenberg, 1992).

Food contaminants. Staple foodstuffs such as cereals and grains are frequently contaminated by fungi that produce highly toxic secondary

Table 1.2. The mutagenic activity of agents identified as being human carcinogens

Carcinogen	Mutagenicity in Salmonella	Cytogenetic effects in vivo
Organic compounds		
Aflatoxins	+	+
4-Aminobiphenyl	+	+
Analgesics containing phenacetin	+	+
Azathioprine	+	+
Benzene	−	+
Benzidine	+	+
Betel quid with tobacco	+	+
Bis(chloromethyl)ether and chloromethyl methyl ether	+	I
Chlorambucil	+	+
Chlornaphazine	+	+
Cyclophosphamide	+	+
Ethylene oxide	+	+
Melphalan	+	+
Methyl-CCNU[a]	+	+
MOPP[b] (and other combined therapies)	+	+
Mustard gas	+	+
Myleran	+	+
2-Naphthylamine	+	+
Tobacco products (smokeless)	+	+
Tobacco smoke	+	ND
Treosulphan	+	+
Vinyl chloride	+	+
Complex mixtures		
Alcoholic beverages	−	+
Coal-tar pitches	+	ND
Coal tars	+	ND
Mineral oils	+	ND
Salted fish, Chinese-style	+	ND
Shale oils	?	?
Soots	+	ND
Strong inorganic acid mists and vapours	−	ND
Wood dusts	?	?
Hormones		
Diethylstilboestrol	−	+
Oestrogen replacement therapy	ND	ND
Oestrogen, non-steroidal	ND	ND
Oestrogen, steroidal	ND	ND
Oral contraceptives, combined	ND	ND
Oral contraceptives, sequential	ND	ND
Metals		
Arsenic compounds	−	+
Beryllium and beryllium compounds	−	−
Cadmium and cadmium compounds	+	+
Chromium compounds (hexavalent)	+	+
Nickel and nickel compounds	−	ND

Table 1.2. The mutagenic activity of agents identified as being human carcinogens (continued)

Carcinogen	Mutagenicity in Salmonella	Cytogenetic effects in vivo
Fibres		
Asbestos	–	ND
Erionite	ND	ND
Talc containing asbestiform fibres	–	–
Other		
8-Methoxypsoralen + UV	+	ND
Solar radiation	+	+
Radon		–
Infection with *Schistosoma haematobium*	ND	ND
Infection with *Opisthorchis viverrini*	ND	ND
Infection with *Helicobacter pylori*	ND	ND
Infection with hepatitis B and C	ND	ND

Adapted and updated from Barrett (1992).

a 1-(2-Chloroethyl)-3-(4-methylcyclohexyl)-1-nitrosourea.

b Mustine hydrochloride, vincristine, procarbazine, prednisone.

c +, positive response; –, negative response; ?, some types showed activity; ND, no data; I, inconclusive data.

metabolites. Aflatoxins, produced by *Aspergillus* species, are probably the most notorious examples of natural but nasty genotoxic carcinogens. The fungi that produce them are ubiquitous in hot, humid parts of the world, and contaminate staples such as maize and groundnuts to an extent that results in lifelong dietary exposure of large human populations to these powerful mutagenic carcinogens. Naturally occurring mixtures of aflatoxins are categorized as being causally linked to hepatocellular carcinoma, one of the commonest of all human cancers (IARC, 1993).

Mutagens generated during cooking. The best studied mutagens produced by cooking are the heterocyclic amines, formed by pyrolysis of amino acids and proteins at temperatures that cause the food to brown (Wakabayashi *et al.*, 1992). They are extremely mutagenic in bacterial systems, possibly because the tester strains are uniquely efficient at activating them. Their high mutagenic activity is not reflected in such exceptionally high carcinogenic activity in rodents and primates, although they do induce tumours in a variety of organs. At present their impact, either carcinogenic or mutagenic, on human health is unknown. Other mutagens associated with cooking include nitrosamines formed in gas flames, polycyclic aromatic hydrocarbons formed in overcooked or barbecued food, and furans formed by heating sugars. In addition, mutagenic compounds are generated by heating fats.

1.4.2 Endogenous processes

DNA is an inherently unstable macromolecule and mutations can occur naturally (or 'spontaneously') through depurination, deamination and through errors in replication. Reactive oxygen species generated in normal endogenous metabolic processes also cause DNA damage and mutation (Lindahl, 1993). It has been argued by Ames and colleagues that it is this 'natural' source of mutagens that is primarily responsible for human cancer, heart disease and ageing (Ames *et al.*, 1995). Indeed, the complexicity of DNA repair and proof-reading processes that have evolved in both prokaryotic and eukaryotic organisms bears witness to the importance of error-free replication of DNA (Kunkel, 1992).

Other mechanisms by which mutagens are generated by endogenous processes have been postulated; for example lipid peroxidation may lead to the formation of mutagenic species; DNA adducts formed by one such species, malondialdehyde, have been detected in human DNA (see Chapter 18). Mutagenic nitrosamines can also be generated endogenously from secondary amines and nitrite, the latter being formed from nitrate (Tricker *et al.*, 1992).

Chronic infection with certain viruses, bacteria and parasites has been found to cause cancer in man or predispose to it (*Table 1.2*). It has been proposed that nitric oxide (NO), a short-lived chemical messenger produced by many different cell types for a variety of physiological functions, and which is found at elevated concentrations in chronic infections, could be a crucial element in the mechanism by which such infections cause cancer (Liu and Hotchkiss, 1995; Ohshima and Bartsch 1994). At least three mechanisms are suggested, based on evidence obtained from experiments conducted *in vitro* and in some cases from studies of humans. The mechanisms are: formation of mutagenic and carcinogenic *N*-nitroso compounds by reaction of NO with secondary amines; direct deamination of DNA bases; and direct oxidation of DNA after formation of peroxynitrite or hydroxy radicals.

1.4.3 Radiation

Because ionizing and ultraviolet radiation occur naturally, life has evolved in their presence – indeed it could be argued that evolution has been driven by the mutagenic effects of these physical mutagens. However, there is renewed concern regarding both, with the depletion of the ozone layer leading to a predicted increase in exposure to ultraviolet light, and the activities of the civil and military nuclear industries that result in increased exposure to ionizing radiation.

1.4.4 Tobacco

Tobacco smoke is a complex mixture of over 4000 chemicals. It is estimated to contain 400–500 gaseous components and the particulate phase more than 3500. At least 43 of these are known animal carcinogens and, of course, tobacco smoking is a well established cause of many human cancers, as well as cardiovascular and respiratory diseases. Tobacco smoke also contains many free-radical species with mutagenic potential. Among those chemicals present in tobacco smoke that are known, in other exposures, to be human carcinogens are 2-naphthylamine, 4-aminobiphenyl, benzene, arsenic, chromium, vinyl chloride, and possibly nickel, cadmium, acrylonitrile, benzo[a]pyrene and polonium-210 (IARC, 1986).

1.4.5 Man-made chemicals

As already discussed, man-made chemicals are the major preoccupation of genetic toxicologists and regulators for the obvious reason that they may represent the most preventable source of mutation. They encompass both manufactured chemicals and also by-products of human activity, such as traffic and urban pollution. Again using the analogy of carcinogenesis, many of the known human carcinogens are also mutagens (*Table 1.2*). Many of the industrial processes that are known to result in occupational cancer involve their workers in exposure to mutagens; for example, polycyclic aromatic hydrocarbons in aluminium, iron, steel and coke production, and aromatic amines in the rubber and dye-stuffs industries.

1.5 Somatic mutations and their causative agents

The mechanisms of carcinogenesis and mutagenesis appear to be inextricably linked. Mutation is a possible consequence of DNA damage, and DNA damage is also thought to be an early stage in the process by which the majority of chemical carcinogens initiate tumours. Mutations in several critical genes have been found in tumours. The first types of such genes to be identified were the proto-oncogenes, in which mutations at one of a few critical codons can produce an activated gene product that causes cell transformation (Bishop, 1991). In contrast, tumour suppressor genes encode proteins essential for control of cell growth; when these genes are inactivated by loss or mutation, aberrant growth can result (Marx, 1993). The tumour suppressor gene *p53* has been found to be mutated in about half of all human tumours (Greenblatt *et al.*, 1994). The protein that it encodes is a transcription factor that provides a cell-cycle checkpoint, preventing cells with unrepaired DNA damage from entering S phase and stimulating them to undergo apoptosis (programmed cell death). Inactivation of the p53 protein by deletion or loss allows survival of cells that carry unrepaired DNA damage, leading ultimately to gene amplification, aneuploidy and other chromosomal aberrations (Smith and Fornace, 1995). Germline mutations in the *p53* gene have been detected in families afflicted by the Li–Fraumeni syndrome – an autosomal dominant syndrome which confers a high risk of diverse cancers at many sites (Malkin *et al.*, 1990; Srivastava *et al.*, 1990). In the Li–Fraumeni syndrome, sufferers have a germline mutation in one copy of the *p53* gene, and thus the probability that the remaining copy becomes inactivated by chromosome loss or by mitotic recombination is considerably greater than in individuals with two functional copies of the gene (Vogelstein, 1990).

A third class of critical gene is involved in maintenance of the integrity of the genome. Hereditary non-polyposis colon cancer (HNPCC) is the result of germ line mutations in genes that encode proteins involved in the process of detecting DNA replication mismatches and correcting them (Eshleman and Markowitz, 1995). Failure of this mechanism generates a 'mutator phenotype' by which the probability of mutations occurring from replication errors is greatly enhanced and the affected individuals accumulate somatic mutations that lead to tumours.

On the basis of the *p53* mutation spectra seen with different human tumours, conclusions have been drawn about the relative importance of exogenous mutagens and endogenous processes in causing these mutations. Mutations occurring at CpG dinucleotides, in which the cytosine is biomethylated, can occur through the natural process of deamination whereby 5-methylcytosine is converted to thymine, resulting in C→T and G→A transitions. Such mutations predominate in germline mutations in the haemophilia B gene, and in germline mutations in the *p53* gene, strongly suggesting that C→T transitions result from endogenous processes (Biggs *et al.*, 1993; Greenblatt *et al.*, 1994). Such mutations in *p53* predominate in colorectal cancer. With lung cancer, however, G→T transversions in the *p53* gene predominate, consistent with mutation being caused by the mutagenic carcinogens present in tobacco smoke (Biggs *et al.*, 1993; Greenblatt *et al.*, 1994). Breast cancer is an interesting case, in which the spectrum of mutations is intermediate between that for colorectal cancer and that for lung cancer. This implies that an environmental agent or agents could be responsible for a proportion of human breast cancer (Biggs *et al.*, 1993). The studies demonstrating changing breast cancer incidences among migrant populations support this possibility (Higginson *et al.*, 1992). Many environmental carcinogens induce mammary tumours in experimental animals, including PAHs, nitro-PAHs, and heterocyclic amines, yet classical epidemiological studies have yet to implicate an initiating (i.e. genotoxic) agent for human breast cancer.

1.6 Genotoxicity

Because some short-term screening tests for mutagens and carcinogens employ endpoints – for example, DNA repair and chromosomal damage – which do not measure mutation *per se* the term 'genotoxicity', rather than mutagenicity, is now generally used when referring to the biological property detected in such tests. 'Genotoxicity' is a useful term whose precise definition is elusive. Most workers in the field know what they mean when they use the term, but attempts to come up with a universally acceptable definition have failed. For example, it has been defined as:

> "any deleterious change in the genetic material regardless of the mechanism by which the change is induced" (D'Arcy and Harron, 1993),

or:

"a rather loose term that in a broad sense may refer to the property of a substance as being harmful to the genetic material. [...] it is used used in a narrow sense to refer to the ability of a substance to react with DNA either directly or after metabolic activation" (COM, 1989).

A Working Group convened by IARC (Vainio *et al.*, 1992) clearly struggled to produce a definition and came up with the following:

"Genotoxicity
The generalization that (organic) chemical carcinogens are metabolized, or spontaneously degrade, to produce reactive electrophilic intermediates which become bound to cellular macromolecules, and in particular to DNA, has influenced much subsequent research [...] – in particular, the development of 'short-term' tests for carcinogenic activity [...]. From the present perspective of carcinogen evaluation, genotoxicity is indicative of initiating activity. Such data are currently sought, however, as a primary indicator of chemical reactivity within biological processes, being relevant to mutagenesis, carcinogenesis and teratogenesis ... ."

However, Ashby (1995) has pointed out that Druckrey (who probably coined the term 'genotoxic') concluded, after extensive study of the relationships between mutagenicity, carcinogenicity and teratogenicity, that:

"carcinogenesis undoubtedly is far more complex than mutagenesis, and any generalization as to a 'mutation theory of cancer' cannot be considered as satisfactory ... on the other hand, the existence of 'genotoxic' substances is beyond dispute."

Druckrey's definition of 'genotoxic' is:

"any agent which, by virtue of its physical or chemical properties, can induce or produce heritable changes in those parts of the genetic apparatus that exercise homeostatic control over somatic cells, thereby determining their malignant transformation" (Druckrey, 1973).

1.7 The rise and rise of environmental mutagenesis

Environmental mutagenesis reached a wider scientific audience when 'mutation research' became a subject in its own right, with its own journal (*Mutation Research*, founded in 1964) and when geneticists working on the effects of ionizing radiation and chemicals founded the Environmental Mutagen Society in the USA in 1969. This was soon followed by a European Branch in 1970 (now known as the European Environmental Mutagen Society) and there are now Environmental Mutagen Societies throughout the world. The formation of the International Commission for Protection against Environmental Mutagens and Carcinogens (ICPEMC) in 1977 gave environmental mutageneticists a global platform from which to influence opinion on the need to safeguard the environment against mutagenic and carcinogenic cataclysm. The activities of these organizations and the extensive and rapidly growing scientific literature on environmental mutagenesis awoke governments to the potential harm that mutagens could do to the human population and to the biosphere. The need to take steps to recognize and regulate environmental mutagens was given extra urgency by the accumulating data that suggested that mutagenicity was a reasonable predictor of carcinogenicity. Thus it was that in the early 1980s public health and environmental agencies in several industrialized countries added mutagenicity to the list of toxic properties to be evaluated before chemicals such as food and feed additives, medicines and industrial could enter commerce. A list of published guidelines is given by Kirkland (1994). This demand was made practicable by the advent of simple, short-term tests that could detect the genetic toxicity of compounds within days or weeks (see Chapters 6, 7, 11 and 12).

1.8 Harmonization of screening for genetic toxicity

Screening of chemicals for genotoxicity, under statutory control, is now accepted as a routine component of toxicological evaluation, and measures to standardize and harmonize protocols for performing such tests throughout the industrialized world are now at an advanced stage under the auspices of, for example, the Organization for Economic Co-operation and Development (OECD), the European Union, the Japanese Ministry of Health and Welfare and the Food and Drug Adminstration (FDA) in the USA (see, for example, D'Arcy and Harron, 1993; Galloway, 1994). Harmonization operates at two levels. The first level, which is covered, for example, by the OECD, dictates general agreement as to how each of the various tests should be performed in order to avoid the numerous pitfalls of experimental design, technique and interpretation known to the screening community (see Galloway, 1994; UKEMS, 1989, 1990, 1993). Once these OECD protocols have been agreed they will be embodied in the statutory regulations of the 26 signatories to the OECD. The second level of harmonization is at the level of strategy – the way in which test batteries are deployed. Guidelines will be agreed that will govern not only minimum protocols but also what tests should be used, in what order, and under what circumstances. In the immediate future this second level of harmonization will apply only to pharmaceutical products (medicines) for use in humans (D'Arcy and Harron, 1993). When the strategy and protocols have been agreed (probably by 1996), the final document will form the basis of legislation by the participants in the process (EU, Japan and USA).

1.9 Hazard identification, risk assessment and risk management

The use of short-term screening tests for detecting mutagens and carcinogens is essentially a method for detecting *hazards*. Showing that a substance is a potent genotoxin in a variety of different organisms and cell lines *in vitro* provides evidence of a qualitative property, namely genotoxicity. Short-term tests can also be used for detecting the presence of genotoxins in the environment and in human populations (MacGregor *et al.*, 1994). Even before submitting chemicals to such tests, some information may be gained by examining the molecule for 'structural alerts' to its potential carcinogenicity (Tennant and Ashby, 1991). Such an approach has been elaborated to provide (it is claimed) quantitative predictions of carcinogenic hazard using computerized expert systems in an activity known as quantitative structure–activity relationships (QSAR; Parry, 1994).

Estimating the extent to which a compound or exposure would pose a *risk* of germ line or somatic mutation to the human population is a much more complicated undertaking that requires quantitative data from a variety of sources (see Section 1.3 and Sankaranarayanan, 1994). The problems of hazard identification, risk assessment and risk management, in human populations and in the biota are reviewed extensively in Brusick (1994). Information on numbers of people at risk, and their exposure to and absorption of the suspect substance is essential in any programme of risk assessment and management. The risk is zero if there is no exposed population. If there is evidence of human exposure, the number of people at risk and the nature of the exposure must be taken into account when attempting to assess the risk of germ line or somatic mutation. Studies conducted in mammals *in vivo* can be undertaken to produce dose–reponse curves for somatic or germ line mutation, and data on absorption, distribution, metabolism, excretion and DNA-binding. Such studies also provide valuable clues to mechanism, which must also be taken into account in risk evaluation. Some of these studies (for example, biomonitoring using white blood cells and body fluids) can also be undertaken in human populations or in human volunteers. It would also be useful to have validated models by which quantitative data obtained from studies in animals could be applied to human populations. Such models have yet to be developed to a state universally acceptable to the scientific community and to regulatory agencies.

References

Ames BN, Gold LS, Willett WC. (1995) The causes and prevention of cancer. *Proc. Natl Acad. Sci. USA* **92:** 5258–5265.

Ashby J. (1995) Druckrey's definition of genotoxic. *Mutat. Res.* **329:** 225.

Ashford NA. (1994) Monitoring the worker and the community for chemical exposure and disease: legal and ethical considerations in the US. *Clin. Chem.* **40:** 1426–1437.

Auerbach C, Robson JM. (1946) Chemical production of mutations. *Nature* **157:** 302.

Barrett JC. (1992) Mechanism of action of known human carcinogens. In: IARC Scientific Publications No. 116, *Mechanisms of Carcinogenesis in Risk Identification* (eds H Vainio, PN Magee, DB McGregor, AJ McMichael). International Agency for Research on Cancer, Lyon, pp. 115–134.

Biggs PJ, Warren W, Venitt S, Stratton MR. (1993) Does a genotoxic carcinogen contribute to human breast cancer? The value of mutational spectra in unravelling the aetiology of cancer. *Mutagenesis* **8:** 275–283.

Bishop JM. (1991) Molecular themes in oncogenesis. *Cell* **64**: 235–248.

Block G, Patterson B, Subar A. (1992) Fruit, vegetables and cancer prevention: a review of the epidemiological literature. *Nutr. Cancer* **18**: 1–29.

Bromwich P, Cohen J, Stewart I, Walker A. (1994) Decline in sperm counts: an artefact of changed reference range of "normal"? *Br. Med. J.* **309**: 19–22.

Brusick DJ. (ed.) (1994) *Methods for Genetic Risk Assessment.* Lewis Publishers, Boca Raton, FL.

COM (Committee on Mutagenicity of Chemicals in Food, Consumer Products and the Environment) (1989) *Report on Health and Social Subjects 35. Guidelines for the Testing of Chemicals for Mutagenicity.* Her Majesty's Stationery Office, London, pp. 34–64.

Cooper DN, Krawczak M. (1993) *Human Gene Mutation.* BIOS Scientific Publishers, Oxford.

D'Arcy PF, Harron DWG. (eds) (1993) *Proceedings of the 2nd International Conference on Harmonization, Orlando 1993.* Greystone Books Ltd, Antrim.

Doll R, Peto R. (1981) The causes of cancer: quantitative estimates of avoidable risks of cancer in the United States today. *J. Natl Cancer Inst.* **66**: 1191–1308.

Druckrey H. (1973) Specific carcinogenic and teratogenic effects of indirect alkylating methyl and ethyl compounds, and their dependency on stages of ontogenic developments. *Xenobiotica* **3**: 271–303.

Dunnick JK, Hailey JR. (1992) Toxicity and carcinogenicity of quercetin, a natural component of foods. *Fund. Appl. Toxicol.* **19**: 423–431.

Efron E. (1984) *The Apocalyptics.* Simon and Schuster, New York.

Eshleman JR, Markowitz SD. (1995) Microsatellite instability in inherited and sporadic neoplasms. *Curr. Opin. Oncol.* **7**: 83–89.

Galloway SM. (ed.) (1994) Report of the International Workshop on standardisation of genotoxicity test procedures. *Mutat. Res.* **312**: 322.

Giwercman A, Carlsen E, Keiding N, Skakkebaek NE. (1993) Evidence for increasing incidence of abnormalities of the human testis: a review. *Environ. Hlth Perspect.* **101** (Suppl 2): 65-71.

Greenblatt MS, Bennett WP, Hollstein M, Harris CC. (1994) Mutations in the p53 tumor suppressor gene: clues to cancer etiology and molecular pathogenesis. *Cancer Res.* **54**: 4855–4878.

Higginson J, Muir CS, Muñoz N. (1992) *Human Cancer: Epidemiology and Environmental Causes.* Cambridge University Press, Cambridge, UK.

IARC. (1975) In: IARC Monographs on the Evaluation of the Carcinogenic Risk of Chemicals to Man, Vol. 9. *Some Aziridines. N-, S- and O-mustards and Selenium.* International Agency for Research on Cancer, Lyon, pp. 181–192.

IARC. (1986) IARC Monographs on the Evaluation of Carcinogenic Risks to Humans, *Tobacco Smoking.* Vol. 38. International Agency for Research on Cancer, Lyon.

IARC. (1993) Aflatoxins. In: IARC Monographs on the Evaluation of Carcinogenic Risks to Humans. Some Naturally Occurring Substances, *Food Items and Constituents: Heterocyclic Aromatic Amines and Mycotoxins.* Vol. 56. International Agency for Research on Cancer, Lyon, pp. 245-395.

Kirkland DJ. (1994) Preface. *Mutat. Res.* **312**: 195–199.

Kunkel TA. (1992) DNA replication fidelity. *J. Biol. Chem.* **267**: 18251–18254.

Lawley PD. (1994a) From fluorescence spectra to mutational spectra, a historical overview of DNA-reactive compounds. In: *DNA Adducts: Identification and Biological Significance* (eds K Hemminki, A Dipple, DEG Shuker, FF Kadlubar, D Segerback, H Bartsch). International Agency for Research on Cancer, Lyon, pp. 3–22.

Lawley PD. (1994b) Historical origins of current concepts of carcinogenesis. *Adv. Cancer Res.* **65**: 17–111.

Lindahl T. (1993) Instability and decay of the primary structure of DNA. *Nature* **362**: 709–715.

Little J, Vainio H. (1994) Mutagenic lifestyles? A review of evidence of associations between germ-cell mutations in humans and smoking, alcohol consumption and use of 'recreational' drugs. *Mutat. Res.* **313**: 131–151.

Liu RH, Hotchkiss JH. (1995) Potential genotoxicity of chronically elevated nitric oxide: a review. *Mutat Res.* **339:** 73–89.

MacGregor JT, Claxton LD, Lewtas J, Jensen R, Lower WR, Pesch GG. (1994) Monitoring environmental genotoxicants. In: *Methods for Genetic Risk Assessment* (ed. DJ Brusick). Lewis Publishers, Boca Raton, FL, pp. 171–243.

Malkin D, Li FP, Strong LC *et al.* (1990) Germ-line p53 mutations in a familial syndrome of breast cancer, sarcomas, and other neoplasms. *Science* **250:** 1233–1238.

Maltoni C, Selikoff IJ. (1988) Preface. Living in a Chemical World: Occupational and Environmental Significance of Industrial Carcinogens. *Ann. NY Acad. Sci.* **534:** xv–xvi.

Marx J. (1993) Learning how to suppress cancer. *Science* **261:** 1385–1387.

Masood E. (1995) MRC rejects call for radiation tests enquiry. *Nature* **376:** 107.

Muller HJ. (1928) Production of mutations by X-rays. *Proc. Natl Acad. Sci. USA* **14:** 714–726.

Mulvihill JJ. (1990) Sentinel and other mutational effects in offspring of cancer survivors. *Prog. Clin. Biol. Res.* **340:** 179–186.

Neel JV, Schull WJ, Awa AA, Satoh C, Kato H, Otake M, Yoshimoto Y. (1990) The children of parents exposed to atomic bombs: estimates of the genetic doubling dose of radiation for humans. *Am. J. Hum. Genet.* **46:** 1053–1072.

Ohshima H, Bartsch H. (1994) Chronic infections and inflammatory processes as cancer risk factors: possible role of nitric oxide in carcinogenesis. *Mutat. Res.* **305:** 253–264.

Olsen GW, Bodner KM, Ramlow JM, Ross CE, Lipshultz LI. (1995) Have sperm counts been reduced 50% in 50 years? A statistical model revisited. *Fertil. Steril.* **63:** 887–893.

Oram R, Tomkins R. (1995) FT Guide to the tobacco business. *Financial Times,* 6 March.

Parry JM. (1994) Detecting and predicting the activity of rodent carcinogens. *Mutagenesis* **9:** 3–5.

Proctor RN. (1995) *Cancer Wars. How Politics Shapes What We Know and Don't Know About Cancer.* BasicBooks, New York.

Rogers AE, Zeisel SH, Groopman J. (1993) Diet and carcinogenesis. *Carcinogenesis* **14:** 2205–2217.

Ross WCJ. (1953) The chemistry of cytotoxic alkylating agents. *Adv. Cancer Res.* **1:** 397–449.

Ruhl CM, Park SJ, Danisa O, Morgan RF, Papirmeister B, Sidell FR, Edlich RF, Anthony LS, Himel HN. (1994) A serious skin sulfur mustard burn from an artillery shell. *J. Emerg. Med.* **12:** 159–166.

Sankaranarayanan K. (1994) Estimation of genetic risks of exposure to chemical mutagens: relevance of data on spontaneous mutations and experience with ionizing radiation. *Mutat. Res.* **304:** 139–158.

Smith ML, Fornace AJ. (1995) Genomic instability and the role of *p53* mutations in cancer cells. *Curr. Opin. Oncol.* **7:** 69–75.

Srivastava S, Zou Z, Pirollo K, Blattner W, Chang EH. (1990) Germ-line transmission of a mutated *p53* gene in a cancer-prone family with Li–Fraumeni syndrome. *Nature* **348:** 747–749.

Tennant RW, Ashby J. (1991) Classification according to chemical structure, mutagenicity to Salmonella and level of carcinogenicity of a further 39 chemicals tested for carcinogenicity by the US National Toxicology Program. *Mutat. Res.* **257:** 209–227.

Tricker AR, Pfundstein B, Kalble T, Preussman R. (1992) Secondary amine precursors to nitrosamines in human saliva, gastric juice, blood, urine and faeces. *Carcinogenesis* **13:** 563–568.

UKEMS. (1989) *Statistical Evaluation of Mutagenicity Test Data: UKEMS Sub-Committee on Guidelines for Mutagenicity Testing: Report: Part III* (ed. DJ Kirkland). Cambridge University Press, Cambridge, UK.

UKEMS. (1990) *Basic Mutagenicity Tests: UKEMS Recommended Procedures: UKEMS Sub-Committee on Guidelines for Mutagenicity Testing. Report. Part I revised* (ed. DJ Kirkland). Cambridge University Press, Cambridge, UK.

UKEMS. (1993) *Supplementary Mutagenicity Tests: UKEMS Recommended Procedures: UKEMS Sub-Committee on Guidelines for Mutagenicity Testing. Report. Part II revised* (eds DJ Kirkland, M Fox). Cambridge University Press, Cambridge, UK.

Vainio H, Magee PN, McGregor DB, McMichael AJ. (eds) (1992) Introduction. IARC
  Scientific Publications, no. 116, *Mechanisms of Carcinogenesis in Risk Identification.*
  International Agency for Research on Cancer, Lyon, p. 3.
Vogelstein B. (1990) A deadly inheritance. *Nature* **348:** 681–682.
Wakabayashi K, Nagao M, Esumi H, Sugimura T. (1992) Food-derived mutagens and
  carcinogens. *Cancer Res.* **52** (Suppl.): 2092s–2098s.
Wattenberg LW. (1992) Inhibition of carcinogenesis by minor dietary constituents. *Cancer Res.*
  52 (Suppl.): 2085s–2091s.

Role of mutagenesis and mitogenesis in carcinogenesis

J. Carl Barrett

2.1 Introduction

Carcinogenesis is a complex, multistep process, and cancer usually requires several decades to develop in humans. Although most cancers are derived from a single cell (i.e. are clonal in origin), the transformation of a normal cell into a malignant cell requires multiple mutations within the cancer cell. The carcinogenic process can be induced by both mutagenic and non-mutagenic chemicals and is also influenced by non-carcinogenic chemicals. Thus, cancer development involves multiple mutations, multiple steps, multiple causes and multiple mechanisms. Consequently, both mutagens and mitogens can influence the carcinogenic process.

2.2 Multistep carcinogenesis

Cancer arises as the result of multiple mutations in critical target genes controlling normal cell division and cell death. There is now substantial experimental evidence in humans and in rodents to support the multistep model of carcinogenesis (Barrett, 1993; Foulds, 1975; Nowell, 1976) as well as the somatic mutation theory of carcinogenesis (Barrett, 1993; Boveri, 1929). A generalized scheme of multistep carcinogenesis is shown in *Figure 2.1*, and the molecular basis for the multiple steps in colorectal cancer as defined by Vogelstein and colleagues (1988) is illustrated in *Figure 2.2*. According to the clonal evolution theory of cancer proposed by Nowell (1976), a malignant cancer is the consequence of genetic or epigenetic alterations of multiple, independent genes involved in growth control. A cancer evolves from a cell with the first critical mutation (initiation), which confers a selective growth advantage and thus clonal expansion (tumour promotion). An increase

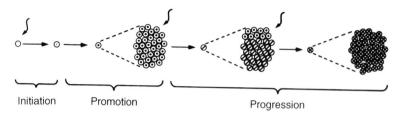

Initiation   Promotion   Progression

Figure 2.1. Representation of the clonal evolution model of cancer proposed by Nowell (1976).

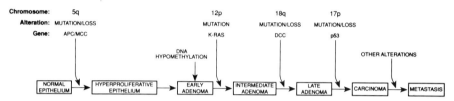

Figure 2.2. The multistep model of colorectal cancer developed by Vogelstein and colleagues (1988). Reproduced from Vogelstein *et al.* (1988) with permission from the Massachusetts Medical Society.

in the number of cells with the first mutation increases the probability of a second mutation in another critical target gene (tumour progression) due to the increased number of cells at risk (*Figure 2.1*). Thus, cancer is a progressive process, as described by Foulds (1975), in which a series of mutations and clonal expansions occur resulting in an accumulation of mutations in critical genes in cells with increasing dysplastic and malignant growth properties. If one or more of the early mutations also increase genomic instability, the rate of neoplastic progression will accelerate (Loeb, 1989; Nowell, 1976).

Studies by Vogelstein and colleagues (Fearon and Vogelstein, 1990; Vogelstein *et al.*, 1988, 1989) of colorectal cancer, and later by investigators studying other cancers, have clearly documented that human and rodent cancers result from multiple mutations in oncogenes and tumour suppressor genes and that these mutations accumulate as cells progress from a precancerous to a malignant state, providing clear experimental documentation of the Foulds and Nowell hypotheses. Based upon these findings and other experimental and epidemiological evidence (Barrett, 1993), the multistep model of cancer is now generally accepted.

The critical target genes for mutations in cancer development are proto-oncogenes and tumour suppressor genes (Barbacid, 1986; Boyd and Barrett, 1990; Weinberg, 1989), which are involved in regulation of cell division, cell death and/or genomic stability. The number of genes involved in neoplastic development is not known with certainty. Most colorectal cancers have three or more altered genes (Vogelstein *et al.*, 1988, 1989), and estimates of as many as 10 or more mutational changes have been proposed to occur in adult human cancers (Boyd and Barrett, 1990). These findings are consistent with multihit models developed on the basis of specific incidence rates of cancers increasing exponentially with the 5th to 7th power of age (Kaldor and Day, 1987). Analysis of multistep carcinogenesis at the molecular level, therefore, indicates that the process of neoplastic evolution is significantly more complicated than the relatively simple two-stage, initiation and promotion, model of carcinogenesis or even a three-stage model of initiation, promotion and progression (Pitot *et al.*, 1981). As an example, the model described by Vogelstein and coworkers for colorectal cancers (*Figure 2.2*) shows that multiple genetic changes must occur after the promotion or clonal growth of the initiated cells (Vogelstein *et al.*, 1988, 1989). Thus, the progression phase of carcinogenesis represents multiple stages at which chemicals might influence the neoplastic process (Hennings *et al.*, 1983).

There are three general mechanisms by which an agent can influence the multistep, carcinogenic process (*Table 2.1*). An agent can induce a heritable alteration in one or more critical genes in the multistep process by a direct genetic alteration. A second general mechanism involves a heritable, epigenetic alteration in one or more critical genes. Although much is known about how chemicals induce genetic changes, little is known about the mechanisms of carcinogen-induced epigenetic, heritable changes (Rubin, 1994). A third mechanism by which an agent can influence multistep carcinogenesis is the facilitation of clonal expansion of an initiated or intermediate cell, which increases the probability of additional, spontaneous (mutational or epigenetic) heritable changes.

Table 2.1. Mechanisms by which an agent can influence multistep carcinogenesis

• By inducing heritable mutation in a critical gene
• By inducing heritable, epigenetic change in a critical gene
• By increasing clonal expansion of a cell with a heritable alteration in a critical gene, allowing for increased probability of additional events

2.3 Mutagenesis as a mechanism of carcinogenesis

The origin of the somatic mutation theory of carcinogenesis is generally credited to Theodor Boveri, who in 1914 published his book (Boveri, 1914) entitled *Zur Frage der Entstehung Maligner Tumoren (On the Problem of the Origin of Malignant Tumours)*. The English translation of this book by his wife, Marcella Boveri, was published in 1929 (Boveri, 1929). Boveri's hypothesis on the origin of malignant tumours was extraordinarily comprehensive and included many predictions which have turned out to be correct. For these reasons, Boveri is generally acknowledged as the father of the somatic mutation theory of carcinogenesis. There is now considerable evidence to support this theory as discussed elsewhere (Barrett, 1991, 1993).

Genetic changes can be classified either as gene mutations, which include point mutations, deletions and frameshift mutations, chromosome rearrangements, gene amplification, or as aneuploidy. There are now many examples of each of these mutational changes in different tumours (*Table 2.2*), which provide strong support for the somatic mutation theory of carcinogenesis. Point mutations have been observed to activate proto-oncogenes and to inactivate tumour suppressor genes in certain cancers. Chromosome rearrangements of oncogenes are also well documented. Gene amplification as well as numerical chromosome changes are important in a number of different cancers (Barrett, 1991). Therefore, chemicals that induce any one of these types of genetic event can heritably alter a critical target gene, resulting in neoplastic development.

Table 2.2. Examples of molecular, genetic and cytogenetic changes in tumours

Type of genetic change	Examples
Gene mutation	Point mutation (G→T) in codon 12 of the c-Ha-*ras* gene in EJ/T24 bladder carcinoma
	Point mutation (A→G) in the splice acceptor sequence of exon 21 in the retinoblastoma gene of J82 bladder carcinoma cells
	Deletions of tumour suppressor genes
Chromosome rearrangement	Philadelphia translocation t(9;22) in chronic myelogenous leukaemia t(8;14) in Burkitt's lymphoma
Gene amplification	N-*myc* gene in neuroblastomas c-*myc* gene in lung carcinomas *neu* gene in mammary carcinomas
Aneuploidy	+12 in chronic lymphocytic leukaemia +8 in ANLL, blast phase of CML +15 in murine T-cell leukaemias −22 in meningiomas −15 in Syrian hamster tumours induced by transfection of v-Ha-*ras* and v-*myc*

2.4 The role of mitogenesis in carcinogenesis

The failure to detect mutagenic activity of certain carcinogens indicates that these chemicals may act by alternative mechanisms of action, one possibility being by increasing cell proliferation (Ames and Gold, 1990; Cohen and Ellwein, 1990). This hypothesis is supported by the fact that most, if not all, types of cancers can arise spontaneously in at least one species. Normal cell division results in a low level of spontaneous errors during DNA replication. Endogenous DNA damage can result from depurination or cytosine deamination under physiological conditions, and from oxidative damage associated with normal cellular metabolism. Thus, mutations can occur 'spontaneously' from normal cellular processes. Exogenous mutagens in food, air, or water will add to the endogenous mutagenic burden (Loeb, 1989).

There is strong evidence that cell proliferation is a risk factor for cancer in humans and rodents (Preston-Martin et al., 1990). Certain human carcinogens (e.g. hormonal carcinogens) are known to influence the rate of cell proliferation in target tissues (Preston-Martin et al., 1990). However, mechanisms in addition to cell proliferation should be considered for hormonal carcinogens (Barrett, 1993; Huff et al., 1995).

Cell proliferation can influence the carcinogenic process by a number of mechanisms (Table 2.3). Cell proliferation is necessary for fixation of DNA damage and expression of mutations and for mutagens to exhibit carcinogenic activity. Increased cell replication in the absence of an exogenous mutagen will increase the number of spontaneous mutations in a population of cells. This may increase the frequency of initiated cells in a target tissue, depending on the rate of cell death in the tissue. If increasing mitogenesis does not increase the size of a tissue, then cell renewal is balanced by cell death and spontaneously arising mutant cells may also die at an increased rate. It might be assumed that initiated or preneoplastic cells are less likely to die than normal cells but this is not true in many cases (Grasl-Kraupp et al., 1994; Preston et al., 1994). Preneoplastic and cancer cells may have an increased rate of cell death as well as cell replication. Many tumour promoters function by blocking cell death of preneoplastic cells with little effect on cell replication (Schulte-Hermann et al., 1983). The importance of this mechanism in chemical carcinogenesis is illustrated by the recent findings of Schulte-Hermann and coworkers studying the mechanism of dietary restriction and carcinogenesis. Restriction of dietary calories reduces cancer development in experimental animals and possibly also in humans. Grasl-Kraupp et al. (1994) demonstrated that dietary restriction eliminates initiated and preneoplastic cells by apoptosis. Preneoplastic cells are more susceptible to apoptosis when caloric intake is reduced. Therefore, promotion of chemically induced cancers by dietary factors involves modulation of cell death. Further studies on the role of inhibition or enhancement of apoptosis by chemical carcinogens may yield new insights into mechanisms of carcinogenesis.

Table 2.3. Mechanisms by which mitogens might influence carcinogenesis

- Increase fixation and expression of premutagenic DNA lesions
- Increase the number of spontaneous, mutated cells occurring during cell replication
- Promote clonal expansion of initiated cells
- Increase the number of preneoplastic cells by blocking cell death
- Increase the number of mutated cells by indirect mutagenic processes

Mitogenic chemicals, such as phorbol ester tumour promoters, can influence the carcinogenic process by promoting clonal expansion of initiated cells (Barrett, 1985, 1987a; Hennings and Yuspa, 1985). Whether this is due to increased mitogenesis, decreased cell death of the initiated cell, or both is not clear.

Before cell proliferation per se can be accepted as the causative mechanism for certain carcinogens, several facts should be considered (Table 2.4). Firstly, many toxic and/or hyperplastic stimuli are not carcinogenic (Hoel *et al.*, 1988; Huff, 1995; Ledda-Columbano *et al.*, 1989). A review of the literature in this field and further studies of noncarcinogenic toxic agents are needed (Huff, 1993). Secondly, cell division occurs repeatedly in all organisms (*Table 2.4*); therefore, it is not clear when cell division is causative in the carcinogenic process. This, of course, depends on the target tissue. A distinction must also be made between proliferation of stem cells that may give rise to malignant cells and proliferation of cells that are committed to a terminal differentiation process. Furthermore, cell division of initiated or intermediate cells may occur at rates which differ from those of normal cells. Finally, the observation that multiple mutations are involved in the development of many neoplasms suggests that even a weak mutagenic response, which is below the level of detection of current assays, is sufficient to influence the neoplastic process in a specific target tissue. This is a plausible explanation for certain non-genotoxic carcinogens, some of which may act by indirect mutagenic processes.

Table 2.4. Evidence against cell proliferation *per se* being carcinogenic

• Many toxic and/or hyperplastic stimuli are non-carcinogenic
• Cell division occurs repeatedly in all organisms[a]
For humans:
1 egg→10^{14} cells in adult organism
10^{13} cells still capable of cell division
10^7 cell divisions/sec occur in adult organism
10^6 cell divisions/sec in intestine
• Multiple mutations (3–10?) are required for a normal cell to evolve into a cancer cell

[a] David Prescott, personal communication.

2.5 Non-genotoxic carcinogens

Interaction of electrophiles with DNA to produce DNA damage and adducts and subsequent mutations is a well-established mechanism of action for many chemical carcinogens (Brookes and Lawley, 1964; McCann and Ames, 1976; Miller and Miller, 1976). Failure to detect DNA damage and mutation induction by some chemicals led to the hypothesis that these chemicals act by different mechanisms (Weisburger and Williams, 1981; Williams, 1987). Weisburger and Williams proposed the classification of carcinogenic chemicals as either genotoxic or epigenetic on the basis of two distinct mechanisms of action. Although this proposal was criticized for various reasons, it served an important role in defining mechanisms of action of carcinogens. Two major criticisms of this classification scheme have been discussed (Barrett, 1987b). First, any classification scheme is problematic because classification implies exclusivity, which is probably rarely the case with chemical carcinogens that can operate through multiple mechanisms of action, which are not mutaully exclusive (Barrett, 1987a). Mutagenic carcinogens also have non-mutagenic epigenetic mechanisms important to their carcinogenic activity, and certain chemicals that do not damage DNA are mutagenic by indirect mechanisms.

The terminology originally used by Weisburger and Williams (genotoxic and epigenetic carcinogens) was also problematic. Genotoxic is now generally accepted to indicate a chemical that induces DNA damage (Williams, 1987), although it was originally used by Druckrey in a broader context (Barrett, 1987b). 'Epigenetic' has been replaced by most investigators with the term 'non-genotoxic', which describes a chemical that lacks the ability to damage DNA directly. The term 'epigenetic' was first used by Waddington in 1940 to denote 'the science concerned with causal analysis of development' (Waddington, 1940). With the understanding of the molecular biology of gene expression, the term epigenetic has also been used to describe processes related to the expression of genetic material by transcriptional and translational control mechanisms (Rieger *et al.*, 1976). Thus, the following definitions have been proposed (Barrett, 1987b):

(i) *Epigenetic change.* Any change in a phenotype which does not result from an alteration in DNA sequence. This change may be stable and heritable, and includes alterations in DNA methylation, transcriptional activation, translational control and post-translational modifications.

(ii) *Genetic change.* Any change in a phenotype which results from an alteration in primary DNA sequence. This change may be a single base pair change, a deletion, an insertion, a rearrangement or duplication of one or more base pairs, or loss or gain of an entire chromosome (genetic change = mutation).

If non-genotoxic carcinogens are defined as chemicals that do not directly induce DNA damage, then some non-genotoxic carcinogens can, in fact, be mutagenic. Many genotoxicity assays measure only the activity of a chemical to induce point mutations or DNA damage. However, chemicals can also induce genetic changes at the chromosomal level without causing gene mutations or directly damaging DNA. These chemicals, therefore, would be negative in some genotoxicity assays (Barrett, 1987b; Jackson *et al.*, 1993; Toman *et al.*, 1980). Certain exceptions to the correlation between carcinogenesis and mutagenicity based on results in the Ames test (e.g. benzene, arsenic, diethylstilboestrol and asbestos) may relate to the ability of certain chemicals to act specifically as chromosome mutagens (i.e. clastogens and/or aneuploidogens). These examples have been discussed elsewhere (Barrett, 1993).

True epigenetic mechanisms are important in carcinogenesis (Rubin, 1985, 1994). Methylation of DNA at the 5-position of cytosine is important in the regulation of gene expression and is one possible epigenetic mechanism for the heritable change in cancer cells (Jones, 1986, 1987). Chemicals, such as 5-azacytidine and ethionine, may affect DNA methylation through an interaction with the DNA methyltransferase enzyme. It has also been suggested that DNA-alkylating agents heritably alter DNA methylation patterns (Jones, 1986, 1987). This provides an epigenetic mechanism for heritable alterations in expression of genes involved in carcinogenesis. Other epigenetic mechanisms for carcinogens can be proposed (Barrett, 1987b). A number of these may involve tumour-promoting activity of carcinogenic chemicals. In cancers, loss of imprinting (gene silencing) due to DNA methylation is commonly observed (Makos et al., 1992). Some chemicals, including genotoxic chemicals, may produce epigenetic changes in DNA expression by altering DNA methylation.

2.6 Mechanisms of indirect mutagenesis

A number of non-DNA targets can be perturbed by chemicals leading to mutagenic events in the cell and possibly contributing to carcinogenesis (Table 2.5). The progression of the cell cycle is controlled by the action of both positive and negative growth regulators. The key players in this activity include a family of cyclins and cyclin-dependent kinases, which are themselves regulated by other kinases and phosphatases. Maintenance of balanced cell-cycle controls may be linked directly to genomic stability. Loss of the checkpoints involved in the cell-cycle control may result in unrepaired DNA damage during DNA synthesis or mitosis leading to genetic mutations and contributing to carcinogenesis (Afshari and Barrett, 1993; Hartwell and Kastan, 1994). Repair enzymes are also potential targets for chemicals. Arsenic, a human carcinogen, may act by interfering with enzymes involved in DNA repair and recombination (Barrett and Lee, 1992). Blocking cell death (apoptosis) may allow mutated, preneoplastic cells to survive and progress to neoplasia (Marsman and Barrett, 1994).

Table 2.5. Processes involving non-DNA targets resulting in indirect mutagenesis or heritable cellular changes

•   Disruption of cell-cycle checkpoint
•   Interference of repair enzymes
•   Blocking cell death (apoptosis)
•   Stress-induced mutagenesis
•   Loss of imprinting/DNA demethylation (heritable epigenetic change)
•   Stimulation of spontaneous and oxidative mutagenesis

Cairns et al. (1988) described the ability of bacteria to mutate specific genes at a high frequency without cell division, in response to environmental challenges. This phenomenon has been referred to as adaptive mutation, directed mutation, and even Cairnsian mutation (Hall, 1991; Harris et al., 1994; MacPhee, 1993). Although this phenomenon was challenged as a methodological artefact (Culotta, 1994; Hall, 1991; MacPhee, 1993), recent findings of specific mutational spectra associated with this experimental model (Foster and Trimarchi, 1994; Rosenburg et al., 1994) provide strong support for adaptive mutation in bacteria. It remains to be determined if adaptive mutations can also occur in mammalian cells. An important question that arises is whether other novel mechanisms of mutagenesis exist by which cells under environmental stress can generate specific mutations. These phenomena, if they exist, may be highly relevant to the carcinogenesis process.

2.7 Conclusions

Most, if not all, tumours have genetic changes, which may or may not result from mutagenic exposures. It is important to understand the mechanisms by which carcinogenic chemicals induce genetic changes that arise in chemically induced tumours. Rather than simply dividing the possible mechanisms into two categories (i.e. genotoxic and non-genotoxic), a number of mechanisms of chemically induced mutations in tumours can be envisioned (*Table 2.6*). If a chemical induces a cancer and that cancer has genetic changes, it is possible that the chemical directly induced the genetic change, for example, through formation of DNA adducts. At the other extreme, the chemical may induce the cancer by a non-genetic mechanism, the tumour becomes genetically unstable and mutations arise due to the nature of the tumour rather than the mutation causing the tumour. A number of mechanisms exist between these two extremes. The chemical may induce mutations by indirect mutational mechanisms (e.g. disruption of spindle function or generation of reactive oxygen radicals). These reactive oxygen radicals may arise due to the intrinsic properties of the chemical or due to receptor-mediated production of enzymes increasing rates of oxygen metabolism in cells. Other receptor-mediated changes can also lead indirectly to mutations, as outlined in *Table 2.6*. Any attempt to classify chemical carcinogens using the mechanism of mutation induction must consider the complexity and the multitude of possible mechanisms.

Table 2.6. Mechanisms of chemically induced mutations in tumours

Chemical→DNA (adduct)→mutation

Chemical→microtubule (spindle dysfunction)→DNA (aneuploidy)→mutation

Chemical→O_2 (activated)→DNA→mutation

Chemical→receptor→enzyme→O_2 (activated)→DNA→mutation

Chemical→receptor→protein (e.g. recombinase)→DNA→mutation

Chemical→receptor→protein→DNA synthesis/cell division (normal mutation rate)→mutation

Chemical→receptor→protein→DNA synthesis/cell division (loss of checkpoint/increased mutation rate)→mutation

Chemical→receptor→protein→DNA synthesis/cell division →tumour→mutation

Many chemical carcinogens operate via a combination of mechanisms, and even their primary mechanism of action may vary depending on the target cells. For example, some chemicals are complete carcinogens in one tissue, promoters in another, and initiators in another. Classification of carcinogens into mutually exclusive categories may be misleading and hinder our comprehension of the complexity of chemical carcinogenesis which involves both mutagenesis and mitogenesis.

References

Afshari CA, Barrett JC. (1993) Cell cycle controls: potential targets for chemical carcinogens? *Environ. Health Perspect.* **101** (Suppl. 5): 9–14.

Ames BN, Gold SL. (1990) Too many rodent carcinogens: mitogenesis increases mutagenesis. *Science* **249:** 970–971.

Barbacid M. (1986) Mutagens, oncogenes and cancer. *Trends Genet.* **2:** 188–192.

Barrett JC. (1985) Tumour promotion and tumour progression. In: *Carcinogenesis – A Comprehensive Survey*, Vol. 8, *Cancer of the Respiratory Tract: Predisposing Factors* (eds MJ Mass, DG Kaufman, JM Siegfried, VE Steele, S Nesnow). Raven Press, New York, pp. 423–429.

Barrett JC. (1987a) A multistep model for neoplastic development: role of genetic and epigenetic changes. In: *Mechanisms of Environmental Carcinogenesis*, Vol. 2, *Multistep Models of Carcinogenesis* (ed. JC Barrett). CRC Press, Boca Raton, FL, pp. 117–126.

Barrett JC. (1987b) Genetic and epigenetic mechanisms in carcinogenesis. In: *Mechanisms of Environmental Carcinogenesis*, Vol. 1, *Role of Genetic and Epigenetic Changes* (ed. JC Barrett). CRC Press, Boca Raton, FL, pp. 1–15.

Barrett JC. (1991) Relationship between mutagenesis and carcinogenesis. In: *Origins of Human Cancer* (eds J Brugge, T Curren, E Harlow, F McCormick). Cold Spring Harbor Laboratory Press, Cold Spring Harbor, NY, pp. 101–112.

Barrett JC. (1993) Mechanisms of multistep carcinogenesis and carcinogen risk assessment. *Environ. Health Perspect.* **100:** 9–12.

Barrett JC, Lee T-C. (1992) Mechanisms of arsenic-induced gene amplification. In: *Gene Amplification in Mammalian Cells: a Comprehensive Guide* (ed. RE Kellems). Marcel Dekker, New York, pp. 441–446.

Boveri T. (1914) *Zur Frage der Entstehung Maligner Tumouren*. Gustave Fischer, Jena, Germany.

Boveri TH. (1929) *The Origin of Malignant Tumours*. Williams and Wilkins, Baltimore, MD.

Boyd JA, Barrett JC. (1990) Genetic and cellular basis of multistep carcinogenesis. *Pharmacol. Ther.* **46:** 469–486.

Brookes P, Lawley PD. (1964) Evidence for the binding of polynuclear aromatic hydrocarbons to the nucleic acids of mouse skin: relation between carcinogenic power of hydrocarbons and their binding to deoxyribonucleic acid. *Nature* **202:** 781–784.

Cairns J, Overbaugh J, Miller S. (1988) The origin of mutants. *Nature* **335:** 142–145.

Cohen SM, Ellwein LB. (1990) Cell proliferation in carcinogenesis. *Science* 249: 1007-1011.

Culotta E. (1994) A boost for 'adaptive' mutation. *Science* **265:** 318–319.

Fearon ER, Vogelstein B. (1990) A genetic model for colorectal tumorigenesis. *Cell* **61:** 759–767.

Foster PL, Trimarchi JM. (1994) Adaptive reversion of a frameshift mutation in *Escherichia coli* by simple base deletions in homopolymeric runs. *Science* **265:** 407–409.

Foulds L. (1975) *Neoplastic Development*. Academic Press, New York.

Grasl-Kraupp B, Bursch W, Turrkay-Nedecky B, Wagner A, Lauer B, Schulte-Hermann R. (1994) Food restriction eliminates preneoplastic cells through apoptosis and antagonizes carcinogenesis in rat liver. *Proc. Natl Acad. Sci. USA* **91:** 9995–9999.

Hall BG. (1991) Is the occurrence of some spontaneous mutations directed by environmental challenges? *New Biologist* **3:** 729–733.

Harris RS, Longerich S, Rosenberg SM. (1994) Recombination in adaptive mutation. *Science* **264:** 258–260.

Hartwell LH, Kastan MB. (1994) Cell cycle control and cancer. *Science* **266:** 1821–1828.

Hennings H, Yuspa SH. (1985) Two-stage tumour promotion in mouse skin: an alternative interpretation. *J. Natl Cancer Inst.* **74:** 735–740.

Hennings H, Shores R, Wenk ML, Spangler EF, Tarone R, Yuspa SH. (1983) Malignant conversion of mouse skin tumours is increased by tumour initiators and unaffected by tumour promoters. *Nature* **304:** 67–69.

Hoel DG, Haseman JK, Hogan MD, Huff J, McConnell EE. (1988) The impact of toxicity on carcinogenicity studies: implications for risk assessment. *Carcinogenesis* **9**: 2045–2052.

Huff J. (1995) Mechanisms, chemical carcinogenesis, and risk assessment: cell proliferation and cancer. *Am. J. Indus. Med.* **27**: 292–300.

Huff J, Boyd J, Barrett JC. (1995) Environmental influences on hormonal carcinogenesis: prologue. In: *Cellular and Molecular Mechanisms of Hormone Carcinogenesis: Environmental Influences* (eds J Huff, J Boyd, JC Barrett). Wiley-Liss, New York, in press.

Jackson MA, Stack HF, Waters MD. (1993) The genetic toxicology of putative nongenotoxic carcinogens. *Mutat. Res.* **296**: 241–277.

Jones PA. (1986) DNA methylation and cancer. *Cancer Res.* **46**: 461–466.

Jones P. (1987) Role of DNA methylation in regulating gene expression, differentiation, and carcinogenesis. In: *Mechanisms of Environmental Carcinogenesis*, Vol. 1, *Role of Genetic and Epigenetic Changes* (ed. JC Barrett). CRC Press, Boca Raton, FL, pp. 17–29.

Kaldor JM, Day NE. (1987) Interpretation of epidemiological studies on the context of the multistage model of carcinogenesis. In: *Mechanisms of Environmental Carcinogenesis*, Vol. 2, *Mechanisms of Multistep Carcinogenesis* (ed. JC Barrett). CRC Press, Boca Raton, FL, pp. 21–57.

Ledda-Columbano GM, Columbano A, Curto M, Coni MGE, Sarma DSR, Pani P. (1989) Further evidence that mitogen-induced cell proliferation does not support the formation of enzyme-altered islands in rat liver by carcinogens. *Carcinogenesis* **10**: 847–850.

Loeb LA. (1989) Endogenous carcinogenesis: molecular oncology into the twenty-first century – presidential address. *Cancer Res.* **49**: 5489–5496.

MacPhee DG. (1993) Is there evidence for directed mutation in bacteria? Discussion Forum. *Mutagenesis* **8**: 3–5.

Makos M, Nelkin DB, Lerman MI, Latif F, Zbar B, Baylin SB. (1992) Distinct hypermethylation patterns occur at altered chromosome loci in human lung and colon cancer. *Proc. Natl Acad. Sci. USA* **89**: 1929–1933.

McCann J, Ames BN. (1976) Detection of carcinogens as mutagens in the Salmonella/microsome test: assay of 300 chemicals: discussion. *Proc. Natl Acad. Sci. USA* **73**: 950–954.

Marsman DS, Barrett JC. (1994) Apoptosis and chemical carcinogenesis. *J. Risk Analysis* **14**: 321–326.

Miller EC, Miller JA. (1976) The mutagenicity of chemical carcinogens: correlations, problems and interpretations. In: *Chemical Mutagenesis Principles and Methods for their Detection*, Vol. 1 (ed. A Hollaender). Plenum Press, New York, pp. 83–119.

Nowell P. (1976) The clonal evaluation of tumour cell populations. *Science* **194**: 23–28.

Pitot HC, Goldsworthy T, Moran S. (1981) The natural history of carcinogenesis: implication of experimental carcinogenesis in the genesis of human cancer. *J. Supramol. Struct. Cell. Biochem.* **17**: 133–146.

Preston GA, Lang J, Maronpot R, Barrett JC. (1994) Regulation of apoptosis during neoplastic progression: enhanced susceptibility after loss of senescence gene and decreased susceptibility after loss of a tumour suppressor gene. *Cancer Res.* **54**: 4214–4223.

Preston-Martin S, Pike MC, Ross RK, Jones PA, Henderson BE. (1990) Increased cell division as a cause of human cancer. *Cancer Res.* **50**: 7415–7421.

Rieger R, Michaelis A, Green MM. (1976) *A Glossary of Genetics and Cytogenetics – Classical and Molecular*. Springer-Verlag, New York.

Rosenberg SM, Longerich S, Gee P, Harris RS. (1994) Adaptive mutation by deletions in small mononucleotide repeats. *Science* **265**: 405–407.

Rubin H. (1985) Cancer as a dynamic developmental disorder. *Cancer Res.* **45**: 2935–2942.

Rubin H. (1994) Incipient and overt stages of neoplastic transformation. *Proc. Natl Acad. Sci. USA* **91**: 12076–12080.

Schulte-Hermann R, Schuppler J, Timmermann-Trosiener I, Ohde G, Bursch W, Berger H. (1983) The role of growth of normal and preneoplastic cell populations for tumour promotion in rat liver. *Environ. Hlth Perspect.* **50**: 185–194.

Toman Z, Dambly C, Radman M. (1980) Induction of a stable, heritable epigenetic change by mutagenic carcinogens: a new test system. In: *Molecular and Cellular Aspects of Carcinogen Screening Tests, IARC Scientific Publications no. 27.* International Agency for Research on Cancer, Lyon, pp. 243–255.

Vogelstein B, Fearon ER, Hamilton SR, Kern SE, Presinger AC, Leppert M, Nakamura Y, White R, Smits AMM, Bos JL. (1988) Genetic alterations during colorectal-tumour development. *New Engl. J. Med.* **319:** 525–532.

Vogelstein B, Fearon ER, Kern SE, Hamilton SR, Preisinger AC, Nakamura Y, White R. (1989) Allelotype of colorectal carcinomas. *Science* **244:** 207–211.

Waddington CH. (1940) *Organizers and Genes.* Cambridge University Press, London.

Weinberg RA. (1989) Oncogenes, antioncogenes, and the molecular bases of multistep carcinogenesis. *Cancer Res.* **49:** 3713–3723.

Weisburger JH, Williams GM. (1981) Carcinogen testing: current problems and new approaches. *Science* **214:** 401–407.

Williams GM. (1987) DNA reactive and epigenetic carcinogens. In: *Mechanisms of Environmental Carcinogenesis*, Vol. 1, *Role of Genetic and Epigenetic Changes* (ed. JC Barrett). CRC Press, Boca Raton, FL, pp. 113–127.

3

Molecular mechanisms of mutagenesis and mutational spectra

Barry W. Glickman, Gopaul Kotturi, Johan de Boer and
Wolfgang Kusser

3.1 Introduction

Mutation has become increasingly the subject of scientific scrutiny. Initial
interest in mutagenesis was restricted to those biologists who recognized the
potential contribution that mutation could make to enabling evolution and
creating biodiversity. With the recognition that ionizing radiation, chemicals
and ultraviolet radiation (UV) could induce mutation, the association between
the environment and the origin of mutation became clear. The relevance of
these observations was further supported as the association between mutation
and cancer became evident. Indeed, it is now well established that most human
carcinogens are also mutagens. In the past few years the link between mutation
and cancer has been strengthened further by the discovery of proto-oncogenes
and tumour suppressor genes which are often mutated during carcinogenesis.

Not only is mutation linked to cancer and birth defects, but it is also
implicated in a number of human diseases ranging from atherosclerosis to
diabetes. Thus mutation is directly implicated in human health. As a
consequence, a significant effort has gone into the development of screening tools
designed to identify potential mutagens and to reduce human exposure to them.
This chapter provides a basis for understanding the mechanisms of mutation and
describes some of the newer tools used in studying mutational events.

3.1.1 The relevance of mutagenesis

Industrial development has added to the natural burden of mutagens in the
environment, and the presence of mutagens is now an important indicator of
environmental contamination. Where human health is involved, the choice has

been either to eliminate the source of exposure, or to remove people from the area of exposure, the latter doing little to improve the state of the environment. From an environmental point of view, screening for mutagens provides an early warning of the potentially tragic effects of both cancer and birth defects. In addition, the analysis of mutation at the molecular level may supply information about the sources of mutation and the metabolic processes involved.

The importance of mutagenesis is not restricted to environmental issues. An understanding of mutagenesis is also likely to provide insights into the process of ageing, evolution and development. Finally, mutation is the single most important analytical tool in the hands of geneticists and developmental biologists.

3.1.2 Selected definitions

In broad terms, any heritable genetic alteration can be viewed as a mutation. At the level of the organism, mutation might be manifested as an alteration in phenotype affecting the whole organism. Such changes generally reflect germ-cell events that are passed from one generation to the next. In some cases the alteration affects only a portion of the organism (i.e. is mosaic) and may reflect a somatic mutation occurring at some later stage of development. Alternatively, such mosaicism may reflect a loss of heterozygosity (LOH) which may be the result of a gene conversion, recombination or mitotic non-disjunction event. While the latter pathways may not involve the creation of new mutations, these represent a class of genetic changes which may play a predominant role in the expression of mutation in mammalian systems (Klinedienst and Drinkwater, 1991).

More classical definitions reflect the nature of the molecular event. Macroscopic alterations can be described by the nature of the visible chromosomal alteration. Smaller events including deletions, duplications and insertions may be detected by Southern blotting or PCR analysis, while the term 'point mutation' covers mutations ranging in size from single base substitutions to those involving 50 or more base pairs. The definition of point mutation refers to molecular events that cannot be detected by the sizing of restriction fragments or amplified PCR fragments where the limits of detection vary with the procedure being used. In this way base substitutions, frameshifts and small deletions and insertions are classified as point mutations.

Base substitutions involve either *transitions* (the substitution of a purine by a purine or a pyrimidine by a pyrimidine) or *transversions* (the substitution of a purine by a pyrimidine or a pyrimidine by a purine). Base substitutions predominantly cause *missense* mutations which involve a single amino acid substitution. *Nonsense* mutations result when a base substitution produces a termination codon in place of an amino acid. Because of the number of available nonsense suppressors, amber (TAG) and ochre (TAA) mutations have been more commonly recovered than opal (TGA) mutations. *Frameshift* events involve the gain or loss of a small number of base pairs (e.g. 1–4), with larger events generally being cited as deletions and insertions. It should be remembered that the term 'frameshift' refers to the alteration in codon reading frame that was originally used to detect these events, and not to the proposed mechanism for their occurrence (Streisinger *et al.*, 1966). As will be seen later, this is important only in that there is no clear mechanistic distinction between frameshift and deletion events.

3.2 Mutational specificity

Not long ago researchers had to be content with data limited to the kinetics of mutation induction and the shape of killing curves. The application of molecular techniques has dramatically altered our capabilities. This chapter stresses that details of the molecular nature of mutations can provide insights into the premutagenic lesions and the processes responsible for mutation. We can now relate the consequences of mutation to its cause. In theory, differences in mutational spectra in people *in vivo* can reveal differences in lifestyle, occupational exposure or the genetically determined capacity to repair DNA damage. It is this possibility that makes the study of mutation at the DNA sequence level so attractive.

3.2.1 Defining mutational specificity

Mutational specificity is the description of a collection of mutations at the molecular level. Mutational spectra thus usually involve the collection of mutants by selection for a phenotypic change. As a consequence, mutational spectra are strongly influenced by the genetic marker being studied. An optimal target for mutational studies requires a locus which has a wide range of selectable mutations. The production of a mutational spectrum requires information on both the nature and precise location of each mutation. By the nature of a mutation we mean the specific class of event. Information concerning the location of the mutation provides the context in which the mutation occurred. Location is not restricted to the regional DNA sequence: it includes additional components such as strand orientation with respect to replication, the direction and level of transcription, the presence of DNA binding proteins and the general genetic background of the organism.

3.2.2 Mutational specificity *in vivo and* in vitro

A number of novel techniques are providing insights into the mechanisms of mutation. It is now possible to collect mutational specificity data *in vivo*, and compare these results with data obtained using *in vitro*-damaged plasmids which are introduced. For example, a plasmid can be damaged by UV light *in vitro*, modified by photoreactivation, and introduced into cells with defined repair capabilities. Moreover, it is also possible to circumvent the problem that mutagenic treatments almost always produce a plethora of DNA lesions through the use of site-specific modifications either *in vivo* by transformation or transfection assays, or *in vitro,* using purified polymerases or replication complexes (Basu and Essigmann, 1988; Bradley *et al.,* 1993; Comess *et al.,* 1992; Lindsley and Fuchs, 1994). We have drawn from these kinds of experiments to supplement the mutational specificity data determined by *in vivo* analyses.

3.3 Mutations are targeted, non-random events

It is essential to recognize that most mutation is *targeted* (i.e. the result of DNA damage), whether that damage be a result of endogenous or exogenous insult. This has extremely important consequences, as mutations reflect both the nature and the position of DNA damage. We discuss later how the nature of a DNA lesion dictates the mutational event observed. We will also indicate how the location, and DNA context, of a lesion has a profound effect on the mutational consequences. The combination of the type and location of DNA lesions is dependent upon the nature of the mutational treatment. It thus follows that the *mutational spectrum* produced by each physical or chemical mutagen will be unique.

The first effective demonstration of the targeting of mutation by DNA damage comes from the work of J.H. Miller (1970) who demonstrated distinct mutational spectra for different mutagens in the *lac*I gene of *Escherichia coli.* For example, the S_N1 alkylating agent N-methyl-N'-nitrosoguanidine (MNNG) produced almost exclusively G:C→A:T transitions with a preference for target sites preceded by a guanine. In contrast, the S_N2 alkylating agent ethylmethanesulphonate (EMS) produced a slightly broader spectrum of base substitutions although still predominately G:C→A:T transitions *(Table 3.1),* but without the same nearest-neighbour preference (Pienkowska *et al.,* 1993). Despite the difference in specificity, G:C→A:T transitions induced by both alkylating agents are thought to be mediated by the miscoding properties of the O^6-alkylguanine adduct (Swann, 1990). As can be seen in *Table 3.1,* G:C→A:T transitions also predominate in the spectrum of UV-induced mutation in the *lac*I gene (Schaaper *et al.,* 1987), but these occur almost exclusively at cytosine-containing dipyrimidine sites as would be expected if cyclobutane pyrimidine dimers (CPD) or (6–4) pyrimidine–pyrimidone lesions ((6–4) photoproduct) were responsible for the mutagenic effects of UV light (see *Table 3.2*). *Figure 3.1* offers a comparison of the mutational spectra of UV and EMS. Despite the fact that both treatments produce predominantly G:C→A:T transitions, the differences in distribution are quite stunning. The context of the nearest-neighbour effects can be determined by comparing the sequences surrounding the more common sites of mutation in *Table 3.2.* The data in *Table 3.3* demonstrate, at least in the broader sense, that mutational spectra obtained in bacteria are quite similar to those recovered in mammalian cells.

Table 3.1. Comparison of mutational spectra

Class	Spontaneous[a] %	anti-BPDE[b] %	UV[c] %	EMS[d] %
GC→AT	33.3	4.8 (2.6)	56.9	97.9
AT→GC	9.2	0	9.7	0.6
GC→TA	5.6	23.8 (15.6)	8.3	0.4
GC→CG	2.9	4.8 (2.6)	6.9	0
AT→TA	8.5	19.0 (6.5)	6.9	0.3
AT→CG	11.7	0 (1.3)	1.4	0.2
+1 fs	0	0	0	0
−1 fs	4.4	19.0 (27.3)	4.2	0.4
del	16.7	0 (15.6)	5.6	0.1
ins	7.8	9.5 (22.1)	0	0
complex	0	19.4 (6.5)	0	0
Total	412	21 (77)	72	1129

These data are for the *lac*I gene in *E. coli*. Only the DNA binding region (nucleotides 29–206) is considered, but the data between brackets for *anti*-BPDE are for the whole gene.
[a] Schaaper and Dunn (1987); [b] Bernelot-Moens *et al.* (1990); [c] Schaaper *et al.* (1987); [d] Pienkowska *et al.* (1993).

Table 3.2. Sequence context of mutational events in *lac*I gene in *E. coli* following ethylmethanesulphonate (EMS; Pienkowska *et al.*, 1993) or UV (Schaaper *et al.*, 1987) treatment for the sites with 5% or more of the mutations

EMS-induced mutants		UV-induced mutants	
42	TAA C GTT	75	GTT T CCC
56	GTC G CAG	89	GTT T CCC
57	TCG C AGA	90	TTT C CCG
75	TCT C TTA	120	TTT C TGC
92	TCC C GCG		
93	CCC G CGT		
120	TTT C TGC		

Table 3.3. Comparison of the numbers of mutants recovered from the *lac*I gene of *E. coli* and the *aprt* gene of Chinese hamster cells after exposure to UV and benzo[*a*]pyrene diolepoxide (*anti*-BPDE)

Class	UV		anti-BPDE	
	aprt CHO[a]	lacI E. coli[b]	aprt CHO[c]	lacI E. coli[d]
GC→AT	17	41	1	2
AT→GC	1	7	0	0
GC→TA	0	6	13	12
GC→CG	4	5	3	2
AT→TA	2	0	2	5
AT→CG	0	1	0	1
+1 fs	1	0	1	0
−1 fs	0	3	1	21
del/dupl	0	0	0	29
dbl subst	7	0	0	0
complex	0	0	0	5
Total	34	72	21	77

[a] Drobetsky *et al.* (1987); [b] Schaaper *et al.* (1987) (DNA binding region only, nucleotides 29–206); [c] Mazur and Glickman (1988); [d] Bernelot-Moens *et al.* (1990) (entire *lac*I gene, Uvr⁻, deletions/duplications includes +/−4 bp hotspot).

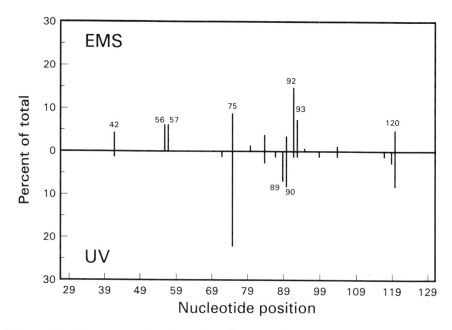

Figure 3.1. Percentage of total number of recovered mutations in the *lac*I gene at sites between the GTG translation initiation codon (position 29) and position 129, after exposure to EMS (Pienkowska *et al.*, 1993) and UV (Schaaper *et al.*, 1987).

3.4 Spontaneous mutation

Spontaneous mutation has generally been viewed as arising during the replication of DNA. Very soon after the elucidation of the structure of DNA, Watson and Crick (1953) recognized the potential for mispairings of natural isomers and tautomers to produce transition events. The range of possibilities was later expanded to include transversion events by Topal and Fresco (1976). An improved understanding of the mechanisms of DNA replication has led to the partitioning of error rates into dNTP selection, incorporation and proof-reading stages. Careful *in vitro* modelling has led to error rates approaching one misincorporation per 10^6 base pairs (Kunkel, 1992). This is far greater than the *in vivo* error rates which approach one per 10^9 or even per 10^{10} (Drake, 1969). This difference has been accounted for by mismatch repair (Glickman and Radman, 1980) which operates following replication to correct errors of incorporation as well as frameshift and deletion slippage intermediates (Schaaper and Dunn, 1987). More recently mismatch repair has been found to play an important role in mammalian cells and a defect in this process has been linked to the familial predisposition for colorectal cancer in Lynch II families (Altonen *et al.*, 1993; Ionov *et al.*, 1993).

3.4.1 Defining spontaneous versus background mutation

Spontaneous mutation is a term that would be best restricted to control values obtained under laboratory conditions where environmental factors such as nutrients, temperature and pH can be controlled. It is not that the effects of these factors are well understood, but reflects the fact that all the organisms being studied experience the same environment and share a common genetic background. In contrast, investigations of natural populations involve diverse genetic backgrounds and different exposures. Consider, for example, a human population where any mutational effects will be confounded by age, occupation and lifestyle considerations. For this reason, the control data from such populations should be denoted as background mutation.

3.4.2 The origins of spontaneous mutation

In addition to mutations arising from errors in DNA replication, mutations can also arise from inaccurate attempts at DNA repair and the incorporation of modified bases having ambiguous base-pairing properties. In addition, at least four endogenous processes are likely to contribute to spontaneous mutation. These are oxidation (Ames, 1983; Harmon, 1981), methylation, deamination and depurination (Saul and Ames, 1986). The spectrum of glycosylases and repair enzymes operational in a cell indicates the challenge these processes most likely present to the integrity of genetic material (Lindahl, 1982).

While each of the above-mentioned pathways can lead to mutation, the key issue is their relative contribution to spontaneous mutation. This is not a simple question. The answer will, to a great extent, depend upon the chosen gene target and the cell type being studied. For example, in the case of the *lac*I gene harboured on an F-prime in *E. coli*, fully two-thirds of the forward *lac*I$^+$ to *lac*I$^-$ mutations involve the gain or loss of 4 base pairs (5'-CTGG-3') within a triple repeat of this sequence (Farabaugh *et al.*, 1978). This frameshift hotspot is also observed when the *lac*I gene is resident on the chromosome (Holliday and Glickman, 1991), but not when it is cloned into the bacteriophage M13 genome (Yatagai *et al.*, 1991). However, in the absence of this type of sequence in the target gene, frameshifts become relatively rare events. Indeed, we have shown that the disruption of the repeat at the frameshift hotspot eliminates these events from the spectrum and results in a 65% reduction in the spontaneous mutation frequency.

The next most predominant spontaneous mutation recovered using this system is the G:C→A:T transition. In the past, speculation has been that this reflects the natural deamination of cytosine to uracil and/or the consequence of O^6-alkylguanine resulting from endogenous alkyl-donors. Both pathways are possible, as exemplified by the raised levels of spontaneous mutation in *E. coli* strains defective in uracil-*N*-glycosylase (Ung$^-$) or alkyltransferase (AGT; Fix and Glickman, 1986; Domoradzki *et al.*, 1984). The fact that the spontaneous mutation frequency increases in strains defective in these DNA repair pathways indicates the potential contribution of these lesions to mutagenesis, although this information alone does not reveal their relative contribution under normal circumstances.

The contribution of deamination to mutagenesis is also evident from the prevalence of G:C→A:T transitions at sites of potential cytosine methylation. In the *E. coli lac*I system, the three amber mutations (6,15 and 34) involving 5'-CC5MeAGG-3' sequences are spontaneous hotspots (Coulondre *et al.*, 1977). This is thought to be related to the deamination of 5-methylcytosine producing thymine rather than uracil, and hence not being removed from DNA by uracil-*N*-glycosylase (Duncan and Miller, 1980). The relevance of this mutational pathway is also evident in mammalian and human cells where G:C→A:T transitions predominate. Indeed, transition events at 5'-CpG-3' sites are disproportionately involved in mutational events leading to a variety of human genetic disorders and cancer involving the *p53* gene (Cooper and Youssoufian, 1988; Hollstein *et al.*, 1991; Rady *et al.*, 1992; Schorderet and Gartler, 1992).

3.4.3 A role for DNA sequence in directing deletion and frameshift events

The observation by Streisinger *et al.* (1966) that frameshift mutations occurred within repeated sequences was the first to suggest a role for DNA sequence in mutation. The mechanism proposed is simple. During replication the reiterated sequences provide an opportunity for slippage to occur. When the strand being replicated slips back on the template, further synthesis results in the insertion of extra bases, whereas when the template slips, bases are lost (see *Figure 3.2a*).

This *misalignment* model has been extended to the formation of deletions and insertions (Albertini *et al.*, 1982). Deletions were found to occur between direct repeats and involve the loss of one of the repeats and all of the sequence between the repeats. Deletions could thus be explained by a model in which the first copy of the repeat became paired with the complement of the second copy of the repeat (*Figure 3.2b*).

The analysis of 729 spontaneous *E. coli lac*I mutants revealed that 46 of the 72 deletion events recovered were flanked by direct repeats (Holliday and Glickman, 1991). Further analysis revealed not only a role for direct repeats, but also for inverted repeats or 'quasipalindromic' sequences as in the model proposed by Glickman and Ripley (1984). Perhaps more important in this study of a large number of spontaneous mutations is the observation that almost all multiply recovered events (15/17) involved sequences where the potential secondary structure of the displaced strand was stabilized by a combination of direct and inverted repeats.

We have recently examined the nature of mutation in the *lac*I gene carried in a transgenic mouse known as Big Blue^R (see Chapter 15). While deletions only account for 2.1% ($n=6/1282$) of spontaneous mutations recovered from a range of tissues, five of these deletions were found to be flanked by direct repeats of 1–8 bp (unpublished results).

(a)
```
                                A-C
                                | |
        5'-T-A-G-A-C-T-T A-C.......
        3'-A-T-C-T-G-A-A-T-G-T-G-T-G-C-A-T-C-T-5'
```

(b)
```
        5'-A-T-G-G-C-A-T-A-C-T-G-G-C-A-T-C-T-T-G-C-A-C-3'
        3'-T-A-C-C-G-T-A-T G-A-C-C-G-T-A-G-A-A-C-G-T-G-5'
                          | |
                          G A
                        A   C
                        C   C
                        C   G
                        G   T
                        T   C
                        A-T-G
```

(c)
```
        5'-T-A-A-G-C-G-T-A-C-T-T-A-A-C-G-C-T-A-G-3'
        3'-A-T-T-C-G-C-A-T-G-X-A-T-T-G-C-G-A-T-C-5'
```

Figure 3.2. DNA sequence can direct mutational events. (a) A repeated sequence resulting in a frameshift mutation. In the example chosen, the extending strand has slipped back and extra bases are being added. This will produce a +2 frameshift. (b) A deletion involves the misalignment of a repeated sequence. In the case illustrated, a 6 base pair direct repeat (in bold) permits the slippage of one copy in the extending strand to pair incorrectly with the complement of the second copy of the repeat. Because of the extended homology, DNA synthesis can continue but the newly synthesized strand has lost one of the two copies of the repeated sequence and all of the sequence in between. (c) A novel model by Kunkel and Soni (1988) demonstrates how a slippage event can bypass a non-instructive DNA lesion and result in a base substitution mutation. In this model a non-instructive lesion has blocked DNA synthesis but the strand is extended by a slippage event where the next base has supplied the template. When the strands reorientate in the proper alignment, the erroneously inserted base supplies the 3' terminus to bypass the lesion.

3.4.4 A role for DNA sequence in directing base substitution events

Sequence-directed events are not restricted to frameshift, deletion and duplication events. A number of examples have been reported where base substitutions were likely to have been templated by misaligned DNA sequences (Golding and Glickman, 1985; Ripley and Glickman, 1982). Fowler *et al.* (1974) proposed a model for dislocation mutagenesis in which base substitution errors would result from mistemplating by transient misalignments of the template–primer complex. Kunkel and his colleagues (Boosalis *et al.*, 1989; Kunkel and Soni, 1988) provided experimental evidence for dislocation mutagenesis and thereby also explained the prevalence of certain base substitutions within specific DNA sequence contexts. According to this model, a transient misalignment would permit DNA polymerase to continue past a non-instructive lesion by allowing the adjacent base to template the strand extension. The subsequent realignment of the strand would then permit the bypass of the lesion (*Figure 3.2c*). The transient misalignment bypass model dictates the base that will be inserted opposite the lesion, and hence the base substitution. This model predicts that altering the base adjacent to a mutational hotspot involving dislocation mutagenesis would change the base substitution observed. This experiment has been tested in the *E. coli lac*I gene and the results can be taken as strong support for the occurrence of this mutational mechanism *in vivo* (W.D. Sedwick, personal communication).

3.5 Induced mutation

Examining the effects of environmental mutagens requires an understanding of the basic mechanisms of mutation. This section describes how mutational specificity can be a useful tool for understanding the nature of mutation. It can be especially powerful when accompanied by some knowledge of the potential DNA lesions. Ultimately, for the examination of weak mutagens, or exposures to low levels of mutagens, the comparisons of mutational spectra of background versus exposed organisms will likely prove to be extremely important.

3.5.1 Ultraviolet light

Several studies have been carried out on the mutational specificity of UV light. Early indications from the study of nonsense mutations in the *E. coli lac*I gene suggested that G:C→A:T mutations predominated at dipyrimidine sites (Coloundre and Miller, 1977). A similar conclusion was reached in one of the first sequencing studies using the single-strand bacteriophage M13 as the target (Brandenburger *et al.*, 1981). The first extensive study of UV-induced mutational specificity involving the sequencing of mutations in the *lac*I gene was that of Schaaper *et al.* (1987) in which mutation was examined in both a repair-proficient (Uvr$^+$) and a repair-deficient (UvrB$^-$) strain of *E. coli*. While the induced mutation frequency per unit dose was three- to four-fold greater in the UvrB$^-$ strain, the distribution of mutations in both strains was rather similar. About 80% of the recovered mutations were base substitutions, 10% were frameshifts and 5% deletions.

The vast majority of mutations recovered were G:C→A:T transitions (*Table 3.1*). These occurred preferentially at dipyrimidine sites (*Table 3.2*). Tandem double events, CC→TT ('CC' are adjacent bases), involving dipyrimidine sites were also recovered, these being considered to be the hallmark of UV mutagenesis. Numerous studies have reported an increase in G:C→A:T, or tandem events, at dipyrimidine sites (Armstrong and Kunz, 1990, 1992; Drobetsky *et al.*, 1987, 1989, 1994; Ivanov *et al.*, 1983; McGregor *et al.*, 1991; Wang *et al.*, 1993; Vrieling *et al.*, 1992).

An important observation is the preferential repair of lesions in the transcribed strand. This occurs in both bacterial and mammalian systems and reflects the coupling of transcription to nucleotide excision repair (NER; Mellon *et al.*, 1987; Mellon and Hanawalt, 1989). The preferential repair of lesions in the transcribed strand is reflected in the specificity of mutation which in transcribed genes occurs primarily in the non-transcribed strand (NTS). Consequently there is a shift of mutations from a bias towards the transcribed strand (TS) in normal human fibroblasts irradiated in early S-phase, when there is little opportunity for repair, to a bias towards the NTS when cells are irradiated in G_1 and there is an opportunity for repair (McGregor *et al.*, 1991).

The premutational lesions for UV light remain unclear. The two primary lesions are the cyclobutane pyrimidine dimer (CPD) and the (6–4) pyrimidine–pyrimidone [(6–4) photoproduct]. Early studies concentrated on the potential role of the CPD in UV mutagenesis. The recognition that the (6–4) photoproduct, which is produced at 5–10% the rate of CPD, may also be a premutagenic lesion has provoked considerable interest (Haseltine, 1983). Data from *E. coli* indicate that the (6–4) photoproduct could be an important contributor to UV mutagenesis (Glickman *et al.*, 1986). However, a broad range of UV-induced lesions has been shown to be mutagenic in engineered plasmid vectors transfected into *E. coli*, although to differing degrees (see *Table 3.4*). Other factors contributing to UV mutagenesis include the deamination of cytosine and the possible photoreactivation of CPDs and (6–4) photoproduct that can occur at significant rates (Tessman and Kennedy, 1991). While the contribution of the different DNA lesions remains controversial, the hallmark of UV-induced mutation remains the G:C→A:T transition and the tandem transitions at pyrimidine sites.

Table 3.4. Summary of the mutagenic and bypass effect obtained with site-specific and stereospecific UV-induced DNA lesions transfected into *E. coli*

Photoproduct	Percentage bypass	Percentage error	Predominant mutation
TT [(6–4) photoproduct, UVC][a]	22.1	91	3'T→C (89%)
TT [(6–4) photoproduct, UVB][a]	12.3	53	3'T→C (25%)
TC [(6–4) photoproduct, UVC][b]	24.5	34	C→T (28%), T→A (5%)
TC [(6–4) photoproduct dewar, UVB])[b]	12.5	79	C→T (36%), T→A (15%)
TT *cis-syn* CPD[c]	19	6	3'T→A (5%)
TT *trans-syn* CPD[d]	29	11	5'T→A (2.5%)
OT (O=abasic site)[e]	7	50	5'T→A (23%),5'T→C (18%)
TO (O=abasic site)[e]	5	23	3' T→C (14%)
UU *cis-syn* (uracil–uracil CPD)[f]	19	5	UU→TA (3.3%)
UU *trans-syn* (uracil–uracil CPD)[f]	9	15	UU→CT (8.8%)
			UU→AT (3.9%)

The bypass rate was calculated by percentage of vectors that were replicated when compared with controls. The percentage error was determined as the ratio of plasmids carrying the altered and the original sequence. The fraction of the predominant mutation reflected the fraction of the total mutants recovered. In the case of the TT (6–4) photoproduct, 89 of 91, or 98% of the mutants were the transition, 3' T→C).
[a] LeClerc *et al.* (1991); [b] Horsfall and Lawrence (1994); [c] Banerjee *et al.* (1988); [d] Banerjee *et al.* (1990); [e] Lawrence *et al.* (1990); [f] Gibbs *et al.* (1993).

Sequence context plays a major role in the deposition of DNA damage. Using 'average' DNA, the relative formation of UV-induced cyclobutane dimers at different dipyrimidine sites has been estimated by Mitchell *et al.* (1992) and Kotturi *et al.* (unpublished) as TT>TC≈CT>CC in the ratio of 52:21:19:7 for UVB light (280–320 nm) and (68:16:13:3) for UVC light (240–280 nm). In general, there is a trend towards increased levels of UV-induced damage in regions rich in pyrimidines (Brash *et al.*, 1987; Koehler *et al.*, 1991) and significant site-to-site variation is observed (Brash *et al.*, 1987; Koehler *et al.*, 1991; Kotturi *et al.*, unpublished; Pfeifer *et al.*, 1991; Sage *et al.*, 1992). There is a good correlation between the high frequency of CPD and (6–4) photoproducts at 5'-TCC-3' sites and the G:C→A: Ttransition mutational hotspots in the adenine phosphoribosyltransferase (*aprt*) locus in Chinese hamster ovary (CHO) cells (Drobetsky and Sage, 1993).

One reason for the lack of clarity is the possibility that both lesions are capable of contributing to UV mutagenesis. A second consideration is that while the initial deposition of damage within a DNA sequence can be determined, DNA repair is both strand- and sequence-specific. The current models for mutagenesis and carcinogenesis suggest that the rate of repair at a given site is more important than the amount of DNA damage (Kunala and Brash, 1992; Tornaletti and Pfeifer, 1994). In studies at modest UV fluences (20 J m^{-2}), Tornaletti and Pfeifer (1994) and Gao *et al.* (1994) measured the rate of repair in the human *p53* and *PGK1* genes and found that the site-to-site variation in the rate of excision of CPD lesions was as great as 15-fold. Although no rules can yet be formulated to predict the sites of slow repair, considerable data from the studies of mutational specificity indicate the relevance of repair rates to mutagenesis (Kunala and Brash, 1992).

3.5.2 Polyaromatic hydrocarbons – benzo[a]pyrene

The first documented case of induced cancer was the description of scrotal cancers in chimney sweeps (Pott, 1775) and over 150 years elapsed before the polycyclic aromatic hydrocarbons (PAHs) were recognized as the likely source of the elevated risk of cancer. Typical PAHs include such compounds as benzo[a]pyrene (B[a]P), 5-methylchrysene and 7,12-dimethylbenz[a]anthracene (DMBA).

PAHs require activation to their ultimate mutagenic form and, as has been shown for B[a]P and DMBA, are often potent carcinogens (Mane *et al.*, 1990). Activation results in dihydrodiol epoxides of these compounds that react and form bulky adducts primarily with exocyclic amino groups of guanine (N^2) and adenine (N^6) which open the epoxide ring.

Benzo[a]pyrene is converted into its reactive metabolites, the diol epoxides (Weinstein *et al.*, 1976), by the enzymatic action of P450s and epoxide hydrolase (Thakker *et al.*, 1985). The racemic mixtures of the diol epoxide stereoisomers that have the potential to bind to DNA are: (1) (±)-r-7,t-8-dihydroxy-9,t-10-epoxy-7,8,9,10-tetrahydro-benzo[a]pyrene (*anti*-BPDE); (2)

($\pm$)-r-7,t-8-dihydroxy-c-9,10-epoxy-7,8,9,10-tetrahydro-benzo[a]pyrene (*syn*-BPDE); and (3) ($\pm$)-r-t-9,10-dihydroxy-c-7,8-epoxy-7,8,9,10-tetrahydro-benzo[a]pyrene (Tang *et al.*, 1992). These epoxides alkylate nucleic acids and predominantly (95% of the adducts) form a covalent bond with the exocyclic N^2 amino group of guanine (Sayer *et al.*, 1991). Adducts at the N^6 position of adenine are formed, but to a lesser degree (Cheng *et al.*, 1989; Harvey, 1979; Sayer *et al.*, 1991). Other minor adducts form at N–7 amino group of guanine (King *et al.*, 1979), deoxycytosine bases (Meehan *et al.*, 1977) and those which are present due to alkylation of denatured DNA (Sayer *et al.*, 1991).

The various reactive diol epoxides have different mutagenic potential and much effort has been devoted to their study. Each of the three racemic diol epoxides has two optically active isomeric forms. The reaction of each of these with DNA can result in either *cis* or *trans* addition. Thus a wide spectrum of different stereostructural adducts is possible. The (+)-enantiomer of *anti*-BPDE is considered the most mutagenic in mammalian cells and only this adduct was shown to be strongly carcinogenic when applied to mouse skin (Buening *et al.*, 1978; Slaga *et al.*, 1979). On the other hand, the (–)-enantiomer is more mutagenic in bacteria, when similar levels of adduct formation are compared (Carothers *et al.*, 1988; King and Brookes, 1984; Stevens *et al.*, 1985; Wood *et al.*, 1977).

The complexity of all the different possible adducts makes it an interesting problem to deduce which adducts are formed and where. The replication of plasmids engineered to contain an adduct whose structure and position are strictly defined allows the recovery of mutations induced by specific lesions. Mackay *et al.* (1992) determined the mutagenic specificity of the (+)-*anti*-BPDE, by incorporating the (+)-*anti*-BPDE-N^2-Gua adduct as part of the sequence 5'-CT$\underline{G}$CA-3' in a plasmid. Replication resulted almost exclusively (57/58) in G:C→T:A transversions (Mackay *et al.*, 1992). When the 5' flanking T was replaced with any other base, the contribution of G:C→T:A decreased to 65% (Rodriguez and Loechler, 1993a). This change in spectrum was attributed to 'adduct structural polymorphism', in which the adduct conformation is modulated by the local sequence context. A similar effect is observed with other bulky adducts, such as the AAF-C8-Gua adduct, which have been shown to adopt different conformations (Belguise-Valladier and Fuchs, 1991; Veaute and Fuchs, 1991). The influence of local sequence context on the conformation of an adduct was further investigated by Rodriguez and Loechler (1993b) by comparing the mutational spectra of an (+)-*anti*-BPDE-adducted plasmid by varying the 5' base and the adducted plasmid treatment prior to transformation and analysing mutation with respect to the 5' sequence context. The mutational alterations at 5'-T$\underline{G}$-3' sequences were predominantly G:C→T:A transversions (27/29) while the spectrum at 5'-G$\underline{G}$-3' sequences consistently yielded a lower percentage of G:C→T:A mutations (31/52). This indicated two possible adduct conformations. An interesting observation was made regarding the mutational pattern of one of the 'hotspots' for base substitution, at a single 5'-C$\underline{G}$-3' site (5'-C$\underline{G}_{115}$-3'). Before the adducted plasmid was heated, the mutational pattern resembled that of 5'-TG-3'sequences, predominantly G:C→T:A mutations (13/15). After heating, the pattern shifted to that of the 5'-GG-3' sequence context resulting in only 15/33 G:C→T:A transversions. Heating had no significant effect on the mutational pattern of 5'-TG-3', or 5'-GG-3' sites. Further investigation revealed that apurinic sites were not formed at an appreciable rate concurrent with the shift in mutational spectra (Drouin and Loechler, 1993). The difference in mutational specificity seemed to result from adduct conformation which was influenced both by heat and by local sequence context.

These results with *anti*-BPDE indicate the importance of adducts and mutations at G:C base pairs. The sequence specificity of adduct formation has been examined in polymerase-pause studies with modified T7 DNA polymerase (Thrall *et al.*, 1992). Polymerase-pause sites are taken as an indication of adduct formation. Such sites predominantly involve G:C base pairs, especially at runs of two or more guanines. Differences at the various guanine residues indicate sequence-specific context effects which may reflect either sequence-specific properties or the presence of different conformations of the bulky adduct (Rodriguez and Loechler, 1993b).

Consistent with the importance of G:C sites for adduct formation is the observation that the G:C→T:A transversion is the hallmark of *anti*-BPDE-induced mutations. This transversion represents about 60% of the induced base substitutions (Bernelot-Moens *et al.*, 1990; Chen *et al.*, 1990; Mazur and Glickman, 1988; Rodriguez and Loechler, 1993b; Yang *et al.*, 1987). However, the complete mutational spectrum is quite complex, much more so than for example, after treatment with an alkylating agent such as EMS (*Table 3.1*). Following *anti*-BPDE treatment mutations such as G:C→A:T transitions, G:C→C:G transversions and frameshifts also occur (Mazur and Glickman, 1988; Rodriguez and Loechler, 1993b; Zhu *et al.*, 1994). The specificity of *anti*-BPDE is quite similar in both mammalian and bacteria systems with some cases of an increased recovery of G:C→C:G transversions in mammalian cells.

Base substitutions recovered in the *lacI* gene of *E. coli* (Bernelot-Moens *et al.*, 1990) following BPDE treatment were predominantly G:C→T:A transversions (*Table 3.3*). However, 50% of the mutations recovered were −1 bp frameshifts which occurred mostly in runs of guanines. BPDE-induced mutations have also been analysed in the *aprt* gene of CHO cells by Mazur and Glickman (1988). Again, the predominant mutations were found to be G:C→T:A transversions (*Table 3.3*). Further analysis suggested that their occurrence was biased towards runs of guanines, especially when flanked 5' by adenine. BPDE also induced frameshift mutations in runs of guanine in the human *aprt* gene (Zhu *et al.*, 1994) and in the *E. coli lacI* gene in a transgenic mouse model by Kohler *et al.* (1991).

It is not surprising, considering the overall similarity of mutational spectra of *anti*-BPDE in bacterial and mammalian systems, that following exposure of animals to B[*a*]P the Ha-*ras* mutations recovered from animal tumours were consistent with expectations for this mutagen (Bizub *et al.*, 1986; Quintanilla *et al.*, 1986).

3.5.3 Acetylaminofluorene

Acetylaminofluorene (AAF), is a prototype carcinogenic and mutagenic aromatic amine that produces DNA adducts *in vivo* (Beland and Kadlubar, 1985). A range of DNA adducts are produced, but the C8-adduct of guanine is considered to be the prime premutagenic lesion. Work by Shibutani and Grollman (1993a), using site-specific lesions as the template for strand extension, indicates that the favoured insertion opposite this lesion *in vitro* is adenine. This predicts a G:C→T:A transversion *in vivo* and this is the mutational specificity observed in the *lacI* gene in *E. coli* (Schaaper *et al.*, 1990) and in M13 plasmid targets (Gupta *et al.*, 1988).

Frameshift mutations have also been recovered *in vitro* (Shibutani and Grollman, 1993b). The frameshift events are context-dependent, reflecting the bypass of lesions by a dislocation translesion mechanism as predicted by Kunkel (see Section 3.4.4). Similar frameshifts were observed *in vivo* in the AAF-induced spectrum in *lacI* (Schaaper *et al.*, 1990; Burnhouf *et al.*, 1989); each of the three guanines in the *NarI* restriction sequence 5'-GGCGCC-3' were found to be targets for frameshifts. The repetitive nature of the target sequence is consistent with the role of DNA sequence context in determining mutational outcomes.

Interestingly, differences in adduct structure can be shown to have dramatic consequences on mutation spectra. The loss of an acetyl group from *N*-2-acetylaminofluorene (AAF) to *N*-2-aminofluorene (AF) changes the primary mutational event from frameshift to base substitution (Fuchs *et al.*, 1981; Koffel-Schwartz *et al.*, 1984). This has been attributed to a difference in the amount of helix distortion. It is also worth noting that this magnitude of a change in spectra will have dramatic consequences for the biological effects of these agents. In mutation research this will be reflected in differences in mutational potency in different target genes and selection systems.

3.5.4 Cis-*diamminedichloroplatinum and carboplatin*

Cis-diamminedichloroplatinum (cisplatin, *cis*-DDP) is a well-characterized mutagenic antitumour drug which is used in the clinical treatment of ovarian, head and neck, and testicular tumours. Its antitumour activity is based on cytoxicity resulting from interactions with DNA (Lepre and Lippard, 1990; Roberts and Thomson, 1979). As a consequence there is some evidence suggesting that there may be some long-term side-effects. Rats and mice develop leukaemia, skin tumours and other carcinomas after exposure to *cis*-DDP (Barnhart and Bowden, 1985; Kempf and Ivankovic, 1986; Leopold *et al.*, 1979) and studies with human lymphocytes provide evidence of chromosome breakage and rearrangements (Morin-Faure and Marcollet, 1983).

Cis-DDP binds with the *N*7 of purines (Pinto and Lippard, 1985) to form intrastrand adducts in a two-step process in which *cis*-DDP preferentially binds with the *N*7 of Gua and then links with a neighbouring purine

(Fichtinger-Schepman *et al.*, 1985). The distribution of adducts types at low *cis*-DDP concentrations is: 65% *cis*-(Pt(NH$_3$)$_2$ (d(GpG)-N7(1),-N7(2))) (*cis*-DDP GG12); 25% *cis*-(Pt(NH$_3$)$_2$ (d(ApG)-N7(1),-N7(2))) (*cis*-DDP AG12); 5% *cis*-(Pt-(NH$_3$)$_2$ (d(GpNpG)-N7(1),-N1(3))) (*cis*-DDP GNG13); and 1% *cis*-(Pt(NH$_3$)$_2$ (d(GpC)-N7(1))/(d(CpG)-N7(4))) (*cis*-DDP GC/CG14). The percentage of interstrand cross-links (*cis*-DDP GC/CG14) has been reported as lying between 5 and 10% of the total adducts in human extracts (Calsou *et al.*, 1992; Hansson and Wood, 1989) and CHO cells (Jones *et al.*, 1991).

In general, the sequence specificity of mutations parallels the deposition of DNA damage. A study by Brouwer *et al.* (1981) of the induction of nonsense mutation in the *lac*I gene of *E. coli* found the guanines in the 5'-GAG-3' and 5'-GCG-3' sequences to be hotspots of base substitutions. This could be taken as an indication that the 25% of *cis*-DDP-AG12 lesions and the 5% *cis*-DDP-GNG are the most potent lesions. However, an examination of the available sites in the *lac*I gene reveals that a restricted number of available sites may have influenced the mutations recovered. A study by Burnhouf *et al.* (1989) in which the tet^R plasmid, pBR322, was treated *in vitro* and transfected into *E. coli* showed that *cis*-DDP induced SOS-dependent mutations. Mutations occurred most frequently at 5'-AG-3' and 5'-GG-3' sequences which correlates closely with the distribution of DNA damage sites. Mutations were also found at the 5'-GAG-3' site as in the *lac*I study.

Cis-DDP has been reported to be mutagenic in both CHO and V79 cells (O'Neill *et al.*, 1977; Zwelling *et al.*, 1979) and the specificity of *cis*-DDP was determined in the *aprt* gene of CHO cells (de Boer and Glickman, 1989). This study revealed a broad range of mutational events including transversions, transitions, frameshifts, and short deletions and duplications occurring at, or proximal to, 5'-AGG-3' and 5'-GAG-3' sequences (Cariello *et al.*, 1992).

An exciting example of the use of mutational spectra to determine mutational mechanisms can be drawn from a study of another Pt compound using the CHO *aprt* system. In this case the mutational specificity of a related Pt compound, carboplatin, was examined (de Boer and Glickman, 1992). The mutational spectrum produced by carboplatin was found to be identical to that recovered following treatment with *cis*-DDP. This suggests that both compounds share a common mutational mechanism.

3.6 The molecular aetiology of cancer and the *p53* tumour suppressor gene

The molecular nature of mutation in human tumours can be assessed *in vivo* by the study of mutations in tumours. The best studied case is the tumour suppressor gene *p53* which has been implicated in all major human malignancies. A *p53* database with over 2500 entries has been compiled by Hollstein *et al.* (1994) and Cariello *et al.* (1994a) who developed software for its analysis. The gene is well suited for mutational analysis as over 90% of these mutations are concentrated within a conserved region of the gene. Moreover, the 100 and more different codons involved make *p53* a good target for molecular epidemiology which we define as the distribution of molecular characteristics in a defined population. In this sense, the study of *p53* mutations is of particular relevance. The knowledge of the precise nature of the changes in DNA at the sequence level provides insights into the nature of the mutational process in people *in vivo*.

3.6.1 Characteristics of the p53 *gene as a mutational target*

The human tumour suppressor gene *p53* encodes a 393 amino acid nuclear phosphoprotein which has been established as an important control factor in the G_1/S check-point of cell division and a regulator of controlled cell death, apoptosis. The sequence of this protein in mammals, chicken, *Xenopus* and rainbow trout exhibits five highly conserved protein domains (Soussi *et al.*, 1989). The DNA sequence of the human *p53* gene spans 20 kb and contains 11 exons. The conserved protein region corresponds to exons 5 to 9 of the genomic DNA, or about 1.6 kb, which makes it an approachable target for DNA sequencing and mutational analysis. The crystal structure of the central conserved domain of the human P53 protein has been elucidated and the results show a fascinating array of structural motifs involved in DNA binding, metal ion binding and stabilization (Cho *et al.*, 1994). Although more than 100 different sites are affected by codon mutations in human cancers, mutational hotspots at codons 175, 248, 249 and 273 participate in DNA binding and stabilization (Cho *et al.*, 1994). Interactions of the P53 protein with DNA as transcriptional regulator and with other proteins are important in our current understanding of its function as an inhibitor of cell division and in apoptosis (reviewed in Zambetti and Levine, 1993). The importance of *p53* in non-human carcinogenesis is shown by the P53 null mouse model: mice lacking the *p53* gene develop multiple cancers at an early age and all die of cancer prematurely (Donehower *et al.*, 1992). Mice hemizygous for the *p53* locus also develop cancer earlier and more frequently that normal animals (Donehower *et al.*, 1992).

It is essential to bear in mind, however, that mutation represents the consequence of a series of processes. These include the original deposition of damage, the action of diverse repair systems, gender-specific factors, cell-cycle status, as well as developmental processes that all play a role in determining the mutational spectrum. It is also important to recognize that mutations are selected in specific target genes so that the observed spectra depend upon the nature of the gene and its protein product.

As the mutations in *p53* are taken from a large number of studies using a variety of techniques, any meta-study on the distributions of mutations in this target has to take into account the potential bias introduced by the technique used to detect and sequence the base changes in this gene. Many studies focus on the conserved exon 5–8 region of the coding sequence of *p53*. This region includes 181 codons or 540 bp, of which 332 or 61% have already been found mutated in human tumours. This region of the *p53* gene thus represents a target comparable in size to the *hprt* gene which includes 219 codons or 657 bp of which 238 or 36% have been found to confer 6-thioguanine resistance (Cariello, 1994a). The detection method used most commonly is single-strand conformation polymorphism analysis (SSCP; Orita *et al.*, 1989); other methods used include denaturing gradient gel electrophoresis (DGGE) and constant denaturing capillary electrophoresis (CDCE). A comparison of methods has shown them able to detect at least 90% of the mutations occurring in the target studied (Condie *et al.*, 1993).

The analysis of mutations in the *p53* gene by type of tissue reveals characteristics which can help to contribute to our understanding of the identity of the aetiological agent (see also Greenblatt *et al.*, 1994). Here we outline four examples: (i) skin cancer, (ii) colon cancer, (iii) liver cancer, and (iv) bladder cancer.

3.6.2 Skin cancer

Analysis of *p53* mutations in invasive squamous cell carcinomas demonstrated a mutational specificity consistent with exposure to UV light (Brash *et al.*, 1991). Specifically, *p53* mutations often occur at dipyrimidine sites and include several tandem double-base substitutions (most often CC→TT), a feature characteristic of UV-induced mutation (Drobetsky *et al.*, 1987). An analysis of the *p53* database shows that tandem base substitutions in the *p53* gene comprise 18% of the mutations found in skin cancer. In contrast, less than 0.1% of the mutations found in all other cancer sites combined involve tandem double events (Hollstein *et al.*, 1994). The origin of this highly specific UV light fingerprint reflects the nature of the UV-specific premutagenic DNA lesions, the cyclobutane pyrimidine dimer and the (6–4) photoproduct (see Section 3.5.1).

3.6.3 Colon cancer

The mutations in *p53* detected in colon cancer are dominated by G:C→A:T transitions (63%, $n=960$), and most of them (47% of all *p53* mutations found in colon cancer) occur at CpG sites (Hollstein *et al.*, 1994). As described earlier, transitions at CpG sites are characteristic of mutations arising spontaneously in mammalian cells (de Jong *et al.*, 1988) and are thought to result from the spontaneous deamination of 5-methylcytosine. The high incidence of G:C→A:T transitions suggests that mutational events in colon cancer might be dominated by spontaneous promutagenic events including deamination and replication errors. Replication errors might be of particular importance considering the high turnover rate of colonic epithelial cells. This would fit quite well with the epidemiological evidence indicating that DNA-mismatch repair is an important process in the control of spontaneous mutation in the colon. This is supported by the observation that a defect in mismatch repair is characteristic of hereditary, non-polyposis colon cancer (HNPCC; Aaltonen *et al.*, 1993; Ionov *et al.*, 1993).

3.6.4 Liver cancer

A strong correlation has been observed between the dietary intake of aflatoxin B_1, the excretion of aflatoxin-B_1-$N7$-guanine in urine, and the incidence of liver cancer (Donahue *et al.*, 1982). High-risk regions are also often characterized by high rates of hepatitis B infection. One such region is Qidong Province in China where the mutational spectrum of *p53* mutations detected in liver tumours has been investigated. The mutational spectrum is typical for exogenous mutagen exposure with G:C→T:A transversions at codon 249 of *p53* accounting for over 50% (12/21) of the mutations (Hsu *et al.*, 1991). The predominance of this event and the appearance of a hotspot are consistent with induction by chemical exposure. Aflatoxin B_1 induces predominantly G:C→T:A transversions in *E. coli lac*I (Foster *et al.*, 1983) and *supF* (Courtemanche and Anderson, 1994) as well as in the human *hprt* gene exon 3, the latter exhibiting a distinct G:C→T:A transversion hotspot at position 209 (Cariello *et al.*, 1994b). This specific transversion is also consistent with expectations for a polyaromatic compound such as aflatoxin B_1.

3.6.5 Bladder cancer

The spectrum of *p53* mutation found in bladder tumours has proved unique. The spectrum contains a high proportion of G:C→C:G transversions and reveals double mutations both in smokers and non-smokers (Kusser *et al.*, 1994; Spruck *et al.*, 1993). The G:C→C:G transversion is normally a relatively rare event. It represents only 8% of all the *p53* mutations recovered so far and just 3% in colon cancer (n=960). In contrast, in bladder cancer G:C→C:G transversions account for 21% of the mutations. Their origin is still somewhat speculative. A possible connection to oxidative DNA damage has been suggested (Spruck *et al.*, 1993), and indeed this rare mutation has been induced by the singlet oxygen-generating mutagen methylene blue plus light (McBride *et al.*, 1992). It is also of interest to note that several aromatic compounds (e.g. benzene) generate oxygen radicals during their metabolism. However, other factors in addition to oxidative damage must also be considered. Endogenous factors and bladder-specific metabolism may contribute. For example, allelic differences in the gene encoding a glutathione-*S*-transferase (see Chapter 5) have been identified as a bladder cancer risk factor (Bell *et al.*, 1993). We have found that the spontaneous mutational spectrum of the *lac*I gene in Big Blue[R] transgenic mice bladder is slightly different from that observed for any other of the tissues tested to date (de Boer *et al.*, unpublished results).

As *p53* mutations appear to occur late in the development of bladder cancer (Kusser *et al.*, 1994), the spectrum of mutation in this gene may reflect the microenvironment in the cell rather than specific exogenous mutagens. The presence of two *p53* mutations within one tumour suggests a possible positive selection effect for these mutations and/or the possible occurrence of a more general genetic instability. The latter would not be surprising considering the requirement for several sequential mutations during progression from a normal cell to a tumour.

3.7 The future implications of mutational specificity

Mutational specificity has spear-headed a major advance in the field of genetic toxicology. The demonstration that mutation is almost exclusively determined by DNA damage and the development of rapid cloning and sequencing techniques have altered the field of mutation research. Our ability to characterize background mutation at the level of DNA sequence presents the promise of the detection of very low levels of mutation, resulting from exposure either to low levels of mutagens or to weak mutagens. In addition, a knowledge of the nature of the mutational alterations and DNA sequence context will have a significant impact on our understanding of premutational lesions and their repair.

Future developments are likely to build on our knowledge of mutational specificity. Transgenic animals including bivalves, fish and rodents will continue to be developed as will techniques to monitor the genetic health of populations. Perhaps new techniques will eventually permit the assessment of mutation without phenotypic selection, complementing the current transgenic models. More likely, new techniques such as sequencing by hybridization will relieve the technical challenge of assessing mutation in large numbers of individuals.

In the more immediate future, however, effective statistical approaches are still required for the meaningful comparison of mutational spectra, as are new algorithms for the storage and analysis of mutational data. The ability to interpret mutational spectra is accompanied by a broad range of demands and reflects advances in many fields besides biology.

References

Aaltonen LA, Peltomaki P, Leach FS, Sistonen P, Rylkkanen L, Mecklin JP, Jarvinen H, Powell SM, Jen J, Hamilton SR, Petersen GM, Kinzler KW, Vogelstein B, de la Chapelle A. (1993) Clues to the pathogenesis of familial colorectal cancer. *Science* 260: 812–816.

Albertini AM, Hoffer M, Calos MP, Miller JH. (1982) On the formation of spontaneous deletions: the importance of short sequence homologies in the generation of large deletions. *Cell* 29: 319–328.

Ames BN. (1983) Dietary carcinogens and anti-carcinogens: oxygen radicals and degenerative diseases. *Science* 221: 1256–1264.

Armstrong JD, Kunz BA. (1990) Site and strand specificity of UV-B mutagenesis in the SUP4-o gene of yeast. *Proc. Natl Acad. Sci. USA* 87: 9005–9009.

Armstrong JD, Kunz BA. (1992) Photoreactivation implicates cyclobutane dimers as the major promutagenic UVB lesions in yeast. *Mutat. Res.* 268: 83–94.

Banerjee SK, Borden A, Christensen RB, LeClerc JE, Lawrence CW. (1990) SOS-dependent replication past a single *trans-syn* T–T cyclobutane dimer gives a different mutation spectrum and increased error rate compared with replication past this lesion in uninduced cells, *J. Bacteriol.* 172: 2105–2112.

Banerjee SK, Christensen RB, Lawrence CW, LeClerc JE. (1988) Frequency and spectrum of mutations produced by a single *cis-syn* thymine–thymine cyclobutane dimer in a single-stranded vector. *Proc. Natl Acad. Sci. USA* **85:** 8141–8145.

Barnhart KM, Bowden GT. (1985) Cisplatin as an initiating agent in two-stage mouse skin carcinogenesis. *Cancer Lett.* **29:** 101–105.

Basu AK, Essigmann JM. (1988) Site-specifically modified oligodeoxynucleotides as probes for the structural and biological effects of DNA-damaging agents. *Chem. Res. Toxicol.* **1:** 1–18.

Beland FA, Kadlubar FF. (1985) Formation and persistence of arylamine DNA adducts *in vivo. Environ. Hlth Perspect.* **62:** 19–30.

Belguise-Valladier P, Fuchs RPP. (1991) Strong sequence-dependent polymorphism in adduct-induced DNA structure analysis of single *N*-2-acetylaminofluorene residues bound within the *NarI* mutation hot spot. *Biochemistry* **30:** 10091–10100.

Bell DA, Taylor JA, Paulson DF, Robertson CN, Mohler JL, Lucier GW. (1993) Genetic risk and carcinogen exposure: a common inherited defect of the carcinogen-metabolism gene glutathione *S*-transferase M1 (GSTM1) that increases susceptibility to bladder cancer. *J. Natl Cancer Inst.* **85:** 1159–1164.

Bernelot-Moens C, Glickman BW, Gordon AJE. (1990) Induction of specific frameshift and base substitution events by benzo[*a*]pyrene diol epoxide in excision-repair-deficient *Escherichia coli. Carcinogenesis* **11:** 781–785.

Bizub D, Wood AW, Skala AM. (1986) Mutagenesis of the Ha-*ras* oncogene in mouse skin tumors induced by polycyclic aromatic hydrocarbons. *Proc. Natl Acad. Sci. USA* **83:** 6048–6052.

Boosalis MS, Mosbaugh DW, Hamatake R, Sugino A, Kunkel TA, Goodman MF. (1989) Kinetic analysis of base substitution mutagenesis by transient misalignment of DNA and by miscoding. *J. Biol. Chem.* **264:** 11360–11366.

Bradley LJN, Yarema KJ, Lippard SJ, Essigmann JM. (1993) Mutagenicity and genotoxicity of the major DNA adduct of the antitumor drug *cis*-diamminedichloroplatinum (II). *Biochemistry* **32:** 982–988.

Brandenburger A, Godson GN, Radman M, Glickman BW, van Sluis CA, Doubleday OP. (1981) Radiation-induced base substitution mutagenesis in single-stranded DNA phage M13. *Nature* **294:** 180–182.

Brash DE, Rudolph AR, Simon JA, Lin A, McKenna GJ, Baden HP, Halperin AJ, Ponten J. (1991) A role for sunlight in skin cancer: UV-induced *p53* mutations in squamous cell carcinoma. *Proc. Natl Acad. Sci. USA* **88:** 10124–10128.

Brash D, Seetharam S, Kraemer KH, Seidman MM, Bredberg A. (1987) Photoproduct frequency is not the major determinant of UV base substitution hot spots or cold spots in human cells. *Proc. Natl Acad. Sci. USA* **84:** 3782–3786.

Brouwer J, Van der Putte J, Fichtinger-Schepman AMJ, Van Reedijk J. (1981) Base pair substitution hotspots in GAG and GCG nucleotide sequences in *Escherichia coli* K-12 induced by *cis*-diamminedichloroplatinum(II). *Proc. Natl Acad. Sci. USA* **78:** 7010–7014.

Buening MD, Wislocki PG, Levin W, Yagi H, Thakker DR, Akagi H, Koreeda M, Jerina DM, Conney AH. (1978) Tumorigenicity of the optical enantiomers of the diastereomeric benzo[*a*]pyrene-7,8-diol-9,10-epoxides in newborn mice: exceptional activity of (+)-7-β,8-α-dihydroxy-9,10-α-epoxy-7,8,9,10-tetrahydrobenzo[*a*]pyrene. *Proc. Natl Acad. Sci. USA* **75:** 5358–5361.

Burnhouf D, Koehl P, Fuchs RPP. (1989) Single adduct mutagenesis: strong effect of the position of a single acetylaminofluorene adduct within a mutation hot spot. *Proc. Natl Acad. Sci. USA* **86:** 4147–4151.

Burnouf D, Daune M, Fuchs RPP. (1987) Spectrum of cisplatin-induced mutations in *Escherichia coli. Proc. Natl Acad. Sci. USA* **84:** 3758–3762.

Calsou P, Frit P, Salles B. (1992) Repair synthesis by human cell extracts in cisplatin-damaged DNA is preferentially determined by minor adducts. *Nucleic Acids Res.* **20:** 6363–6368.

Cariello NF. (1994) Database and software for the analysis of mutations at the human *hprt* gene. *Nucleic Acids Res.* **22:** 3547–3548.

Cariello NF, Beroud C, Soussi T. (1994a) Database and software for the analysis of mutations at the human *p53* gene. *Nucleic Acids Res.* **22:** 3549–3550.

Cariello NF, Cui L, Skopek TR. (1994b) *In vitro* mutational spectrum of Aflatoxin B₁ in the human hypoxanthine guanine phosphoribosyl transferase gene. *Cancer Res.* **54:** 4436–4441.

Cariello NF, Swenberg JA, Skopek TR. (1992) *In vitro* mutational specificity of cisplatin in the human hypoxanthine 3 guanine phosphoribosyltransferase gene. *Cancer Res.* **52:** 2866–2873.

Carothers AM, Urlaub G, Grunberger D, Chasin LA. (1988) Mapping and characterization of mutations induced by benzo[*a*]pyrene diol epoxide at dihydrofolate reductase locus in CHO cells. *Somat. Cell Mol. Genet.* **14:** 169–183.

Chen RH, Maher VM, Mccormick JJ. (1990) Effect of excision repair by diploid human fibroblasts on the kinds and locations of mutations induced by (racemic)-7-β,8-α-dihydroxy-9,10-α-epoxy-7,8,9,10-tetrahydrobenzo[*a*]pyrene in the coding region of the HPRT gene. *Proc. Natl Acad. Sci. USA* **87:** 8350–8354.

Cheng SC, Hilton BD, Roman JM, Dipple A. (1989) DNA adducts from carcinogenic and noncarcinogenic enantiomers of benzo[*a*]pyrene dihydrodiol epoxide. *Chem. Res. Toxicol.* **2:** 334–340.

Cho Y, Gorina S, Jeffrey PD, Pavletich NP. (1994) Crystal structure of a *p53* tumor suppressor–DNA complex: understanding tumorigenic mutations. *Science* **265:** 346–355.

Comess KM, Burstyn JN, Essigmann JM, Lippard SJ. (1992) Replication inhibition and translesion synthesis on templates containing site-specifically placed *cis*-diamminedichloroplatinum(II) DNA adducts. *Biochemistry* **31:** 3975–3990.

Condie A, Eeles R, Borresen AL, Coles C, Cooper C, Prosser J. (1993) Detection of point mutations in the *p53* gene: comparison of single-strand conformation polymorphism. Constant denaturant gel electrophoresis, and hydroxylamine and osmium tetroxide techniques. *Hum. Mutat.* **2:** 58–66.

Cooper D, Youssoufian H. (1988) The CpG dinucleotide and human genetic disease. *Hum. Genet.* 151–155

Coulondre C, Miller JH. (1977) Genetic studies of the lac repressor. IV. Mutagenic specificity in the *lac*I gene of *Escherichia coli*. *J. Mol. Biol.* **117:** 577–606.

Courtemanche C, Anderson A. (1994) Shuttle-vector mutagenesis by aflatoxin B₁ in human cells: effect of sequence context on the *supF* mutational spectrum. *Mutat. Res.* **306:** 143–151.

De Boer JG, Glickman BW. (1989) Sequence specificity of mutation induced by the anti-tumor drug cisplatin in the CHO *aprt* gene. *Carcinogenesis* **10:** 1363–1367.

De Boer JG, Glickman BW. (1992) Mutations recovered in the Chinese hamster *aprt* gene after exposure to carboplatin: a comparison with cisplatin. *Carcinogenesis* **13:** 15–17.

de Jong PJ, Grosovsky AJ, Glickman BW. (1988) Spectrum of spontaneous mutations at the *APRT* locus of Chinese hamster ovary cells: an analysis at the DNA sequence level. *Proc. Natl Acad. Sci. USA* **85:** 3499–3503.

Domoradzki J, Pegg AE, Dolan ME, Maher VM, McCormick JJ. (1984) Correlation between O⁶-methylguanine, DNA-methyltransferase activity and resistance of human cells to cytotoxic and mutagenic effect of *N*-methyl-*N*′-nitro-*N*-nitrosoguanidine. *Carcinogenesis* **5:** 1641–1647.

Donahue PR, Essigmann JM, Wogan GN. (1982) Aflatoxin DNA adducts: detection in urine as a dosimeter of exposure. In: *Indicators of Genotoxic Exposure* (eds BA Bridges, BE Butterworth, IB Weinstein). Cold Spring Harbor Laboratory Press, Cold Spring Harbor, NY, pp. 221–229.

Donehower LA, Harvey M, Slagle BL, Arthur MJ, Montgomery Jr CA, Butel JS, Bradley A. (1992) Mice deficient in *p53* are developmentally normal but susceptible to spontaneous tumors. *Nature (Lond.)* **356:** 215–221.

Drake JW. (1969) Comparative rates of spontaneous mutagenesis. *Nature (Lond.)* **221:** 1132.

Drobetsky EA, Sage E. (1993) UV-induced G:C to A:T transitions at the aprt locus of Chinese hamster ovary cells cluster at frequently damaged 5′-TCC-3′ sequences. *Mutat. Res.* **289:** 131–136.

Drobetsky EA, Grosovsky AJ, Glickman BW. (1987) The specificity of UV-induced mutations at an endogenous locus in mammalian cells. *Proc. Natl Acad. Sci. USA* **84:** 9103–9107.

Drobetsky EA, Grosovsky AJ, Skandalis A, Glickman BW. (1989) Perspectives on UV light mutagenesis: investigation of the CHO *aprt* gene carried on a retroviral shuttle vector. *Somat. Cell Mol. Genet.* **15:** 401–409.

Drobetsky EA, Moustacchi E, Glickman BW, Sage E. (1994) The mutational specificity of simulated sunlight at the *aprt* locus in rodent cells. *Carcinogenesis* **15:** 1577–1583.

Drouin EE, Loechler EL. (1993) AP sites are not significantly involved in mutagenesis by the (+)-anti diol epoxide of benzo[*a*]pyrene: the complexity of its mutagenic specificity is likely to arise from adduct conformational polymorphism. *Biochemistry* **32:** 6555–6562.

Duncan B, Miller JH. (1980) Mutagenic deamination of cytosine residues in DNA. *Nature (Lond.)* **287:** 560–561.

Farabaugh PJ, Schmeissner U, Hoffer M, Miller JH. (1978) Genetic studies of the *lac* repressor VII. On the molecular nature of spontaneous hotspots in the *lac*I gene of *Escherichia coli. J. Mol. Biol.* **126:** 847–857.

Fichtinger-Schepman AMJ, van der Veer JL, den Hartog JHJ, Lohman PHM, Reedijk J. (1985) Adducts of the antitumour drug *cis*-diamminedichloroplatinum(II) with DNA: formation, identification, and quantitation. *Biochemistry* **24:** 707–713.

Fix D, Glickman BW. (1986) Differential enhancement of spontaneous transition mutations in the *lac*I gene of an *ung* strain of *Escherichia coli. Mutat. Res.* **175:** 41–45.

Foster PL, Eisenstadt E, Miller JH. (1983) Base substitution mutations induced by metabolically activated aflatoxin B_1. *Proc. Natl Acad. Sci USA* **80:** 2695–2698.

Fowler RG, Degnen GE, Cox EC. (1974) Mutational specificity of a conditional *Escherichia coli* mutator, mutD5. *Mol. Gen. Genet.* **133:** 179–191.

Fuchs RP, Schwartz N, Daune MP. (1981) Hot spots of frameshift mutations induced by the ultimate carcinogen *N*-acetoxy-*N*-2-acetylaminofluorene. *Nature (Lond.)* **294:** 657–659.

Gao S, Drouin R, Golmquist GP. (1994) DNA repair rates mapped along the human *PGK1* gene at nucleotide resolution. *Science* **263:** 1438–1440.

Gibbs PE, Kilbey BJ, Banerjee SK, Lawrence CW. (1993) The frequency and accuracy of replication past a thymine–thymine cyclobutane dimer are very different in *Saccharomyces cerevisiae* and *Escherichia coli. J. Bacteriol.* **175:** 2607–2612.

Glickman BW, Radman M. (1980) *Escherichia coli* mutator mutants deficient in methylation-instructed DNA mismatch correction. *Proc. Natl Acad. Sci. USA* **77:** 1063–1067.

Glickman BW, Ripley LS. (1984) Structural intermediates of deletion mutagenesis: a role for palindromic DNA. *Proc. Natl Acad. Sci. USA* **81:** 512–516.

Glickman BW, Schaaper RM, Haseltine WA, Dunn RL, Brash DE. (1986) The C–C (6–4) UV photoproduct is mutagenic in *Escherichia coli. Proc. Natl Acad. Sci. USA* **83:** 6945–6949.

Golding GB, Glickman BW. (1985) Sequence-directed mutagenesis: evidence from a phylogenetic history of human alpha-interferon genes. *Proc. Natl Acad. Sci. USA* **82:** 8577–8581.

Greenblatt MS, Bennett WP, Hollstein M, Harris CC. (1994) Mutations in the *p53* tumor suppressor gene: clues to cancer etiology and molecular pathogenesis. *Cancer Res.* **54:** 4855–4878.

Gupta PK, Lee MS, King CM. (1988) Comparison of mutagenesis induced in single- and double-stranded M13 viral DNA by treatment with *N*-hydroxy-2-aminofluorene. *Carcinogenesis* **9:** 1337–1345.

Hansson J, Wood RDR, Wood D. (1989) Repair synthesis by human extracts in DNA damaged by *cis*- and *trans*-diamminedichloroplatinum(II). *Nucleic Acids Res.* **17:** 8073–8091.

Harmon P. (1981) The aging process. *Proc. Natl Acad. Sci.* USA **78:** 7124–7128.

Harvey RG. (1979) Benzo[*a*]pyrene-7,8-dihydrodiol 9,10-oxide adenosine and deoxyadenosine adducts: structure and stereochemistry. *Science* **206:** 1309–1311.

Haseltine WA. (1983) Ultraviolet light repair and mutagenesis revisited. *Cell* **33:** 13–17.

Holliday J, Glickman BW. (1991) Mechanisms of spontaneous mutation in DNA repair proficient *Escherichia coli. Mutat. Res.* **250:** 55–71.

Hollstein M, Rice K, Greenblatt MS, Soussi T, Fuchs R, Sorlie T, Hovig E, Smith-Sorensen B, Montesano R, Harris CC. (1994) Database of *p53* somatic mutations in human tumors and cell lines. *Nucl. Acids Res.* **22:** 3551–3555.

Hollstein M, Sidransky D, Vogelstein B, Harris CC. (1991) *p53* mutations in human cancers. *Science* **253:** 49–53.

Horsfall MJ, Lawrence CW. (1994) Accuracy of replication past the T–C (6–4) adduct. *J. Mol. Biol.* **235:** 465–471.

Hsu IC, Metcalf RA, Sun T, Welsh JA, Wang NJ, Harris CC. (1991) *p53* gene mutational hot spot in human hepatocellular carcinoma from Qidong, China. *Nature (Lond.)* **350:** 427–428.

Ionov Y, Peinado MA, Malkbosyan S, Shibata D, Perucho M. (1993) Ubiquitous somatic mutations in simple repeated sequences reveal a new mechanism for colonic carcinogenesis. *Nature (Lond.)* **363:** 558–561.

Ivanov EL, Koval'tsova SV, Korolev VG. (1983) Molecular nature of direct gene mutations induced by gamma and ultraviolet irradiation in *Saccharomyces cerevisiae* yeasts. *Genetika* **19:** 1063–1069.

Jones JC, Zehn W, Reed E, Parker RJ, Sancar A, Bohr VA. (1991) Gene-specific formation and repair of cisplatin intrastrand adducts and interstrand cross-links in Chinese hamster ovary cells. *J. Biol. Chem.* **266:** 7101–7107.

Kempf SR, Ivankovic S. (1986) Carcinogenic effect of cisplatin (*cis*-diamminedichloro-platinum(II), CDDP) in BD IX rats. *J. Cancer Res. Clin. Oncol.* **111:** 133–136.

King HWS, Osborne ME, Brookes P. (1979) The *in vitro* and *in vivo* reaction at the N[7]-position of guanine of the ultimate carcinogen derived from benzo[*a*]pyrene. *Chem. Biol. Interact.* **24:** 345–353.

King HWS, Brookes P. (1984) The nature of mutation induced by the diol epoxide of benzo[*a*]pyrene in mammalian cells. *Carcinogenesis* **5:** 965–970.

Klinedienst DK, Drinkwater NR. (1991) Reduction to homozygosity is the predominant spontaneous mutational event in cultured human lymphoblastoid cells. *Mutat. Res.* **250:** 365–374.

Koehler DR, Awadallah SS, Glickman BW. (1991) Sites of preferential induction of cyclobutane pyrimidine dimers in the nontranscribed strand of *lacI* correspond with sites of UV-induced mutation in *Escherichia coli*. *J. Biol. Chem.* **266:** 11766–11773.

Koffel-Schwartz N, Verdier JM, Bichara M, Freund AM, Daune MP, Fuchs RPP. (1984) Carcinogen-induced mutation spectrum in wild-type, uvrA and umuC strains of *Escherichia coli*. *J. Mol. Biol.* **177:** 33–51.

Kohler SW, Provost GS, Fieck A, Kretz PL, Bullock WO, Sorge JA, Putman PL, Short J. (1991) Spectra of spontaneous and mutagen-induced mutations in the *lacI* gene in transgenic mice. *Proc. Natl Acad. Sci. USA* **88:** 7958–7962.

Kunala S, Brash DE. (1992) Excision repair at individual bases of the *Escherichia coli lacI* gene: relation to mutation hot spots and transcription coupling activity. *Proc. Natl Acad. Sci. USA* **89:** 11031–11035.

Kunkel TA. (1992) DNA replication fidelity. *J. Biol. Chem.* **267:** 18251–18254.

Kunkel TA, Soni A. (1988) Mutagenesis by transient misalignment. *J. Biol. Chem.* **263:** 14784–14789.

Kusser WC, Miao X, Glickman BW, Friedland JM, Rothman N, Hemstreet GP, Mellot J, Swan DC, Schulte PA, Hayes R. (1994) *p53* mutations in human bladder cancer. *Environ. Mol. Mutagen.* **24:** 156–160.

Lawrence CW, Borden A, Banerjee SK, LeClerc JE. (1990) Mutation frequency and spectrum resulting from a single abasic site in a single-stranded vector. *Nucl. Acids Res.* **18:** 2153–2157.

LeClerc JE, Borden A, Lawrence CW. (1991) The thymine–thymine pyrimidine–pyrimidone(6–4) ultraviolet light photoproduct is highly mutagenic and specifically induces 3′ thymine-to-cytosine transitions in *Escherichia coli*. *Proc. Natl Acad. Sci. USA* **88:** 9685–9689.

Leopold WR, Miller EC, Miller JA. (1979) Carcinogenicity of antitumour *cis*-platinum(II) coordination complexes in mouse and rat. *Cancer Res.* **39**: 913–918.

Lepre CA, Lippard SJ. (1990) Interaction of platinum antitumour compounds with DNA. In: *Nucleic Acids and Molecular Biology* (eds F. Eckstein and D.M.J. Lilley). Springer-Verlag, Berlin, pp. 3–9.

Lindahl T. (1982) DNA repair enzymes. *Annu. Rev. Biochem.* **51**: 61–87.

Lindsley JE, Fuchs RPP. (1994) Use of single-turnover kinetics to study bulky adduct bypass by T7 DNA polymerase. *Biochemistry* **33**: 764–772.

Mackay W, Benasutti M, Drouin E, Loechler EL. (1992) Mutagenesis by (dextro)-anti-B(a)P-N-2-Gua, the major adduct of activated benzo[*a*]pyrene, when studied in an *Escherichia coli* plasmid using site-directed methods. *Carcinogenesis* **13**: 1415–1425.

Mane SS, Purnell DM, Hsu IC. (1990) Genotoxic effects of five polycyclic aromatic hydrocarbons in human and rat mammary epithelial cells. *Environ. Mol. Mutagen.* **15**: 78–82.

Mazur M, Glickman BW. (1988) Sequence specificity of mutations induced by benzo[*a*]pyrene-7,8-diol-9,10-epoxide at endogenous *aprt* gene in CHO cells. *Somat. Cell Mol. Genet.* **14**: 393–400.

McBride TJ, Schneider JE, Floyd RA, Loeb LA. (1992) Mutations induced by methylene blue plus light in single-stranded M13mp2. *Proc. Natl Acad. Sci. USA* **89**: 6866–6870.

McGregor WG, Chen RH, Lukash L, Maher VM, McCormick JJ. (1991) Cell cycle-dependent strand bias for UV-induced mutations in the transcribed strand of excision repair-proficient human fibroblasts but not in repair-deficient cells. *Mol. Cell. Biol.* **11**: 1927–1934.

Meehan T, Straub K, Calvin M. (1977) Benzo[alpha]pyrene diol epoxide covalently binds to deoxyguanosine and deoxyadenosine in DNA. *Nature (Lond.)* **269**: 725–727.

Mellon I, Hanawalt PC. (1989) Induction of the *Escherichia coli* lactose operon selectively increases repair of its transcribed strand. *Nature (Lond.)* **342**: 95–98.

Mellon I, Spivak G, Hanawalt PC. (1987) Selective removal of transcription-blocking DNA damage from the transcribed strand on the mammalian DHFR gene. *Cell* **51**: 241–249.

Miller JH. (1970) Carcinogenesis by chemicals: an overview – G. H. A. Clowes memorial lecture. *Cancer Res.* **30**: 559–576.

Mitchell DL, Nairn RS. (1989) The biology of the (6–4) photoproduct. *Photochem. Photobiol.* **49**: 805–819.

Morin-Faure J, Marcollet M. (1983) Immunocytochemical study of the action of *cis*-dichlorodiamino platinum on the human metaphase chromosome. *Eur. J. Cell Biol.* **30**: 316–319.

O'Neil JP, Couch DB, Machanoff R, SanSebastian JR, Brimer PA, Hsie AW. (1977) A quantitative assay of mutation induction at the hypoxanthine-guanine phosphoribosyl transferase locus in Chinese hamster ovary cells (CHO/HGPRT system): utilization with a variety of mutagenic agents. *Mutat. Res.* **45**: 103–109.

Orita M, Iwahana H, Kanazawa H, Hayashi K, Sekiya T. (1989) Detection of polymorphisms of human DNA by gel electrophoresis as single-strand conformation polymorphisms. *Proc. Natl Acad. Sci. USA* **86**: 2766–2770.

Pfeifer GP, Drouin R, Riggs AD, Holmquist GP. (1991) *In vivo* mapping of a DNA adduct at nucleotide resolution: detection of pyrimidine (6–4) pyrimidine photoproducts by ligation-mediated polymerase chain reaction. *Proc. Natl Acad. Sci. USA* **88**: 1374–1378.

Pienkowska M, Glickman BW, Ferreira A, Anderson M, Zielenska M. (1993) Large-scale mutational analysis of EMS-induced mutation in the *lacI* gene of *Escherichia coli. Mutat. Res.* **288**: 123–131.

Pinto AL, Lippard SJ. (1985) Binding of the antitumor drug *cis*-diamminedichloroplatinum(II) (cisplatin) to DNA. *Biochim. Biophys. Acta* **780**: 167–180.

Pott P. (1775) Chirurgical observations relative to the cataract, the polypus of the nose, the cancer of the scrotum, the different kinds of ruptures, and the mortification of the toes and feet. Clarke and Collins, London.

Quintanilla M, Brown K, Ransden M, Balmain A. (1986) Carcinogenic-specific mutation and amplification of Ha-ras during mouse skin carcinogenesis. *Nature (Lond.)* **322:** 78–83.

Rady PF, Scinicariello F, Wagner F, King S. (1992) *p53* mutations of basal cell carcinomas. *Cancer Res.* **52:** 3804–3806.

Ripley LS, Glickman BW. (1982) Unique, self-complementarity of palindromic sequences provides structural intermediates for mutation. *Cold Spring Harbor Symp. Quant. Biol.* **47:** 851–861.

Roberts JJ, Thomson AJ. (1979) The mechanism of action of antitumor platinum compounds. *Prog. Nucleic Acids Res. Mol. Biol.* **22:** 71–133.

Rodriguez H, Loechler EL. (1993a) Mutagenesis by the dextro-anti-diol epoxide of benzo[a]pyrene: what controls mutagenic specificity? *Biochemistry* **32:** 1759–1769.

Rodriguez H, Loechler EL. (1993b) Mutational specificity of the dextro anti-diol epoxide of benzo(alpha)-pyrene in a *supF* gene of an *Escherichia coli* plasmid: DNA sequence context influences hotspots, mutagenic specificity and the extent of SOS enhancement of mutagenesis. *Carcinogenesis* **14:** 373–383.

Sage E, Cramb E, Glickman BW. (1992) The distribution of UV damage in the *lac*I gene of *Escherichia coli:* correlation with mutation spectrum. *Mutat. Res.* **269:** 285–299.

Saul RL, Ames BN. (1986) Background levels of DNA damage in the population. In: *Mechanisms of DNA Damage and Repair: Implications for Carcinogenesis and Risk Assessment.* (eds MG Simic, L Grossman, AC Upton). Plenum Press, New York, pp.529–536.

Sayer JM, Chadha A, Agarwal SK, Yeh HJC, Yagi H, Jerina DM. (1991) Covalent nucleoside adducts of benzo[a]pyrene 7,8-diol 9,10-epoxides: structural reinvestigation and characterization of a novel adenosine adduct on the ribose moiety. *J. Organ. Chem.* **56:** 20–29.

Schaaper RM, Dunn RL. (1987) Spectra of spontaneous mutations in *Escherichia coli* strains defective in mismatch correction: the nature of *in vivo* DNA replication errors. *Proc. Natl Acad. Sci. USA* **84:** 6220–6224.

Schaaper RM, Dunn RL, Glickman BW. (1987) Mechanisms of UV-induced mutation: mutational spectra in the *Escherichia coli lac*I gene for a wild-type and excision-deficient strain. *J. Mol. Biol.* **198:** 187–202.

Schaaper RM, Koffel-Schwartz N, Fuchs RPP. (1990) *N*-acetoxy-*N*-acetyl-2-aminofluorene-induced mutagenesis with the *lac*I gene of *Escherichia coli. Carcinogenesis* **11:** 1087–1095.

Schorderet D, Gartler S. (1992) Analysis of CpG suppression in methylated and non-methylated species. *Proc. Natl Acad. Sci. USA* **89:** 957–981.

Shibutani S, Grollman AP. (1993a) Nucleotide misincorporation on DNA templates containing *N*-(deoxyguanosin-N^2-yl)-2-(acetylamino)fluorene. *Chem. Res. Toxicol.* **6:** 819–824.

Shibutani S, Grollman AP. (1993b) On the mechanism of frameshift (deletion) mutagenesis *in vitro. J. Biol. Chem.* **268:** 11703–11710.

Slaga TJ, Bracken WJ, Gleason G, Levin W, Yagi H, Jerina DM, Conney AH. (1979) Marked differences in skin tumor-initiating activities of the optical enantiomers of the diastereomeric benzo[a]pyrene-7,8-diol-9,10-epoxides. *Cancer Res.* **39:** 67–71.

Soussi T, de Fromentel CC, Stürzbecher H, Ullrich S, Jenkins J, May P. (1989) Evolutionary conservation of the biochemical properties of *p53:* specific interaction of *Xenopus laevis p53* with simian virus 40 large T antigen and mammalian heat shock proteins 70. *J. Virol.* **63:** 3894–3901.

Spruck III CH, Rideout III WM, Olumi AF, Ohneseit PF, Yang AS, Tsai YC, Nichols PW, Horn T, Hermann GG, Steven K, Ross RK, Yu MC, Jones PA. (1993) Distinct pattern of *p53* mutations in bladder cancer: relationship to tobacco usage. *Cancer Res.* **53:** 1162–1166.

Stevens CW, Bouck N, Burgess JA, Fahl WE. (1985) Benzo[a]pyrene diol-epoxides: different mutagenic efficiency in human and bacterial cells. *Mutat. Res.* **152:** 5–14.

Streisinger G, Okada Y, Emrich J, Newton J, Tsugita A, Terzhagi E, Inouye M. (1966) Frameshift mutations and the genetic code. *Cold Spring Harbor Symp. Quant. Biol.* **31:** 77–84.

Swann PF. (1990) Why do O^6-alkylguanine and O^4-alkylthymine miscode? The relationship between the structure of DNA containing O^6-alkylguanine and O^4-alkylthymine and the mutagenic properties of these bases. *Mutat. Res.* **233:** 81–94.

Tang MS, Pierce JR, Doisy RP, Nazimiec ME, Macleod MC. (1992) Differences and similarities in the repair of two benzo[a]pyrene diol epoxide isomers induced DNA adducts by uvrA, uvrB, and uvrC gene products. *Biochemistry* **31:** 8429–8436.

Tessman I, Kennedy MA. (1991) The two-step model of UV mutagenesis reassessed: deamination of cytosine in cyclobutane dimers as the likely source of the mutations associated with photoreactivation. *Mol. Gen. Genet.* **227:** 144–148.

Thakker DR, Yagi H, Levin W, Wood AW, Conney AH, Jerina DM. (1985) Bay-region diol epoxides. In: *Bioactivation of Foreign Compounds* (ed. AH Anders). Academic Press, New York, pp. 177–242.

Thrall BD, Mann DB, Smerdon MJ, Springer DL. (1992) DNA polymerase, RNA polymerase and exonuclease activities on a DNA sequence modified by benzo[a]pyrene diolepoxide. *Carcinogenesis* **13:** 1529–1534.

Topal MD, Fresco JR. (1976) Complementary base pairings and the origin of substitution mutations. *Nature (Lond.)* **263:** 285–289.

Tornaletti S, Pfeifer GP. (1994) Slow repair of pyrimidine dimers at *p53* mutation hotspots in skin cancer. *Science* **263:** 1436–1438.

Veaute X, Fuchs RPP. (1991) Polymorphism in *N*-2-acetylaminofluorene induced DNA structure as revealed by DNase I footprinting. *Nucleic Acids Res.* **19:** 5603–5606.

Vrieling H, Zhang LH, van Zeeland AA, Zdienicka MZ. (1992) UV-induced *hprt* mutations in a UV-sensitive hamster cell line from complementation group 3 are biased towards the transcribed strand. *Mutat. Res.* **274:** 147–155.

Wang YC, Maher VM, Mitchell DL, McCormick JJ. (1993) Evidence from mutation spectra that the UV hypermutability of xeroderma pigmentosum variant cells reflects abnormal, error-prone replication on a template containing photoproducts. *Mol. Cell. Biol.* **13:** 4276–4283.

Watson JD, Crick FHC. (1953) The structure of DNA. *Cold Spring Harbor Symp. Quant. Biol.* **18:** 123–131.

Weinstein IB, Jeffrey AM, Jeanette KW, Blobstein SH, Harvey RG, Harris C, Autrup H, Kasai H, Nakanishi K. (1976) Benzo[a]pyrene diol epoxide intermediates in nucleic acid binding *in vitro* and *in vivo*. *Science* **193:** 592–595.

Wood AW, Chang RL, Levin W, Yagi H, Thakker DR, Jerina DM, Conney AH. (1977) Differences in mutagenicity of the optical enantiomers of the diastereoisomeric benzo[a]pyrene 7,8-diol-9,10-epoxides. *Biochem. Biophys. Res. Commun.* **77:** 1389–1396.

Yang JL, Maher VM, McCormick JJ. (1987) Kinds of mutations formed when a shuttle vector containing adducts of (+-) - 7-β, 8-α-dihydroxy- 9, 10-α-epoxy-7,8,9,10-tetrahydrobenzo[a]pyrene replicates in human cells. *Proc. Natl Acad. Sci. USA* **84:** 3787–3791.

Yatagai F, Horsfall M, Glickman BW. (1991) Specificity of SOS mutagenesis in native M13*lac*I phage. *J. Bacteriol.* **173:** 7996–7999.

Zambetti GP, Levine AJ. (1993) A comparison of the biological activities of wild-type and mutant *p53*. *FASEB J.* **7:** 855–865.

Zarbl H, Sukumar S, Arthur AV, Martin-Zanca D, Barbacid M. (1985) Direct mutagenesis of Ha-*ras*-1 oncogenes by *N*-nitroso-*N*-methylurea during initiation of mammary carcinogenesis in rats. *Nature (Lond.)* **315:** 382–385.

Zhu Y, Bye S, Stambrook PJ, Tischfield JA. (1994) Single-base deletion induced by benzo[a]pyrene diol epoxide at the adenine phosphoribosyltransferase locus in human fibrosarcoma cell lines. *Mutat. Res.* **321:** 73–79.

Zwelling LA, Bradley MO, Sharkey NA, Anderson T, Kohn KW. (1979) Mutagenicity, cytotoxicity and DNA crosslinking in V79 Chinese hamster cells treated with *cis*- and *trans*-Pt(II) diaminedichloride. *Mutat. Res.* **67:** 271–280.

Inherited susceptibility to mutation

D.G. Harnden

> It is axiomatic to modern biology that mutation is a wholly random process in so far as the nature of the resulting change in genetic 'information' is concerned. Any *control* of mutation can only be in the form of modification in one direction or the other of the rate of mutation. It becomes inescapable, then, for us to postulate that the rate of germline mutation in the species, and of somatic mutation in the individual, must have optimal values for each particular species. If those two statements are, as I suggest, axiomatic, the two sets of mutation rates must be determined genetically: their regulation is just one of the responsibilities laid on the nucleotide sequences of the genome.
>
> [Macfarlane Burnet, 1974]

4.1 Introduction – gene–environment interaction

The mutation rate in human populations has long been a matter of major concern. To a large extent, the focus has been on the impact of environmental factors on the occurrence of new mutations in the population and the possibility of monitoring mutation rate on a regular basis. Neel (1977) concludes that the cost of monitoring would be great but the cost of not monitoring even greater. However, as Burnet points out in the above quotation, the mutation rate for germ cells, which is the principal concern of most population geneticists, is only part of the problem. One must also consider the somatic mutation rate in the individual. These two parameters are not independent variables. Any mutation in the germline which alters the mutation rate in somatic cells will have an impact on the fate of the individual but may also have far-reaching evolutionary implications. A simple example will illustrate the point. There are two distinct reservoirs of influenza A viruses evolving in birds and mammals, respectively. Genetic analysis suggests that most, perhaps all, mammalian influenza A viruses originated from an avian ancestor. Not only must the cross-species infection take place but the virus must also adapt to its new host. Scholtissek *et al.*

(1993) provide evidence to show that the necessary variability is provided by a mutator mutation in the polymerase complex. This produces a large number of variants, some of which will adapt rapidly to the new host, thus providing an evolutionary advantage. While the situation is clearly more complex in higher organisms, the principle is the same. The emergence of a mutator phenotype will influence the individual and ultimately the species. Environmental events could have an impact on both the origin of mutator mutations and the increase in somatic mutation arising from such mutations, since many mutator mutations confer a susceptibility rather than an inevitability of somatic genetic events. Thus, the interaction between environmental factors and the genetic constitution of the individual is what will eventually determine impact. Indeed, it is hard to see how the influence of environmental mutagens can be measured unless the enormous variation in the human population is taken into consideration. Average values may not be an appropriate measure.

4.2 Mechanisms of inherited susceptibility to mutation

Mutation rates and the mechanism of mutation have been the subject of intense investigation in many different species for the larger part of a century. Gradually an overall picture is beginning to emerge and it is clear that some, at least, of the mechanisms identified in lower organisms also apply to man. There are three ways in which an unusual frequency of somatic mutations could occur (*Table 4.1*):

(i) Errors in the enzymic machinery required for DNA replication and mismatch correction can act as mutator mutations. Once such a mutation has been induced, possibly by an environmental agent, and stabilized in the genome, the cell with the mutant enzyme will give rise to further errors without further external influence.

(ii) Transposition of transposable elements occurs spontaneously, but the rate and type of transposition leading to mutation may be influenced by specific environmental stimuli. The number of elements and the rate of intrinsic and induced transpositions are, however, characteristics of the genome.

(iii) Environmentally induced damage may be influenced by the genetic constitution in a variety of ways including access to DNA, configuration of chromatin, genomic surveillance and DNA repair.

4.2.1 Replication errors

While environmentally induced lesions will constitute an important component in the burden of somatic mutations, there are a number of different mechanisms by which errors may be created in DNA which are not the direct result of environmental agents but which reside in faults in the DNA

Table 4.1. Mechanisms of susceptibility to mutation

1. Replication errors
   a. Polymerase malfunction
   b. Unstable sequences

2. Transposable elements

3. Environmentally induced damage
   a. Access to DNA
   b. Configuration of chromatin
   c. Genomic surveillance
     (i)   Monitoring
     (ii) Signal transduction
   d. DNA repair
     (i)   Transcription factors
     (ii) Repair enzymes

replication and mismatch correction mechanisms. Such mechanisms have been well explored in virus, bacterial and yeast systems. From an extensive literature a few examples must suffice. Mutations in DNA polymerase genes have been found to be responsible for a mutator phenotype in several situations. In a series of vaccinia mutants selected for aphidicolin resistance, Taddie and Traktman (1991) found lesions in the DNA polymerase gene in each mutant. One of these mutants caused a 20- to 40-fold increase in the frequency of spontaneous mutations within the virus stock.

Many mutator strains of *Escherichia coli* are known, though the precise nature of the defect is not always clear. Isbell and Fowler (1989) using the temperature-dependent dna Q49 allele, which is a strong mutator at 37°C, found that not only was the frequency of mutations altered at 37°C (as compared with 30°C) but so also was the spectrum of mutations. They concluded that the epsilon subunit of DNA polymerase III, which is encoded by dna Q49, malfunctions at the higher temperature leading to defective proof-reading and post-replicative mismatch repair. The same conclusion was reached by Jonczyk *et al.* (1988) using an entirely different system. Similarly, in *Saccharomyces cerevisiae* Morrison and Sugino (1994) have studied mutants deficient in the exonuclease function of the DNA polymerase II (epsilon) and III (delta) genes and found increased spontaneous mutation rates at specific loci. These two exonucleases act in series with the PMS1 mismatch correction system.

Strand *et al.* (1993) show that three yeast genes, known to be involved in DNA mismatch repair, cause destabilization of tracts of simple repetitive DNA. The three genes *PMS1*, *MLH1* and *MSH2* are homologues of bacterial mismatch repair genes and cause 100- to 700-fold increases in tract instability. Mutator phenotypes of a very similar kind have now been observed in members of families susceptible to colorectal cancer (see later).

4.2.2 Transposable elements

The classic work of McClintock (1950) showed that somatic mutation in maize could be due to the mobilization and transposition of specific genetic elements. In plants, somatic mutations may become part of a meristem and, therefore, be incorporated into the germlines. Transposable elements play a major part in such mutations and preferentially affect 'regions that are more or less labile depending on the physiological status of the cell or organism' (Cullis, 1990).

Such elements have also been recognized in many different animal species and, moreover, they are often, though not invariably, associated with a mutator phenotype. They are most studied in species that have been subject to detailed genetic analysis for other reasons (e.g. *Drosophila melanogaster, Caenorhabditis elegans* and *Mus musculus*). Indeed, the capacity of these elements to cause mutation can be harnessed to facilitate classical genetic analysis. Cooley *et al.* (1988) describe an elegant system of controlled mutagenesis using crosses between two strains of *Drosophila,* one containing a non-autonomous transposon and the other a P element transposase. The resulting crosses yield a library of single-element insertions which provide a resource for *Drosophila* molecular genetics.

In the present context, the most important conclusion that emerges from such studies is that the mutator phenotype is dependent not only upon the nature of the transposable element but also on the genotype of the organisms harbouring the element. Kim *et al.* (1994) described a series of experiments in *Drosophila* involving a genetic instability system caused by a transposable element GYPSY. A mutator strain (MS) derived from a stable strain (SS) is shown to differ in two respects. First, genetic crosses yielding one MS chromosome in an SS genetic environment showed elevated frequencies of spontaneous mutation, suggesting that the MS strain contained transpositionally active GYPSY elements. Second, microinjection of single active elements into the SS strain showed amplification and enhanced transposition, while injection into a different stable strain has shown no such effect, indicating that the SS genotype contains a mutation or mutations in genes regulating GYPSY transposition. Similar conclusions can be drawn from the studies of Collins *et al.* (1989) on *C. elegans.* Members of a family of transposable elements, Tc3, are found in several different wild-type varieties of *C. elegans* but are apparently not active. In a mutant strain TR679, Tc3 transposition and excision occur at high frequencies and are responsible for several spontaneous mutations in specific genes. In this particular strain, another family of transposable elements, Tc1, is also highly active but other mutator mutants with high Tc1 activity do not show Tc3 activity. Thus, the nature of the transposable element and the genotype of the host organism are both important variables in causing the mutator phenotype – an example of genetic control of mutation.

Dispersed repetitive sequence elements with structural similarities to retrovirus proviruses are found in many eukaryotic species. In rodents, endogenous retroviral elements include murine leukaemia virus (MLV), murine mammary tumour virus, intracisternal A particles and virus-like 30S elements (VL30). The importance of the long terminal repeats (LTRs) of exogenous retroviruses in activation of cellular oncogenes is well-documented. There is also some evidence that LTRs of endogenous virus-like particles may play a part in activation of cellular oncogenes following transposition and insertion. Spontaneous rates of transposition are not high but these rates are greatly enhanced by the provision of MLV functions in *trans*. Carter *et al.* (1986) found that newly integrated copies of VL30 are at the very high frequency of one or two new insertions per cell, raising the possibility that transposition and insertional deregulation by VL30 elements may be a novel mechanism of mutagenesis and oncogenesis.

There are, however, few reports of transposable elements in man. Paulson *et al.* (1985) describes a transposon-like element in human DNA with a retrovirus-like structure but with only limited homology to known viruses. Other reports also suggest a retrovirus-like structure (Hehlamann *et al.*, 1988) but their biological relevance is so far unknown. It seems improbable, therefore, that transposable elements play a major role in mutation-prone phenotypes in human populations.

4.2.3 Susceptibility to environmental agents

The probability of DNA damage occurring following exposure to an external mutagenic influence depends, in the first instance, on whether or not the damaging stimulus reaches the genome. Potten *et al.* (1993) show quite clearly that the induction of thymine dimers in human skin and the associated repair process is related to the skin type of the individual, a genetically determined characteristic and, moreover, that simulation of the more heavily pigmented skin types, using an artificial tanning agent, protects the DNA of fair-skinned individuals from primary damage. Similarly, there is a vast literature on the metabolism of chemical carcinogens (reviewed briefly by Harnden, 1990) which shows that the probability of a potentially mutagenic or carcinogenic chemical interacting with DNA is dependent on the metabolizing enzymes, which may either detoxify the compound or convert it into the ultimate carcinogen which damages DNA. These metabolizing enzymes are, of course, under genetic control and evidence is accumulating to show that many of the loci controlling production of these enzymes are highly polymorphic, thus conferring considerable variation in the capacity of individuals to handle carcinogens and mutagens (see Chapter 5).

While most studies correlate variation within the population with cancer incidence, some address the question of alterations in the rate of induction of genetic change. For example, exposure of lymphocytes to diepoxybutane reveals a bimodal distribution for the induction of sister chromatid exchanges with 20% of healthy workers being twice as sensitive to the mutagen as the remaining 80% (Wiencke and Kelsey, 1993). While it is not proven that this is due to a polymorphism in a metabolizing enzyme, the authors suggest that this kind of study may help to elucidate such population variations.

A second level at which genetic constitution may influence the probability of mutation following an environmental insult is the suggestion by Hittelman and Pandita (1994) that the configuration of the chromatin, for example in cells from ataxia telangiectasia (AT) patients, may influence the probability of initial damage occurring in the DNA. Using a premature chromosome condensation technique (see Chapter 7), they show an excess of initial damage in AT cells as compared with controls and suggest that structural relationships enhance the probability that a damaging event will be converted into a chromosomal lesion even before replication or repair has occurred.

Once DNA damage occurs, a whole battery of repair functions is available, depending on both the origin of the cell and the nature of the induced lesion. Variation within populations in their repair response is well documented from prokaryotes to man and, in many cases, this leads to variation in the frequency with which DNA lesions are converted into mutations. The basis of the repair defect will range from errors in the mechanisms of genomic surveillance and recognition of errors through to simple absence or malfunctioning of specific enzymes (reviewed by Defais, 1990).

In prokaryotes this point scarcely needs emphasis since the whole basis of the development of 'tester strains' of *Salmonella typhimurium* and *E. coli* is the selection of strains which have particular mutability characteristics dependent upon the genotype of the strain (see Chapter 6). Thus, the inheritance of susceptibility to mutation is taken as axiomatic by everyone working in the field of environmental mutagenesis. It is somewhat surprising, therefore, that such susceptibility in man is regarded as something unusual. This may arise from an almost intuitive understanding that larger, longer-lived animals must have more stable genomes.

4.3 Variation in mutation frequency in specific regions of the genome

There are a number of reports suggesting that mutations are not randomly distributed throughout the genome. The presence of 'hot spots' within specific genes is well documented [e.g. in p53 mutations in tumour cells (Cerutti *et al.*, 1994)]. This does not necessarily mean that mutations, either spontaneous or induced, occur more frequently at these locations since their recognition may be simply a measure of their preferential survival. However, in the case of immunoglobulin gene rearrangement, it is clear that specific and clearly defined regions within the coding sequence are subject to unusually high levels of mutation. Wabl *et al.* (1989) have measured the spontaneous mutation rate in these hypermutable regions in a pre-B-lymphocyte cell line at 10^{-5}

mutations per base pair per generation. This compares with less than 10^{-9} in the control cell line. The mechanisms of hypermutation are by no means clear but several studies show that the hypermutability does not extend into the region immediately upstream of the coding sequence of the hypermutable sites. Rothenfluh *et al.* (1993) sequenced the 5' flanking region in a series of mouse antibodies and found 97% of mutations in the transcribed region of the gene. They conclude that the data are consistent with a mutational model which requires transcription of the target gene and direct mutation of RNA or cDNA (i.e. a reverse transcription model). Similarly, Azuma *et al.* (1993) and Rogerson (1994) agree that there is a sharp 5' boundary to the hypermutable region and suggest precise control by *cis*-acting elements. This indicates tight genetic control over the hypermutation even though the mechanism remains unclear.

It also appears that some DNA sequences are inherently unstable and consequently lead to replication errors. Some of these errors may manifest only as DNA polymorphisms, without obvious phenotypic effect, but others lead to a major change and even pathological consequences. Some such errors result from anomalous recombination events between highly homologous sequences. For example, Weil *et al.* (1994) have shown that highly homologous loci on the human X and Y chromosomes are hot spots for ectopic recombination which is responsible for a substantial proportion of XX maleness. Similarly, homologous regions next to the growth hormone gene constitute hot spots for deletion in that region (Vnencak-Jones and Phillips, 1990).

4.4 Spontaneous germline mutation in man

The estimation of mutation rate for an inherited mutation at a specific locus is not easy and the measurement of an increase in that rate is daunting. Neel (1977) estimates that, using a battery of 20 protein variations, it would be necessary to study populations of several hundred thousand to identify an increase of 50% in germline mutations with a probability of more than 0.01 (assuming an observed spontaneous frequency of 20 per 100 000 births). Using 'sentinel phenotypes' – well-defined clinical syndromes – the numbers would be slightly less, but still in the range of hundreds of thousands. It would, therefore, be difficult, if not impossible, to demonstrate that in a human subpopulation there was an increased susceptibility to germline mutation.

It might be considered that those syndromes where there is an instability of a direct repeat sequence within a gene, where the length of the repeat might vary quite considerably between families and between individuals within a family, are examples of susceptibility to germline mutation. In the fragile-X syndrome, for example, the length of the repeat is directly related to the severity of the clinical phenotype (Oberlé *et al.*, 1991). Thus, males carrying a short repeat may be asymptomatic, while the male offspring of daughters of these males may have an increased insert length and a severe phenotype.

Atypically for an X-linked syndrome, carrier females with a long insert may be affected. Mosaic individuals are also identified, suggesting that the instability occurs in somatic cells as well as in germ cells.

Similarly, in Huntingdon's chorea, myotonic dystrophy and several other disorders, instability of the repeat sequence leads to variation in the expression of the syndrome (Richards and Sutherland, 1994). However, these are examples of variations in a pre-existing mutation rather than susceptibility to *de novo* mutational events.

4.5 Induced germline mutation in man

There is good evidence that treatment of rodents with alkylating agents can lead to an increased incidence of cancer in the offspring and in the second and third generation (Tomatis *et al.*, 1975, 1977). The descendants of rats treated during pregnancy showed an increased incidence of tumours, especially of the kidney and nervous system, down to the third generation. This implies an induced germline mutation leading to genomic instability causing predisposition to spontaneous malignancy. Evidence that ionizing radiation can similarly induce germline mutations, leading to increased cancer susceptibility, comes from the work of Nomura (1983, 1986), who demonstrated an increase in large lung nodules in urethane-treated offspring of parents who were X-irradiated before mating. While the occurrence of tumours does not prove increased mutability in somatic cells of these offspring, it is highly suggestive. There is, however, no confirmation, as yet, of Nomura's work.

Evidence from man is also scanty. A major discussion surrounds the hypothesis put forward by Gardner *et al.* (1987a and b) that the data from the area around the nuclear plant at Sellafield in Cumbria, England, can be interpreted as indicating that paternal exposure to ionizing radiation before conception leads to an increase in the incidence of leukaemia in the children of the exposed workers. In a careful critique, Doll *et al.* (1994) argue against the notion that preconceptional irradiation has caused a germline mutation which, in turn, leads to the somatic mutations known to be associated with childhood leukaemia. In particular, they point out that for the hypothesis to hold good, the leukaemia gene(s) would have to be 'uniquely and inordinately' sensitive to ionizing radiation since no other cancers or inherited syndromes are increased in association with the increased leukaemia and, second, the gene(s) would have to be selectively susceptible to occupational radiation since there is only a minor inherited component in spontaneous childhood leukaemia in the general population which is exposed to low doses of background radiation. Further, the mutation rate would have to be about 2000 times the background rate.

Doll and his colleagues (Doll *et al.*, 1994) conclude that the association between paternal irradiation and leukaemia is largely or wholly a chance finding. The case they put forward is quite compelling. However, as we have seen, there are mutational hot spots in the genome and the work of Kadhim *et al.* (1992), which demonstrates continuing instability of the genome many cell divisions after exposure to alpha particle irradiation, shows that quality of radiation may be an important factor and could possibly discriminate between occupational and background radiation since the balance of radionuclide exposure in an occupational situation will be quite different from that in the normal environment. More telling is Kinlen's (1993) observation that the excess of leukaemia at Seascale extends to children who were not born at Sellafield. At present, therefore, direct evidence that environmental agents may induce germline mutations which lead to genomic instability is weak.

4.6 Syndromes and families

There are, however, a number of specific situations in which it is clear that susceptibility to somatic mutation is inherited in classical Mendelian fashion. There are several well-defined syndromes where susceptibility to mutation is an integral part of the syndrome and also a number of situations in which susceptibility to mutation appears to influence the occurrence of cancer without there being any other inherited phenotype associated with the hypermutability.

4.6.1 Rb1 *and* p53

Some of the genes that confer susceptibility to cancer are also found to be mutated in certain types of tumour (e.g. *Rb1* in retinoblastoma and *p53* in a whole range of different cancers). In these cases it is not necessary to postulate that these genes inevitably cause further mutation. In the case of *Rb1* the observed elevated frequency of retinoblastoma in carriers of a mutant gene can be adequately explained by the occurrence of a second mutation at the same locus on the unaffected chromosome at a frequency no different from normal somatic mutation rates (Knudson, 1971). Similarly, it could be that the mere presence of a *p53* mutation in cells of subjects from Li–Fraumeni families (Birch *et al.*, 1994; Malkin *et al.*, 1990), is one step along the multistage progression towards malignancy and that this is all that is required to explain the increased frequency of cancer in subjects carrying *p53* germline mutation. However, the role of *p53* in cell cycle control strongly suggests that defects in *p53* are likely to be associated with increased mutation at other loci, since premature entry of damaged cells into S phase in the absence of a normal *p53* response will inevitably lead to mutational events. Specific studies on the genetic stability of cells from Li–Fraumeni patients with a *p53* germline mutation have not yet been reported.

Comparison has been made between Li–Fraumeni families and *p53* knockout mice. Mice carrying either a mutant *p53* gene or a *p53* null allele are more susceptible to the development of cancer than are wild-type mice of the same strain (Donehower *et al.*, 1992; Harvey *et al.*, 1993; Jacks *et al.*, 1994). Homozygotes show a higher incidence and a more rapid tumour development than do heterozygotes, though the latter also have an increased incidence of tumours. The type of tumour, however, differs; the heterozygotes develop predominantly sarcomas while the homozygotes develop predominantly thymic lymphomas (Jacks et al., 1994). Mice heterozygous for p53 mutations are more susceptible to the effects of the liver carcinogen, dimethylnitrosamine, in that survival time after exposure is greatly reduced (Harvey et al., 1993), but the incidence of tumours was high in both the controls and p53 mutants. It cannot be assumed from the data that germline p53 mutations are causing additional somatic mutation. Indeed, Kemp et al. (1993) find that p53 null mice show no increase in initiation or promotion events following a skin tumour induction regimen using 7,12-dimethylbenz[a]anthracene as the initiator and 12-0-tetradecanoyl-phorbol-13-acetate as the promoter. However, the rate of progression from papilloma to carcinoma was greatly increased, the effect being more dramatic in the homozygotes than in the heterozygotes. Kemp et al. show that the conversion to malignancy in the heterozygotes is accompanied by loss of the normal p53 allele, but again this does not necessarily imply an increased frequency of mutation, since mutations occurring at a normal frequency in the normal p53 allele would be subject to selection. Direct evidence of increased mutation frequency in either homozygotes or heterozygotes for a p53 mutation is lacking.

4.6.2 Werner's syndrome

Werner's syndrome is a rare autosomal recessive disorder which is characterized by premature signs of ageing in young adults, and which closely resembles the normal ageing process. In culture, cells from Werner's patients show a greatly reduced lifespan and a spontaneous tendency to develop both chromosome abnormalities and mutations in specific genes. Fukuchi et al. (1989) show that the frequency of mutations in the hypoxanthine phosphoribosyltransferase (HPRT) gene is roughly two orders of magnitude greater in SV40-transformed Werner's fibroblasts than in similarly transformed control fibroblasts. It was necessary to use the SV40 transformants because of the short lifespan of the untransformed Werner's fibroblasts, but the authors consider that this has not materially altered the result. Fukuchi and colleagues also show that a surprisingly high proportion of the HPRT mutants are deletions, giving rise to the speculation that recombination-like mechanisms might be involved. However, Monnat et al. (1992) find that the deletion mechanism in Werner's fibroblasts, which involves non-homologous recombination, does not differ significantly from the mechanism observed in control cells. Although the location of the Werner's gene at 8p11.2–p12 has been determined (Nakura et al., 1993) the underlying mechanism is still unknown. The similarity of the features of Werner's syndrome to normal ageing could suggest that a mutational process akin to that seen in Werner's syndrome underlies the ageing process in normal individuals. It is highly probable that other genes have effects similar to, but less dramatic than, the Werner's gene. Indeed, Burnet's (1974) treatise is based on this contention.

4.6.3 Xeroderma pigmentosum

Xeroderma pigmentosum (XP) is an autosomal recessive condition with extreme ultraviolet (UV) sensitivity, a defect of excision repair and a susceptibility to skin cancer. It is quite clear that XP cells in culture are unusually susceptible to the induction of mutations by UV radiation and by specific chemical carcinogens (Maher and McCormick, 1976; Tatsumi *et al.*, 1987). Tatsumi and colleagues examined the mutability of two XP cell lines, one from complementation group A and one from complementation group C. Both showed greatly enhanced mutation rates following exposure to UV when compared with normal control and XP heterozygote cell lines. The group A cell line was, however, dramatically more sensitive showing an increase in mutant fraction from 2×10^{-6} to 1.1×10^{-3}, or roughly a 500-fold increase over the spontaneous level. The heterozygote cell line showed no increase in mutagenic response to UV. It seems probable, therefore, that the DNA repair defect leads to the hypermutability, which in turn provides a range of mutant cells from amongst which clones with a selective advantage emerge to become the precursors of malignant lesions. While mutant cells with other phenotypes must arise, they are unlikely to be identified *in vivo* though it has been suggested that areas of abnormal pigmentation on sun-exposed skin may represent mutant clones (Burnet, 1974).

4.6.4 Ataxia telangiectasia

Ataxia telangiectasia (AT) is a rare, recessive, multisystem disorder characterized by progressive cerebellar degeneration, ocular telangiectasia, immune deficiency, DNA repair deficiency and susceptibility to specific malignant diseases. The occurrence of an unusually high level of spontaneous chromosome abnormalities in both lymphocytes and fibroblasts is well documented (Taylor, 1982). Moreover, AT cells are unusually susceptible to the induction of chromosome damage by ionizing radiation. Cytogenetically marked clones of lymphocytes are found in the peripheral blood of AT patients and the frequency of these abnormal cells increases with time. These cells are, therefore, true mutants and not simply a reflection of an ongoing chromosome instability.

On the other hand, attempts to demonstrate mutability at specific loci in AT cells have proved difficult. Arlett and Harcourt (1982) found that AT cells were not mutable with ionizing radiation. However, Tatsumi and Takebe (1984) found that mutations could be induced in SV40-transformed fibroblasts and EBV-transformed lymphocytes, but that the mutation frequency was lower than in the appropriate controls. Thus, AT cells appear to be hypomutable following exposure to X-rays. Cole *et al.* (1989) found an increased frequency of mutations in the circulating lymphocytes of AT patients as compared with controls. This could reflect the disturbance of the immune system rather than a higher frequency of the original mutation.

While the existence of a DNA repair defect is clear (Cox, 1982), the nature of that defect has still not been defined. Recent studies have, however, thrown more light on the nature of the mutational events in AT cells. Tatsumi-Miyajima *et al.* (1993) have shown, using a shuttle vector plasmid propagated in AT cells, that there is a lack of fidelity in the repair of double-strand breaks and, moreover, that this results in a high proportion of deletion mutations. Using similar systems, Runger *et al.* (1992) demonstrated an increase in spontaneously occurring insertions and complex mutations, while Meyn (1993) found a dramatic increase in intrachromosomal recombination rates. Possibly the most interesting finding is that the misjoin mechanism involves deletions at the sites of short direct repeats at various distances from the initial break site (Ganesh *et al.*, 1993; Thacker *et al.*, 1992). The specific nature of these events suggests that while AT cells are not hypermutable in numerical terms, they have an inherited defect in a repair system which involves a recombination-like process. Thus, AT patients have an inherited predisposition to a specific class of mutation.

The nature of the AT defect is becoming clearer. Kastan *et al.* (1992) and Khanna and Lavin (1993) found a defect in the radiation-induced increase in p53 levels in AT cells as compared with normal cells. Lu and Lane (1993), however, failed to detect a substantive defect in the p53 response. It is clear, however, that AT cells show a defect in cell cycle control and also have an increased susceptibility to DNA damage-induced apoptosis (Meyn *et al.*, 1994). Meyn and his colleagues propose that the AT gene product plays a crucial role in a signal transduction network that activates multiple cellular functions in response to DNA damage. The model implies that there is no basic defect in DNA repair systems in AT but, rather, that the process is disrupted by disturbances to progression through the cell cycle.

A novel gene, termed *ATM* and mapping at the known locus of the AT gene, has been found to be mutated in AT patients from all complementation groups (Savistky *et al.* 1995). This gene encodes a putative protein which has domains homologous with phosphatidylinositol 3-kinases, with the product of the *rad 3*[+] gene which is required for G2/M cell cycle checkpoint control in yeast and also with the TOR1 and TOR2 proteins of yeast which are involved in controlling the G1 phase of the cell cycle. This finding strengthens the above hypotheses and opens the way to a full understanding of the AT defect.

4.6.5 Cockayne's syndrome

Cockayne's syndrome (CS) is a rare condition showing severe physical and mental retardation, cachexia, microcephaly and a characteristic face with beaked nose and sunken eyes. The patients are sun-sensitive and cultured cells show hypersensitivity to the lethal effects of UV light. Arlett and Harcourt (1982) suggest that CS fibroblasts are hypermutable by UV light but the data are limited.

4.6.6 Fanconi's anaemia

Fanconi's anaemia (FA) is characterized by a progressive pancytopaenia, short stature, microcephaly, abnormal skin pigmentation and skeletal abnormalities, particularly of the radius and thumb. The patients are prone to developing malignancies, especially leukaemia.

FA cells are particularly sensitive to the cytotoxic action of bifunctional alkylating agents such as mitomycin-C and this sensitivity is used as a diagnostic tool. A variety of other studies have been reported (reviewed by Lehmann and Dean, 1990) which show that FA cells are sensitive to the lethal effects of DNA cross-linking agents, but not to any marked extent, if at all, to monofunctional agents, or to X-rays or UV light.

FA lymphocytes show an increase in the incidence of spontaneous chromosome aberrations and also chromosome aberrations induced by bifunctional agents (Cohen et al., 1982; Schroeder-Kurth et al., 1989). There is also some evidence that FA cells are sensitive to chromosome damage by ionizing radiation (Duckworth-Rysiecki and Taylor, 1985) which is at variance with the toxicity data, since chromosome damage is normally a good index of cytotoxicity.

There is little information on specific locus mutation in FA cells. Takebe et al. (1987) report briefly an increased sensitivity to mutation by diepoxybutane. Thioguanine-resistant mutants were increased by approximately 10-fold. They failed, however, to demonstrate increased mutation by mitomycin-C, which they attribute to its acute toxicity.

There are four complementation groups of Fanconi's anaemia. One gene has been cloned by Strathdee et al. (1992a) complementing the C group defect (FACC). All four genes, including FACC, appear to be on chromosome 9 (Strathdee et al., 1992b). This gives a means by which the nature of the gene can be further explored and the role of the gene in inducing chromosome breakage and mutation clarified.

4.6.7 Bloom's syndrome

Bloom's syndrome (BS) is a rare genetic disorder, inherited in an autosomal recessive manner; the main clinical features are short stature, a characteristic face with prominent nose, sun-sensitivity and a progressive skin rash which eventually leads to blistering and scarring. Sufferers also have impaired immunity and an elevated risk of developing cancer, particularly leukaemia, at a very early age. Bloom's lymphocytes and fibroblasts are characterized by an unusual frequency of specific types of cytogenetic abnormalities. Most common are quadriradial configurations involving chromatid interchanges between homologous chromosomes (Ray and German, 1981). Non-homologous exchanges are also seen occasionally. Curiously, quadriradials have not been observed in direct bone marrow preparations. A second striking cytogenetic abnormality in Bloom's cells is an unusually high frequency of sister chromatid exchanges (SCE). The 10- to 14-fold increase is considered diagnostic for Bloom's syndrome. The elevated SCE frequency is not always found in all cells and it appears that there may be differential expression of the phenotype in different subpopulations of cells. The repeated presence of chromosomally marked clones of lymphocytes provides evidence for *in vivo* chromosome mutation in Bloom's syndrome.

The primary defect has been identified as a reduced or anomalous activity of DNA ligase I (Willis *et al.*, 1987). It is also of interest that a unique patient (known as 46BR), who is hypersensitive to a range of DNA damaging agents, has been found to have a mutation in the DNA ligase I gene (Barnes *et al.*, 1992). The authors suggest that ligase I is essential for the joining of Okazaki fragments and the completion of DNA repair.

4.7. Genomic instability and mutator phenotype

It has been well known for the best part of a century that the chromosomes of cancer cells show a wide range of aberrations. Some of these chromosome changes have proved to be quite specific and associated with specific molecular rearrangements and with specific disease types. However, these specific rearrangements are often masked by a plethora of apparently non-specific rearrangements and aneuploidies. Genetic instability has, therefore, become an accepted feature of the vast majority of tumour cells, especially in advanced disease. The mechanism for generating such diversity and the surprising viability of these grossly aberrant cells have, until recently, received scant attention. An understanding of the mechanism of programmed cell death (apoptosis) and the role played by failure of the apoptotic mechanism in permitting continued survival of grossly damaged cells, have provided insight into one of the mechanisms for maintaining the genetic integrity of somatic cells (e.g. Potten *et al.*, 1994). In determining mutation rates (induced or spontaneous) in somatic cells, one must, therefore, continually bear in mind the possibility that certain types of mutation will be recognized as harmful and thus activate the genetically controlled mechanisms of cellular suicide.

4.7.1 Cell lines

Specific mechanisms of generating genetic diversity in somatic mammalian cells have also been recognized recently. Not surprisingly, they show a clear relationship with mechanisms already well understood from prokaryotes and lower eukaryotes. Some lines of cultured mammalian cells show a spontaneous mutator phenotype. Liu *et al.* (1993), for example, describe HPRT mutations occurring in a line of V79 Chinese hamster cells which contains a mutant DNA polymerase-α and a mutator phenotype (Aphr-4-2 cells). The rates of specific mutation types (base substitutions and deletions) were found to be elevated in the Aphr-4-2 cells and the authors suggest that DNA polymerase-α may play a role in determining the rate of different molecular types of spontaneous mutations *in vivo*. DNA polymerase-α has also been implicated in mutator mutants of the mouse cell line DM3A (Hyodo and Susuki, 1990) but mutations in other replication-associated enzymes also show a mutator phenotype [e.g. CTP synthetase (Yamauchi *et al.*, 1990) and ribonucleotide reductase (Caras and Martin, 1988)].

Boyer *et al.* (1993) set out specifically to test the hypothesis that a mutator phenotype may be associated with carcinogenesis using a diploid fibroblast strain and its transformed derivative as well as HeLa cells. Using cellular extracts, they measured deletions and mutations in a plasmid construct. They conclude that genetic instability, associated with transformation, does not involve reduced fidelity of replication of undamaged DNA or reduced mismatch repair. However, others working along the same lines have come to quite different conclusions. Studies on *E. coli* have previously shown that loss of mismatch correction confers a mutator phenotype and it has now been shown that mammalian cells also show the phenomenon. Kat *et al.* (1993) used a mutant lymphoblastoid cell line (MT1) resistant to *N*-methyl-*N'*-nitro-*N*-nitrosoguanidine (MNNG) and displaying a mutator phenotype. They showed that MNNG-induced mutations in the HPRT gene are almost exclusively GC → AT transitions, while spontaneous mutations in this cell line are single nucleotide insertions, transversions and AT → GC transitions. From these, and other observations, they conclude that MT1 cells are deficient in strand-specific mismatch repair. Similarly, Branch *et al.* (1993) show that human and hamster cell lines that lack O^6-methylguanine-methyltransferase activity, but also display resistance to the cytotoxic effect of alkylating agents such as *N*-methyl-*N*-nitrosourea (MNU), display a mutator phenotype. These cells are shown to be defective in a mismatch repair function, probably components of a GT binding complex. It is suggested that this GT binding activity helps prevent spontaneous mutation and that tolerance of methylation damage is associated with absence of this activity and a mutator phenotype.

4.7.2 Colon cancer and HNPCC

More directly, Parsons *et al.* (1993) show that cells from a subset of colorectal cancers and from most tumours developing in hereditary non-polyposis colorectal cancer (HNPCC) show a profound defect in strand-specific mismatch repair. Cells from these tumours show a 100-fold increase in the mutation rate of microsatellite sequences $(CA)_n$. Similarly, Thibodeau *et al.* (1993) and Aaltonen *et al.* (1993) found an increased instability of microsatellite sequences in cells from carcinoma of the colon. Thibodeau *et al.* showed the effect to be associated particularly with the proximal colon and inversely correlated with loss of heterozygosity on chromosomes known to be involved in colonic cancers, suggesting a mechanism of cancer induction quite different from classic tumour suppressor genes. Aaltonen *et al.* again stress the absence of loss of heterozygosity and show that the effect is demonstrated in a proportion of spontaneous cancers as well as in those from familial cases and come to an identical conclusion. Ionov *et al.* (1993) agree that about 12% of spontaneous colorectal cancers show microsatellite instability and suggest that cells from such cases may carry up to 100 000 mutations of this type.

Fishel *et al.* (1993) and Leach *et al.* (1993) reported simultaneously that the gene responsible for this genomic instability in HNPCC is the human equivalent of the bacterial *MutS* gene and the yeast *MSH* gene. Both groups identify a mutation in a splice site and associate the constitutional mutation with HNPCC families. Fishel *et al.* propose the name *hMSH2* (the human *MutS* homologue). The gene maps to chromosome 2p22–21 and they have demonstrated that expression of the human gene in *E. coli* causes a dominant mutator phenotype, confirming that this is indeed the gene causing instability in the tumour cells. This instability is similar to that reported by Strand *et al.* (1993) in yeast cells (Section 4.2.1). Palombo *et al.* (1994) point out that a GT binding protein which they had recognized in the HeLa cells some years earlier is homologous with the *hMSH2* gene and suggest that many similar pathways remain to be discovered.

It had been known that a proportion of HNPCC families did not show linkage to chromosome 2. Bronner *et al.* (1994) exploited the known linkage of some families to chromosome 3 and the prior discovery of the *hMSH2* gene in the chromosome 2 families to identify another human homologue of a mismatch repair gene. This new gene, *hMLH1*, is the human *MutL* homologue and is located on chromosome 3p21.3–23. They demonstrate nonsense mutations in affected individuals from a chromosome 3-linked HNPCC family and suggest that this is a second gene for hereditary carcinoma of the colon, not associated with adenomatous polyposis.

Thus, specific inherited mechanisms for generating genomic instability have now been associated with a human cancer and it can be anticipated that more such examples will be recognized in other cancers in the future. Indeed, such mutator phenotypes may have far-reaching consequences for other human disorders.

References

Aaltonen LA, Peltomaki P, Leach FS, Sistonen P, Pylkkanen L, Mecklin J-P, Jarvinen H, Powell SM, Jen J, Hamilton SR, Petersen GM, Kinzler KW, Vogelstein B, de la Chapelle A. (1993) Clues to the pathogeneses of familial colorectal cancer. *Science* 260: 812–818.

Arlett CF, Harcourt SA. (1982) Variation in response to mutagens amongst normal and repair deficient human cells. In: *Induced Mutagenesis* (ed. CW Lawrence). Plenum Press, New York, pp. 249–266.

Azuma T, Motoyama N, Fields LE, Loh DY. (1993) Mutation of the chloramphenicol acetyl transferase transgene driven by the immunoglobulin promoter and intron enhancer. *Int. Immunol.* 5: 121–130.

Barnes DE, Tomkinson AE, Lehmann AR, Webster DB, Lindahl T. (1992) Mutations in the DNA ligase I gene of an individual with immunodeficiencies and cellular hypersensitivity to DNA damaging agents. *Cell* 69: 495–503.

Birch JM, Hartley AL, Tricker KJ, Prosser J, Condie A, Kelsey AM, Harris M, Morris-Jones PM, Binchy A, Crowther D, Craft AW, Eden OB, Evans GR, Thompson E, Mann JR, Martin J, Mitchell ELD, Santibanez-Koref M. (1994) Prevalence and diversity of constitutional mutations in the *p53* gene among 21 Li–Fraumeni families. *Cancer Res.* **54:** 1298–1304.

Boyer JC, Thomas DC, Maher VM, McCormick JJ, Kunkel TA. (1993) Fidelity of DNA replication by extracts of normal and malignantly transformed cells. *Cancer Res.* **53:** 3270–3275.

Branch P, Aquilina G, Bignami M, Karran P. (1993) Defective mismatch binding and a mutator phenotype in cells tolerant to DNA damage. *Nature* **362:** 652–654.

Bronner CE, Baker SM, Morrison PT, Warren G, Smith LG, Lescoe MK, Kane M, Earabino C, Lipford J, Lindblom A, Tannergard P, Bollag RJ, Godwin AR, Ward DC, Nordenskjold M, Fishel R, Kolodner R, Liskay RM. (1994) Mutation in the DNA mismatch repair gene homologue hMLH 1 is associated with hereditary non-polyposis colon cancer. *Nature* **368:** 258–266.

Burnet M. (1974) *Intrinsic Mutagenesis.* Medical Aid Technical Publishing, Lancaster, UK.

Caras IW, Martin DW. (1988) Molecular cloning of the cDNA for a mutant mouse ribonucleotide reductase M1 that produces a dominant mutator phenotype in mammalian cells. *Mol. Cell Biol.* **8:** 2698–2704.

Carter AT, Norton JD, Gibson Y, Avery RJ. (1986) Expression and transmission of a rodent retrovirus-like (VL30) gene family. *J. Mol. Biol.* **188:** 105–108.

Cerutti P, Hussain P, Pourzand C, Aguilar F. (1994) Mutagenesis of the *H-ras* protooncogene and the *p53* tumour suppressor gene. *Cancer Res.* **54:** 1934s–1938s.

Cohen MM, Simpson SJ, Honig GR, Maurer HS, Nicklas JW, Martin AO. (1982) The identification of Fanconi anemia genotypes by clastogenic stress. *Am. J. Hum. Genet.* **34:** 794–810.

Cole J, Arlett CF, Green MHL, James SE, Henderson L, Cole H, GalATrepat M, Benzi R, Price ML, Bridges BA. (1989) Measurement of mutant frequency to 6-thioguanine resistance in circulating T-lymphocytes for human population monitoring. In: *New Trends in Genetic Risk Assessment* (ed. G Jollis). Academic Press, London, pp. 175–203.

Collins J, Forbes E, Anderson P. (1989) The Tc3 family of transposable elements in *Caenorhabditis elegans. Genetics* **121:** 47–55.

Cooley L, Kelley R, Spradling A. (1988) Insertional mutagenesis of the *Drosophila* genome with angle P elements. *Science* **239:** 1121–1128.

Cox R. (1982) A cellular description of the repair defect in ataxia telangiectasia. In: *Ataxia-Telangiectasia – a Cellular and Molecular Link Between Cancer, Neuropathology and Immune Deficiency* (eds BA Bridges, DG Harnden). John Wiley & Sons, Chichester, pp. 141–153.

Cullis CA. (1990) DNA rearrangements in response to environmental stress. In: *Genomic Responses to Environmental Stress,* Vol. 28, *Advances in Genetics* (ed. JG Scandalios). Academic Press, San Diego, pp. 72–97.

Defais M. (1990) Mechanisms of repair in mammalian cells. In: *Chemical Carcinogenesis and Mutagenesis II* (eds CS Cooper, PL Grover). Springer-Verlag, Berlin, pp. 51–70.

Doll R, Evans HJ, Darby S. (1994) Paternal exposure not to blame – commentary on the Gardner hypothesis. *Nature* **367:** 378–380.

Donehower LA, Harvey M, Slagle BL, McArthur MJ, Montgomery CA, Butel JS, Bradley A. (1992) Mice deficient for p53 are developmentally normal but susceptible to spontaneous tumours. *Nature* **356:** 215–221.

Duckworth-Rysiecki G, Taylor AMR. (1985) Effects of ionising radiation on cells from Fanconi's anaemia. *Cancer Res.* **45:** 416–420.

Fishel R, Lescoe MK, Rao MRS, Copeland NG, Jenkins NA, Garber J, Kane M, Kolodner R. (1993) The human mutator gene homologue MSH2 and its association with hereditary nonpolyposis colon cancer. *Cell* **75:** 1027–1036.

Fukuchi K, Martin GM, Monnat RJ. (1989) Mutator phenotype of Werner syndrome is characterised by extensive deletions. *Proc. Natl Acad. Sci. USA* **86:** 5893–5897.

Ganesh A, North P, Thacker J. (1993) Repair and misrepair of site specific DNA double strand breaks by human cell extracts. *Mutat. Res.* **299:** 251–259.

Gardner MJ, Hall AJ, Downes S, Terrell JD. (1987a) Follow up study of children born elsewhere but attending schools in Seascale, West Cumbria (schools cohort). *Br. Med. J.* **295:** 819–822.

Gardner MJ, Hall AJ, Downes S, Terrell JD. (1987b) Follow up study of children born to mothers resident in Seascale, West Cumbria (birth cohort). *Br. Med. J.* **295:** 822–827.

Harnden DG. (1990) Genetic susceptibility to chemical carcinogens. In: *Chemical Carcinogenesis and Mutagenesis II* (eds CS Cooper, PL Grover). Springer-Verlag, Berlin, pp. 225–248.

Harvey M, McArthur MJ, Montgomery CA, Butel JS, Bradley A, Donehower LA. (1993) Spontaneous and carcinogen induced tumorigenesis in p53-deficient mice. *Nature Genetics* **5:** 225–229.

Hehlamann R, Brack-Werner R, Leib-Mosch C. (1988) Human endogenous retroviruses. *Leukaemia* **2:** 167–177.

Hittelman WN, Pandita TK. (1994) The possible role of chromatin alteration in the radiosensitivity of ataxiATelangiectasia. *Int. J. Rad. Biol.* **66:** s109–s113.

Hyodo M, Susuki K. (1990) Fidelity of DNA polymerase alpha partially purified from a mutator mutant and wild type mouse FM3A cells. *Tokai J. Exp. Clin. Med.* **15:** 5–11.

Ionov Y, Peinado MA, Malkhosyan S, Shibata D, Perucho M. (1993) Ubiquitous somatic mutations in simple repeated sequences reveal a new mechanism for colonic carcinogenesis. *Nature* **363:** 558–561.

Isbell RJ, Fowler RG. (1989) Temperature-dependent mutational specificity of an *Escherichia coli* mutator dna Q49 defective in 3′–5′ exonuclease (proof reading). *Mutat. Res.* **213:** 149–156.

Jacks T, Remington L, Williams BO, Schmitt EM, Halachmi S, Bronson R, Weinberg R. (1994) Tumour spectrum analysis in p53-deficient mice. *Current Biol.* **4:** 1–7.

Jonczyk P, Fijalkowska I, Ciesla Z. (1988) Overproduction of the epsilon subunit of DNA polymerase III counteracts the SOS mutagenic response in *Escherichia coli*. *Proc. Natl Acad. Sci. USA* **85:** 9124–9127.

Kadhim MA, Macdonald DA, Goodhead DT, Lorimore SA, Marsden SJ, Wright EG. (1992) Transmission of chromosomal instability after plutonium α particle irradiation. *Nature* **355:** 738–740.

Kastan MB, Zhan Q, el Deiry WS, Carrier F, Jacks T, Walsh WV, Plunkett BS, Vogelstein B, Fornace AJ. (1992) A mammalian cell cycle checkpoint pathway utilising p53 and GADD45 is defective in ataxia telangiectasia. *Cell* **71:** 587–597.

Kat A, Thilly WG, Fang WH, Longley MJ, Li GM, Modrich P. (1993) An alkylation tolerant, mutator human cell line is deficient in strand specific mismatch repair. *Proc. Natl Acad. Sci. USA* **90:** 6424–6428.

Kemp CJ, Donehower LA, Bradley A, Balmain A. (1993) Reduction of *p53* gene dosages does not increase initiation or promotion but enhances malignant progression of chemically induced skin tumours. *Cell* **74:** 813–822.

Khanna KK, Lavin MF. (1993) Ionising radiation and UV induction of p53 protein by different pathways in ataxia telangiectasia cells. *Oncogene* **8:** 3307–3312.

Kim AI, Lyubomirskaya NV, Belyaeva ES, Shostack NG, Ilyin YV. (1994) The introduction of a transpositionally active copy of retrotransposon GYPSY into the stable strain of *Drosophila melanogaster* causes genetic instability. *Mol. Gen. Genet.* **242:** 472–477.

Kinlen LJ. (1993) Can paternal preconceptual radiation account for the increase in leukaemia of non-Hodgkin's lymphoma in Seascale. *Br. Med. J.* **306:** 1718–1721.

Knudson AG. (1971) Mutation and cancer: statistical study of retinoblastoma. *Proc. Natl Acad. Sci. USA* **68:** 820–823.

Leach FS, Nicolaides NC, Papadopoulos N and 32 others. (1993) Mutations of a unit S homolog in hereditary non polyposis colorectal cancer. *Cell* **75:** 1215–1225.

Lehmann AR, Dean SW. (1990) Cancer prone disorders with defects in DNA repair. In: *Chemical Carcinogenesis II* (eds CS Cooper, PL Grover). Springer-Verlag, Berlin, pp. 71–101.

Liu PK, Trujillo JM, Monnat RJ. (1993) Spectrum of spontaneous mutation in animal cells containing an aphidicolin resistant DNA polymerase-alpha. *Mutat. Res.* **288:** 229–236.

Lu X, Lane DP. (1993) Differential induction of transcriptionally active p53 following UV or ionising radiation: defects in chromosome instability syndromes? *Cell* **75**: 765–778.

Maher VM, McCormick JJ. (1976) Effect of DNA repair on the cytotoxicity and mutagenicity of UV irradiation and of chemical carcinogenesis in normal and xeroderma pigmentosum cells. In: *Biology of Radiation Carcinogenesis* (eds JM Yuhars, RJ Tennant, ID Regan). Raven Press, New York, pp. 129–145.

Malkin D, Li FP, Strong LC, Fraumeni JF, Nelson CE, Kim DH, Kassel J, Gryka MA, Bischoff FZ, Tainsky MA, Friend SH. (1990) Germline mutations in a familial syndrome of breast cancer, sarcoma and other neoplasms. *Science* **250**: 1233-1238.

McClintock B. (1950) The origins and behaviour of mutable loci in maize. *Proc. Natl Acad. Sci. USA* **36**: 344–355.

Meyn MS. (1993) High spontaneous intrachromosomal recombination rates in ataxia telangiectasia. *Science* **260**: 1327–1330.

Meyn MS, Strasfeld L, Allen C. (1994) Testing the role of p53 in the expression of genetic instability and apoptosis in ataxia telangiectasia. *Int. J. Rad. Biol.* **66** (suppl.): s141–s149.

Monnat RJ, Hackmann AF, Chiaverotti TA. (1992) Nucleotide sequence analysis of human hypoxanthine phosphoribosyltransferase (HPRT) gene deletions. *Genomics* **13**: 777–787.

Morrison A, Sugino A. (1994) The 3′–5′ exonucleases of both DNA polymerases delta and epsilon participate in errors of DNA replication in *Saccharomyces cerevisiae*. *Mol. Gen. Genet.* **242**: 289–296.

Nakura J, Miki T, Nagano K, Kihara K, Ye L, Kamino K, Fujiwara Y, Yoshida S, Murano S, Fukuchi K. (1993) Close linkage of the gene for Werner's syndrome to ANK1 and D8587 on the short arm of chromosome 8. *Gerontology* **39** (suppl.1): 11–15.

Neel JV. (1977) The monitoring of human populations for mutations affecting protein structure and/or function: problems and prospects. In: *Conference on Population Monitoring Methods for Detecting Increased Mutation Rates.* Harald Boldt Verlag, Boppard, pp. 91–109.

Nomura T. (1983) X-ray induced germ line mutation leading to tumours: its manifestation in mice given urethan post natally. *Mutat. Res.* **121**: 59–65.

Nomura T. (1986) Further studies on X-ray and chemically induced germ line alterations causing tumours and malformations in mice. *Proc. Clin. Biol. Res.* **209B**: 13–20.

Oberlé I, Rousseau F, Hertz D, Kretz C, Devys D, Hanauer A, Boue J, Bertheas MF, Mandel JL. (1991) Instability of a 550 base pair DNA segment and abnormal methylation in fragile X-syndrome. *Science* **252**: 1097–1102.

Palombo F, Hughes M, Jiricny J, Truong O, Hsuan J. (1994) Mismatch repair and cancer. *Nature* **367**: 417.

Parsons R, Li GM, Longley MJ, Fang WH, Papadopoulos N, Jen J, de la Chapelle A, Kinzler KW, Vouelstein B, Modrich P. (1993) Hypermutability and mismatch repair deficiency in RER + tumour cells. *Cell* **75**: 1227–1236.

Paulson KE, Deka N, Schmid CW, Misra R, Schindler DW, Rush MG, Kadky L, Leinward L. (1985) A transposon-like element in human DNA. *Nature* **316**: 359–361.

Potten CS, Chadwick CA, Cohen AJ, Nikaido O, Matsunaga T, Schipper NW, Young AR. (1993) DNA damage in irradiated human skin *in vivo:* Automated direct measurement by image analysis (thymine dimers) compared with indirect measurement (unscheduled DNA synthesis) and protection by 5-methoxypsoralen. *Int. J. Radiat. Biol.* **63**: 313–323.

Potten CS, Merritt AJ, Hickman J, Hall P, Faranda A. (1994) The characterisation of radiation-induced apoptosis in the small intestine and its biological implications. *Int. J. Rad. Biol.* **65**: 71–79.

Ray JH, German J. (1981) The chromosome changes in Bloom's syndrome, ataxia telangiectasia and Fanconi's anaemia. In: *Genes, Chromosomes and Cancer* (eds FE Arrighi, PN Rao, E Stabblefield). Raven Press, New York, pp.351–378.

Richards RI, Sutherland GR. (1994) Simple repeat DNA is not replicated simply. *Nature Genetics* **61**: 114–116.

Rogerson BJ. (1994) Mapping the upstream boundary of somatic mutations in rearranged immunoglobulin transgenes and endogenous genes. *Mol. Immunol.* **31**: 83–98.

Rothenfluh HS, Taylor L, Rothwell AL, Both GW, Steele EJ. (1993) Somatic hypermutation in 5′ flanking regions of heavy chain antibody variable regions. *Eur. J. Immunol.* **23**: 2152–2159.

Runger TM, Poot M, Kraemer KH. (1992) Abnormal processing of transfected plasmid DNA in cells from patients with ataxia telangiectasia. *Mutat. Res.* **293**: 47–54.

Savitsky K, Bar-Shira A, Gilad S et al. (1995) A single Ataxia Telangiectasia gene with a product similar to P1-3 kinase. *Science* **268**: 1749–1753.

Scholtissek C, Ludwig S, Fitch WM. (1993) Analysis of influenza A virus nucleoproteins for the assessment of molecular genetic mechanisms leading to new phylogenetic virus lineages. *Arch. Virol.* **131**: 237–250.

Schroeder-Kurth TM, Auerbach AD, Obe G. (1989) *Fanconi Anaemia: Clinical Cytogenetic and Experimental Aspects.* Springer-Verlag, Berlin.

Strand M, Prolla TA, Liskay RM, Petes TD. (1993) Destabilisation of tracts of simple repetitive DNA in yeast by mutations affecting mismatch repair. *Nature* **365**: 274–276.

Strathdee CA, Gavish H, Shannon WR, Buchwald M. (1992a) Cloning of cDNAs for Fanconi's anaemia by functional complementation. *Nature* **256**: 763–767.

Strathdee CA, Duncan AMV, Buchwald M. (1992b) Evidence for at least four Fanconi anaemia genes including FACC on chromosome 9. *Nature Genetics* **1**: 196–198.

Taddie JA, Traktman P. (1991) Genetic characterisation of the vaccinia virus DNA polymerase: identification of point mutations conferring altered drug sensitivities and reduced fidelity. *J. Virol.* **65**: 869–879.

Takebe H, Tatsumi K, Tachibana A, Nishigori C. (1987) High sensitivity to radiation and chemicals in relation to cancer and mutation. In: *Radiation Research,* Vol. 2 (eds EM Fielden, JF Fowler JH Hendry, D Scott). Taylor and Francis, London, pp. 443–448.

Tatsumi K, Takebe H. (1984) Gamma-irradiation induces mutation in ataxia Telangiectasia lymphoblastoid cells. *Gann* **75**: 1040–1043.

Tatsumi K, Toyoda M, Hashimoto T, Furuyama JI, Kurihara T, Inoue M, Takebe H. (1987) Differential hypersensitivity of xeroderma pigmentation lymphoblastoid cell lines to ultraviolet light mutagenesis. *Carcinogenesis* **8**: 53–57.

Tatsumi-Miyajima J, Yagi T., Takebe H. (1993) Analysis of mutations caused by DNA double strand breaks produced by a restriction enzyme in shuttle vector plasmids propagated in ataxia telangiectasia cells. *Mutat. Res.* **294**: 317–323.

Taylor AMR. (1982) Cytogenetics of ataxia telangiectasia. In: *Ataxia Telangiectasia: a Cellular and Molecular Link Between Cancer, Neuropathology and Immune Deficiency* (eds BA Bridges, DG Harnden). Wiley, Chichester, pp. 53–81.

Thacker J, Chalk J, Ganesh A, North P. (1992) A mechanism for deletion formation in DNA by human cell extracts: the involvement of short sequence repeats. *Nucl. Acids Res.* **20**: 6183–6188.

Thibodeau SN, Bren G, Schaid D. (1993) Microsatellite instability in cancer of the proximal colon. *Science* **260**: 818–821.

Tomatis L, Hilfrich J, Turusov V. (1975) The occurrence of tumours in F1, F2 and F3 descendents of BD rats exposed to *N*-nitrosomethylurea during pregnancy. *Int. J. Cancer* **15**: 385–390.

Tomatis L, Ponomarko V, Turusov V. (1977) Effects of ethyinitrosourea administration during pregnancy on three subsequent generations of BDV1 rats. *Int. J. Cancer* **19**: 240–248.

Vnencak-Jones CL, Phillips JA. (1990) Hot spots for growth hormone gene deletions in homologous regions outside of *Alu* repeats. *Science* **250**: 1745–1748.

Wabl MR, Jack HM, von Borstel RC, Steinberg CM. (1989) Scope of action of the immunoglobulin mutator system. *Genome* **31**: 118–121.

Weil D, Wang I, Dietrich A, Popustka A, Weissenbach J, Petit C. (1994) Highly homologous loci on the X and Y chromosomes are hot spots for ectopic recombinations leading to XX maleness. *Nature Genetics* **7**: 414–419.

Wiencke JK, Kelsey KT. (1993) Susceptibility to induction of chromosomal damage by metabolites of 1,3-butadiene and its relationship to "spontaneous" sister chromatid exchange frequencies in human lymphocytes. In: *Butadiene and Styrene: Assessment of Health Hazards* (eds M Sorsa, K Peltonen, H Vainio, K Hemminke). IARC, Lyon, pp. 265–273.

Willis AE, Weksberg R, Tomlinson S, Lindahl T. (1987) Structural alterations of DNA ligase I in Bloom's syndrome. *Proc. Natl Acad. Sci. USA* **84**: 8016–8020.

Yamauchi M, Yamauchi N, Meuth M. (1990) Molecular cloning of the human CTP synthetase gene by functional complementation with purified metaphase chromosomes. *EMBO J.* **9**: 2095–2099.

Pharmacogenetic polymorphisms

Gillian Smith, C.A. Dale Smith and C. Roland Wolf

5.1 Introduction

A relationship between the inheritance of defective genes and the aetiology of diseases such as albinism and alkaptonuria was first recognized at the beginning of this century (Garrod, 1909). These 'inborn errors of metabolism' were subsequently shown to be the result of deficiencies in, or lack of expression of, enzymes encoded by specific defective or mutant genes. For example, patients with alkaptonuria were shown to have elevated levels of homogentisic acid, produced during the metabolism of the amino acid tyrosine. It was another 50 years, however, before it was reported that the enzyme responsible for the oxidative metabolism of homogentisic acid, homogentisic acid oxidase, was not expressed in the livers of affected patients (La Du *et al.*, 1958). These patients who were therefore unable to metabolize the compound further. Similarly, the observation that isoniazid metabolism exhibited marked interindividual variation led to the identification of 'slow' and 'rapid' metabolizers/acetylators of the drug, 30 years before the molecular basis for the polymorphism at the *NAT2* gene locus was reported (Blum *et al.*, 1991, Price-Evans *et al.*, 1960). These and similar observations led to the birth of pharmacogenetics, defined as the study of individual variation in the rate of drug metabolism, or the determination of the genetic basis of idiosyncratic responses to drugs or foreign chemicals (Nebert, 1990).

Many subsequent studies have demonstrated that phenotypic variation in the metabolism of many clinically important drugs is the consequence of genetically determined differences (polymorphisms) in the activities of a significant number of human drug-metabolizing enzymes. Polymorphic alleles are defined as allelic variants at a single genetic locus which are present in more than 1% of the population. The enzymes for which the genetic basis for their polymorphic expression has been studied in most detail are the cytochromes P450, glutathione *S*-transferases (GST) and the *N*-acetyl transferases (NAT).

Drug metabolism in man is thought to occur in two distinct phases. On entering the body, drugs are first subjected to phase I metabolism, where functionalization reactions are performed, creating a 'reactive-centre' (for example, -OH, -NH$_2$, -COOH) in the molecule. The most ubiquitous of phase I catalysts are the cytochrome P450 monooxygenases, which are capable of metabolizing a wide range of structurally diverse substrates. Other enzymes involved in phase I metabolism are described in *Table 5.1.* Following the creation of an electrophilic reactive centre, phase II enzymes such as the glutathione *S*-transferases and UDP–glucuronyl transferases (*Table 5.1*) are responsible for conjugation reactions, involving the incorporation of, for example, a glutathione or glucuronic acid moiety into the molecule. Phase II reactions are, in most cases, responsible for the ultimate excretion of drug from the body. As can be seen from *Table 5.1*, the expression of many of the enzymes involved in drug metabolism in man is genetically polymorphic.

Table 5.1. Some enzymes involved in drug metabolism in man

Phase I: functionalization reactions	Phase II: conjugation reactions
P450 monooxygenases	*Glutathione S-transferases*
Flavoprotein monooxygenases	*N-acetyl transferases*
Monoamine oxidases	*N*-acyl transferases
Alcohol dehydrogenases	*UDP-glucuronyl transferases*
Aldehyde dehydrogenases	*Sulphotransferases*
Arylesterases	*Methyl transferases*
Cholinesterases	*Epoxide hydrolases*
Epoxide hydrolases	
Amidases	
Nitroreductases	

Enzymes in italic have been shown to, or are thought to, exhibit genetic polymorphism in man.

The existence of multiple alleles at loci which encode drug-metabolizing enzymes can result in differential susceptibilities of individuals within a population to the mutagenic or carcinogenic effects of drugs or environmental chemicals. For example, the ability to metabolize the antihypertensive drug debrisoquine to its 4-OH metabolite shows a bimodal distribution within the population (*Figure 5.1*), which has been shown to result from the polymorphic expression of the cytochrome P450 gene active in its metabolism, *CYP2D6* (Mahgoub *et al.*, 1977). The majority of individuals (extensive metabolizers, EM) have at least one intact copy of the *CYP2D6* gene and can metabolize debrisoquine efficiently, resulting in most, if not all of the administered drug being converted to its primary (4-OH) metabolite. The metabolic ratio for debrisoquine (i.e. the ratio of the concentration of unchanged drug to the concentration of the 4-OH metabolite) is therefore low. Individuals with a mutated, inactive form of CYP2D6 (poor metabolizers, PM), are unable to metabolize the drug, and have a correspondingly high metabolic ratio.

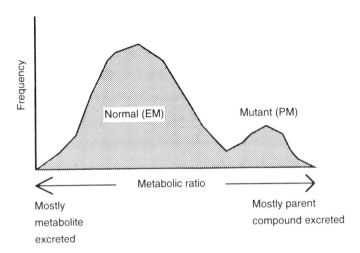

Figure 5.1. Bimodal distribution of drug clearance by CYP2D6 in a population study.

Several factors are important in determining the clinical significance of the polymorphic expression of a particular drug-metabolizing enzyme (Tucker, 1994). If the affected enzyme is responsible for primary 'first pass' metabolism of the drug, then the phenotypic consequences of altered expression are more pronounced than if it is active at a later stage of the metabolic pathway. If, however, minor metabolites of the parent drug are particularly toxic, then the polymorphic expression of a relatively minor enzyme can have significant phenotypic consequences. The relative level of expression of detoxifying enzymes (phase II metabolism) is also an important consideration, as is the possibility of compensation by other non-polymorphically expressed enzymes when the preferred route of metabolism is no longer active. This is particularly relevant in multigene families, such as the cytochromes P450, where many unique isozymes have overlapping substrate specificities.

The therapeutic index of a drug is also an important determinant of whether polymorphic expression of the drug-metabolizing enzymes active in its metabolism are likely to have phenotypic consequences – a wide variation in plasma concentration is not so important for drugs with a high therapeutic index but, if the therapeutic index of the drug is low, then subtle variations in the route or rate of metabolism can have serious clinical consequences.

Significant differences in the allele frequencies of polymorphically expressed human drug-metabolizing enzymes have been observed between populations of different racial origin (Kalow, 1982). For example, while the *CYP2D6* PM genotype is seen in between 5 and 10% of the Caucasian population, it is relatively rare in Orientals, occurring with a frequency of only

1% in the Chinese population (Johansson *et al.*, 1991). A similar variation in allele frequency is seen for *CYP2C19* (de Morais *et al.*, 1994), which catalyses the hydroxylation of mephenytoin, and at the *NAT2* gene locus (Grant, 1993). It is therefore vitally important to recognize these racial differences when screening a population in order to determine accurately the allele frequencies for a given polymorphism.

Many enzymes which are active in drug metabolism in man also play an important role in the metabolic activation or detoxification of a wide range of xenobiotic toxins which occur as environmental pollutants or which are ingested in food. For example, it has been demonstrated that the potent carcinogen NNK [4-(methylnitrosamino)-1-(3-pyridyl)-1-butanone], a component of cured tobacco and tobacco smoke, is metabolically activated by CYP2D6 (Crespi *et al.*, 1991). It seems reasonable, therefore, that polymorphisms in human drug-metabolizing enzymes may be linked to the incidence of diseases such as cancer and Parkinson's disease, the aetiologies of which have been linked to exposure to environmental toxins.

Advances in molecular cloning techniques have allowed many allelic variants of several human drug-metabolizing enzymes to be characterized at a genetic level, and specific nucleotide changes have been identified as the basis for altered protein structure and/or function. Certain genes, for example the gene encoding cytochrome P450 CYP2D6, carry mutations within introns or at intron/exon boundaries which can result in splice-junction defects, frame-shift mutations and/or defective processing of genomic DNA. While many mutations are not accompanied by a change in protein function (altered phenotype), single nucleotide substitutions can alter the nature and function of the protein product produced or, alternatively, alter the regulatory mechanisms of the gene.

Until relatively recently, determination of an individual's drug-metabolizing capacity was achieved by phenotypic assessment , for example by administration of a probe drug such as caffeine, and measuring the relative concentrations of unchanged drug and its metabolites in the urine. Such studies have inherent limitations; they do not allow for drug–drug interactions or the effect of external factors such as alcohol consumption or cigarette smoking. In addition, no allowance is made for the state of health of the patient – impaired renal function, for example, can have a marked effect on the rate of drug clearance from the body. Development of DNA-based polymerase chain reaction (PCR), single-stranded conformational polymorphism (SSCP) or restriction fragment length polymorphism (RFLP) methods of analysis allows the genotype of an individual to be determined easily and relatively quickly in a minimally invasive procedure. As a result, it has become much more straightforward to relate a particular genotype to disease incidence or to predict whether an individual may show an idiosyncratic response to a prescribed drug.

To date, the genetic polymorphisms within human drug-metabolizing enzymes that have been best characterized are those identified in genes encoding cytochromes P450 *CYP2D6* and *CYP2C19* and those at the *GSTM1* and *NAT2* gene loci.

5.2 Cytochromes P450

Cytochrome P450 enzymes are a multigene superfamily of monomeric mixed-function monooxygenases, responsible for the phase I metabolism of a wide range of structurally diverse substrates (Gonzalez, 1990). All P450s have the ability to insert an atom of molecular oxygen into their substrates which, in most cases, leads to an increase in hydrophilicity of the substrate and facilitates its excretion from the cell. Certain substrates, however, are metabolically activated by P450 metabolism, resulting in the formation of chemically reactive electrophiles. Many drugs that are commonly prescribed are substrates for one or more P450 isozyme - examples are given in *Table 5.2*.

Table 5.2. Some drugs that are metabolized by human cytochromes P450

CYPIA	– phenacetin, theophylline, acetaminophen, phenacetin, caffeine
CYP2B	– thio-TEPA, cyclophosphamide, BCNU
CYP2C	– tolbutamide, mephenytoin, warfarin, hexobarbital, ifosfamide, diazepam, omeprazole, propranolol, imipramine
CYP2D	– debrisoquine, sparteine, encainide, bufuralol, metoprolol, amitryptiline, dextromethorphan, amiflamine, codeine, ecstasy, timolol, haloperidol, perhexiline
CYP2E	– chlorzoxazone, acetaminophen, ethanol
CYP3A	– nifedipine, cyclosporine, erythromycin, dapsone, lidocaine, ethynylestradiol, vinblastine, morpholinodoxorubicin, taxol

P450s are haem-containing proteins, with molecular weights in the range 45–55 kDa, and which are located in the endoplasmic reticulum or mitochondrial membrane of mammalian cells. In conjunction with the flavoprotein NADPH cytochrome P450 oxidoreductase, they function as the terminal electron acceptors of an electron transport chain. Based on sequence homology, the P450 superfamily is divided into 10 subfamilies, each member of which has at least 40% homology at the amino acid level (Nelson *et al.*, 1993). Nomenclature and the chromosomal organization of human P450 genes are described in *Table 5.3*.

Subfamilies CYP1, CYP2 and CYP3, the 'microsomal' P450s, are primarily involved in xenobiotic (drug) metabolism – the diversity of these genes, particularly those within Family 2, is thought to have arisen as an adaptive response to environmental challenge (Gonzalez and Nebert, 1990; Wolf, 1986). In agreement with this, the level of expression of many of these genes has been shown to be substrate-inducible, usually regulated at the level of transcription (Gonzalez, 1990). The remaining P450 subfamilies are located in the mitochondria and are important catalysts in the regulation of cellular homoeostasis, with many isoforms responsible for catalysing regio- and stereo-specific steroid hydroxylation reactions. CYP4 proteins, for example, are

Table 5.3. Nomenclature and chromosomal localization of human P450 genes involved in xenobiotic metabolism

Gene family	No. of genes	Chromosomal location	Regulated by
CYP1A	2	15q22–qter	Polycyclic aromatic hydrocarbons, dioxins
CYP1B	>1	–	Steroid hormones (?)
CYP2A	2 or 3	19q13.1–13.3	Pyrazole, ethanol, hormones
CYP2B	2 or 3	19q13.1–13.3	Phenobarbital, PCBs, dexamethasone
CYP2C	5–10	10q24.1–24.3	Hormones, phenobarbital
CYP2D	3 or 4	22q11.2–qter	Unknown
CYP2E	1	10	Ethanol, diabetes, starvation, solvents
CYP2F	1 or 2	19	Unknown
CYP3A	3–5	7q21.3–q22	Hormones
CYP4A	2–4	1	Phenobarbital, clofibrate, fatty acids
CYP4B	1	1p12–q34	Phenobarbital

involved in the ω and ω-1 oxidation of fatty acids, leukotrienes and prostaglandins, while cholesterol 7-hydroxylase, the product of the *CYP7* gene, catalyses the rate-limiting step in the formation of bile acid from cholesterol (Karan and Chang, 1992).

Individual P450 isozymes, in general, have unique substrate specificities, although a certain amount of overlap is observed. There is some evidence that certain P450 genes may be coordinately regulated – this may occur, for example, as the result of the chromosomal proximity of the CYP2A and CYP2B genes on human chromosome 19 (Miles *et al.*, 1989a), and the fact that the divergence of these genes was relatively recent. There are many examples of compounds, for example the antioestrogenic drug tamoxifen, which is commonly prescribed in the treatment of breast cancer (Mani and Kupfer, 1991), and aflatoxin B_1, a potent hepatocarcinogen, which can be metabolized, often to different products, by alternative P450-mediated pathways (Aoyama *et al.*, 1990; Forrester *et al.*, 1990). Thus, an individual's response to the effects of administered drug or carcinogen is governed by the relative activities of the individual P450 isozymes active in its metabolism.

Cytochromes P450 are predominantly expressed in the liver, but there are many reports of isozyme-specific expression in extrahepatic tissues, which has been associated with tissue- or cell-specific substrate activation and resulting toxicities. The levels of expression of specific P450 isozymes in human liver have been shown to vary widely between individuals (Forrester *et al.*, 1990).

Several human P450 genes are now known to be polymorphically expressed and, for some of these, the genetic basis for the polymorphism is now understood. While many of the polymorphic sites present in human P450s have no functional significance, for example because the mutant form has only a single nucleotide change that does not alter the amino acid composition of the corresponding protein, other mutants can have significant phenotypic consequences, leading, in the extreme case, to a complete absence of functional enzyme.

5.2.1 CYP1A

The *CYP1A* gene locus in man has two members: *CYP1A1*, which is predominantly expressed in extrahepatic tissues, including lung (Antilla *et al.*, 1992), lymphocytes (Kouri *et al.*, 1982) and placenta (Song *et al.*, 1985); and the liver-specific *CYP1A2*. Substrates for, and inducers of, *CYP1A1* include polycyclic aromatic hydrocarbons such as 3-methylcholanthrene and benzo[*a*]pyrene (Gonzalez, 1990), while *CYP1A2* is active in the metabolism of nitrosamines and arylamines (Boobis *et al.*, 1994). The expression of both genes is regulated by the binding of an inducing agent to a cytosolic receptor, the Ah (aryl hydrocarbon) or dioxin receptor (Whitlock, 1990)

An *Msp1* restriction fragment length polymorphism has been identified in the 3′ non-coding region of the *CYP1A1* gene as a result of a C→T mutation, 250 base pairs (bp) downstream of the polyadenylation site (Spurr *et al.*, 1987; Petersen *et al.*, 1991). The phenotypic consequence of this mutation, manifested in approximately 10% of the human population, has been shown to be increased inducibility of *CYP1A1* expression in response to PAH substrates (Kawajiri *et al.*, 1990). A further point mutation (A→G, Ile→Val) at position 4889 in the *CYP1A1* gene has been linked with the *Msp1* polymorphism. This mutation lies within the haem-binding region of the protein, and would therefore be expected to have an effect on catalytic activity. Indeed, it was subsequently demonstrated that the mutant form of the protein has a catalytic activity which is seven times higher than the wild-type enzyme (Hayashi *et al.*, 1991).

A further polymorphism, which is only found in African Americans has recently been reported in the *CYP1A1* gene (Crofts *et al.*, 1993). This new mutant allele, which has an AT→GC transition in the 3′ non-coding region of the gene is also detectable by an *Msp1* RFLP, but does not cosegregate with either of the previously described allelic variants. The functional consequences, if any, of this mutant allele are presently unknown.

Although no genetic polymorphism has yet been discovered in the *CYP1A2* gene, considerable interindividual variations have been reported in both the level of expression of this isozyme in human liver and in the rate of metabolism of CYP1A2 substrates, for example caffeine. However, as many CYP1A2 substrates can also be metabolized by other drug-metabolizing enzymes [for example NAT2 (Butler *et al.*, 1992)], it is not possible at present to determine which enzyme, if any, is responsible for the variation in metabolic rate.

5.2.2 CYP2A and CYP2B

Considerable interindividual variation in the level of expression of both CYP2A and CYP2B proteins in human liver has been reported (Forrester *et al.*, 1992). An inactive variant of *CYP2A6*, *CYP2A6v*, has been described (Yamano *et al.*, 1990), which has an allele frequency of 0.02 in a Caucasian population (Daly *et al.*, 1993) and which may result in impaired metabolism of coumarin and other compounds that have been shown to be substrates for CYP2A6, including NNK (Crespi *et al.*, 1991). Alternative splice sites have now been identified in both CYP2B6 and CYP2A7 (Miles *et al.*, 1989b; Yamano *et al.*, 1989; Ding *et al.*, 1995) which may also contribute to the interindividual variation in the level of expression of these proteins. To date, substrates for CYP2A7 have not been identified.

5.2.3 CYP2C

It has been known for some time that there is considerable interindividual variation in the metabolism of *S*-mephenytoin, and that the metabolism of this compound is performed by a member of the CYP2C subfamily (Wilkinson *et al.*, 1989). Approximately 5% of the Caucasian population and up to 23% of Orientals are poor metabolizers (PM) and are unable to metabolize the drug. This PM phenotype has been shown to be inherited as an autosomal recessive trait (Inaba *et al.*, 1986). As a result of an intense period of study, a genetic polymorphism within the *CYP2C19* gene, responsible for catalysing the 4-hydroxylation of *S*-mephenytoin, is the latest metabolic defect for which the genetic basis has been elucidated (de Morais *et al.*, 1994). The PM phenotype has been attributed to a 40 bp deletion (bp 643–682) at the start of exon 5 of the *CYP2C19* gene, leading to the creation of a cryptic splice site, altered reading frame and the generation of a premature stop codon. The resulting truncated protein lacks the haem-binding region and is therefore inactive. This splice junction defect is thought to account for approximately 75% of mutant alleles in both Caucasian and Japanese populations.

5.2.4 CYP2D

The *CYP2D* gene locus in man is comprised of three genes, arranged in tandem on human chromosome 22 (Gough *et al.*, 1993). Two of these, *CYP2D7P* and *CYP2D8P* contain inactivating mutations and are thought to be pseudogenes (Heim and Meyer, 1992; Kimura *et al.*, 1989). The expression of the third gene, *CYP2D6*, has been shown to be genetically polymorphic (Mahgoub *et al.*, 1977), and characterization of allelic variants at this locus has been a major focus of research in this area in recent years. More than 30 commonly prescribed drugs are substrates for CYP2D6 (*Table 5.2*), including those which act on the cardiovascular and central nervous systems. Altered expression of CYP2D6 protein can therefore have severe clinical consequences, leading to marked drug side effects and even, in extreme cases, death (Kroemer *et al.*, 1994).

A number of mutant alleles have now been characterized at the *CYP2D6* locus – these are described in *Table 5.4*. While many of these allelic variants have no phenotypic consequences, several gene-inactivating mutations have now been reported (Evert *et al.*, 1994; Gaedigk *et al.*, 1991; Gough *et al.*, 1990; Kagimoto *et al.*, 1990; Saxena *et al.*, 1994). These mutations result in complete absence of CYP2D6 protein, and an inability in affected individuals (PMs) to metabolize compounds that are substrates for this enzyme. A simple and rapid PCR-based assay is now available by which the major gene-inactivating mutations can be determined (Gough *et al.*, 1990, Smith *et al.*, 1992a). A minority of individuals has been shown to inherit multiple copies of wild-type *CYP2D6* gene (Johansson *et al.*, 1993). These individuals, termed 'ultrarapid metabolizers' require doses of medication which vastly exceed the normal range in order to achieve a therapeutic effect (Bertilsson *et al.*, 1993).

5.2.5 CYP2E

The CYP2E family in man is represented by a single gene, *CYP2E1* (Song *et al.*, 1986), the expression of which has also been shown to be genetically polymorphic. CYP2E1 metabolizes, and is induced by ethanol, catalysing the oxidative metabolism of ethanol to acetaldehyde. Several polymorphisms have been identified in CYP2E1. Two of these, leading to RFLPs with the

Table 5.4. Alleles at the human *CYP2D6* gene locus

Allele	*Xba*I RFLP	Mutation	Functional significance
CYP2D6 wt	29 kb	–	Wild-type allele
CYP2D6-A	29 kb	A2637 deletion (exon 5)	Inactive
CYP2D6-B	29, 44, 9 + 16 kb	G→A (intron 3/exon 4)	Inactive
CTP2D6-C	29 kb	Lys 281 deletion	Decreased activity
CYP2D6-Ch1	29, 44 kb	–	Decreased activity
CYP2D6-D	11.5, 13 kb	Gene deletion	Inactive
CYP2D6-E	29 kb	–	Inactive
CYP2D6-F	29 kb	–	Inactive
CYP2D6-G	29 kb	–	Inactive
CYP2D6-J	29, 44 kb	C→T (exon 1) G→C (exon 9)	Decreased activity
CYP2D6-L	29 kb	–	Wild-type activity
CYP2D6-L2	42 kb	Gene amplification	Increased activity
CYP2D6-L12	175 kb	–	Increased activity
CYP2D6-T	–	Base deletion T^{1795}, exon 3	Inactive
CYP2D6-W	29, 44 kb	–	Decreased activity

restriction enzymes *Pst*I and *Rsa*I have been shown to lie in the 5′ promoter region of the gene and lead to altered interaction with the transcription factor HNF-1 (Watanabe *et al.*, 1990). A further RFLP (*Dra*I) is associated with a mutation in intron 2 of the gene (Uematsu *et al.*, 1991).

5.3 Glutathione *S*-transferases

Glutathione *S*-transferase (GST) genes encode a family of enzymes (mol. wt 17–28 kDa) which catalyse the conjugation of glutathione to electrophilic substrates, most of which are formed during phase I metabolism (Rushmore and Pickett, 1993). There are four classes of dimeric cytosolic GST – α (A), μ (M), π (P) and θ (T). Members of the same gene family exhibit a minimum of 65% amino acid identity and can exist as either homodimers or heterodimers (Mannervik *et al.*, 1992). A fifth form, the trimeric microsomal GST is membrane bound (De Jong *et al.*, 1988). The primary function of the cytosolic enzymes is thought to be the detoxification of reactive electrophiles (Jacoby, 1978), but these proteins, via their glutathione-dependent peroxidase activities, also have an important role in free-radical scavenging, thus protecting the cell from the deleterious effects of oxidative stress (Sies and Ketterer, 1988). In addition, GST expression has been shown to be significantly increased on exposure to antioxidant compounds such as the common food preservatives butylated hydroxyanisole (BHA, E 320) and butylated hydroxytoluene (BHT, E 321). An 'antioxidant response element' (ARE), present in the 5′ promoter region of GST genes, has been shown to be regulated by these chemoprotective compounds (Rushmore *et al.*, 1990).

Like the cytochromes P450, GSTs are thought to have evolved as an adaptive response to environmental insult and are capable of metabolizing a wide variety of structurally diverse substrates, many of which are also inducers of GST expression (Rushmore and Pickett, 1993). Unlike P450s, however, GSTs are constitutively expressed at much higher levels in a wide variety of tissues, with different tissues having characteristic patterns of GST isozyme expression. Increased expression of specific GST isozymes in tumour cells which have become resistant to anticancer drugs has suggested a role for these proteins in the development of resistance to chemotherapy (Hayes and Wolf, 1990).

Variation in the metabolism of *trans*-stilbene oxide in peripheral blood leukocytes led to the elucidation of a genetic polymorphism at the *GSTM* locus (Seidegard *et al.*, 1985). The polymorphic gene, *GSTM1*, is homozygous nulled in approximately 40–50% of the Caucasian population, as a result of a gene deletion (Seidegard *et al.*, 1986). A total of three alleles is present at this locus, the gene deletion and two further alleles, *GSTM1-a* and *GSTM1-b*, which confer an identical phenotype, but which have lysine or asparagine, respectively, at amino acid 172 (Fryer *et al.*, 1993).

Up to 60% of the human population is unable to conjugate and hence detoxify monohalomethanes (Peter *et al.*, 1989). The genetic basis for this polymorphism has recently been reported (Pemble *et al.*, 1994), and has been shown to result from a deletion, in affected individuals, of the *GSTT1* gene. Conjugation with glutathione, catalysed specifically by *GSTT1*, has been demonstrated to be the major route of detoxification of monohalomethane compounds in human erythrocytes. A PCR-based genotyping assay for the *GSTT1(0)* genotype has now been reported (Pemble *et al.*, 1994), the results of which correlate well with phenotyping studies.

5.4 *N*-acetyl transferases

The *N*-acetyl transferases (NATs) are phase II cytosolic enzymes which catalyse the transfer of acetate from acetyl CoA to primary amine and hydrazine groups, forming acetamides and hydrazides. Substrates for these enzymes include drugs such as isoniazid, procainamide, sulphamethazine and caffeine, and occupational carcinogens, such as benzidine and 4-aminobiphenyl, and amino acid pyrolysates, produced as a result of charring food (Grant, 1993; Sim *et al.*, 1992).

Clinical interest in the regulation of drug acetylation was aroused 40 years ago, following the observation that response to isoniazid, greatly used in the treatment of tuberculosis, varied widely among patients and, in certain individuals, was accompanied by neurological side effects (Bonicke and Reif, 1953). Based on these observations, patients were designated either 'fast' or 'slow' acetylators, depending on their ability to metabolize isoniazid to its *N*-acetyl derivatives (Mitchell and Bell, 1957). It became further apparent that NAT substrates fell into two distinct classes – certain 'polymorphic' substrates, such as isoniazid and procainamide exhibited wide interindividual variation in their rates of metabolism, while other 'monomorphic' substrates, *p*-aminobenzoic acid and *p*-aminosalicylic acid, were metabolized at the same rate. Partial purification of NAT proteins from human liver revealed the presence of two distinct isozymes, NAT1, the monomorphic enzyme and the polymorphic homologue, NAT2 (Grant *et al.*, 1989). More recent studies have, however, revealed that the *NAT1* locus is itself polymorphic (Grant, 1994).

The slow acetylator phenotype is inherited as a trimodal autosomal recessive trait in up to 70% of Caucasians (Grant, 1993). Like other polymorphically expressed drug-metabolizing enzymes, however, there is considerable inter-racial variation in the prevalence of the slow-acetylator (PM) phenotype. Unlike the CYP2D6 polymorphism, Chinese populations and Caucasians have similar numbers of individuals with the mutant phenotype, while Japanese and Korean Orientals have a lower frequency (10–30%). The polymorphism is particularly rare in Inuits, with only 5% of individuals possessing the 'slow acetylator' phenotype (Lin *et al.*, 1993, 1994).

In man, three distinct *NAT* loci have been identified and are localized on human chromosome 8 (pter q11) - the functional genes *NAT1* and *NAT2*, and a third gene, *NATP*, which contains several premature stop codons and is thought to be a non-expressed pseudogene (Blum *et al.*, 1990; Hickman *et al.*, 1994). Both *NAT1* and *NAT2* are intronless genes which encode 33–34 kDa proteins, the expression of which is observed in a variety of human tissues, including liver, lung and colon. A total of six allelic variants at the *NAT2* gene locus have now been isolated (*Table 5.5*), most of which have arisen from point mutations (Grant, 1993; Hickman *et al.*, 1992). Using a combination of PCR and RFLP analysis with the restriction enzymes *Bam*HI, *Dde*I, *Fok*I, *Kpn*I and *Taq*I, the *NAT2* genotype of an individual can be determined and, by extrapolation, his or her response to drugs which are substrates for this enzyme can be predicted.

Table 5.5. Mutant alleles at the *NAT2* locus

Allele	Amino acid substitution	Diagnostic restriction enzymes	Functional significance
R1	–	–	Wild-type allele
R2	None	*Fok*I, *Kpn*I, *Bam*HI	–
S1A	Ile^{114}→Thr	*Fok*I, *Taq*I, *Dde*I, *Bam*HI	Decreased V_{max}
S1B	None	*Fok*I, *Taq*I, *Bam*HI	–
S1C	Lys^{268}→Arg	*Fok*I, *Kpn*I, *Taq*I, *Dde*I, *Bam*HI	–
S2	Arg^{197}→Gln	*Kpn*I, *Bam*HI	Decreased stability
S3	Gly^{286}→Arg	*Kpn*I, *Taq*I	Decreased stability
S4	None	*Kpn*I, *Taq*I, *Bam*HI	–

Recent evidence suggests that *NAT1* expression is also genetically polymorphic (Grant, 1994), with affected individuals having a compromised ability to metabolize the classical 'monomorphic' substrate, *p*-aminosalicylic acid. Two variant alleles at the *NAT1* locus have been identified, the first of which contains a point mutation (G^{560}→A) which results in the substitution of a glutamic acid residue for arginine at position 187. The second allele has a point mutation (C^{559}→T), resulting in a premature stop codon at residue 187, and the loss of functional protein.

5.5 Other drug-metabolizing enzymes

Much less is known about the molecular basis of genetic polymorphisms within other human drug-metabolizing enzymes. However, as the technology required to isolate allelic variants of a given gene is now readily available, further mutant alleles will undoubtedly be identified within the population. Examples of human drug-metabolizing enzymes which are known to be, or are thought to be genetically polymorphic are given in *Table 5.1*. These include thiopurine methyl transferase, responsible for the *S*-methylation of 6-mercaptopurine (Lennard *et al.*, 1989) and the pseudocholinesterases, seven allelic variants of which have been identified by RFLP analysis (La Du *et al.*, 1991). Inborn errors in glucuronic acid conjugation (Crigler-Najjar and Gilbert's syndromes) are due to genetic polymorphisms at the UDP–glucuronyl transferase locus (Burchell and Coughtrie, 1992).

The polymorphic expression of two aldehyde dehydrogenase genes, *ALDH1* and *ALDH2*, has been relatively well characterized. Up to 50% of the Chinese and Japanese population have no hepatic expression of the mitochondrial *ALDH2* gene (Agarwal and Goedde, 1992), resulting in acetaldehyde accumulation and the 'alcohol flush' syndrome common in Orientals. A similar phenotype is achieved by the polymorphic expression of two genes at the alcohol dehydrogenase locus, *ADH2* and *ADH3*. The major phenotypic consequence of this polymorphism is increased conversion of ethanol to acetaldehyde. Again, the polymorphism is particularly prevalent in Japanese (Agarwal and Goedde, 1992).

Two allelic variants of the ubiquitously expressed microsomal epoxide hydrolase gene have recently been identified (Hassett *et al.*, 1994). This enzyme plays an important role in catalysing the detoxification of reactive epoxides which are generated by the action of, for example, P450 metabolism of polycyclic aromatic hydrocarbons. Each mutant allele has a single amino acid substitution (tyrosine[113]→histidine or histidine[139]→arginine) compared with the wild-type protein. Although the effects of these amino acid substitutions have not yet been fully evaluated, neither appears to lead to a marked alteration in the catalytic activity of the protein. The frequency of each mutant allele within the population remains to be established.

5.6 Pharmacogenetic polymorphisms and disease susceptibility

Recent advances in pharmacogenetics have made it possible to establish whether drug side effects or idiosyncratic responses to drugs are associated with the polymorphic expression of a drug-metabolizing enzyme. In certain cases, the evidence that these polymorphisms affect drug metabolism *in vivo* is unequivocal. As many of these enzymes are also implicated in the detoxification of mutagens, or in the metabolic activation or detoxification of environmental toxins, it seems reasonable that a link should exist between the level of expression of particular isozymes (the allele frequency of the mutant forms) and the incidence of cancer, and of diseases where exposure to environmental chemicals has been implicated in the aetiology.

Many P450-catalysed reactions result in the formation of chemically reactive electrophiles. These reactive intermediates are often inherently mutagenic and can bind to DNA, inducing mutations which result in oncogene activation or, alternatively, loss of function of a tumour suppressor gene. This sequence of events is thought to be fundamental to the initiation of environmentally linked cancers (Weinberg, 1990).

The polymorphic expression of a number of cytochrome P450 genes has been linked to cancer susceptibility. The resulting variation in the levels of P450 protein expressed in the liver can lead to marked differences in the ability of these enzymes to activate or detoxify a wide range of (pro)carcinogens, resulting in an increase or decrease in cancer susceptibility, respectively.

For example, it has been demonstrated that *CYP1A1* expression is induced in the bronchial airways of more than 80% of lung cancer patients who are smokers (Kawajiri *et al.*, 1990; McLemore *et al.*, 1990), presumably due to the presence of the many polycyclic aromatic hydrocarbons (PAH) and other carcinogens present in tobacco smoke. As a result, the relative level of *CYP1A1* expression in the bronchus has been linked to the pathogenesis of tobacco smoke-induced peripheral pulmonary carcinoma. The *CYP1A1* 'high inducibility' phenotype (*Msp*I RFLP) has also been been associated with lung cancer incidence (Kreyberg type I) in a Japanese population (Hayashi *et al.*, 1992), but the same correlation was not observed in Caucasians (Tefre *et al.*, 1991).

Many studies have been performed in an attempt to establish a link between the polymorphic expression of *CYP2D6* and the incidence of various types of cancer (Caporaso *et al.*, 1989, 1990; Roots *et al.*, 1988,1992; Wolf *et al.*, 1992). The results of many of these studies have been equivocal, due, in part at least, to the analytical methods used. Many investigators used phenotypic rather than genotypic assessment of PM status and this, for reasons discussed previously, can lead to the production of anomalous results. Several studies did not produce correlations which were statistically significant, owing to inadequate population sizes and the lack of adequate case-matched controls. For example, there is significant variation in *CYP2D6* allele frequency between different racial groups, and it is therefore essential to ensure that all study participants are of the same racial origin.

Susceptibility to certain cancer types, most notably bladder cancer, melanoma and leukaemia has, however, been shown to correlate with the *CYP2D6* PM genotype (Wolf *et al.*, 1992) *(Table 5.6)*. Several studies, mostly using phenotypic assessment of PM status, have reported a link between lung cancer incidence and the CYP2D6 PM phenotype. A number of potent carcinogens, including NNK, have been shown to be metabolized by CYP2D6 (Crespi *et al.*, 1991). PM individuals would be unable to metabolize these agents (by a CYP2D6 mediated pathway, at least) and would therefore be less likely to develop chemically linked cancers. The combined results of several genotyping studies (Wolf *et al.*, 1994) indicated that this was indeed the case, although the reduction in lung cancer risk was relatively small (odds ratio = 0.67).

A highly significant correlation was also observed between the PM genotype and the incidence of Parkinson's disease (PD; Kurth and Kurth, 1993; Smith *et al.*, 1992b). Individuals who have the *CYP2D6* PM genotype have been shown to be more than twice as likely to develop PD than case-matched controls. *In vitro* studies (Fonne-Pfister *et al.*, 1987) have shown that MPTP (1-methyl-4-phenyl-1,2,3,6-tetrahydropyridine) can be metabolized by CYP2D6. This compound, a contaminant of synthetic meperidine narcotics, can induce

Parkinsonism when oxidatively metabolized by the mitochondrial monooamine oxidase to its active form MPP^+, the 1-methyl-4-phenyl pyridinium ion (Calne and Lanston, 1983), which is selectively toxic to the *substantia nigra*. MPP^+ can also bind to and be metabolized by CYP2D6, the expression of which has been identified in this region of the brain (Nisnik *et al.*, 1990).

Table 5.6. Distribution of *CYP2D6* alleles in cancer and PD patients (Wolf *et al.*, 1992)

Patients	No. of samples	Genotype[a]			% MA[b]	χ^2	$p<$
		%EM	%H	%PM			
Random controls	720	66.1	29.6	4.3	19.1	–	–
Emphysema	151	61.6	35.1	3.3	20.9	2.35	0.50
All cancers	1759	63.4	31.6	5.0	20.8	6.38	0.05*
Lung cancer[c]	361	64.8	31.6	3.6	19.4	0.96	0.90
Breast cancer	437	66.8	29.3	3.9	18.5	0.22	0.90
Colon cancer	115	63.8	29.6	6.9	21.7	2.88	0.50
Leukaemia[d]	312	65.7	26.9	7.4	30.8	7.65	0.02*
Teratoma[e]	169	65.7	30.8	3.5	18.9	0.32	0.90
Melanoma	127	54.3	39.4	6.3	26.0	7.91	0.02*
Bladder cancer	184	53.8	41.8	4.4	25.3	13.50	0.005*
Prostate cancer	54	59.2	31.5	9.3	25.0	3.62	0.5
Parkinson's disease	229	60.7	27.5	11.9	–	22.3	0.0001*

[a] EM, extensive metabolizer; H, heterozygous; PM, poor metabolizer.
[b] Mutant allele frequency.
[c] 145 squamous cells carcinomas, 74 adenocarcinomas, 52 small cell lung cancers, 90 other lung cancers.
[d] 141 acute myeloid, 117 chronic myeloid, 37 myelodysplastic syndrome, 17 other leukaemias.
[e] Data from teratocarcinoma and seminoma patients.
* Statistically significant.

More recent studies have attempted to link the *CYP2D6* PM genotype to the incidence of Alzheimer's disease (Benitez *et al.*, 1993) and motor neurone disease (James *et al.*, 1994), but no statistically significant correlations have been observed. A further study (Sindrup *et al.*, 1993) investigated a possible link between the *CYP2D6* PM genotype and tolerance of pain, as a result of a compromised ability in PM individuals to catalyse the endogenous conversion of codeine to morphine. The experimental evidence did not, however, confirm this hypothesis. Indeed, the results of a later study (Mikus *et al.*, 1994) suggest that the conversion of endogenous codeine to morphine may not be catalysed by CYP2D6.

CYP2E1 is known to catalyse the metabolism of several compounds which are potent human carcinogens (Yang *et al.*, 1990). These include benzene, nitrosamines and ethanol, which has itself been implicated in the pathogenesis of cancers of the oral cavity, oesophagus, breast and stomach (Driver and Swan, 1987). No definitive link between the polymorphic expression of

CYP2E1 and cancer incidence has, however, yet been identified. Individuals with mutations in *CYP2E1* were shown to be significantly underrepresented in a group of Japanese lung cancer patients, compared with a control population. Similar results were not found, however, in lung cancer patients in a Caucasian population (Kato *et al.*, 1992) and marked ethnic differences were reported in the results obtained in lung cancer populations of different racial origin (Hirvonen *et al.*, 1993; Uematsu *et al.*, 1994). Further studies have demonstrated linkage between the *Dra*I polymorphism in intron 2 and the polymorphisms in the promoter region of the gene, and reported an association between mutations in *CYP2E1* and the development of liver cirrhosis in alcoholics (Maezawa *et al.*, 1994).

The incidence of various cancers has also been associated with the glutathione *S*-transferase GSTM1 polymorphism (Zhong *et al.*, 1993). A particularly strong association has been demonstrated for colon cancer, which was further strengthened when the data were analysed according to the site of the tumour. Over 70% of patients with tumours in the proximal colon were found to be nulled at the GSTM1 locus, compared with 54% with tumours in the distal colon and 42% of a control population. Statistically significant correlations between overexpression of the GSTM1 genotype and susceptibility to cancers of the lung (Howie *et al.*, 1990; Seidegard *et al.*, 1986), stomach and skin (Heagerty *et al.*, 1994) have also been observed. These results strongly suggest a role for the GST, particularly GSTM1, in the detoxification of a wide range of environmental agents which, if allowed to accumulate in the body, could directly contribute to tumour initiation.

Fewer studies have been reported in which cancer susceptibility has been linked to the polymorphic expression of the *N*-acetyl transferase genes. NATs are known to metabolize several classes of chemical carcinogen, including the aromatic amines 4-aminobiphenyl and β-naphthylamine (Grant, 1993). These compounds have been shown to cause bladder cancer in man (Ladero *et al.*, 1985). Individuals classed as 'slow acetylators' at the polymorphic *NAT2* gene locus are presumed to have a compromised ability to detoxify these compounds and, as a consequence, have been shown to have an increased susceptibility to developing bladder cancer (Lin *et al.*, 1993). In contrast, the slow acetylator phenotype appears to be associated with decreased susceptibility to colon cancer (Rodriquez *et al.*, 1993). While this apparent dichotomy has not yet been fully rationalized, it is known that NAT genes are expressed at a relatively high level in the colon, where they are active in the metabolic activation of heterocyclic amines ingested in food. These compounds have been implicated in the aetiology of colon cancer. 'Slow acetylators' are, therefore, afforded some degree of protection from the deleterious effects of these compounds (Kadlubar, 1994). In contrast, both NAT1 and NAT2 proteins catalyse the detoxification of bicyclic aromatic amines, compounds which are known to be potent bladder carcinogens. NAT1 'slow acetylators' therefore have increased susceptibility to bladder cancer, presumably as a consequence of their compromised ability to detoxify these compounds.

There have been recent reports of associations between the NAT2 slow acetylator phenotype and susceptibility to type I diabetes mellitus (Price-Evans, 1984), rheumatoid arthritis (Ladero *et al.*, 1993) and multiple sclerosis (Ladero *et al.*, 1994). No statistically significant associations with the incidence of these diseases were, however, found – due, in part at least to inadequate population sizes. However, with the recent advances in the molecular characterization of the *NAT* gene locus, and the development of simple PCR-based assays for the determination of *NAT* genotype, it should soon be possible to make an accurate assessment of the association, if any, between polymorphisms at the *NAT* locus and disease susceptibility.

5.7 Concluding remarks

The preceding discussion illustrates the recent advances which have been made in understanding the genetic basis for many of the abnormal responses seen in response to certain medications. Development of allele-specific PCR and RFLP analyses have made the determination of individual genotypes relatively straightforward, and this has eliminated many of the problems associated with previous phenotypic assessment of drug-metabolizing capacity. The ability to predict an individual's response to a particular drug, before that drug is administered, will avoid the need to expose patients to drugs which may have harmful side effects. It has become more and more apparent that, in addition to conventional medical practice where adjustments are made to a dosing schedule to allow for age, weight and the medical condition of the patient, it can be equally important to allow for genetic factors which will influence the metabolism and disposition of the prescribed drug.

The ultimate metabolic fate of an administered drug is a balance between the relative activities of enzymes active in its activation and detoxification. The expression of several of these enzymes (cytochromes P450, glutathione *S*-transferases, *N*-acetyl transferases) has been shown to be genetically polymorphic, and the metabolic consequences of these polymorphisms have been described. Other factors, however, may also contribute to the rate of drug elimination from the body. P-glycoprotein (P-gp) is an ATP-dependent membrane-bound glycoprotein which functions as a transporter of many structurally diverse chemicals from the cell. These include anticancer drugs such as adriamycin and vincristine, the exclusion of which from the cell has been shown to lead to the MDR (multidrug resistant) phenotype (Hayes and Wolf, 1990). Although polymorphic expression of genes at the MDR locus has not yet been characterized, any genetically determined variation in P-gp protein expression would have important toxicological consequences.

It has been extensively documented that gene expression at the *CYP1A* locus is mediated by a cytosolic receptor (the Ah receptor), a ligand-dependent transcription factor which binds to a xenobiotic response element (XRE) sequence in the 5′ promoter region of *CYP1A* genes (Burbach *et al.*, 1992). It has recently been reported that wide interindividual variation exists, both in the expression of the Ah receptor protein itself (Hayashi *et al.*, 1994) and in the levels of aryl hydrocarbon nuclear translocator (Arnt), an accessory protein which binds to the Ah receptor and causes its translocation to the nucleus (Hoffman *et al.*, 1991). Genetic polymorphisms in the genes which encode these proteins could significantly affect the inducibility of *CYP1A* genes and, by extrapolation, could influence lung cancer susceptibility.

A number of polymorphisms in human drug-metabolizing enzymes have been linked to cancer susceptibility. It has not yet, however, been unequivocally demonstrated whether these enzymes themselves are involved in the pathogenesis of cancer or whether, for example, they are in linkage disequilibrium with an oncogene at the same chromosomal location. As carcinogenesis is thought to be a multistep process (Weinberg, 1990), inheritance of mutant alleles at more than one polymorphic locus may well lead to an increased susceptibility to disease. Studies published to date suggest that this is indeed the case. For example, in a study of lung cancer incidence among smokers in a Japanese population, individuals with the 'high inducibility' *CYP1A1* allele and who were also nulled at the *GSTM1* locus, showed a marked excess of squamous cell lung carcinoma (Nakachi *et al.*, 1993). There also appears to be an association between the *GSTM1(0)* locus in a series of lung tumours and mutations in the tumour suppressor gene, *p53* (Kawajiri *et al.*, 1993).

Mutations in genes which encode human drug-metabolizing enzymes have therefore been shown to have a profound influence, not only on the metabolism and disposition of a wide range of commonly prescribed drugs, but also on the incidence of 'environmental' diseases such as cancer. Although the technology is now available to determine the 'genetic blueprint' of an individual relatively easily and quickly, the relevance and applicability of this technology to routine clinical medicine has yet to be determined.

References

Antilla S, Vainio H, Hietanen E, Camus AM, Malaveille C, Bron G, Husgafvel-Pursiainen K, Heikkila L, Karjaleinen A, Bartsch H. (1992) Immunohistochemical detection of pulmonary cytochrome P450 1A and metabolic activities associated with P4501A1 and P4501A2 isozymes in lung cancer patients. *Environ. Hlth Perspect.* **98:** 179–182.

Agarwal DP, Goedde HW. (1992) Pharmacogenetics of alcohol metabolism and alcoholism. *Pharmacogenetics* **2:** 48–62.

Aoyama T, Yamano S, Guzelian PS, Gelboin HV, Gonzalez FJ. (1990) Five of 12 forms of vaccinia virus-expressed human hepatic cytochrome P450 metabolically activate aflatoxin B_1. *Proc. Natl Acad. Sci. USA* **87:** 4790–4793.

Benitez J, Barquero MS, Coria F, Molina JA, Jiminez-Jiminez FJ, Ladero JM. (1993) Oxidative polymorphism of debrisoquine is not related to the risk of Alzheimer's disease. *J. Neurol. Sci.* **117:** 8–11.

Bertilsson L, Dahl ML, Sjoqvist F, Aberg-Wistedt A, Humble M, Johansson I, Lundqvist E, Ingelman-Sundberg M. (1993) Molecular basis for rational megaprescribing in ultrarapid hydroxylators of debrisoquine. *Lancet* **341:** 63.

Blum M, Grant DM, McBride DW, Heim M, Meyer UA. (1990) Human arylamine *N*-transferase genes: Isolation, chromosomal localization and functional expression. *DNA Cell Biol.* 9: **193–203**.

Blum M, Demierre A, Grant DM, Heim M, Meyer UA. (1991) Molecular mechanism of slow acetylation of drugs and carcinogens in humans. *Proc. Natl Acad. Sci. USA* **88:** 5237–5241.

Bonicke R, Reif W. (1953) Enzymatic inactivation of isonicotinic acid hydrazide in humans and animals. *Arch. Exp. Pathol. Pharmacol.* 220: 321–333.

Boobis AR, Lynch AM, Murray S, De la Torre R, Solano A, Farre M, Segura J, Gooderman NJ, Davies D. (1994) CYP1A2 catalyzed conversion of dietary heterocyclic amines to their proximate carcinogens is their major route of metabolism in humans. *Cancer Res.* 54: 89–94.

Burbach KMA, Poland A, Bradfield CA. (1992) Cloning of the Ah receptor cDNA reveals a distinctive ligand-activated transcription factor. *Proc. Natl Acad. Sci. USA* **89:** 8185–8189.

Burchell B, Coughtrie MWH. (1992) UDP-Glucuronosyltransferases. In: *Pharmacogenetics of Drug Metabolism* (ed. W Kalow). Pergamon Press, New York, pp. 195–226.

Butler MA, Lang NP, Young JF, Caparaso NE, Vinels P, Hayes RB, Teitel CH, Massengill JP, Lawson MF & Kadlubar FF. (1992) Determination of CYP1A2 and *N*-acetyl transferase-2 phenotypes in the human population by analysis of caffeine urinary metabolites. *Pharmacogenetics* 2: 116–127.

Calne DB, Lanston JW. (1983) Etiology of Parkinson's disease. *Lancet* 2: 1457–1459.

Caporaso NG, Hayes RB, Dosmeci M, Hoover RN, Ayesh R, Hetzel M, Idle JR. (1989) Lung cancer risk, occupational exposure and the debrisoquine metabolic phenotype. *Cancer Res.* 49: 3675–3679.

Caporaso NG, Tucker MA, Hoover RN *et al.* (1990) Lung cancer and the debrisoquine metabolic phenotype. *J. Natl Cancer Inst.* 82: 1264–1272.

Crespi CL, Penman BW, Gelboin HV, Gonzalez FJ. (1991) A tobacco smoke-derived nitrosamine, 4-(methylnitrosamino)-1-(3-pyridyl)-1-butanone, is activated by multiple human cytochrome P450s, including the polymorphic cytochrome P4502D6. *Carcinogenesis* 12: 1197–1201.

Crofts F, Cosma GN, Currie D, Taioli E, Toniolo P, Garte SJ. (1993) A novel CYP1A1 gene polymorphism in African-Americans. *Carcinogenesis* 14: 1729–1731.

Daly AK, Cholerton S, Armstrong M, Idle JR. (1993) Genotyping for polymorphisms in xenobiotic metabolism as a predictor of disease susceptibility. *Environ. Hlth Perspect.* **101:** 117–120.

DeJong R, Morgenstern R, Jornvall H, DePierre JW. (1988) Gene expression of rat and human microsomal GST. *J. Biol. Chem.* 263: 8430–8436.

deMorais SMF, Wilkinson GR, Blaisdeu J, Nakamura K, Meyer UA, Goldstein JA. (1994) The major genetic defect responsible for the polymorphism of *S*-mephenytoin metabolism in humans. *J. Biol. Chem.* 265: 15419–15422.

Ding S, Lake BG, Friedberg T, Wolf CR. (1995) Expression and alternative splicing of cytochrome P450 CYP2A7. *Biochem. J.* 306: 161–166.

Driver HE, Swan RF. (1987) Alcohol and human cancer: a review. *Anticancer Res.* 7: 309–320.

Evert B, Griese EU, Eichelbaum M. (1994) A missense mutation in exon-6 of the cyp2d6 gene leading to a histidine-324 to proline exchange is associated with the poor metabolizer phenotype of sparteine. *Naunyn-schmied Arch. Pharmacol.* 350: 434–439.

Fonne-Pfister R, Bargetzi MJA, Meyer UA. (1987) MPTP the neurotoxin inducing Parkinson's disease, is a potent inhibitor of human and rat P450 enzymes (P450bufl, P450dbl) catalysing debrisoquine 4-hydroxylation. *Biochem. Biophys. Res. Comm.* **148:** 1144–1150.

Forrester LM, Neal GE, Judah DJ, Glancey MG, Wolf CR. (1990) Evidence for involvement of multiple forms of cytochrome P-450 in aflatoxin B_1 metabolism in human liver. *Proc. Natl Acad. Sci. USA* 87: 8306–8310.

Forrester LM, Henderson CJ, Glancey MG, Back DJ, Park BK, Ball SE, Kitteringham NR, McLaren AW, Miles JS, Skett P, Wolf CR. (1992) Relative expression of cytochrome P450 isoenzymes in human liver and association with the metabolism of drugs and xenobiotics. *Biochem. J.* **281:** 359–368.

Fryer AA, Zhao L, Alldersea J, Pearson WR, Strange RC. (1993) Use of site-directed mutagenesis of allele-specific PCR primers to identify the GSTM1A, GSTM1B, GSTM1AB and GSTM1 null polymorphisms at the glutathione *S*-transferase GSTM1 locus. *Biochem. J.* **295:** 313–315.

Gaedigk A, Blum M, Gaedigk R, Eichelbaum M, Meyer UA. (1991) Deletion of the entire cytochrome P450 CYP2D6 gene as a cause of impaired drug metabolism in poor metabolizers of the debrisoquine/sparteine polymorphism. *J. Hum. Genet.* **48:** 943–950.

Garrod AE. (1909) Inborn errors of metabolism. *Lancet* **2:** 142–214.

Gonzalez FJ. (1990) The molecular biology of cytochrome P450s. *Pharmacol. Rev.* **40:** 243–288.

Gonzalez FJ, Nebert DW. (1990) Evolution of the P450 superfamily: animal–plant 'warfare', molecular drive and human genetic differences in drug oxidation. *Trends Genet.* **6:** 182–186.

Gough AC, Miles JS, Spurr NK, Moss JE, Gaedigk A, Eichelbaum M, Wolf CR. (1990) Identification of the primary gene defect at the cytochrome P450 CYP2D locus. *Nature* **347:** 773–776.

Gough AC, Smith CAD, Howell SM, Wolf CR, Bryant SP, Spurr NK. (1993) Localisation of the CYP2D gene locus to human chromosome 22q 13.1 by polymerase chain reaction, *in situ* hybridisation and linkage analysis. *Genomics* **15:** 430–432.

Grant DM. (1993) Molecular genetics of the *N*-acetyl transferases. *Pharmacogenetics* **3:** 45–50.

Grant DM. (1994) Molecular genetics of the acetyltransferases. In: *Proceedings of the 10th International Symposium on Microsomes and Drug Oxidations.* Toronto, Canada, July 1994.

Grant DM, Lottspeich F, Meyer UA. (1989) Evidence for two closely related isozymes of arylamine *N*-acetyltransferase in human liver. *FEBS Lett.* **244:** 203–207.

Hasset C, Aicher L, Sidhu JS, Omiecinski CJ. (1994) Human microsomal epoxide hydrolase: genetic polymorphism and functional expression *in vivo* of amino acid variants. *Hum. Mol. Genet.* **3:** 421–428.

Hayashi SI, Watanabe J, Nakachi K, Kawajiri K. (1991) Genetic linkage of lung cancer associated Msp1 polymorphisms with amino acid replacement in the heme binding region of the human cytochrome P4501A1 gene *J. Biochem.* **110:** 407–411.

Hayashi S, Watanabe J, Kawajiri K. (1992) High susceptibility to lung cancer analysed in terms of combined genotypes of P4501A1 and mu class glutathione *S*-transferase genes. *Jpn. J. Cancer Res.* **83:** 866–870.

Hayashi SI, Watanabe J, Nakachi K, Eguchi H, Gotoh O, Kawajiri K. (1994) Interindividual differences in expression of human Ah receptor and related P450 genes. *Carcinogenesis* **15:** 801–806.

Hayes JD, Wolf CR. (1990) Molecular mechanisms of drug resistance. *Biochem. J.* **272:** 281–295.

Heagerty AHM, Fitzgerald D, Smith A, Bowers B, Jones P, Fryer AA, Zhao L, Alldersea J, Strange RC. (1994) Glutathione *S*-transferase GSTM1 phenotypes and protection against cutaneous tumours. *Lancet* **343:** 266–268.

Heim M, Meyer UA. (1992) Evolution of a highly polymorphic human cytochrome P450 gene cluster CYP2D6. *Genomics* **14:** 49–58.

Hickman D, Risch A, Camilleri JP, Sim E. (1992) Genotyping human polymorphic arylamine *N*-acetyltransferase – identification of new slow allotypic variants. *Pharmacogenetics* **2:** 217–226.

Hickman D, Risch A, Buckle V, Spurr NK, Jeremiah SJ, McCarthy A, Sim E. (1994) Chromosomal localisation of human genes for arylamine *N*-acetyltransferase. *Biochem. J.* **297:** 441–445.

Hirvonen A, Husgaivel-Pursainen K, Antilla S, Karalainen A, Vainio H. (1993) The human CYP2E1 gene and lung cancer. Dra I and Rsa I restriction fragment length polymorphism in a Finnish study population. *Carcinogenesis* **14:** 85–88.

Hoffman EC, Reyes H, Chu FF, Sander F, Conley LH, Brooks BA, Hankinson O. (1991) Cloning of a factor required for activity of the Ah (dioxin) receptor. *Science* **252**: 954–958.

Howie AF, Bell D, Hayes PC, Hayes JD, Beckett GJ. (1990) Glutathione *S*-transferase isoenzymes in human broncheolar lavage: a possible early marker for detection of lung cancer. *Carcinogenesis* **11**: 295–300.

Inaba T, Jurima M, Kalow W. (1986) Family studies of mephenytoin hydroxylase deficiency. *Am. J. Hum. Genet.* **38**: 768–772.

Jacoby WB, (1978) The glutathione S-transferases, a group of multi-functional detoxification proteins. *Adv. Enzymol. Rel. Areas Mol. Biol.* **46**: 383–414.

James CM, Daniels J, Wiles CM, Owen MJ. (1994) Debrisoquine hydroxylase gene polymorphism in motor-neuron disease. *Neurodegeneration* **3**: 149–152.

Johansson I, Yue QY, Dahl ML, Heim M, Sawe J, Bertilsson I, Meyer UA, Sjoquist F, Ingelman-Sundberg M. (1991) Genetic ananlysis of the inter-ethnic differences between Chinese and Caucasians in the polymorphic metabolism of debrisoquine and codeine. *Eur. J. Clin. Pharmacol.* **40**: 553–556.

Johansson I, Lundquist E, Bertilsson L, Dahl ML, Sjoqvist F, Ingelman-Sundberg M. (1993) Inherited amplification of an active gene in the cytochrome P450 CYP2D locus as a cause of ultra rapid metabolism of debrisoquine. *Proc. Natl Acad. Sci. USA* **90**: 11825–11829.

Kadlubar FF. (1994) Biochemical individuality and its implications for drug and carcinogen metabolism: recent insights from acetyltransferase and cytochrome P450IA2 phenotyping and genotyping in humans. *Drug Metab. Rev.* **26**: 37–46.

Kagimoto M, Heim M, Kagimoto K, Zeugin T, Meyer UA. (1990) Multiple mutations of the human cytochrome P450IID6 gene (CYP2D6) in poor metabolizers of debrisoquine. Study of the functional significance of individual mutations by expression of chimeric genes. *J. Biol. Chem.* **265**: 17209–17214.

Kalow W. (1982) Ethnic differences in drug metabolism. *Clin. Pharmacokinet.* **7**: 373–400.

Karam WG, Chiang JYL. (1992) Polymorphisms of human cholesterol 7α hydroxylase. *Biochem. Biophys. Res. Comm.* **185**: 588–595.

Kato S, Shields PG, Caporaso NG, Hoover RN, Trump BF, Sugimura H, Weston A, Harris CC. (1992) Cytochrome P450IIE1 genetic polymorphism, racial variation and lung cancer risk. *Cancer Res.* **52**: 6712–6715.

Kawajiri K, Nakachi K, Imai K, Yoshii A, Shinoda N, Watanabe J. (1990) Identification of genetically high risk individuals to lung cancer by DNA polymorphisms of the cytochrome P4501A1 gene. *FEBS Lett.* **263**: 131–133.

Kawajiri K, Nakachi K, Imai K, Watanabe J, Hayashi SI. (1993) Germ line polymorphisms of p53 and CYP1A1 genes involved in lung cancer. *Carcinogenesis* **14**: 1085–1089.

Kimura S, Umeno M, Skoda RC, Meyer UA, Gonzalez FJ. (1989) The human debrisoquine 4-hydroxylase (CYP2D) locus: sequence and identification of the polymorphic CYP2D6 gene, a related gene and a pseudogene. *Am. J. Hum. Genet.* **45**: 889–904.

Kouri RE, McKinney CE, Slomany DJ, Snodgrass DR, Wray NP, McLemore TL. (1982) Positive correlation between high aryl hydrocarbon hydroxylase activity and primary lung cancer as analysed in cryopreserved lymphocytes. *Cancer Res.* **42**: 5030–5037.

Kroemer HK, Mikus G, Eichelbaum M. (1994) Clinical relationship of pharmacogenetics. In: *Handbook of Experimental Pharmacology* (eds Welling PG, Balant LP). Springer, Berlin, pp. 265–288.

Kurth MC, Kurth JH. (1993) Variant cytochrome P450 CYP2D6 allelic frequencies in Parkinson's disease. *Am. J. Med. Genet.* **48**: 166–168.

Ladero JM, Kwok CK, Jara C. *et al.* (1985) Hepatic acetylator phenotype in bladder cancer patients. *Ann. Clin. Res.* **17**: 96–99.

Ladero JM, Andres MP, Banares A, Fernandez B, Hernandez C, Benitez J. (1993) Acetylator polymorphism in rheumatoid arthritis. *Eur. J. Clin. Pharmacol.* **45**: 279–281.

Ladero JM, Arroyo R, De Andres C, Jimenez-Jiminez FJ, Molina JA, Varela D, Seijas E, Giminez-Roldan S, Benitez J. (1994) Acetylator polymorphism in multiple sclerosis. *Acta Neurol. Scand.* **89:** 102–104.

La Du BN, Zannovi VA, Laster L, Seegmiller JE. (1958) The nature of the defect in tyrosine metabolism in alcaptonuria. *J. Biol. Chem.* **230:** 251.

La Du BN, Bartels CF, Nogueira CP, Arpagans M., Lockridge O. (1991) Proposed nomenclature for human butyrylcholinesterase genetic variants identified by DNA sequencing. *Cell. Mol. Neurobiol.* **11:** 79–89.

Lennard L, van Loon JA, Weinshilboum RM. (1989) Pharmacogenetics of acute azathioprine toxicity: relationship to thiopurine methyltransferase genetic polymorphism. *Clin. Pharmacol. Ther.* **46:** 149–154.

Lin HJ, Han CY, Lin BK, Hardy S. (1993) Slow acetylator mutations in the human polymorphic *N*-acetyltransferase gene in 786 Asians, blacks, hispanics and whites: application to metabolic epidemiology. *Am. J. Hum. Genet.* **52:** 827–834.

Lin HJ, Han CY, Lin BK, Hardy S. (1994) Ethnic distribution of slow acetylator mutations in the polymorphic *N*-acetyltransferase (NAT2) gene. *Pharmacogenetics* **4:** 125–134.

Maezawa Y, Yamauchi M, Toda G. (1994) Association between RFLP of the human cytochrome P450IIE1 gene and susceptibility to alcoholic liver cirrhosis. *Am. J. Gastroenterol.* **89:** 501–565.

Mahgoub BA, Idle JR, Dring LG, Lancaster R, Smith RL. (1977) Polymorphic hydroxylation of debrisoquine in man. *Lancet* **1:** 584–586.

Mani C, Kupfer D. (1991) Cytochrome P-450-mediated activation and irreversible binding of the antioestrogen tamoxifen to proteins in rat and human liver: possible involvement of flavin-containing monooxygenases in tamoxifen metabolism. *Cancer Res.* **51:** 6052–6058.

Mannervik B, Aswasthi YC, Board PG, Hayes JD, Di Ilio C, Ketterer B, Lutowsky L, Morgenstern R, Muramatsu M, Pearson WR, Pickett CB, Sato K, Wirdstern M, Wolf CR. (1992) Nomenclature for human glutathione transferases. *Biochem. J.* **282:** 305–308.

McLemore TL, Adelberg S, Liu MC, McMahon NA, Yu SJ, Hubbard WC, Czerwinski M, Wood TG, Storeng R, Lubet RA, Eggleston JC, Boyd MR, Hines RN. (1990) Expression of CYP1A1 gene in patients with lung cancer: evidence for cigarette smoke-induced gene expression in normal lung tissue and for pulmonary carcinomas. *J. Natl Cancer Inst.* **82:** 1333–1339.

Mikus G, Bochner F, Eicheibaum M, Horak P, Somogyi AA, Spector S. (1994) Endogenous codeine and morphine in poor and extensive metabolizers of the CYP2D6 (debrisoquine/sparteine) polymorphism. *J. Pharmacol. Exp. Ther.* **268:** 546–551.

Miles JS, Bickmore W, Brock JD, McLaren AW, Meehan RR, Wolf CR. (1989a) Close linkage of the human cytochrome P450 IIA and P450 IIB gene families, implication for the assignment of a substrate specificity. *Nucleic Acids Res.* **17:** 2907–2917.

Miles JS, McLaren A, Wolf CR. (1989b) Alternative splicing in the human cytochrome IIB6 gene generates a high level of abberant messages. *Nucleic Acids Res.* **20:** 8241–8255.

Mitchell RS, Bell JC. (1957) Clinical implications of isoniazid, PAS and streptomycin blood levels in pulmonary tuberculosis. *Trans. Am. Clin. Chem. Assoc.* **69:** 98–105.

Nakachi K, Imai K, Hagashi S, Kawajiri SI. (1993) Polymorphism of the CYP1A1 and glutathione S-transferase gene associated with susceptibility to lung cancer in relation to cigarette dose in a Japanese population. *Cancer Res.* **53:** 2994–2999.

Nebert DW. (1990) Polymorphism of human CYP2D genes involved in drug metabolism: possible relationship to individual cancer risk. *Cancer Cells* **3:** 93–96.

Nelson DR, Kamataki T, Waxman DJ, Guengerich FP, Estabrook RW, Feyereisen R, Gonzalez FJ, Coon MJ, Gunsalus IC, Gotoh O, Okuda K, Nebert DW. (1993) The P450 superfamily: Update on new sequences, gene mapping, accession numbers, early trivial names of enzymes, and nomenclature. *DNA Cell Biol.* **12:** 1–51.

Niznik HB, Tyndale RF, Sallee FR, Gonzalez FJ, Hardwick JP, Inaba T, Kalow W. (1990) The dopamine transporter and cytochrome P450IID1 (debrisoquine 4-hydroxylase) in brain: resolution and identification of two distinct [^{3}H] GBR-12935 binding proteins. *Arch. Biochem. Biophys.* **276:** 424–432.

Pemble S, Schroeder KR, Spencer SR *et al.* (1994) Human glutathione *S*-transferase theta (GSTT1): cDNA cloning and the characterisation of a genetic polymorphism. *Biochem. J.* **300:** 271–276.

Peter H, Deutschmann S, Reichel C, Hallier E. (1989) Metabolism of methyl chloride by human erythrocytes. *Arch. Toxicol.* **63:** 351–355.

Petersen DD, McKinney CE, Ikeya K *et al.* (1991) Human CYP1A1 gene: cosegregation of the enzyme inducibility phenotype and an RFLP. *Am. J. Hum. Genet.* **48:** 720–725.

Price Evans DA, Manley KA, McKusick VA. (1960) Genetic control of isoniazid metabolism in man. *Br. Med. J.* **1:** 485–491.

Price-Evans DA. (1984) Survey of the human acetylator polymorphism in spontaneous disorders. *J. Med. Genet.* **21:** 243–253.

Rodriquez JW, Kirlin WG, Ferguson RJ, Dall MA, Gray K, Rutson TD, Lee ME, Kemp K, Urso P, Hein DW. (1993) Human acetylator genotype–relationship to colorectal cancer incidence and arylamine *N*-acetyltransferase expression in colon cytosol. *Arch. Toxicol.* **67:** 445–452.

Roots, I, Drakoulis N, Ploch M, Heinemeyer G, Loddenkimper R, Minks T, Nitz M, Otte F, Koch M. (1988) Debrisoquine hydroxylation phenotype, acetylation phenotype and ABO blood groups as genetic host factors of lung cancer risk. *Klin. Woschenscr.* **66:** 87–97.

Roots I, Brockmoller J, Drakoulis N, Kerb R. (1992) Mutant alleles of cytochrome P450 IID6 in lung cancer. *Clin. Pharmacol. Ther.* **51:** 181.

Rushmore TH, King RG, Poulson KE, Pickett CB. (1990) Identification of a unique xenobiotic response element controlling inducible expression by planar aromatic compounds. *Proc. Natl Acad. Sci. USA* **87:** 3826–3830.

Rushmore TH, Pickett CB. (1993) Glutathione *S*-transferases, structure, regulation and therapeutic implications. *J. Biol. Chem.* **268:** 11475–11478.

Saxena R, Shaw GL, Relling MV, Frame JN, Moir DT, Evans WE, Caporaso N, Weiffenbach B. (1994) Identification of a new variant CYP2D6 allele with a single base deletion in exon 3 and its association with the poor metabolizer phenotype. *Hum. Mol. Genet.* **3:** 923–926.

Seidegard J, DePierre JW, Pero RW. (1985) Hereditary interindividual differences in the glutathione transferase activity towards *trans*-stilbene oxide in resting mononuclear leukocytes due to a particular isozyme(s). *Carcinogenesis* **6:** 1211–1216.

Seidegard J, Pero RW, Miller DW, Beattle EJ. (1986) A glutathione transferase in human leukocytes as a marker for the susceptibility to lung cancer. *Carcinogenesis* **7:** 751–753.

Sies H, Ketterer B. (eds) (1988) *Glutathione Conjugation: Mechanisms and Biological Significance.* Academic Press, New York.

Sim E, Hickman D, Coroneos E, Kelly SL. (1992) Arylamine *N*-acetyltransferase. *Biochem. Soc. Trans.* **20:** 304–309.

Sindrup SH, Poulsen L, Arendt-Nielsen L, Gram LF. (1993) Are poor metabolizers of sparteine/debrisoquine less pain tolerant than extensive metabolizers? *Pain* **53:** 335–349.

Smith CAD, Moss JE, Gough AC, Spurr NK, Wolf CR. (1992a) Molecular genetic analysis of the cytochrome P450 debrisoquine hydroxylase locus and association with cancer susceptibility. *Env. Hlth Perspect.* **98:** 107–112.

Smith CAD, Gough AC, Leigh PN, Summers BA, Harding AG, Maranganore DM, Sturman SG, Schapira AHV, Williams AL, Spurr NK, Wolf CR. (1992b) Debrisoquine hydroxylase gene polymorphism and susceptibility to Parkinson's disease. *Lancet* **339:** 1365–1372.

Song BJ, Friedman FK, Park SS, Tsokos GC, Gelboin HV. (1985) Monoclonal antibody-directed radioimmunoassay detects cytochrome P-450 in human placenta and lymphocytes. *Science* **228:** 490–492.

Song BJ, Gelboin HV, Park SS, Yang CS, Gonzalez FJ. (1986) Complementary cDNA and protein sequences of ethanol inducible rat and human P450s: transcriptional and post-transcriptional regulation of the rat enzyme. *J. Biol. Chem.* **261:** 16689–16697.

Spurr NK, Gough AC, Stevenson K, Wolf CR. (1987) Msp-1 polymorphism detected with a cDNA probe for the P450 I family on chromosome 15. *Nucleic Acids Res.* **15:** 5901.

Tefre T, Ryborg D, Haugen A, Nebert DW, Skaug V, Brogger, Borreson AL. (1991) Human CYP1A1 (cytochrome P450) gene: lack of association between the Msp1 restriction fragment length polymorphism and the incidence of lung cancer in a Norwegian population. *Pharmacogenetics* **1**: 20–25.

Tucker GT. (1994) Clinical implications of genetic polymorphisms in drug metabolism. *J. Pharm. Pharmacol.* **46**: 417–424.

Uematsu F, Kikuchi H, Motomiya M, Abe T, Sagam I, Ohmachi T, Wakoi A, Konamaro R, Watanabe M. (1991) Association between restriction fragment length polymorphism of the cytochrome P450IIE1 gene and susceptibility to lung cancer. *Jpn. J. Cancer Res.* **82**: 254–256.

Uematsu F, Ikawa S, Kitochi H *et al.* (1994) Restriction fragment length polymorphism of the human CYP2E1 (cytochrome P450IIE1) gene and susceptibility to lung cancer: possible relation to low smoking exposure. *Pharmacogenetics* **4**: 58–65.

Watanabe J, Hayashi SI, Nakachi K, Imai K, Suda Y, Sekine K, Kawajiri K. (1990) Pst I and Rsa I RFLPs in complete linkage disequilibrium at the CYP2E gene. *Nucleic Acids Res.* **18**: 7194.

Weinberg RA, (1990) *Oncogenes and the Molecular Origins of Cancer.* Cold Spring Harbor Laboratory Press, Cold Spring Harbor, NY.

Whitlock JP. (1990) Genetic and molecular aspects of 2,3,7,8-tetrachlorodibenzo-*p*-dioxin. *Annu. Rev. Pharmacol. Toxicol.* **30**: 251–277.

Wilkinson GR, Guengerich FP, Branch RA. (1989) Genetic polymorphism of *S*-mephenytoin hydroxylation. *Pharmacol. Ther.* **43**: 53–76.

Wolf CR. (1986) Cytochrome P-450s: polymorphic multigene families involved in carcinogen activation. *Trends Genet.* **2**: 209–214.

Wolf CR, Smith CAD, Gough AL, Moss JE, Vallis KA, Howard G, Carey FJ, Mills K, McNee W, Carmichael J, Spurr NK. (1992) Relationship between the debrisoquine hydroxylase polymorphism and cancer susceptibility. *Carcinogenesis* **13**: 1035–1038.

Wolf CR, Smith CAD, Forman D. (1994) Metabolic polymorphisms in carcinogen metabolising enzymes and cancer susceptibility In: *Genetics of Malignant Disease* (ed BAJ Ponder). Churchill Livingstone, Edinburgh, pp. 718-731.

Yamano S, Nhanburo PT, Aoyama T, Meyer UA, Inaba T, Kalow T, Gelboin HV, McBride OW, Gonzalez FJ. (1989) cDNA cloning, sequence and cDNA-directed expression of human P450 IIB1: identification of a normal and two variant cDNAs derived from the CYP2B locus on chromosome 19 and differential expression of the IIB mRNAs in human liver. *Biochemistry* **28**: 7340–7348.

Yamano S, Tatsuno J, Gonzalez FJ. (1990) The CYP2A6 gene product catalyses coumarin 7-hydroxylation in human liver microsomes. *Biochemistry* **29**: 1322–1329.

Yang CS, Jeong-Sook VH, Ishizaki H, Hong Y. (1990) Cytochrome P450 IIE1 roles in nitrosamine metabolism and mechanism of regulation. *Drug Metab. Rev.* **22**: 147–159.

Zhong S, Wyllie AM, Barnes D, Wolf CR, Spurr NK. (1993) Relationship between the GSTM1 genetic polymorphism and susceptibility to bladder, breast and colon cancer. *Carcinogenesis* **14**: 1821–1824.

Mutagenicity tests in bacteria as indicators of carcinogenic potential in mammals

Errol Zeiger

> Gathering and amassing data is easy; the difficult part is deciding what it means and how to use it.
>
> [Zeiger, unpublished observation]

6.1 Introduction

The most widespread use of bacterial mutagenicity tests is in screening chemicals for the identification of mutagens and, by implication, potential carcinogens. These tests have become the first, if not the only, step in regulatory and industrial schemes for identifying mutagens and carcinogens. There are a number of reasons for this, including the ease and cost of performance of the tests, their sensitivity, their well-documented reproducibility within and among laboratories, and their ability to identify correctly chemical carcinogens. The advantages of having a rapid and reliable system for identifying mutagens become apparent when one is attempting to screen large numbers of chemicals for carcinogenic potential, or environmental or biological samples for the presence of potential carcinogens.

Testing using bacterial systems has its advantages and disadvantages. A major advantage of using bacteria is that they contain their DNA in the form of a single linear or circular chromosome. They do not have the chromosome composition and organization of eukaryotic cells, and do not go through mitosis, so that DNA interaction can be directly implicated in chemically induced mutagenic effects. Therefore, bacterial test systems can identify chemicals with the potential to interact with cellular DNA, either directly or indirectly. The cells have no nuclear membrane; the outer cell wall is easily penetrated by chemicals and can be made more permeable. Also the cells divide rapidly, allowing large numbers of organisms to be tested over many generations during different stages of growth and division, in a relatively short time, and at a relatively low cost.

Bacteria also have disadvantages, some of which are similar to their advantages. Their simple, relatively unadorned single chromosome is structurally and functionally different from mammalian cell chromosomes. Therefore, chemicals that produce only such events as chromosome breakage, recombination or non-disjunction will not be detected using bacterial tests. Additionally, as a rule, bacteria do not contain the enzyme complexes found in mammals that metabolize xenobiotic chemicals.

Bacterial test systems for genetic toxicity fall into two main types – those measuring mutation and those measuring DNA damage that leads to induction or expression of specific DNA damage recognition enzymes. Three bacterial mutation systems most widely used for mutagenicity testing are: the *Salmonella typhimurium his⁻* mammalian/microsome reverse mutation test (Ames test) (Claxton *et al.*, 1987; Gatehouse *et al.*, 1994; Kier *et al.*, 1986; Maron and Ames, 1983; Zeiger, 1985); the *E. coli trp⁻* reverse mutation test (Brusick *et al.*, 1980); and the *Salmonella ara* forward mutation test (Hera and Pueyo, 1986). The procedures used for all three tests are similar. The test with the largest publicly available databases is the Ames test. This test forms the basis of most national and international guidelines for mutagenicity testing.

A number of other bacterial systems are available that do not test for mutation *per se*, but for DNA damaging events that can lead to mutation, enzyme induction, or cell death. These include tests that measure differential killing between DNA repair proficient and DNA repair deficient strains (polA; recA tests); and tests that measure the induction of specific enzymes in response to DNA damage (Elespuru, 1990; Mamber *et al.*, 1986; Quillardet and Hofnung, 1985; Quillardet *et al.*, 1982, 1985; Rossman *et al.*, 1991). DNA damage test systems will not be addressed here.

Bacterial mutagenicity studies have found their greatest uses in classifying chemicals according to biological activity, monitoring biological or environmental samples, and for predicting other toxicological effects. It is not the intention of this chapter to provide detailed protocols of the different test systems; they are available from the referenced literature. Instead, this chapter will examine some of the characteristics of bacterial test systems, their utility for detecting mutagens or presumptive mutagens, and the ability of bacterial mutagenicity tests to identify rodent carcinogens.

6.2 Bacterial mutation test systems

The primary differences among the various bacterial test systems (*Salmonella his*; *E. coli trp*; *Salmonella ara*), regardless of the strains or endpoints used, are the amino acid and vitamin requirements (i.e. supplementation added to the different bacterial growth media). Extensive protocols or guidelines have been published for all these test systems, and will not be detailed here, but are referenced later. Different classes/types of chemicals, and different rationales for testing, require different test protocols. Despite differences in the genetics and endpoints scored in the different bacterial test systems, the testing procedures are similar. Many different types of protocols could be used, but the three most widely used are the plate test, the preincubation test (with and without determining mutation frequency), and the fluctuation test. Detailed descriptions of these procedures, and their variations, are available (Brusick *et al.*, 1980; Claxton *et al.*, 1987; Gatehouse and Delow, 1979; Gatehouse *et al.*, 1994; Kado *et al.*, 1983; Kier *et al.*, 1986; Maron and Ames, 1983; Zeiger *et al.*, 1992). The majority of the methodology references provide guidance and recommendations for the appropriate test protocols for different test situations. These procedures can be used with pure chemicals, defined mixtures, and crude mixtures, or extracts. It is also difficult to associate a cost to the testing of single chemicals because of the wide range of protocols that can be used (i.e. number of bacterial strains; number and types of *in vitro* metabolic activation systems, numbers of test doses and plates/dose, numbers of replicate experiments, etc.).

The tester strains used in the reverse mutation tests are auxotrophic for single amino acids [histidine (*his*) for *Salmonella*, and tryptophan (*trp*) for *E. coli*], and cannot grow and form colonies in the absence of the required amino acid. Spontaneous and mutagen-induced mutants can synthesize the required amino acids and can grow to form colonies on minimal medium. In the forward mutation test, cells are used that cannot survive in the presence of the sugar, arabinose (*ara*). When one of the arabinose-metabolizing genes is inactivated by a mutation, the cell can survive in the presence of the sugar and grow to form a colony.

The *Salmonella his* reversion test has been extensively validated in a number of laboratories and the results compiled. The other tests have been validated in fewer laboratories. Therefore, this chapter will centre around the Ames test because it boasts the largest database of tested chemicals and it has been validated in a number of multilaboratory blind studies (Aeschbacher *et al.*, 1983; Cheli *et al.*, 1980; Dunkel *et al.*, 1984, 1985; Grafe *et al.*, 1981; Knuiman *et al.*, 1987; Margolin *et al.*, 1984; Piegorsch and Zeiger, 1991; Zeiger *et al.*, 1990 ; among others).

6.2.1 Salmonella *reverse mutation (Ames test)*

A large number of tester strains are available which contain different genetic targets and different enabling mutations, for example, cell wall permeability (*rfa*); DNA repair deficiency (*uvrB*); error-prone repair (pKM101) (see *Table 6.1*). Most testing guidelines and routine testing practices use four or five strains (Claxton *et al.*, 1987; Gatehouse *et al.*, 1994; Maron and Ames, 1983). Sometimes, fewer strains are used in testing chemicals or samples with known or expected responses or mechanisms of action. The strains most often recommended are TA98, TA100, TA1535, and TA1537 or TA97. Strains TA102 and TA104 have been proposed for testing aldehydes and peroxides (Levin *et al.*, 1982; Maron and Ames, 1983), but their utility for these classes of chemicals has been questioned (Dillon *et al.*; 1992; Gatehouse *et al.*, 1994).

Table 6.1. Bacterial mutagenicity test systems

Name	Type	Strains available	Selected references
Salmonella mutation test (Ames test)	Reverse	TA97a, TA98, TA100, TA102, TA104, TA1535, TA1537, TA1538, and others	Maron and Ames (1983), Claxton *et al.* (1987), Gatehouse *et al.* (1994), Zeiger (1985)
E. coli mutation test	Reverse	WP2, WP2*uvrA*, WP2 (pKM101), WP2*uvrA* (pKM101)	Brusick *et al.* (1980), Gatehouse *et al.* (1994)
Arabinose resistance test (in *Salmonella*)	Forward*	BA3, BA6, BA13, BA14, and others	Hera and Pueyo (1986), Jurado *et al.* (1993), Ruiz-Rubio and Pueyo (1982)

* Also, reverse mutation test in those strains that contain his mutations.

6.2.2 E. coli *reverse mutation test*

Unlike the *Salmonella* tester strains, the *E. coli* strains are generally designated by their genotypes. Typically, one or two tester strains are used (see *Table 6.1*). These strains contain a mutation at the *trp* locus and either a DNA repair deficiency (*uvrA*) and/or a plasmid (pKM101) that enhances error-prone repair. One report, which compared the efficacy of *E. coli* with *Salmonella* recommended the use of two *E. coli* strains (Gatehouse *et al.*, 1994). These strains appear to be able to detect the full range of mutagenic events detected by the *Salmonella* strains.

6.2.3 Salmonella *forward mutation (*ara^r *test)*

Strains have been developed that allow the measurement of forward mutation at the arabinose locus and, in the same strain, reverse mutation at the *hisG46* or *hisD3052* locus (equivalent to *Salmonella* strains TA100 or TA98) (see *Table 6.1*). Two different types of media are used, one to identify the arabinose-resistant colonies and one to identify the his+ colonies. Any one strain is sufficient to identify mutagens by their ability to affect *ara*. As with the *E. coli* or *Salmonella* reverse mutation tests, other mutations have been incorporated into the cells to make them more permeable to large molecules, and to enhance their ability to detect premutagenic events through interference with normal DNA (error-free) repair, and by enhancing error-prone repair processes.

6.3 Metabolic activation

One major shortcoming of using bacterial systems as surrogates for *in vivo* effects in mammals is that the bacteria do not contain xenobiotic metabolizing systems of the kinds found in vertebrates – specifically the cytochrome P450 family of enzymes associated with the endoplasmic reticulum of mammalian cells, primarily the liver (see Chapter 5). To approximate mammalian metabolic activation of xenobiotic chemicals in an *in vitro* bacterial culture, rodent liver homogenates are combined with enzyme cofactors and incubated with the test chemical and indicator bacteria. Typically, rats or hamsters are treated with the polychlorinated biphenyl mixture, Aroclor 1254, to induce the relevant metabolizing enzymes, and pooled liver homogenates are centrifuged at 9000 *g* to remove the unbroken cells, cell nuclei and tissue debris. The resulting supernatant (S-9) contains microsomal-bound enzymes, soluble enzymes, and cofactors, and is able to provide many of the cytochrome P450 mixed-function oxygenase activities present in the intact liver. Other tissue preparations and liver enzyme inducers have been used (Ong *et al.*, 1980; Robertson *et al.*, 1983; Zeiger, 1985), but the majority of testing, and the test guidelines, are based on induced rat liver. The S-9 liver fraction can be stored frozen at -80°C until needed.

Not all chemicals metabolized by the intact liver will be metabolized by this *in vitro* system, and many chemicals metabolized by induced liver S-9 may form different metabolites, or different proportions of metabolites to those found *in vivo*. A number of metabolic pathways *in situ* are not represented by these liver homogenates, such as the prostaglandin synthase system, organ-specific reductive metabolism, or metabolism by the bacterial flora of the gut. These other metabolic pathways can be incorporated into the test protocol by the use of other tissue preparations, or bacterial gut flora (Mehta *et al.*, 1984; Prival *et al.*, 1984; Reid *et al.*, 1983; Robertson *et al.*, 1983; Wise *et al.*, 1984; Zeiger, 1985). Moreover, the S-9 system does not contain active phase II conjugation enzymes, so that the reaction mixture may accumulate active metabolites that, *in vivo*, would for example, be detoxified by conjugation with glutathione (Wright, 1980; see also Chapter 5). There is no evidence that S-9 or other subcellular fractions from other organs offer an advantage over rat or hamster liver for routine testing of chemicals. Other tissues or other types of inducers may be of use for the study of the metabolism and mutagenicity of specific classes of chemicals.

6.4 Uses of bacterial tests and data

There are important questions as to how bacterial mutagenicity data are used. The use of data differs according to whether the tests will be used to classify a chemical, monitor samples, predict effects in other organisms, or study mechanisms of biological activity. The major impetus behind the widespread use of bacterial mutagenicity testing has been the high correlations originally found with carcinogenicity, and the involvement of mutagenic events as crucial steps in the development of a tumour cell.

6.4.1 Hazard assessment — screening for mutagenic activity

It is not possible to approximate a figure that accurately represents the proportion of bacterial mutagens in a population of chemicals, because these proportions vary greatly with chemical class. For example, the percentage of mutagens among all the chemicals tested by the US National Cancer Institute (NCI) and the National Toxicology Program (NTP) was 36%; however among the nitroaromatics in this population the percentage was 76% (Zeiger et al., 1985a). In other compilations of NTP-tested chemicals, 43% of the total population was mutagenic, but the percentages of mutagens were 91%, 69%, 37%, and 0% for nitroaromatics, aromatic amines, chlorinated chemicals and phthalate esters respectively (Zeiger, 1987; Zeiger et al., 1985b).

6.5 Prediction of carcinogenicity from mutagenicity

It is universally recognized that cancer is a multistage process, and that a mutation is the initiating step for many rodent and most human carcinogens (Fearon and Vogelstein, 1990). Of the carcinogens identified by the mid-1970s, approximately 90% are mutagens (Kier et al., 1986; McCann et al., 1975; Purchase et al., 1978; Sugimura et al., 1976a,b; see Table 6.2). However, with changes in carcinogenesis testing protocols [for example, higher doses, longer exposure times, more extensive pathology, different criteria for selection of chemicals and different chemical classes tested (Fung, et al., 1993)], the proportion of carcinogens that are bacterial mutagens has decreased.

Table 6.2. Predictivity and sensitivity of *Salmonella his* reversion mutation test*

	% Predictivity	% Sensitivity	No. Chems[a]	Reference
Human cancer (IARC group 1)[b]	–	95	20	Shelby and Zeiger (1990)
Human cancer (IARC group 2a,b)	–	71	201	Zeiger (unpublished)
Rodent cancer	92	90	273	McCann and Ames (1976)
(literature database)[c]	95	88	239	Sugimura et al. (1976a)
	93	91	120	Purchase et al. (1978)
	98	80	160	Kier et al. (1986)
Rodent cancer	69	54	224	Zeiger (1987)
(NCI/NTP database)	89	48	114	Zeiger et al. (1990)
	72	56	298	Ashby and Tennant (1991)
Electrophilicity[d]	95	77	296	Ashby and Tennant (1991)

* % Predictivity: the per cent of *Salmonella* mutagens that are carcinogenic or electrophilic;
  % Sensitivity: the per cent of carcinogens or electrophilic chemicals that are mutagenic in *Salmonella*.
[a] Number of chemicals in the database that have been tested for mutagenicity in *Salmonella*.
[b] Organic, non-hormonal chemicals only.
[c] Does not include carcinogenicity test results from the NCI/NTP database.
[d] Structural alerts, as defined by Ashby (1985); Ashby and Tennant (1988).

Salmonella mutagenicity tests on NCI- and NTP-tested chemicals during the past 10–15 years have yielded sensitivity values (proportions of carcinogens that are mutagenic) of about 50% (Ashby and Tennant, 1991; Poirier and de Serres, 1979; Shelby *et al.*, 1988; Zeiger *et al.*, 1990; see *Table 6.2*). These values are highly dependent on the classes of chemicals tested and the patterns of responses used to define carcinogens. The earlier (1970s) compilations included a high proportion of chemicals such as alkylating agents, polycyclic aromatic hydrocarbons and nitrosamines, which are potent carcinogens as a result of their ability to modify DNA. Their DNA reactivities also made them effective bacterial mutagens. As a rule, these chemicals were carcinogenic following limited administration over a relatively short period of time. Chemicals such as those that are carcinogenic only following 2 years of exposure at doses near the maximum tolerated dose (as defined by the NCI/NTP protocols), would not have been detected as carcinogens by these earlier, less sensitive, test protocols. The classes of carcinogens that tend to be non-mutagenic (for example, chlorinated aromatics and phthalates) typically are carcinogenic only with chronic (2 year) exposure at maximally tolerated doses. Many of these carcinogens do not have electrophilic structures that would suggest that they would be DNA reactive.

In the NCI/NTP rodent carcinogenesis database, the organ most often affected is the liver (Haseman and Clark, 1990; Haseman *et al.*, 1984a). The mouse liver is the most common site for the single-site, single-species carcinogens. Relatively few (27%) of the chemicals carcinogenic only at this site are mutagens (Zeiger, 1987), which may reflect the fact that determinations of liver cancer in the B6C3F1 mice used in these tests is based on increases above an already high spontaneous frequency in males (30 – 31%) and females (approx. 8%; Haseman *et al.*, 1984b, 1985), as compared with ≤1.0% in rats (Haseman *et al.*, 1984b, 1985, 1990).

Although approximately 50% of the carcinogens identified by rat and mouse 2-year rodent carcinogenicity tests are mutagenic in *Salmonella*, the positive predictivity of the *Salmonella* test for rodent carcinogenicity is about 80–90% (see *Table 6.2*). That is, a positive response in *Salmonella* is highly predictive of cancer. However, a negative response in the *Salmonella* test is not predictive of non-carcinogenicity because of the high proportion (*c.* 50%) of rodent carcinogens that are not mutagenic (Zeiger *et al.*, 1990). The use of test batteries of *in vivo*, or other *in vitro*, genetic toxicity tests in addition to the *Salmonella* test (Auletta *et al.*, 1993; HPBGC, 1993; Kirkland, 1993; Shelby *et al.*, 1993; Sofuni, 1993) has not been shown to improve significantly the positive predictivity over the value obtained by using *Salmonella* alone (Zeiger *et al.*, 1990). However, these other tests do provide pharmacokinetic information and data on other biological properties of the chemicals which may be useful for interspecies comparisons, risk estimation and risk management.

6.6 The relationship of biological potency to predictivity

The biological potency of a substance can be defined by a measure of the magnitude of the response it produces, as a function of the active test chemical concentration, or by the extent of the animal responses to the challenge. When carcinogenic potency is measured as a function of the carcinogenic dose of test chemical, the mutagenic carcinogens, as a group, were of a higher potency (i.e. effective at lower doses) than the non-mutagenic carcinogens (Brown and Ashby, 1990; Haseman and Clark, 1990; Rosenkranz and Ennever, 1990; Tennant *et al.*, 1987). When carcinogenic potency was measured as a function of the numbers of animal species affected, or the multiplicity of tumour sites, the mutagenic carcinogens tended to be positive in more than one species (rats and mice), and at more than one tissue site (Tennant, 1993; Zeiger, 1987).

The human and rodent data show that the predictive ability of the *Salmonella* mutagenicity test increases with the potency of the carcinogen, when potency is measured by the number of species or target sites affected. As can be seen, among the organic non-hormonal chemicals judged by the International Agency for Research on Cancer (IARC; IARC, 1987) to be human carcinogens (group 1), 95% were mutagenic (see *Table 6.2*). Among the chemicals judged 'probable' or 'possible' human carcinogens (i.e. those with less evidence of carcinogenicity for humans, but generally carcinogenic in more than one rodent species; IARC groups 2a,b) the proportion of mutagenic carcinogens drops to 71%. Similar patterns are seen with rodent carcinogens. Among the chemicals that are carcinogenic at a common site in more than one rodent species (rats and mice), 78% are *Salmonella* mutagens (Tennant, 1993), while approximately 40% of the chemicals carcinogenic in only one species are mutagens (Tennant, 1993; Zeiger, 1987).

In summary, a positive mutagenic response in bacterial test systems is highly predictive for carcinogenicity in rodents. The chemicals predicted to be carcinogens are those that are likely to be multispecies carcinogens and, therefore, more likely to be a potential hazard for humans. Therefore, if chemicals are mutagenic in these bacterial test systems, and their structure and metabolism are suggestive of chemicals known to be carcinogenic, it may be sufficient to label such chemicals as potential rodent carcinogens. These chemicals would trigger a rodent carcinogenicity test only to disprove that assumption, or if such a test were required for regulatory action. Alternatively, an abbreviated carcinogenicity test using, perhaps, a single sex from each of two species, could be performed to prove the assumption, thereby minimizing the cost and use of animals. The carcinogenicity test resources conserved by these approaches can be redirected toward testing those chemicals for which a prediction of carcinogenicity cannot be made, or those chemicals whose carcinogenicity test results would be informative regarding mechanisms of tumour formation, or patterns or kinetics of tumour development.

6.7 Mutagenic vs. non-mutagenic carcinogens

On the basis of the results of *Salmonella* mutagenicity studies, carcinogens have been grouped, for convenience, into two general classes – those that are mutagenic by virtue of their ability to interact directly with DNA, and those that are not. In the NCI/NTP database, approximately 50% of the carcinogens are not mutagenic (Ashby and Tennant, 1991; Zeiger, 1987; Zeiger *et al.*, 1990). One explanation for this phenomenon is that the *in vitro* S-9 activation system was ineffective in activating the chemical to a mutagen, although it can be activated *in situ* in the animal. Also, it is possible that the types of mutations, or mutagenic events, produced by the chemical are not detectable in the bacterial systems used. These mechanisms probably account for a small fraction of the carcinogens 'missed' because it can be seen from their structures that some of these chemicals are electrophilic, or can form electrophiles (Ashby, 1985) which can alkylate DNA to form premutagenic lesions.

However, there are more differences between the mutagenic and non-mutagenic carcinogens than their mutagenicity (see *Table 6.3*). The mutagenic carcinogens are generally electrophiles or capable of being metabolized to electrophilic substances (for example, alkylating agents, nitrosamines, aromatic amines, nitroaromatics, azo dyes) while the non-mutagenic carcinogens (for example, chlorinated hydrocarbons, hormones, phthalates, insolubles) do not have electrophilic properties. Among the chemicals tested for carcinogenicity in rats and mice by the NCI and the NTP, the mutagenic carcinogens were more likely to be carcinogenic in two species, and at multiple tissue sites than the non-mutagens (Tennant, 1993; Zeiger, 1987), whereas the non-mutagens are predominately single-species (usually mouse) carcinogens (Zeiger, 1987). Moreover, the mutagenic carcinogens tend to be carcinogenic at lower concentrations than the non-mutagens (Haseman and Clark, 1990; Rosenkranz and Ennever, 1990). Ashby and Tennant (1988) showed that the patterns of carcinogenic responses were different for the two types of carcinogens. In addition to the mouse liver, there were organs and tissues (for example, thyroid, kidney and the haematopoietic system) where tumours occurred primarily in response to non-mutagens, whereas tumours at other sites (such as Zymbal's gland, lung and skin) were most frequently associated with mutagens.

Table 6.3. Mutagenic (in *Salmonella*) vs. non-mutagenic carcinogens

Mutagenic carcinogens	Non-mutagenic carcinogens
• Can form electrophiles • Produce tumours at lower effective doses • Usually tumorigenic in rats and mice • Often multiple organ and tissue sites	• Non-electrophilic • Require higher doses, longer exposures • Usually in rats or mice, but not both • Often single organ or tissue
Target organs include: Zymbal's gland; lung; nasal; skin • Tumours in liver plus other tissues • Human carcinogens	Target organs include: thyroid; kidney; haematopoietic system • Many are mouse liver tumorigens only

The obvious explanation that encompasses the larger group of non-mutagenic carcinogens is that these chemicals do not induce cancer via mutation as the primary step. Numerous mechanisms other than mutation have been invoked for the carcinogenic activity of the non-mutagens, including hormonal effects, induction of cell proliferation, and induction of chromosome loss or gain (Butterworth, 1990). As more is learned about their mechanisms of action, the non-mutagenic carcinogens will probably be further subdivided on the basis of those mechanisms. All carcinogens will then be grouped into classes based on their primary carcinogenic mechanisms, with mutagenicity being only one such mechanism.

6.8 Summary

In vitro bacterial mutagenicity tests are the most effective, short-term initial procedures for identifying chemicals likely to be carcinogenic in rodents and, by extension, for identifying potential human carcinogens. However, they cannot, by themselves, be used to identify non-carcinogens. In addition, bacterial mutation tests provide information on the mechanism of the carcinogenic action of chemicals. Further developments of these, and other, bacterial mutation test systems are not likely to improve our ability to identify carcinogens. Future studies should be directed to other (non-mutagenic) molecular and cellular events that can initiate tumorigenesis, and the development of test systems to identify rapidly and effectively the chemicals that act via these other events.

References

Aeschbacher HU, Friederich U, Seiler JP. (1983) Criteria for the standardization of *Salmonella* mutagenicity tests: results of a collaborative study. III. The influence of the composition and preparation of the minimal medium in the *Salmonella* mutagenicity test. *Teratogen. Carcinogen. Mutagen.* **3:** 195–203.

Ashby J. (1985) Fundamental structural alerts to potential carcinogenicity or noncarcinogenicity. *Environ. Mutagen.* **7:** 919–921.

Ashby J, Tennant RW. (1988) Chemical structure, *Salmonella* mutagenicity and extent of carcinogenicity as indicators of genotoxic carcinogenesis among 222 chemicals tested in rodents by the U.S. NCI/NTP. *Mutat. Res.* **204:** 17–115.

Ashby J, Tennant RW. (1991) Definitive relationships among chemical structure, carcinogenicity and mutagenicity for 301 chemicals tested by the US NTP. *Mutat. Res.* **257:** 229–306.

Auletta AE, Dearfield KL, Cimino MC. (1993) Mutagenicity test schemes and guidelines: U.S. EPA Office of Pollution Prevention and Toxics and Office of Pesticide Programs. *Environ. Mol. Mutagen.* **21:** 38–45.

Brown LP, Ashby J. (1990) Correlations between bioassay dose-level, mutagenicity to *Salmonella*, chemical structure and sites of carcinogenesis among 226 chemicals evaluated for carcinogenicity by the U.S. NTP. *Mutat. Res.* **244:** 67–76.

Brusick DJ, Simmon VF, Rosenkranz HS, Ray VA, Stafford RS. (1980) An evaluation of the *Escherichia coli* WP2 and WP2 *uvrA* reverse mutation assay. *Mutat. Res.* 76: 169–190.

Butterworth BE. (1990) Consideration of both genotoxic and nongenotoxic mechanisms in predicting carcinogenic potential. *Mutat. Res.* **239:** 117–132.

Cheli C, DeFrancesco D, Petrullo L, McCoy EC, Rosenkranz HS. (1980) The *Salmonella* mutagenicity assay: reproducibility. *Mutat. Res.* **74**: 145–150.

Claxton LD, Allen J, Auletta A, Mortelmans K, Nestmann E, Zeiger E. (1987) Guide for the *Salmonella typhimurium*/mammalian microsome tests for bacterial mutagenicity. *Mutat. Res.* **189**: 83–91.

Dillon DM, McGregor DB, Combes RD, Zeiger E. (1992) Optimal conditions for detecting bacterial mutagenicity of some aldehydes and peroxides. *Mutat. Res.* **271**: 184.

Dunkel VC, Zeiger E, Brusick D, McCoy E, McGregor D, Mortelmans K, Rosenkranz HS, Simmon VF. (1984) Reproducibility of microbial mutagenicity assays: I. Tests with *Salmonella typhimurium* and *Escherichia coli* using a standardized protocol. *Environ. Mutagen.* **6** (suppl. 2): 1–251.

Dunkel VC, Zeiger E, Brusick D, McCoy E, McGregor D, Mortelmans K, Rosenkranz HS, Simmon VF. (1985) Reproducibility of microbial mutagenicity assays: II. Testing of carcinogens and noncarcinogens in *Salmonella typhimurium* and *Escherichia coli*. *Environ. Mutagen.* **7** (suppl. 5): 1–248.

Elespuru RK. (1990) Automation of screening assays for DNA damaging agents which induce the SOS response. In: *Mutation and the Environment. Part D: Carcinogenesis* (eds ML Mendelsohn, RJ Albertini). Wiley-Liss, New York, pp. 345–354.

Fearon ER, Vogelstein B. (1990) A genetic model for colorectal tumorigenesis. *Cell* **61**: 759–767.

Fung VA, Huff J, Weisburger EK, Hoel DG. (1993) Predictive strategies for selecting 379 NCI/NTP chemicals evaluated for carcinogenic potential: scientific and public health impact. *Fund. Appl. Tox.* **20**: 413–436.

Gatehouse DG, Delow GF. (1979) The development of a 'microtitre®' fluctuation test for the detection of indirect mutagens, and its use in the evaluation of mixed enzyme induction of the liver. *Mutat. Res.* **60**: 239–252.

Gatehouse D, Haworth S, Cebula T, Gocke E, Kier L, Matsushima T, Melcion C, Nohmi T, Ohta T, Venitt S, Zeiger E. (1994) Recommendations for the performance of bacterial mutation assays. *Mutat. Res.* **312**: 217–233.

Grafe A, Mattern IE, Green M. (1981) A European collaborative study of the Ames assay. I. Results and general interpretation. *Mutat. Res.* **85**: 391–410.

Haseman JK, Clark A-M. (1990) Carcinogenicity results for 114 laboratory animal studies used to assess the predictivity of four *in vitro* genetic toxicity assays for genetic toxicity. *Environ. Mol. Mutagen.* **16** (suppl. 18): 15–31.

Haseman JK, Crawford DD, Huff JE, Boorman GA, McConnell EE. (1984a) Results from 86 two-year carcinogenicity studies conducted by the National Toxicology Program. *J. Toxicol. Environ. Hlth.* **14**: 621–639.

Haseman JK, Huff J, Boorman GA. (1984b) Use of historical control data in carcinogenicity studies in rodents. *Toxicol. Pathol.* **12**: 126–135.

Haseman JK, Huff JE, Rao GN, Arnold JE, Boorman GA, McConnell EE. (1985) Neoplasms observed in untreated and corn oil gavage control groups of F344/N rats and (C57BL/6N X C3H/HeN)F_1 (B6C3F_1) mice. *J. Natl Cancer Inst.* **75**: 975–984.

Haseman JK, Arnold J, Eustis SL. (1990) Tumor incidences in Fischer 344 rats: NTP historical data. In: *Pathology of the Fischer Rat. Reference and Atlas* (eds GA Boorman, SL Eustis, MR Elwell, CAJ Montgomery, WF MacKenzie). Academic Press, San Diego, pp. 555–564.

Hera C, Pueyo C. (1986) Conditions for the optimal use of the L-arabinose-resistance mutagenesis test with *Salmonella typhimurium*. *Mutagenesis* **1**: 267–273.

HPGBC, Health Protection Branch Genotoxicity Committee. (1993) The assessment of mutagenicity. Health Protection Branch mutagenicity guidelines. *Environ. Mol. Mutagen.* **21**: 15–37.

IARC. (1987) *IARC Monographs on the Evaluation of Carcinogenic Risks to Humans.* Suppl. 7, Overall Evaluations of Carcinogenicity: an Updating of *IARC Monographs, Vols 1–41*. International Agency for Research on Cancer, Lyon.

Jurado J, Alejandre-Duran E, Pueyo C. (1993) Genetic differences between the standard Ames tester strains TA100 and TA98. *Mutagenesis* **8:** 527–532.

Kado NY, Langley D, Eisenstadt E. (1983) A simple modification of the *Salmonella* liquid-incubation assay. Increased sensitivity for detecting mutagens in human urine. *Mutat. Res.* **121:** 25–32.

Kier LD, Brusick DJ, Auletta AE, Von Halle ES, Simmon VF, Brown MM, Dunkel VC, McCann J, Mortelmans K, Prival MJ, Rao TK, Ray VA. (1986) The *Salmonella typhimurium*/mammalian microsome mutagenicity assay. A report of the U.S. Environmental Protection Agency Gene-Tox Program. *Mutat. Res.* **168:** 67–238.

Kirkland DJ. (1993) Genetic toxicology testing requirements: official and unofficial views from Europe. *Environ. Mol. Mutagen.* **21:** 8–14.

Knuiman MW, Laird NM, Louis TA. (1987) Inter-laboratory variability in Ames assay test results. *Mutat. Res.* **180:** 171–182.

Levin DE, Hollstein M, Christman MF, Schwiers EA, Ames BN. (1982) A new *Salmonella* tester strain (TA102) with A:T base pairs at the site of mutation detects oxidative mutagens. *Proc. Natl Acad. Sci. USA* **79:** 7445–7449.

Mamber SW, Okasinski WG, Pinter CD, Tunac JB. (1986) The *Escherichia coli* K-12 SOS chromotest agar spot test for simple, rapid detection of genotoxic agents. *Mutat. Res.* **171:** 83–90.

Margolin BH, Risko RJ, Shelby MD, Zeiger E. (1984) Sources of variability in Ames *Salmonella typhimurium* tester strains: analysis of the International Collaborative Study on 'Genetic Drift'. *Mutat. Res.* **130:** 11–25.

Maron D, Ames BN. (1983) Revised methods for the *Salmonella* mutagenicity test. *Mutat. Res.* **113:** 173–212.

McCann J, Ames BN. (1976) Detection of carcinogens as mutagens in the *Salmonella*/microsome test: assay of 300 chemicals: discussion. *Proc. Natl Acad. Sci. USA* **73:** 950–954.

McCann J, Choi E, Yamasaki E, Ames BN. (1975) Detection of carcinogens in the *Salmonella*/microsome test: assay of 300 chemicals. *Proc. Natl Acad. Sci. USA* **72:** 5135–5139.

Mehta R, Labuc GE, Urbanski SJ, Archer MC. (1984) Organ specificity in the microsomal activation and toxicity of *N*-nitrosomethylbenzylamine in various species. *Cancer Res.* **44:** 4017–4022.

Ong T-M, Mukhtar M, Wolf CR, Zeiger E. (1980) Differential effects of cytochrome P450-inducers on promutagen activation capabilities and enzymatic activities of S-9 from rat liver. *J. Environ. Pathol. Toxicol.* **4:** 55–65.

Piegorsch WW, Zeiger E. (1991) Measuring intra-assay agreement for the Ames *Salmonella* assay. In: *Lecture Notes in Medical Informatics* (ed. L Hothorn). Springer-Verlag, Heidelberg, pp. 35–41.

Poirier LA, de Serres FJ. (1979) Initial National Cancer Institute studies on mutagenesis as a prescreen for chemical carcinogens: an appraisal. *J. Natl Cancer Inst.* **62:** 919–926.

Prival MJ, Bell SJ, Mitchell VD, Peiperl MD, Vaughan VL. (1984) Mutagenicity of benzidine and benzidine-congenor dyes and selected monoazo dyes in a modified *Salmonella* assay. *Mutat. Res.* **136:** 33–47.

Purchase IFH, Longstaff E, Ashby J, Styles JA, Anderson D, Lefevre PA, Westwood FR. (1978) An evaluation of 6 short-term tests for detecting organic chemical carcinogens. *Br. J. Cancer* **37:** 873–959.

Quillardet P, de Bellecombe C, Hofnung M. (1985) The SOS Chromotest, a colorimetric bacterial assay for genotoxins: validation study with 83 compounds. *Mutat. Res.* **147:** 79–95.

Quillardet P, Hofnung M. (1985) The SOS Chromotest, a colorimetric bacterial assay for genotoxins: procedures. *Mutat. Res.* **147:** 65–78.

Quillardet P, Huisman O, D'Ari R, Hofnung M. (1982) SOS Chromotest, a direct assay of induction of an SOS function in *Escherichia coli* K12 to measure genotoxicity. *Proc. Natl Acad. Sci. USA* **79:** 5971–5975.

Reid T, Morton K, Wang C, King C. (1983) Conversion of congo red and 2-azoxyfluorene to mutagens following *in vitro* reduction by whole-cell rat cecal bacteria. *Mutat. Res.* **117:** 105–112.

Robertson IGC, Sivarajah K, Eling TE, Zeiger E. (1983) Activation of some aromatic amines to mutagenic products by prostaglandin endoperoxide synthetase. *Cancer Res.* **43:** 476–480.

Rosenkranz HS, Ennever FK. (1990) An association between mutagenicity and carcinogenic potency. *Mutat. Res.* **244:** 61–65.

Rossman TG, Molina M, Meyer L, Boone P, Klein CB, Wang Z, Li F, Lin WC, Kinney PL. (1991) Performance of 133 compounds in the lambda prophage induction endpoint of the Microscreen assay and a comparison with *S. typhimurium* mutagenicity and rodent carcinogenicity assays. *Mutat. Res.* **260:** 349–367.

Ruiz-Rubio M, Pueyo C. (1982) Double mutants with both His reversion and Ara forward mutation systems of *Salmonella*. *Mutat. Res.* **105:** 383–386.

Shelby MD , Zeiger E. (1990) Activity of human carcinogens in the *Salmonella* and rodent bone-marrow cytogenetics tests. *Mutat. Res.* **234:** 257–261.

Shelby MD, Zeiger E, Tennant RW. (1988) Commentary on the status of short-term tests for chemical carcinogens. *Environ. Mol. Mutagen.* **11:** 437–441.

Shelby MD, Erexson GL, Hook GJ, Tice RR. (1993) Evaluation of a three-exposure mouse bone marrow micronucleus protocol: results with 49 chemicals. *Environ. Mol. Mutagen.* **21:** 160–179.

Sofuni T. (1993) Japanese guidelines for mutagenicity testing. *Environ. Mol. Mutagen.* **21:** 2–7.

Sugimura T, Sato S, Nagao M, Yahagi T, Matsushima T, Seino Y, Takeuchi M, Kawachi T. (1976a) Overlapping of carcinogens and mutagens. In: *Fundamentals of Cancer Prevention* (eds PN Magee, S Takayama, T Sugimura, T Matsushima). University Park Press, Baltimore, pp. 191–215.

Sugimura T, Yahagi T, Nagao M, Takeuchi M, Kawachi T, Hara K, Yamasaki E, Matsushima T, Hashimoto Y, Okada M. (1976b) Validity of mutagenicity tests using microbes as a rapid screening method for environmental carcinogens. In: *Screening Tests in Chemical Carcinogenesis* (eds R Montesano, H Bartsch, L Tomatis). IARC Scientific Publications, Lyon, pp. 81–104.

Tennant RW. (1993) Stratification of rodent carcinogenicity bioassay results to reflect relative human hazard. *Mutat. Res.* **286:** 111-118.

Tennant RW, Margolin BH, Shelby MD, Zeiger E, Haseman JK, Spalding J, Caspary W, Resnick M, Stasiewicz S, Anderson B, Minor R. (1987) Prediction of chemical carcinogenicity in rodents from *in vitro* genetic toxicity assays. *Science* **236:** 933–941.

Wise RW, Zenzer TV, Kadlubar FF, Davis BB (1984) Metabolic activation of carcinogenic aromatic amines by dog bladder and kidney prostaglandin H synthase. *Cancer Res.* **44:** 1893–1897.

Wright AS. (1980) The role of metabolism in chemical mutagenesis and chemical carcinogenesis. *Mutat. Res.* **75:** 215–241.

Zeiger E. (1985) The Salmonella mutagenicity assay for identification of presumptive carcinogens. In: *Handbook of Carcinogen Testing* (eds HA Milman, EK Weisburger). Noyes Publications, Park Ridge, NJ, pp. 83–99.

Zeiger E. (1987) Carcinogenicity of mutagens: predictive capability of the *Salmonella* mutagenesis assay for rodent carcinogenicity. *Cancer Res.* **47:** 1287–1296.

Zeiger E, Risko KJ, Margolin BH. (1985a) Strategies to reduce the cost of mutagenicity screening using the Salmonella/microsome assay. *Environ. Mutagen.* **7:** 901–911.

Zeiger E, Haworth S, Mortelmans K, Speck W. (1985b) Mutagenicity testing of di(2-ethylhexyl)phthalate and related chemicals in *Salmonella*. *Environ. Mutagen.* **7:** 213–232.

Zeiger E, Haseman JK, Shelby MD, Margolin BH, Tennant RW. (1990) Evaluation of four in vitro genetic toxicity tests for predicting rodent carcinogenicity: confirmation of earlier results with 41 additional chemicals. *Environ. Mol. Mutagen.* **16** (suppl. 18): 1–14.

Zeiger E, Anderson B, Haworth S, Lawlor T, Mortelmans K. (1992) Salmonella mutagenicity tests. V. Results from the testing of 311 chemicals. *Environ. Mol. Mutagen.* **19** (suppl. 21): 2–141.

7

In vitro cytogenetics and aneuploidy

E.M. Parry and J.M. Parry

7.1 Introduction

Exposure to environmental agents may, by various means, produce chromosomal mutations; endpoints of relevance to both heritable defects and to cancer induction. Chromosomal mutations, detectable as changes in the structure or number of chromosomes are the most important cause of human reproductive failure, manifested as sterility, low fertility and a high rate of conceptus mortality (Bond and Chandley, 1983; Jacobs and Hassold, 1980; Jacobs *et al.*, 1974). Studies of the frequencies of structural and numerical chromosome abnormalities in live births are technically demanding and probably less than 50% are detected. Autosomal trisomies and unbalanced autosomal structural rearrangements are the most readily detectable. Down's syndrome is the most easily recognized; the sex chromosome trisomies and balanced structural rearrangements are not likely to be detected phenotypically (Hook, 1987).

Whereas almost all numerical abnormalities are presumably fresh mutations, a significant proportion of structural abnormalities detected in fetuses will be inherited from a carrier parent (Hook and Cross, 1987) and will, moreover, be of a structurally balanced type. Structural cytogenetic abnormalities often result from mechanisms different to those that produce numerical aberrations; they are almost exclusively a consequence of direct damage to DNA, like point mutations. Indeed the distinction between these two events may be difficult to make, for instance the deletion of a single base pair from a gene may occur as a result of a similar molecular process causing the deletion of several thousands of bases (which can be seen as a chromosomal deletion microscopically). Thus the induction of structural chromosome abnormalities by exposure to an environmental agent is also likely to be an indicator of the occurrence of other types of mutational events; however, numerical changes may not always be induced by the same agents but may involve events such as the disturbance of the cell division spindle.

Chromosome changes in somatic cells have been found to be associated with, and implicated in, the progression of cancer (Mitelman, 1983). Carcinogenesis is a multistage process which can vary with the inducing agents, the cell type and species; but generally at least two genetic changes are involved and cell proliferation is required to convert the DNA damage into mutations, chromosomal rearrangements, insertions, deletions, gene amplifications or numerical changes.

Environmental exposures are implicated in the induction of chromosomal damage. Recognized sources have been steadily increased during this century from the testing of nuclear weapons, nuclear energy, the use of X-rays for both diagnostic and therapeutic reasons and from many groups of chemicals used in industry, medicine and agriculture. For this reason it is important to assess what levels of damage exist in human male and female germ cells and in somatic cells, and also to investigate and measure the potential of an environmental agent to induce such damage in human cells. This chapter will concentrate on how this can be achieved in somatic cells *in vitro*.

7.2 Cytogenetic endpoints

The direct or indirect consequences of DNA damage can be seen at cytogenetic endpoints as changes in chromosome structure or number. Their occurrence can be related in most cases to two events: the presence of DNA strand breaks and their erroneous repair, and interference with cell division. Thus cell division and cell cycle progression are essential for the expression of chromosome damage and the type of damage detected will be influenced by the position of the cells in the cell cycle. The cell cycle comprises four stages, the two growth phases G_1 and G_2 separated by a stage, S, when DNA synthesis and chromosome replication occurs, and the dividing phase, mitosis. Mitosis is also subdivided into the four stages: prophase, metaphase, anaphase and telophase (PMAT),when chromosome condensation and separation takes place. Cell cycles can therefore be represented as:

$$G_1 \rightarrow S \rightarrow G_2 \rightarrow PMAT \rightarrow G_1 \rightarrow S \rightarrow G_2 \rightarrow PMAT$$

Interphase   Interphase
Mitosis   Mitosis

The time spent in each phase is a characteristic of particular cell types; generally human cells take about 24 hours for the complete cycle, of which G_1 is the longest phase, about 12 hours, S phase is 6–8 hours, G_2 is about 3–4 hours and 1 hour is spent in mitosis. Chinese hamster cell lines often have a cell cycle of about 15 hours.

During interphase the chromosomes are decondensed structures in the cell nucleus which are invisible by light microscopy, but towards the end of G_2 they condense and become visible as objects whose shape and size are characteristic of the species. They reach their maximum condensation and contraction during the metaphase stage of mitosis when they can be seen clearly with a light microscope with the aid of specific stains and recognized by their distinguishing features: length and position of a primary constriction, the centromere. The centromere is the region of the metaphase chromosome that attaches to the spindle apparatus of cell division and at anaphase the chromosomes separate and move apart, one chromatid segregating to each pole of the spindle thus achieving an equal distribution of chromosomes at mitosis. This disjunction of chromosomes is assisted by kinetochores, plate-like compound structures associated with centromeres on the side of the chromosome, joining it to the microtubules of the spindle. According to an agreed convention, chromosomes with different centromere positions are given different names (e.g. a centrally positioned centromere defines a metacentric chromosome whereas one with its centromere at one end is called telocentric).

This brings us to the natural ends of chromosomes, the telomeres. It has been known for a considerable time that when a chromosome is broken the damaged ends are highly reactive and will tend to join up with any other damaged ends available; conversely non-reactivity is a property associated with natural chromosome ends or telomeres (McClintock, 1939; Muller, 1938). Thus telomeres are essential for the maintainance of chromosome stability and appear to help the cell to distinguish an intact chromosome from damaged DNA. Studies of the DNA from telomeres of mammalian cells have revealed that they comprise blocks of short, simple tandem repeats similar in organization to those identified in lower eukaryotes (Blackburn and Szostak, 1984). Telomeres are thought to resolve the problems of DNA replication at the ends of a linear molecule, generally stabilizing chromosome ends. Additionally, they may play roles in the arrangement of chromosomes in the interphase nucleus by their association and position with respect to the nuclear envelope and may participate in determining the natural lifespan of a cell.

The DNA sequences in centromeric regions are also highly specific, and rich in AT bases, repetitive sequences and heterochromatin. This suggests that in the case of both centromeres and telomeres the underlying DNA sequence plays an important role in the structural organization of the chromosome and in its correct segregation at cell division. In addition, the remaining arms of chromosomes can be differentially stained to reveal characteristic banding patterns by physical or chemical procedures that disclose differences in the way that the DNA/chromatin is ordered along the chromosome. These banding patterns are species-specific and are used to define karyotypes and detect some rearrangements. The distribution of mapped genes, radiation-induced breakpoints and exchange points and cancer-associated exchange points along the chromosome appears to be non-random (Holmquist, 1992). Some of these hot spots may depend upon the particular DNA sequence in that region. For instance Boehm *et al.* (1989) report that a breakpoint on chromosome 11p13 in some leukaemias occurred at a region of alternating purine–pyrimidine sequences, a potential region of Z-DNA that could change the shape of the chromosome, making it more accessible to attack and recombination.

Chromosomal stability is also influenced by a cell's ability to control cell cycle progression. Checkpoints in the cell cycle normally operate to prevent damaged cells from progressing from G_1 to S or G_2 to mitosis until the damage has been repaired. Kung *et al.* (1990) have attributed differences in the stringency of cell cycle control to a tendency for chromosome instability in some cell lines. In terms of species differences it is known that humans have higher stringency than rodents and the degree of stringency decreases along with transformation to the cancer state.

Chromosomal changes that can be detected using a light microscope in cultured mammalian cells by metaphase analysis are either structural, as chromosome aberrations and sister chromatid exchanges, or numerical [i.e. changes in the number of whole chromosomes present in the cell as either multiples of the normal diploid set (polyploidy) or loss or gain of individual chromosomes (aneuploidy)]. Interphase analysis can be used to detect micronucleus induction, that is the formation of a small satellite nucleus adjacent to the main nucleus as a result of chromosome damage. It too can be classified as either 'clastogenic' (resulting from chromosome breakage) or 'aneugenic' (resulting from whole chromosomes) in origin.

7.3 Structural chromosome changes

Exposure to environmental agents may induce, by various means, chromosome breakage. Such breaks may be either rejoined or repaired to their original state, remain unrejoined or be rejoined differently to their original state. Some of the last two cases may be detected by cytogenetic techniques as chromosome abberations. When these gross, readily observable chromosomal aberrations are produced, it is often assumed that many of them will be lethal to the cell (i.e. they will not survive cell division events because of their unbalanced nature). Their presence is used as an indicator that the agent and treatment used was able to induce structural chromosome damage and it is inferred that other chromosome aberrations will also have been induced, such as reciprocal translocations, inversions and small deletions, that would not be seen but will have possible mutational and carcinogenic consequences in surviving cells.

The structural chromosome changes that can be detected by conventional *in vitro* cytogenetics are chromosome aberrations, sister chromatid exchanges and micronuclei.

7.3.1 Metaphase analyses

Chromosome abberrations. The obvious prerequisite for chromosome studies is a source of dividing cells. The cells commonly used in these types of studies are mammalian somatic cells, either human peripheral lymphocytes or established cell lines, usually Chinese hamster fibroblasts. Peripheral lymphocytes in healthy individuals are virtually all in the same G_0 (resting stage) or G_1 stage of mitotic interphase and some of them can be stimulated by addition of mitogens (commonly phytohaemagglutinin, an extract from the red bean) to enter the mitotic cycle in short-term culture (Evans and O'Riordan, 1975). They are very popular cells for *in vitro* studies because they are human primary cells, easily cultured with a stable karyotype ($2n=46$) and low spontaneous rates of chromosome damage. Chinese hamster cell lines are also popular because of their relatively small number of large chromosomes ($2n=22$). They should checked for karyotypic stability. Established lines such as Chinese hamster ovary (CHO) cells often show extensively rearranged chromosomes, but provided that these are stable changes and the spontaneous rates of chromosome change are low (less than 5%) their use is acceptable.

A basic treatment protocol involves exposing cycling cells to the test agent for either 3–6 hours or continuously and harvesting the cells at the subsequent mitosis. Harvest time is usually about 1.5 times the normal cell cycle time, so that cells experiencing cell cycle delay induced by the treatment may have time to recover. There is some evidence that longer treatment and sampling times may be necessary to detect some classes of chemically induced damage in some cell types (Bean *et al*, 1994; Galloway, 1994).

The incorporation of systems to mimic metabolic activation is discussed in Section 7.5.1. In order to obtain sufficient numbers of metaphase cells for analysis it is usual to interfere with the spindle using spindle poisons such as colchicine, colcemid or vinblastine (Tjio and Levan, 1956) and to swell the cells with a hypotonic treatment (Hsu, 1952) before fixing and staining them for examination under the light microscope.

The structural chromosome changes that can be detected by conventional metaphase analysis and simple Giemsa staining have been classified by Savage (1976, 1983) and the methods described in detail by Scott *et al.* (1990) and Galloway (1994).

Seven types of chromosomal damage can be distinguished: five are intrachanges and two are interchanges (*Figure 7.1*). The five intrachanges are terminal deletions, minutes, acentric rings, centric rings and inversions (these last are generally not observable). The two interchanges are reciprocal translocations; either symmetrical interchanges (not generally detectable in metaphase analysis without special staining procedures to identify subchromosomal bands) or asymmetrical interchanges (dicentrics).

Chromosome damage detected cytologically at metaphase can be divided into chromosome and chromatid aberrations. The type of damage induced by exposure to an environmental agent will be dependent upon the properties and mode of action of that agent and upon the cellular reaction and cell cycle position.

Thus exposures to ionizing radiations and agents that cause direct DNA damage during G_0 or G_1 stages of the cell cycle induce chromosome-type damage, because the lesion has been replicated, whereas in S or G_2 stages chromatid-type damage is produced (Evans, 1984).

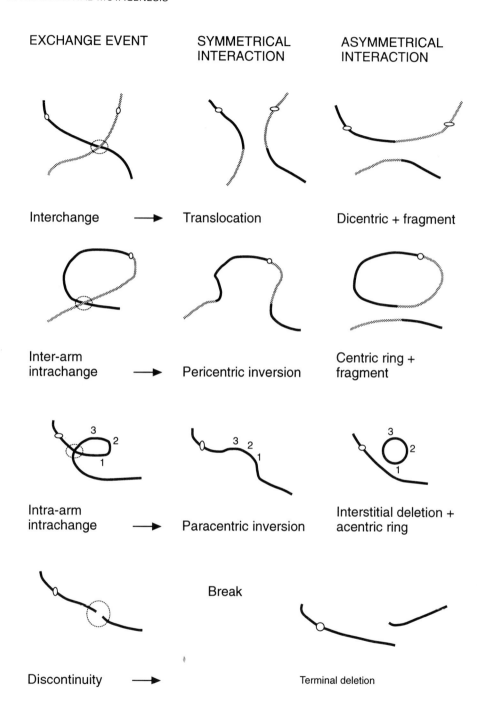

Figure 7.1. Basic aberration categories.

Many chemicals cause only chromatid-type damage even though the cells have been exposed in G_1 and examined in their first subsequent division. These abberations may be the consequence of errors in replication or processing occurring at DNA synthesis (S) following exposure. The scoring of chromatid breaks, particularly gaps (i.e. small regions of discontinuity), is not so reliable an indicator of damage as scoring chromosome-type damage since it can be more subjective and influenced by technical factors.

In vitro cytogenetic analysis usually aims to measure aberrations at the first division after treatment. Preparations made from cells cultured for longer periods will contain cells in their second or subsequent divisions and cells with some chromosome-type damage may fail to undergo repeated division due to mechanical problems at anaphase and/or subsequent genetic imbalance. Some chromosome-type aberrations seen at second or later divisions may result from a duplication of aberrations that were initially of the chromatid type.

If the exposure and treatment conditions delay the normal cell cycle, prolonged culture time may be necessary to reach first division. This can be measured by labelling the DNA with 5-bromo-deoxyuridine (BrdU) and using a stain that distinguishes between sister chromatids. Chromosomes at first division will have equally stained chromatids whereas those at second and subsequent divisions will show differential staining.

Practical aspects of chromosome aberration assays have been discussed by Scott *et al.* (1990), Galloway *et al.* (1985) and Galloway (1994). Perhaps one of the most important aspects of analysis is familiarity with the karyotype of the cell being used. This will undoubtedly help to overcome many of the confounding factors such as gaps being confused with discontinuities of staining or unstained secondary constrictions, dicentrics being confused with secondary constrictions or overlapping chromosomes and satellite associations being confused with chromosome exchanges.

Within the limits of resolution imposed by use of the light microscope and for the reasons mentioned above, it is often impossible to detect all events, especially multilesion events. Chromosome condensation and packaging may also act to disguise many events. Some of the aberrations depicted in *Figure 7.1* that can be seen by conventional methods are shown in *Figure 7.2*.

Sister chromatid exchanges. Sister chromatid exchanges (SCEs) involve breakage of DNA in both chromatids followed by an exchange of whole DNA duplexes. These events occur in S phase and are induced efficiently by mutagenic and carcinogenic agents that form covalent adducts with DNA or interfere directly or indirectly with DNA replication. The induction of SCEs has been correlated with the induction of point mutations (Carrano *et al.*, 1978) and cytotoxicity (Natarajan *et al.*, 1983). The methods of preparation of metaphases for SCE analysis are essentially similar to those used for studying chromosome aberrations except that BrdU is added to the cell culture medium and incorporated for two rounds of replication (Latt, 1973; Perry and Wolff, 1974; Perry and Thomson,1984). When stained by this differential staining technique, chromosomes at first division have equally stained chromatids

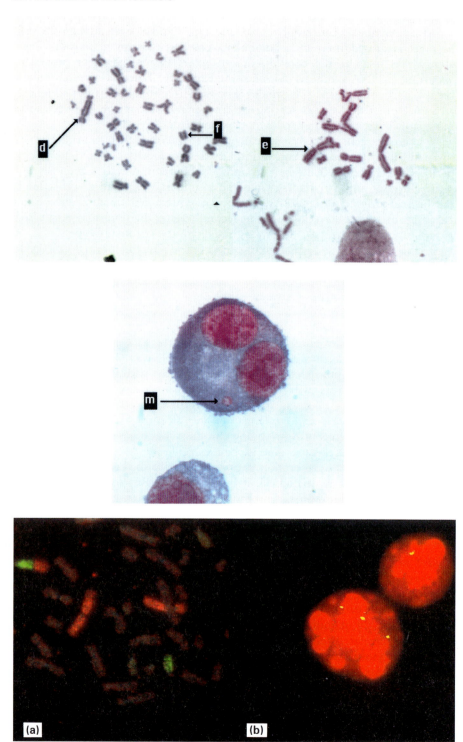

Figure 7.2. Chromosome aberrations detected by Giemsa staining. d, dicentric; e, exchange; f, fragment.

Figure 7.3. Micronucleus (m) in a cytochalasin B blocked binucleate cell.

Figure 7.4. Detection of structural and numerical chromosome aberrations in metaphase and interphase cells by FISH. (a) Human metaphase chromosomes showing a translocation between chromosomes 1(red) and 14 (green). (b) Human cytochalasin B blocked binucleate cell showing non-disjunction of centromeric DNA probe marking chromosome 8.

---

whereas chromosomes at second and subsequent divisons have differentially stained sister chromatids. This is often referred to as harlequin staining.

The frequency of SCE has been correlated with the extent of BrdU incorporation in DNA – the more incorporation, the higher the frequency (Natarajan *et al.*, 1986; Pinkel *et al.*, 1985). This effect can be minimized by detecting SCE, with antibodies against BrdU, considerably lowering the incorporation levels.

7.3.2 Interphase analyses

Micronuclei. Micronuclei can be formed at anaphase by either chromosomal fragments or whole chromosomes failing to segregate normally and being excluded from the two daughter nuclei resulting from cell division. Thus to detect micronuclei the cell must be allowed to progress through cell division and micronuclei can be seen in the resulting interphase cells. Because acentric fragments can form micronuclei, the latter have been used as an indicator of an agent's ability to induce structural chromosome damage, based on the premise that it is unlikely that agents will produce acentric fragments and not also produce heritable aberrations (Heddle, 1994). Adler (1990) found very similar doubling doses when comparing the induction of chromosome aberrations with micronuclei following treatment with six clastogens.

The micronucleus assay was first described for *in vitro* studies of radiation effects by Evans *et al.* (1959) and for *in vivo* bone marrow cell studies by Matter and Schmid (1971) and Heddle (1976): it has been adapted and used widely for *in vitro* studies (Degrassi and Tanzarella, 1988; Eastmond and Tucker, 1989; Fenech and Morley, 1985; Parry *et al.*, 1988). Because it uses interphase cells, it is a more rapid system with the potential to analyse quickly large numbers of cells.

Since it is essential for the cell to divide in order for micronucleus formation to occur, the *in vitro* micronucleus assay has been improved by identification of those cells that have completed a cell division. This can be done by labelling the chromosomes with BrdU (Pincu *et al.*,1984) or [^{3}H]thymidine (Fenech and Morley, 1985) or by treating the cells with cytochalasin B (CB) an inhibitor of cytokinesis (Fenech and Morley, 1985). The CB method is popular at present because there are fewer doubts about its effect on chromosome structure. Cells that have divided in the presence of CB are binucleate and therefore easily distinguished for scoring (*Figure 7.3*).

Micronuclei that have been formed as a result of chromosome breakage can be distinguished from those formed by whole chromosomes roughly on the basis of their size, but this is an unreliable method. It is preferable to classify them on the presence or absence of centromeres. This is usually done by staining with antikinetochore antibodies (Degrassi and Tanzarella, 1988; Eastmond and Tucker, 1989; Thomson and Perry, 1988), sometimes called 'Crest' staining because the antibody is derived from the serum of patients with the autoimmune disease scleroderma, CREST syndrome (Moroi et al., 1980). Crest antibodies are directed against the kinetochore structure associated with the centromere. It is also possible to use DNA pericentromeric sequence probes and in situ hybridization techniques to mark centromeres (see Section 7.5.2).

Premature chromosome condensation. Interphase cells can also be studied using the premature chromosome condensation technique (Hittleman, 1981). This method fuses non-dividing target cells with mitotic CHO or HeLa cells; the chromosomes of the treated cells become condensed in the hybrid, G_1 chromosomes appearing as single chromatin threads and G_2 chromosomes as double structures. Thus, in theory, interphase chromosomes can be visualized and analysed for any induced damage relatively soon after the initiating event and compared with those present after allowing for repair at the first mitosis and possibly later.

7.4 Aneuploidy

Aneuploidy is a change in the number of whole chromosomes from the normal diploid or haploid number that is a characteristic of each species. Unlike structural chromosome aberrations, aneuploidy may be induced by the action of environmental agents upon a variety of cellular targets, all involved in cell division, as well as the chromosomes themselves (Parry and Parry, 1989). Damage to any part of the cell division apparatus such as the spindle, microtubules, microtubular organizing centre or kinetochores may have an effect on correct chromosome segregation, and direct chromosomal damage may also affect correct chromosome orientation and attachment to the spindle.

There are two basic types of aneuploidy: the first type is non-disjunction of chromosomes at anaphase, resulting in one trisomic daughter cell and one monosomic daughter cell. This can be ascribed to a failure of normal separation of sister chromosomes. The other aneugenic event is chromosome loss; here one daughter cell becomes monosomic and the other daughter cell remains normal. The lost chromosome, resulting from a failure of correct spindle attachment or interaction with the spindle, may be excluded from both new daughter nuclei and may form a micronucleus. Alternatively, it may become randomly incorporated into either daughter nucleus, in which case either one daughter cell becomes trisomic and the sister cell monosomic, or both become normal and the aneugenic event is 'repaired'.

Various mammalian *in vitro* methods to detect aneuploidy have been reviewed by Galloway and Ivett (1986), Oshimura and Barrett (1986), Parry *et al.* (1995). Currently, the most commonly used methods to detect aneugenic effects are counting metaphase chromosomes, analysing micronuclei in interphase cells and examining cell division aberrations.

7.4.1 Metaphase analysis by chromosome counting

Mammalian cells cultured and treated as for a chromosome aberration study can be used to measure the induction of aneuploidy by the identification and counting of metaphase chromosomes. However, for any aneugenic event to be expressed it is essential to allow the cells to progress to the second metaphase following treatment. Either the whole chromosome set or specific chromosomes may be counted.

Unless a cell line is being used that has a distinctive karyotype where all chromosomes are morphologically different, total chromosome counting cannot offer conclusive evidence for or against numerical stability. This difficulty may be overcome by chromosome identification with banding techniques, but this is a time-consuming procedure that would not be practical for routine investigations. Thus, it is usual to count unbanded chromosomes in a familiar cell type and assume that a diploid chromosome number can be equated with diploidy.

The preparative techniques used to make metaphase spreads can be adapted so that the possibility of any artefactual chromosome gain or loss can be minimized. Most importantly, this involves preserving the cell membranes and not overswelling the cells but at the same time achieving sufficient chromosome spreading to allow chromosome counting.

The alternative approach, counting a single chromosome recognizable by its distinctive shape, has been used by Tenchini *et al.* (1983) for the quinacrine-stained human Y chromosome. This approach assumes that all chromosomes in a diploid set are equally susceptible to aneuploidy-inducing events, but as yet there is not sufficient experimental evidence either for or against this point of view.

In man, trisomies 13 (Patau syndrome), 18 (Edwards' syndrome), 21 (Down's syndrome) and both loss and gain of the sex chromosomes are the most common forms of aneuploidy found in live births (Hook, 1987). This may be because these chromosomes are more sensitive to aneuploidy-inducing events or may indicate that they are the only ones that are able to survive in an unbalanced condition. It is generally thought that the latter is true: evidence from studies of spontaneously aborted fetuses show that about 50% are chromosomally abnormal, mainly aneuploid (Hassold *et al.*,1980). But it is difficult to establish the true situation since the frequency of aneuploidy for different chromosomes may vary at different stages of embryonic development. Studies using human sperm (Bischoff *et al.*, 1994; Martin and Rademaker, 1990), long-term fibroblast cultures (Wenger, 1989) and cultured human lymphocytes (Martin *et al.*, 1980; Nowinski *et al.*, 1990; Ohtaki *et al.*, 1994; Wenger *et al.*, 1984) all indicate that autosome loss is inversely related to chromosome loss (possibly a technical artefact), whereas it is random for chromosome gain. Sex and age were shown to be contributing factors to aneuploidy.

The unambiguous identification of chromosomes now appears to be possible by the use of *in situ* hybridization methods, which are discussed in Section 7.5.2.

7.4.2 Interphase analysis by micronucleus studies

Micronuclei can be formed as a result of chromosome breakage or aneugenic events; thus this method can be used to detect aneuploidy induction if methods are included to allow the identification of whole chromosomes in cells that have divided. This is discussed in Section 7.3.2. and the development of this approach using *in situ* hybridization techniques and metabolically competent cells is discussed in Sections 7.5.2 and 7.5.1.

7.4.3 Cell division aberrations

Manifestations of aberrant cell division, such as C-mitoses, increases in mitotic metaphases, cell-cycle delay, chromosome clumping or contraction, premature chromatid separation, anaphase and telophase bridges, chromosome lagging, multipolar spindles and diplochromosomes, have been used to indicate the induction of aneuploidy following chemical treatment. Aberrant mitotic effects were first described by Levan (1938) following treatment of *Allium* with the spindle inhibitor colchicine and various authors have suggested that such effects can be indicative of aneuploidy and may also suggest the mechanism responsible for aneuploidy induction (Dulout and Natarajan, 1987; Hsu *et al.*, 1983; Liang and Satya-Prakash, 1985; Onfelt, 1987; Parry *et al.*, 1982).

7.5 Method developments

7.5.1 Metabolic activation

It is known that some chemicals are either activated or deactivated *in vivo* by metabolic processes that may be absent in cells cultured *in vitro*. Attempts to mimic these events are made by the inclusion of liver-derived metabolizing systems into the cell culture. The method most commonly used (Natarajan *et al.*, 1976) is to add a liver homogenate (S-9 + cofactors) obtained from the livers of Aroclor-induced rats. Problems of toxicity and reactivity in culture mean that the exposure times that can be used +S-9 are limited to about 3 hours. Typically, S-9 fractions from rat liver contain phase I enzymes that activate mutagens (see Chapter 5) and also enzymes that detoxify mutagens such as glutathione-*S*-transferases, sulphotransferases and glucuronyl transferases.

In an attempt to overcome the problems associated with exogenous metabolizing sytems and to improve the limited intrinsic capacity for metabolic activation of cultured cell lines, genetic engineering has been used to incorporate some genes to express the necessary enzymes (Crespi *et al.*, 1991; Doehmer *et al.*, 1990). Genetically engineered Chinese hamster cells have been used in micronucleus assays by Ellard *et al.* (1991) and the use of genetically engineered human cells for micronucleus studies has been described by Crofton-Sleigh *et al.* (1993).

7.5.2 *Fluorescence* in situ *hybridization*

Conventional analysis of metaphase chromosome damage is often limited to asymmetrical exchanges as discussed earlier because of the frequent difficulties in distinguishing symmetrical events. Yet it is the symmetrical events that are potentially of biological importance since they are the stable rearrangements of genetic information that may result in the inappropriate gene expression that has been associated with the development of cancer (Mitelman *et al.*, 1991). It is generally assumed that symmetrical and asymmetrical exchanges occur with equal frequency, as demonstrated for G-banded irradiated chromosomes by Buckton (1976). However, recent studies comparing Giemsa and fluorescence *in situ* hybridization staining methods have not always found this equivalence (Columna *et al.*, 1993; Cremer *et al.*, 1990; Lucas *et al.*, 1989; Natarajan *et al.*,1992): excessive numbers of symmetrical exchanges were commonly reported, and complex rearrangements that were not previously detectable have been revealed.

In situ hybridization (ISH), frequently coupled with fluorescent labelling (FISH, see also Chapter 14), allows for the detection of nucleic acid sequences on chromosomes at either metaphase or interphase. The reaction involves a heteroduplex formation between a chemically modified single-stranded DNA probe sequence and a denatured chromosome immobilized on a microscope slide. Initially probes were labelled radioactively, but this has largely been replaced by chemical labelling for reasons of greater efficiency, clarity and safety. Once the probe has hybridized to the chromosome its position can be revealed by an antibody tagged either with a fluorochrome (the usual method) or with an enzyme such as a peroxidase. The signals appear in sharp contrast against a counterstained background. Various probes are now available for human chromosomes. These include:

(i) repetitive sequence probes which may also be chromosome-specific (e.g. alphoid sequences that identify juxtacentromeric region sequences; Cremer *et al.*, 1986, 1988);

(ii) single-copy unique sequences that identify the precise band location of genes on chromosomes; and

(iii) composite probes comprising mixtures of chromosome-specific probes obtained from chromosome-specific libraries that hybridize to an entire chromosome so that it can be 'painted' (Cremer *et al.*, 1986; Lichter *et al.*, 1988; Pinkel *et al.*, 1988).

The probes are usually labelled with either biotin or digoxigenin and are detected with a fluorochrome-linked antibody, using fluorescein (green), rhodamine or Texas red (red) and AMCA (blue). Combined labelling allows for the simultaneous detection of several targets, for instance Ried et al. (1992) used this approach to detect seven different probes simultaneously, although about three may be more suitable for chromosome aberration studies.

The application of these methods to the analysis of chromosome damage is an exciting field of research. Not only does it allow for a more complete analysis of structural chromosome damage, but also for more fundamental research into the underlying mechanisms producing chromosome damage of all types. With suitable probes, FISH can be applied equally well to condensed or uncondensed (interphase) chromosomes and, since these latter are more freely available, the amount of information generated can be increased and extended to non-dividing tissues (Figure 7.4). Thus, painting whole metaphase chromosomes can be used to investigate both structural and numerical chromosome aberrations. Moreover, repetitive DNA probes (often pericentromeric and chromosome-specific) can be used to count chromosomes in interphase nuclei. Because of the complexity of the rearrangements revealed by chromosome painting, new methods of classification have been proposed (Savage and Simpson, 1994; Tucker et al., 1995).

Recently, premature chromosome condensation has been combined with FISH to examine chromosome-specific damage in interphase cells (Brown et al., 1992; Pandita et al., 1994). The aim is to score chromosome aberrations in cells from the G_0 or G_1 phase of the cell cycle, relatively soon after exposure to an agent, so that more information can be obtained about the initial events leading to chromosome damage.

Most of the studies reported so far are confined to human cells because suitable probes are already available. However, painting probes have been developed for the mouse and rat (Breneman et al., 1993; Hoebee et al., 1994) and centromeric mouse probes for gamma satellite DNA (Hörz and Altenburger, 1981; Weier et al., 1991) have been used in micronucleus studies to identify all centromeres (Miller et al., 1991).

7.6 Conclusions

It is hoped that the new methods described here will assist in our understanding of how environmental agents induce chromosomal damage. We know that the endpoints seen as a result of some environmental exposures will depend upon the particular action of that agent, but also depend upon many features of the cell and the specific chromosomal arrangement. For instance, there can be 'hot spots' for the action of agents leading to the non-random distribution of chromosome damage. The outcome of damage induction at a particular region of a chromosome may depend upon the chromosomal organization and the location of that chromosome in the nucleus in relation to other chromosomes. To form a chromosome exchange, double-strand breaks on different chromosomes must be available and adjacent for interaction to occur. The spatial arrangement of chromosomes in the interphase nucleus may influence this, in that chromosome breaks must be sufficiently close to each other for recombination to occur. It is also known that telomeres have an important role to play in chromosome stability.

The availability of the new techniques of FISH means that many more questions may be asked about chromosomal aberrations and the extent to which different chromosomes and subchromosomal regions are involved in their formation at the levels of the initial damage, after repair, at first mitosis and after several generations as stable changes. It is expected that these developments will increase the power of *in vitro* cytogenetics in human health protection.

References

Adler ID. (1990) Chromosome studies in male germ cells, their relevance for the prediction of heritable effects and their role in screening protocols. In: *Chromosome Aberrations: Basic and Applied Aspects* (eds G Obe, AT Natarajan). Springer-Verlag, Heidelberg, pp. 107–119.

Bean CL, Bradt CI, Hill RB, Johnson TE, Stallworth MV, Galloway SM. (1994) Chromosome aberrations: persistence of alkylation damage and modulation by O⁶–alkyguanine-DNA alkytransferase. *Mutat. Res.* **307**: 67–81.

Bischoff FZ, Nguyen DD, Burt KJ, Shaffer LG. (1994) Estimates of aneuploidy using multicolour fluorescence *in situ* hybridisation on human sperm. *Cytogenet. Cell Genet.* **66**: 237–243.

Blackburn EH, Szostak JW. (1984) The molecular structure of centromeres and telomeres. *Annu. Rev. Biochem.* **53**: 163–194.

Boehm T, Mengle-Gaw L, Kees UR, Spurr N, Lavenir I, Forster A, Rabbitts TH. (1989) Alternating purine–pyrimidine tracts may promote chromosomal translocations seen in a variety of human lyphoid tumours. *EMBO J.* **8**: 2621–2631.

Bond DJ, Chandley AC. (1983) *Aneuploidy.* Oxford University Press, Oxford.

Breneman JW, Ramsey MJ, Lee DA, Eveleth GG, Minkler JL, Tucker JD. (1993) The development of chromosome specific composite DNA probes for the mouse and their application to chromosome painting. *Chromosoma* **102**: 591–598.

Brown JM, Evans J, Kovacs MS. (1992) The prediction of human tumour radiosensitivity *in situ:* an approach using chromosome aberrations detected by fluorescence *in situ* hybridisation. *Int. J. Radiat. Oncol. Biol. Phys.* **24**: 279–286.

Buckton KE. (1976) Identification with G and R banding of the position of breakpoints induced in human chromosomes by *in vitro* X-irradiation. *Int. J. Radiat. Biol.* **29**: 475–488.

Carrano AV, Thompson LH, Lindl PA, Minkler JL. (1978) Sister chromatid exchanges as an indicator of mutagenesis. *Nature* **271**: 551–553.

Columna EA, Giaccia AJ, Evans JW, Yates BL, Morgan WF. (1993) Analysis of restriction enzyme-induced chromosomal aberrations by fluorescence *in situ* hybridisation. *Environ. Mol. Mutagen.* **22**: 26–33.

Cremer T, Landegent JE, Brücknev A, Scholl HP, Schardin M, Hager HD, Devilee P, Pearson P, van der Ploeg M. (1986) Detection of chromosome aberrations in human interphase nucleus by visualisation of specific target DNAs with radioactive and nonradioactive in situ hybridisation techniques: diagnosis of trisomy 18 with probe L1.84. *Hum. Genet.* **74**: 346–352.

Cremer T, Lichter B, Borden J, Ward DC, Manuelidis L. (1988) Detection of chromosome aberrations in metaphase and interphase cells by *in situ* hybridisation using chromosome-specific library probes. *Hum. Genet.* **80**: 235–246.

Cremer T, Popp S, Emmerich P, Lichter P, Cremer C. (1990) Rapid metaphase and interphase detection of radiation-induced chromosome aberrations in human lymphocytes by chromosomal suppression *in situ* hybridisation. *Cytometry* **11:** 110–118.

Crespi CL, Gonzalez FJ, Steimel DT, Turner TR, Gelboin HV, Penman BW, Langenbach R. (1991) A metabolically competent human cell line expressing five cDNAs encoding procarcinogen-activating enzymes: application to mutagenicity testing. *Chem. Res Toxicol.* **4:** 566–572.

Crofton-Sleigh C, Doherty A, Ellard S, Parry EM, Venitt S. (1993) Micronucleus assays using cytochalasin-blocked MCL-5 cells a proprietary human cell line expressing five human cytochrome P-450 and microsomal epoxide hydrolase. *Mutagenesis* **4:** 363–372.

Degrassi F, Tanzarella C. (1988) Immunofluorescent staining of kinetochores in micronuclei: a new assay for the detection of aneuploidy. *Mutat. Res.* **203:** 339–345.

Doehmer J, Seidel A, Oesch F, Glatt HR. (1990) Genetically engineered V79 Chinese hamster cells metabolically activate the cytostatic drugs cyclophosphamide and ifosfamide. *Environ. Hlth Perspect.* **88:** 63–65.

Dulout FN, Natarajan AT. (1987) A simple and reliable *in vitro* test system for the analysis of induced aneuploidy as well as other cytogenetic endpoints using Chinese hamster cells. *Mutagenesis 2:* 121–126.

Eastmond DA, Tucker JD. (1989) Identification of aneuploidy-inducing agents using cytokinesis-blocked human lymphocytes and an antikinetochore antibody. *Environ. Mol. Mutagen.* **13:** 34–43.

Ellard S, Mohammad Y, Dogra S, Wölfel C, Doehmer J, Parry JM. (1991) The use of genetically engineered V79 Chinese hamster cultures expressing rat liver CYP1A1, 1A2 and 2B1 cDNAs in micronucleus assays. *Mutagenesis* **6:** 461–470.

Evans HJ. (1984) Human peripheral blood lymphocytes for the analysis of chromosome aberrations in mutagen tests. In: *Handbook of Mutagenicity Test Procedures* (eds BJ Kilbey, M Legator, W Nichols, C Ramel). Elsevier, Amsterdam, pp. 405–428.

Evans HJ, O'Riordan ML. (1975) Human peripheral lymphocytes for the analysis of chromosome aberrations in mutagen tests. *Mutat. Res.* 31: 135–148.

Evans HJ, Neary GJ, Williamson FS. (1959) The relative biological efficiency of single doses of fast neutrons and gamma-rays on *Vicia faba* roots and the effect of oxygen. Part II. Chromosome damage: the production of micronuclei. *Int. J. Rad. Biol.* **3:** 216–229.

Fenech M, Morley AA. (1985) Measurement of micronuclei in lymphocytes. *Mutat. Res.* **147:** 29–36.

Galloway SM. (1994) Chromosome aberrations induced in vitro: mechanisms, delayed expression, and intriguing questions. *Environ. Mol. Mutagen.* 23 (suppl. 24): 44–53.

Galloway SM, Ivett JL. (1986) Chemically induced aneuploidy in mammalian cells in culture. *Mutat. Res.* **167:** 89–105.

Galloway SM, Bloom AD, Resnick M, Margolin BH, Nakamura F, Archer P, Zeigler E. (1985) Development of a standard protocol for *in vitro* cytogenetic testing with Chinese hamster ovary cells: comparison of results for 22 compounds in two laboratories. *Environ. Mutagen.* **7:** 1–51.

Hassold T, Chen N, Funkhouser J, Jooss T, Mannel B, Matsuura A, Wilson C, Yamane JA, Jacobs PA. (1980) A cytogenetic study of 1000 spontaneous abortions. *Annu. Hum. Genet.* **44:** 151–164.

Heddle JA. (1976) Measurement of chromosomal breakage in cultured cells by the micronucleus technique. In: *Mutation-Induced Chromosome Damage to Man* (eds HJ Evans, DC Lloyd). Edinburgh University Press, Edinburgh, pp. 191–200.

Heddle JA. (1994) Revelling in cytogenetics. *Environ. Mol. Mutagen.* **23(24):** 35–38.

Hittleman WN. (1981) Premature chromosome condensation for the detection of mutagenic activity. In: *Cytogenetic Assays of Environmental Mutagens* (ed. P Tsu). Allanhead, Totowa, NJ, pp. 353–384.

Hoebee B, de Stoppelaar JM, Suijkerbuijk RF, Monard S. (1994) Isolation of rat chromosome-specific paint probes by bivariate flow sorting followed by degenerate oligonucleotide primed-PCR. *Cytogenet. Cell Genet.* **66:** 277–282.

Holmquist GP. (1992) Chromosome bands, their chromatin flavors and their functional features. *Am. J. Hum. Genet.* **51:** 17–37.

Hook EB. (1987) Surveillance of germinal human mutations for effects of putative environmental mutagens and utilization of a chromosome registry in following rates of cytogenetic disorders. In: *Cytogenetics; Basic and Applied Aspects* (eds G Obe, A Basler). Springer-Verlag, Berlin, pp. 141–165.

Hook EB, Cross PK. (1987) Rates of mutant and inherited structural cytogenetic abnormalities detected at amniocentesis: results of about 63,000 fetuses. *Ann. Hum. Genet.* **51:** 27–55.

Hörz W, Altenburger W. (1981) Nucleotide sequence of mouse satellite DNA. *Nucleic Acids Res.* **9:** 683–696.

Hsu TC. (1952) Mammalian chromosomes *in vitro*. I. The karyotype of man. *J. Hered.* **43:** 17–23.

Hsu TC, Shirley LR, Takanari H. (1983) Cytogenetic assays for mitotic poisons: the diploid Chinese hamster cell system. *Anticancer Res.* **3:** 155–160.

Jacobs PA, Hassold TJ. (1980) The origin of chromosome abnormalities in spontaneous abortions. In: *Human Embryonic and Fetal Death* (eds EB Hook, IH Porter). Academic Press, New York, pp. 289–298.

Jacobs PA, Melville M, Ratcliffe S, Keay AJ, Syme J. (1974) A cytogenetic survey of 11680 newborn infants. *Annu. Hum. Genet.* **37:** 359–376.

Kung AL, Sherwood SW, Schimke RT. (1990) Cell line-specific differences in the control of cell cycle progression in the absence of mitosis. *Proc. Natl Acad. Sci. USA* **87:** 9553–9557.

Latt SA. (1973) Microfluorometric detection of deoxyribonucleic acid replication in human metaphase chromosomes. *Proc. Natl Acad. Sci. USA* **70:** 3395–3399.

Levan A. (1938) The effect of colchicine on root mitoses in allium. *Hereditas* **24:** 471–486.

Liang JC, Satya-Prakash KL. (1985) Induction of aneuploidy by mitotic arrestants in mouse bone marrow. *Mutat. Res.* **155:** 61–70.

Lichter P, Cremer T, Tang CC, Watkins PC, Manuelidis L, Ward DC. (1988) Rapid detection of human chromosome 21 aberrations by *in situ* hybridisation. *Proc. Natl Acad. Sci. USA* **85:** 9664–9668.

Lucas JN, Tenjin T, Straume T, Pinkel D, Moore D, Litt M, Gray JW. (1989) Rapid human chromosome aberration analysis using fluorescence *in situ* hybridisation. *Int. J. Radiat. Biol.* **56:** 35–44. (Erratum, *Int. J. Radiat. Biol.* **56:** 201.)

Martin JM, Kellet JM, Kahn J. (1980) Aneuploidy in cultured human lymphocytes. I. Age and sex differences. *Age Aging* **9:** 147–153.

Martin RH, Rademaker A. (1990) The frequency of aneuploidy among individual chromosomes in 6,821 human sperm complements. *Cytogenet. Cell Genet.* **64:** 23–26.

Matter B, Schmid W. (1971) Trenimon-induced chromosome damage in bone-marrow cells of six mammalian species evaluated by the micronucleus test. *Mutat. Res.* **12:** 417–425.

McClintock, B. (1939) The behavior in successive nuclear divisions of a chromosome broken at meiosis. *Proc. Natl Acad. Sci. USA* **25:** 405–416.

Miller BM, Zitzelsberger HF, Weier HUG, Adler ID. (1991) Classification of micronuclei in murine erythrocyes: immunofluorescent staining using CREST antibodies compared to *in situ* hybridisation with biotinylated gamma satellite DNA. *Mutagenesis* **6:** 297–302.

Mitelman F. (1983) Catalogue of chromosome aberrations in cancer. *Cytogenet. Cell Genet.* **36:** 1–515.

Mitelman F, Kaneko Y, Trent JM. (1991) Report of the committee on chromosome changes in neoplasia. *Cytogenet. Cell Genet.* **58:** 1053–1079.

Moroi Y, Peebles C, Fritzler MJ, Steigerwald J, Tan EM. (1980) Autoantibody to centromere (kinetochore) in scleroderma sera. *Proc. Natl Acad. Sci. USA* **77:** 1627–1631.

Muller HJ. (1938) The remaking of chromosomes. *Collecting Net* **13:** 181–195.

Natarajan AT, Tates AD, van Buul PPW, Meijers M, de Vogel N. (1976) Cytogenetic effects of mutagens/carcinogens after activation in a microsomal system *in vitro*. *Mutat. Res.* **37:** 83–90.

Natarajan AT, Tates AD, Meijers M, Neuteboom I, de Vogel N. (1983) Induction of sister-chromatid exchanges (SCEs) and chromosome aberrations by mitomycin C and methyl methanesulphonate in Chinese hamster ovary cells. *Mutat. Res.* **121:** 211–223.

Natarajan AT, Rotteveel AHM, van Pieterson J, Schliermann MG. (1986) Influence of incorporated 5-bromodeoxyuridine on the frequencies of spontaneous and induced sister-chromatid exchanges, detected by immunological methods. *Mutat. Res.* **163:** 51–55.

Natarajan AT, Vyas RC, Darroudi F, Vermeulen S. (1992) Frequencies of X-ray induced chromosome translocations in human peripheral lymphocytes as detected by *in situ* hybridisation using chromosome specific libraries. *Int. J. Radiat. Biol.* **61:** 199–203.

Nowinski GP, Van Dyke DL, Tilley BC, Jacobsen G, Babu VR, Worsham MT, Wilson GN, Weiss L. (1990) The frequency of aneuploidy in cultured lymphocytes is correlated with age and gender but not with reproductive history. *Am. J. Hum. Genet.* **46:** 1101–1111.

Ohtaki K, Sposto R, Kodama Y, Nakano M, Awa AA. (1994) Aneuploidy in somatic cells of *in utero* exposed A-bomb survivors in Hiroshima. *Mutat. Res.* **316:** 49–58.

Önfelt A. (1987) Spindle disturbances in mammalian cells. III: Toxicity, C-mitosis and aneuploidy with 22 different compounds. Specific and unspecific mechanisms. *Mutat. Res.* **182:** 135–154.

Oshimura M, Barrett JC. (1986) Chemically induced aneuploidy in mammalian cells: mechanisms and biological significance in cancer. *Environ. Mol. Mutagen.* **8:** 129–159.

Pandita TK, Gregoire V, Dhingra K, Hittelman WN. (1994) Effect of chromosome size on aberration levels caused by gamma radiation as detected by fluorescence *in situ* hybridisation. *Cytogenet. Cell Genet.* **67:** 94–101.

Parry EM, Danford N, Parry JM. (1982) Differential staining of chromosomes and spindle and its use as an assay for determining the effect of diethylstilboestrol on cultured mammalian cells. *Mutat. Res.* **105:** 243–252.

Parry EM, Henderson L, Mackay JM. (1995) Procedures for the detection of chemically induced aneuploidy; recommendations of a UK Environmental Mutagen Society Working Group. *Mutagenesis* **10:** 1–14.

Parry JM, James S, Lynch A, Warr T, Parry EM. (1988) The development and validation of assays for the detection and assessment of aneugenic chemicals. *Genome* **30:** 252–262.

Parry JM, Parry EM. (1989) Induced chromosome aneuploidy: its role in the assessment of the genetic toxicology of environmental chemicals. In: *New Trends in Genetic Risk Assessment* (eds G Jolles, A Cordier). Academic Press, London, pp. 261–296.

Perry PE, Wolff S. (1974) New Giemsa method for the differential staining of sister chromatids. *Nature* **251:** 156–158.

Perry PE, Thomson EJ. (1984) The methodology of sister chromatid exchanges. In: *Handbook of Mutagenicity Test Procedures* (eds BJ Kilbey, M Legator, W Nichols, C Ramel). Elsevier, Amsterdam, pp. 495–529.

Pincu M, Bass D, Norman A. (1984) An improved micronuclear assay in lymphocytes. *Mutat. Res.* **139:** 61–65.

Pinkel D, Thompson LH, Gray JW, Vanderlaan M. (1985) Measurement of sister-chromatid exchanges at very low bromodeoxyuridine substitution levels using monoclonal antibody in Chinese hamster ovary cells. *Cancer Res.* **45:** 5795–5798.

Pinkel D, Landegent J, Collins C, Fuscoe J, Segraves R, Lucas J, Gray JW. (1988) Fluorescence in situ hybridisation with human chromosome-specific libraries: detection of trisomy 21 and translocations of chromosome 4. *Proc. Natl Acad. Sci. USA* **85:** 9138–9142.

Ried T, Baldini A, Rand TC, Ward DC. (1992) Simultaneous visualisation of seven different DNA probes by *in situ* hybrization using combinatorial fluorescence and digital imaging microscopy. *Proc. Natl Acad. Sci. USA* **89:** 1388–1392.

Savage JRK. (1976) Annotation: classification and relationships of induced chromosome structural change. *J. Med. Genet.* **13:** 103–122.

Savage JRK. (1983) Some practical notes on chromosome aberrations. *Clin. Cytogenet. Bull.* **1:** 64–76.

Savage JRK, Simpson P. (1994) FISH painting patterns resulting from complex exchanges. *Mutat. Res.* **312:** 51–60.

Scott D, Danford ND, Dean BJ, Kirkland DJ. (1990) Metaphase chromosome aberration assays *in vitro*. In: *Basic Mutagenicity Tests: UKEMS Recommended Procedures* (ed. DJ Kirkland). Cambridge University Press, Cambridge, pp. 62–86.

Tenchini ML, Mottura A, Velicogna M, Passima M, Rainaldi G, de Carli L (1983) Double Y as an indicator in a test of mitotic non-disjunction in cultured human lymphocytes. *Mutat. Res.* **121:** 139–146.

Thomson EJ, Perry PE. (1988) The identification of micronucleated chromosomes: a possible assay for aneuploidy. *Mutagenesis* **3:** 415–418.

Tjio JH, Levan A. (1956) The chromosome number of man. *Hereditas* **42:** 1–6.

Tucker JD, Morgan WF, Awa AA, Bauchinger M, Blakey D, Cornforth MN, Littlefiejd LG, Natarajan AT, Shasserre C. (1995) A proposed system for scoring structural aberrations detected by chromosome painting. *Cytogenet. Cell Genet.* **68:** 211–221.

Weier HUl, Zitzelsberger HF, Gray JW. (1991) Non-isotopic labelling of murine heterochromatin *in situ* by hybridisation with *in vitro* synthesized biotinylated mouse major satellite DNA. *Biotechniques* **10:** 498–505.

Wenger SL, Golden WL, Dennis SP, Steel MW (1984) Are the occasional aneuploid cells in peripheral blood cultures significant? *Am. J. Med. Genet.* **19:** 715–719.

Wenger SL. (1989) Nonmodal chromosome gain and loss in human fibroblast cultures. *Cytogenet. Cell Genet.* **52:** 201–210.

8

In vivo assays for mutagenicity

John A. Heddle

8.1 Introduction

Genetic toxicology, which has seen little change for many years, is now experiencing the beginnings of a series of rapid changes. The new methods of molecular biology are being used to provide greater insight into the mechanisms involved in both mutagenesis and carcinogenesis and to develop novel and versatile assays. These assays represent a new and needed opportunity for genetic toxicology to prove its usefulness, since it has been established that *in vitro* tests for genotoxicity are not predictive of rodent carcinogenicity (Tennant *et al.*, 1987). This chapter is concerned with general aspects of *in vivo* assays and assay protocols, and contains specific recommendations for their improvement and use. In particular, the importance of the biological mechanisms underlying the assays and carcinogenesis must be considered if the assays are to be used effectively: there is no one-to-one equivalence of cancer and mutation; not all mutations are the same, and the turnover rate of each tissue influences the optimal protocol for the detection of mutation in that tissue. Each of these points will be considered in turn.

In principle, *in vivo* assays should have many advantages over their *in vitro* counterparts. No review of *in vivo* assays could fail to emphasize these advantages, obvious though they are. The reasons for the failure of *in vitro* assays to predict rodent carcinogenicity are less obvious. Suggestions range from those that fault the carcinogenicity assays (Heddle, 1988), to criticism of the interpretation of the *in vitro* results, to criticism of the conduct of the tests. Many attempts have been made to improve upon the utility of *in vitro* tests in various ways, but there is no general acceptance of any of these proposals. Whatever the reasons for the failure of *in vitro* tests, the lack of correlation with rodent carcinogenicity has undoubtedly stimulated further interest in *in vivo* assays and, in the general rush to embrace these assays, their defects are infrequently mentioned. They have defects, nonetheless.

It has been a pervading faith of genetic toxicology that carcinogens (at least initiating carcinogens) are mutagenic and act through a mutational mechanism. Further, it has been accepted that the use of assays for genetic damage would be justified if this hypothesis proves to be correct, as genetic toxicologists expect it will. The discovery of oncogenes (Bishop, 1983; Weinberg, 1989) and tumour suppressor genes (Cavenee et al., 1986; Stanbridge, 1990; Zhu et al., 1992), together with the clonal nature of human cancers (Fialkow, 1977), provides experimental support verging on proof that the faith was and is justified. Indeed the work of Vogelstein and others indicates that multiple mutations of various kinds are important in carcinogenesis (Cho and Vogelstein, 1992; Fearon and Jones, 1992; Fearon and Vogelstein, 1990; Goyette et al., 1992). This does not mean that even in vivo assays are perfect or can easily be made so. The possibility exists, however, that in vivo assays for mutagenicity may prove to be better predictors of human cancer than rodent bioassays, whose utility is much more an article of faith than a demonstrated fact, and whose performance has left a great deal to be desired. It will be impossible for any assay to predict rodent carcinogenicity more reliably than the inherent variability of the rodent bioassay itself (Heddle, 1988).

In genetic toxicology and environmental mutagenesis, 'in vivo' means in mammals, unlike its meaning in other biological fields. Thus this chapter does not include references to tests in plants, insects, or even non-mammalian vertebrates. The critical aspect of 'in vivo' is the exposure of the cells to be analysed and any subsequent expression of the genetic damage, not the analysis itself, which is typically conducted on tissue samples rather than on the intact organism. The reason for this is clear: the advantages of the in vivo assays lie primarily in their ability to model the human situation closely. A culture dish has no digestive tract, kidney, or even liver, for which S-9 is a poor substitute. To capitalize on their relevance, the investigator must make every aspect of the in vivo assays conform as closely as possible to the human situation. Unfortunately, it is not feasible to model the human situation perfectly, which must lead to some caveats on any risk extrapolation. Test animals are not humans: they do not eat human diets, which may include antimutagens, comutagens, enzyme inducers, or other chemicals that can affect the mutagenic response. Furthermore, test animals do not harbour human bacterial flora and do not metabolize all chemicals just like humans. To the extent that these influence either the mutagenicity or carcinogenicity of chemicals, they influence the validity of the extrapolation from test result to human risk. To think that in vivo assays are perfect, or can be made so at present, is to invite serious error. Nevertheless, in vivo assays are obviously better models of the human situation than are in vitro assays. How good they are as predictors of human carcinogenesis, or even of rodent carcinogenesis, is not yet known. Some things are known and will be dealt with in a specific section on risk extrapolation. Some points for consideration are given in Table 8.1. One of the most important considerations for genetic toxicology is the different categories of mutation which are not always detectable by a single technique.

Table 8.1. Propositions that I hold to be true, but which are not self-evident

Proposition	Corollary
1. Chemicals will induce mutations in tissues in which they are not carcinogenic (because the mutation spectrum is inappropriate)	This may produce 'false positives' in a validation against a cancer bioassay
2. Genotoxic chemicals must induce mutations in the cells in which they are carcinogenic	The only satisfactory way to demonstrate that a chemical is a non-genotoxic carcinogen is to test for all types of mutations in the target cells
3. The spectrum of mutations detected must be compared to the spectrum of mutations that are carcinogenic in the target cells and other factors that influence mutagenicity and carcinogenicity taken into account	Mutagenic potency and carcinogenic potency will be only roughly correlated
4. Both chromosomal aberrations and intragenic mutations are involved in the carcinogenic process	Both chromosomal changes and intragenic changes must be measured
5. It will be possible to measure increases in mutation that are too small to be detected as an increase in carcinogenicity by the cancer bioassay	This will produce 'false positives' in a validation against the cancer bioassay
6. It will be difficult to establish the non-carcinogenicity of a chemical from its mutagenicity because of tissue specificity	It will be important to know which tissues are the potential target tissues so as to be able to concentrate the testing on those
7. It is discrimination between carcinogens and non-carcinogens that is wanted, not detection of carcinogens[b]. Some attempts to improve the assays by making them more sensitive will make them worse by reducing discrimination[c]	*In vivo* assays can be made too sensitive and rendered ineffective by taking the same approach that has been used *in vitro*, i.e. forcing the assay to give a positive for known potent carcinogens[a]
8. The greatest achievement of genetic toxicology will be to discover the causes of common human cancers, not of preventing cancer by testing xenobiotic compounds	

[a] In my opinion the addition of S-9 to the bacterial tests reduced their value rather than enhancing it. Undoubtedly more carcinogens are detected by the assays when S-9 is added than without it, but so are more non-carcinogens. The result is that discrimination is reduced.

[b] A simple rubber stamp saying CARCINOGEN! will detect all carcinogens, and is cheaper than any assay, but is useless because it fails to discriminate between carcinogens and non-carcinogens.

[c] There will be a temptation to increase the sensitivity of the *in vivo* tests by using mice defective in DNA repair, treated with sensitizing compounds, and by other means yet to be invented. These should be resisted until it can be shown that such changes merely enhance the sensitivity of the assay without changing the results qualitatively and without making innocuous compounds seem hazardous.

Factors that need to be matched include all of the important experimental variables (species, sex, target tissue, dose, dose rate, route of administration, etc.). There may still be other factors, such as differences between sites within the genome, locus-specific effects that will influence the comparison. For such reasons perfect correlations may be difficult to establish even when chemically induced mutations are the main factor determining the carcinogenicity.

8.2 The nature of mutations

Mutations are of two major kinds, chromosomal aberrations and specific locus mutations, which have traditionally been detected by different techniques. Numerical chromosomal changes are extremely important in inherited mutation but convincing evidence that they are induced by external agents at anything like environmental doses is lacking. The importance of numerical changes in carcinogenesis has been posited, given their frequent occurrence in cancerous cells and their potential for leading to loss of heterozygosity, but no experimental demonstration of this in animals exists. Hence assays that quantify numerical changes are rarely used and are not discussed here. Structural aberrations, in contrast, have been found to be important and to be inducible by environmental agents. Indeed, we have proposed that chromosomal rearrangements are responsible for some loss of heterozygosity in tumours (Halberstadt and Heddle, 1995) and correspond to what has been called mitotic recombination (Lasko and Cavanee, 1991). Nevertheless, our search for loss of heterozygosity induced by ethyl nitrosourea and X-rays at the *Dlb-1* locus failed to produce evidence of this (Vomiero and Heddle, unpublished observations). It has been argued on the basis of the cancer-proneness but hypomutability of the chromosome breakage syndromes that aberrations may be more important than intragenic mutations (Heddle, 1991; Heddle *et al.*, 1983). The dividing line between an aberration and a mutation is nebulous. At what point is a deletion big enough to be called an aberration: 1 bp, 1 kb, 100 kb, 1 Mb? In practice, the genetic changes detected are called whatever the assay is designed to detect and not necessarily what it actually detects. Two examples may illustrate the potential overlap of these events. First, it is possible to conceive of a deletion that would lie entirely within the gene for Duchenne muscular dystrophy, which is approximately 2 Mb long (Worton and Thompson, 1988), and still be large enough to produce a detectable micronucleus (Heddle and Carrano, 1977), albeit a small one. Secondly, a genetic alteration involving a chromosome exchange with one breakpoint within the *hprt* locus would be identified as a translocation by chromosome painting (Pinkel *et al.*, 1988) and a mutant by thioguanine resistance (Albertini *et al.*, 1990).

8.3 The importance of neutral mutations

Robust and reliable protocols must be designed with a knowledge of the nature of the mutations being studied and the biology of the cell populations in which they are being assayed. Neutral mutations are those that are neither detrimental nor advantageous to the cells that contain them. The acid test for neutrality is whether or not the frequency of the cells containing the mutations changes after all of the mutations have been induced. It has recently been suggested that the new transgenic assays for somatic mutation in mice do assay neutral mutations, and strong evidence for this neutrality has been provided in the small intestine where the cells are reproducing rapidly and selection would have been readily detected had it existed (Tao *et al.*, 1993a). The advantage of such mutations is that their frequency is a measure of the number that were induced, not of any secondary selection for or against the mutant cell. If the events are neutral, then their frequency is the integral of the mutation rate over the life time of the cells and their ancestors back to the zygote. This means that effects of multiple treatments will be additive and that the assay may be made more sensitive by the use of chronic or subacute protocols.

Recent work with the peripheral blood micronucleus assay provides a prime example of what can be done. Micronuclei in many cells are lethal events, but in the mammalian erythrocyte there is no nucleus anyway. In the mouse there is virtually no selection against micronucleated erythrocytes, so the erythrocyte population simply accumulates micronuclei over the 30-day lifespan of the erythrocyte (MacGregor *et al.*, 1980). This has been put to dramatic use by Zetterberg and Grawé (1993) and Grawé *et al.* (1993) who were able to reduce the size of the dose to the individual erythroblasts and thus extend the radiation dose–response curve downwards to much lower doses by means of automated analysis. (The frequency of micronuclei in the whole erythrocyte population at 30 days will be no greater than that in the newly formed polychromatic erythrocytes on any of the days, so there is no increase in sensitivity over the traditional micronucleus assay on a per cell basis. But in the mature erythrocyte population, the frequency increases with daily treatments for the 30-day lifespan of the erythrocyte.) In principle the same enhancement of sensitivity through multiple treatments is also possible for other events, especially translocations and neutral mutations.

The new methods of chromosome painting (Pinkel *et al.*, 1988), which can now be applied to the mouse and rat, provide a novel means of detecting chromosomal aberrations (Breneman *et al.*, 1993; see also Chapter 14). Furthermore, many of these changes may be neutral and hence accumulate on chronic treatment (as discussed later for gene mutations). Hence the same protocol that works well for gene mutations *in vivo* (as discussed later) should work well for the detection of chromosomal aberrations. The events detected are more relevant than micronuclei, which are usually cell-lethal events, but the assay is probably not as sensitive, particularly when micronuclei are assessed by flow cytometry. The accumulation of micronuclei is limited to 30 days in the mouse because this is the lifetime of the cells, but the accumulation of translocations should not be limited. If so, longer treatments could reduce or eliminate the sensitivity gap. The mutations detected in the Muta™ Mouse (Gossen *et al.*, 1989) and the Big Blue™ (Kohler *et al.*, 1991) Mouse should also be neutral and accumulate during chronic exposures, since the mice do not need the loci involved (see Chapter 15). Hence, we now find ourselves in a favourable position in which the protocols that produce the greatest sensitivity are similar for three assays, all of which can be performed on a single mouse.

8.4 Treatment protocols

8.4.1 Route of administration

Since the main advantage of *in vivo* assays is their relevance to human exposure, it is foolish to use anything other than the relevant route when testing any chemical. For some agents the route of administration alters the effect. This being so, only the route that corresponds to the human route should be used. This simple and obvious principle is often ignored, thereby reducing the relevance of the assay. The obvious corollary is that more than one route should be used if there is more than one route of human exposure, although this would not seem necessary if it is already known that the metabolism and pharmacokinetics are similar for both routes.

8.4.2 Number of treatments

One of the most important recommendations to make concerning protocols is that multiple or chronic treatment protocols be used. There is insufficient experience with multiple treatments to be certain of this recommendation, but I am confident it will prove correct. The lack of experience is particularly acute because of a failure to consider the time-after-sampling factor. Two papers have addressed the number-of-treatment issue recently and come to similar conclusions.

Shephard *et al.* (1994) have compared data from the literature on the results of acute treatments with their data from chronic treatments. They used this information to extrapolate to the minimum detectable dose and concluded that a large increase in sensitivity could be obtained by the use of the chronic treatments. They concluded that the *lac*I transgenic mouse system will only be as sensitive as a cancer bioassay if they are treated for about 8 months (no one has actually done this) and that the assay can be made about as sensitive as the conventional micronucleus assay. (But the micronucleus assay can be made more sensitive by flow cytometry.)

While the extrapolations of Shephard *et al.* (1994) are extreme and rely on data from different laboratories, we have come to similar conclusions on both theoretical and practical grounds. The practical grounds are that the effects of multiple treatments are additive for both potent mutagens like ethyl nitrosourea (Tao *et al.*, 1993a) and weak mutagens like methyl methane sulphonate and 1,2-dimethylhydrazine (Tao and Heddle, 1994; Tao *et al.*, 1993b). This led directly to the proposal that multiple treatments would be approximately as much more effective than single treatments as the ratio of the total doses that could be delivered. The 'approximately' is required because this depends upon the shape of the dose–response curve: the enhancement obtained by the use of the multiple treatment protocol will be reduced if the dose–response curve curves upwards. The theoretical grounds are that the mutations are neutral (i.e. they provide neither a selective advantage nor disadvantage). Hence the mutant frequency is simply the integral of the mutation rate as a function of time. Obviously this leads to the same conclusion: the more you treat, the more mutants will be present and the more sensitive the assay will be. A direct test has shown that this is the case in that a low daily dose was detectable when repeated for 30, 60 and 90 days (Shaver-Walker *et al.*, in preparation) in the small intestine and the colon.

The effects of chronic and subacute treatment may not be as simple as present knowledge suggests, nor may such treatments always be more relevant to human exposure. In some cases, human exposure may be more like a series of widely or randomly spaced subacute exposures than like a chronic or daily subacute exposure. To the extent that there are stem cell effects, and differences in metabolism and DNA repair, then these may affect the mutagenicity of the treatments. Probably such differences will only affect the quantitative mutagenicity of the treatments and the mutation spectrum, and not obviate mutagenicity altogether at treatment doses. It will require much better theoretical knowledge of DNA repair mechanisms or much more sensitive mutation assays to determine if this is also true at the normal levels of human exposure.

There are data suggesting that both increases and decreases in mutagenic potency will be observed in chronic protocols. In the initial work of Gossen *et al.* (1989) who examined the effect of expression time, the mutant frequency declined at the later times in some cases. This may reflect mutations induced in stem cells. Such a result would be expected if adaptation were to occur or if repair enzymes were to be induced (Samson and Cairns, 1977; Wolff, 1992), or if the stem cells were to be inherently resistant because they are mostly in G_0 phase. The results obtained in most multiple treatment protocols cannot be interpreted in this light because the sampling times after treatment were inadeq'uate to permit complete expression of the mutations and thus the different treatments had different effective expression times, which could give any result.

There is good evidence that chronic treatments can increase the mutagenic potency of some chemicals, as well as decrease the potency of others. Shaver-Walker *et al.* (in preparation) have found enhanced mutagenicity of ethyl nitrosourea at the *Dlb-1* locus when a daily protocol was used for long enough. Initially there is a decrease in the mutagenicity of the chemical: the same dose given over 30 days is less effective than when given acutely. But over longer time periods the mutagenic efficiency increases until it is higher than that of the acute treatment! The generality of the result remains to be determined. The proposed explanation is that the stem cells are ordinarily quiescent and thereby resistant to chemical mutagenesis. Only those stem cells that are in S phase at the time of treatment or soon afterwards will be mutated at a high rate; the quiescent stem cells will have sufficient time to repair most of the lesions by a less error-prone mechanism before they enter S and so will have fewer mutations. Hence, acute treatments would be far less effective than they could be if all the cells were actively dividing. Daily treatments would increase the proportion of the sensitive cycling stem cells because the treatments induce not only mutation but also cell death. The stem cells, which maintain their numbers and respond to any deficiency in their numbers or those of their differentiated progeny, must therefore divide more often to compensate for the cell death that has been induced. In consequence, there will be more cycling stem cells and the sensitivity to mutation will be increased. Such a mechanism would be expected to operate in any cell population with stem cells, for any agent that mutates cycling cells more effectively than non-cycling cells. It is the stem cells that are important for carcinogenesis.

8.5 Sampling times

Inadequate attention has been paid to expression time for mutation *in vivo*. The biological basis of expression time is complex and is specific to the cell type and the locus being studied. 'Expression' *in vivo* involves the whole process of mutagenesis: the pharmacokinetics, DNA repair, expression of the gene, and tissue turnover. Since the first three are relatively rapid events, the last dominates the change of mutation frequency as a function of time, and thus the suitable choice of sampling times. The appropriate sampling time to maximize the sensitivity of the assay can thus be predicted from the biology of the tissue: short times will be adequate for epithelial cell populations and other tissues in which cells are turning over rapidly. Equally, such tissues will be the most sensitive to most chemical mutagens since most are S-active agents. There are exceptions, such as bleomycin, which is X-ray-like as judged by clastogenicity, but they are rare.

Tissues which turn over slowly, such as liver, will have long expression times. Indeed, it would seem likely that the liver could be made a much more sensitive target for mutation if partial hepatectomy before treatment were incorporated into the assay as it has been in some experimental cancer models. While this may make the assays less relevant to some human exposure, it may make them more relevant to the exposures of those populations in which liver cancer is relatively common and liver cell damage by other agents is an important factor so the liver is proliferating anyway. So far there appear to be no data that violate the expectation that the peak mutant frequency will occur much later in tissues that turn over slowly than in tissues that turn over rapidly. It must be remembered that tissues are typically composed of a variety of cell types and that the mutant frequency may be dominated by one or another of these. Ultimately it is the stem cells that are important but they are typically a small fraction of the total cells. Only after the tissue has turned over is the mutation rate of the stem cells revealed. Thus, investigation of the expression time in the tissue of interest is vital for the design of the proper protocol for testing that tissue. Any test can be made negative by sampling too soon after treatment, so the significance and reality of a negative result depend upon the choice of the proper sampling time.

8.6 Statistics

The appropriate statistical method to use for analysing mutant frequency at the transgenic loci is still under debate. Unfortunately that debate has not been properly informed by the knowledge that exists about the basic biology involved nor by much of the data that have been obtained to date. The biology of development dictates that there will be jackpots of mutations. These are clones of mutant cells that represent the progeny of a stem cell mutated early in development. The ultimate jackpot arises from a mutation that exists in the sperm or egg, and this has been observed. The mouse with this jackpot had been treated with mitomycin C. The mutant frequency observed was about 12 500 x 10^{-5}. Since all tissues examined showed the same mutant frequency, and the frequency corresponds to one mutation in one of the 80 copies of the transgene present, the origin was clearly an inherited spontaneous mutation in one of the 80 copies of the *lacZ* gene, not the result of the treatment with mitomycin C. Smaller jackpots are readily observed in other assays (e.g. Heddle *et al.*, 1992). Indeed most of the specific locus assays with endogenous loci rely on the formation of a jackpot to make the mutation evident. For this reason several assays require the treatment of the mouse while *in utero* so that the embryonic stem cells would be hit early enough to result in a mutant clone large enough to be visible (Russell, 1983; Stephenson and Searle, 1986; Winton *et al.*, 1988). If such a developmental jackpot should occur in a stem cell of a treated animal, it could make the treatment seem mutagenic as judged by the mean, even if it is not. So any statistical methods must take this into account: there will be spontaneous jackpots at some frequency in both the control and treated groups. The accumulated experience with these assays indicates that these will be fairly rare, but they will occur, nonetheless.

The second factor that has not been considered properly is sampling time. It seems highly likely that the variability observed among mice will depend upon the sampling time used. At early times, which have been typical so far, the variability will probably be greater than at later times when the mutant frequency stabilizes. I know of no analysis that has addressed this point, but it should be considered. It is also possible that after treatments or sampling times that lead to the measurement of mutations in a small number of stem cells, the variability will be increased by the reduced number of independent loci being examined. In this the transgenic models have a significant advantage over native loci, since each cell contains multiple target loci. On the other hand, in some assays of endogenous loci, mutant clones are recognizable as such and are only counted once (e.g the *Dlb-1* assay, where a mutant ribbon or patch is counted once regardless of size).

If the number of mutable loci examined is large, then it is probable that Poisson statistics will be appropriate at low mutant numbers and normal statistics at high mutant numbers, provided provision is made for outliers representing developmental jackpots.

8.7 Risk extrapolation

The qualitative and quantitative extrapolation of data from *in vivo* mutation assays to cancer rates in experimental animals and people is complicated by at least the following factors: the difference between mutation rate and mutant frequency, the spectrum of mutations induced and the biology of stem cells. The obvious factors of species specificity and the influence of other factors such as other mutagens, promoters, antimutagens and other environmental factors need only to be mentioned briefly. Among these, the diet of the animals and their intestinal flora may be particularly important. Inherent differences are also an obvious concern, differences in P450 metabolizing enzymes and DNA repair enzymes being of particular interest. It may be possible in the future to construct strains of mice and rats with human genes for these and other functions that are important. In the meantime, we depend on the undoubted general similarities of anatomy, physiology, and metabolism among mammals to justify *in vivo* assays. We should be ready, however, to reject extrapolations to humans from animals where these assumptions can be shown to be incorrect. For example, to condemn a chemical merely because it is mutagenic in mice would be inappropriate if it were not metabolized to the mutagenic form in man.

8.7.1 Mutation rate vs. mutant frequency

The spontaneous *mutant frequency* at neutral loci such as the *lac*I and *lac*Z transgenes represents the integral of the *mutation rate* from conception to the time of sampling. On the other hand, the mutant frequency induced after treatment is typically that induced over a short period of time (plus the spontaneous, of course). Hence, a treatment that doubles the *mutation rate* will not double the *mutant frequency* but may increase it only slightly. Conversely, a treatment that doubles the mutant frequency must be increasing the mutation rate many fold. The magnitude of this factor may be estimated in a reasonable manner by comparing the increase in mutant frequency induced by a treatment with the increase that occurs with time spontaneously. It has been shown that this increase with age is measurable in the spleen (Lee *et al.*, 1994) and the small intestine and so, probably, in other tissues as well. If risk estimations are to be made by the doubling dose method, it is the doubling of this rate that is relevant, subject to the considerations given below.

8.7.2 Mutation spectrum

The mutation spectrum induced by chemical carcinogens differs in many cases from the spontaneous spectrum. Whether this will be true of all agents is unclear, but it is likely that at least some agents will mimic the spontaneous spectrum by chance or because the spontaneous spectrum arises from a similar set of DNA lesions: the contribution of replication errors and spontaneous deaminations, etc., to the spontaneous spectrum is unknown and may even differ for different tissues. Oxidative metabolism is a prime suspect as the source of many spontaneous mutations and metabolic activity varies markedly with cell type.

The nature of the mutation spectrum is extremely important for risk extrapolation. It is clear that there are different kinds of mutation involved in carcinogenesis, from highly specific intragenic mutations and chromosomal aberrations, to rather non-specific chromosomal deletions and rearrangements. It also seems that the cancerous state can be reached by more than one pathway in a particular cancer type, for cancers of a specific type are characterized by several sets of mutations and rarely by a unique set. Studies with transgenic animals that inherit one of the oncogenes already mutated show that they are at increased risk of cancer. While this has not been shown for all oncogenes, it is probable that this result is general and that an increased rate of mutation at one locus will increase the cancer rate even if the other loci are unaffected. Loci may be unaffected because the nature of the mutations being induced is not that required for activation.

The consequence of these considerations for risk extrapolation is that the proportionality between mutagenicity and carcinogenicity may depend upon the spectrum and upon the target tissue. If the mutations being produced in the liver are not those that are important for the development of liver cancer, then the risk of liver cancer will not be increased no matter how mutagenic the agent may be in the assay. Equally, if the agent produces all of the mutational types that are important in the development of liver cancer, then it will be far more carcinogenic than another agent, equally mutagenic in the assays that produces only one of the required mutational types. In addition, those mutagens that produce a lot of cell killing may be far more dangerous for a given level of mutagenicity when this is measured under acute protocols. It is to be hoped that chronic protocols, which should reflect the increased sensitivity to cell killing, will reflect this factor. There may, of course, be an additional aspect to cell killing, namely the stimulation of a mutated cell to divide and thereby to become cancerous.

8.7.3 Dose

Finally, it would be inappropriate not to mention dose extrapolations. Even if the use of chronic treatment protocols should bring the sensitivity of the assays for gene mutations to the level of animal carcinogenicity studies, the doses used will still be higher than common human exposures and vastly higher than most. Extrapolation to low doses is therefore necessary and the nature of the dose–response curves becomes vital. Shaver-Walker *et al.* (in preparation) have shown that the use of subacute exposures permits the extension of the dose–response relationship to much lower doses than could ordinarily be used for measurement. The maximum extension, however, falls far short of that required for a knowledge of the dose–response curve in the range of human exposure. This does not mean that the assays are not useful. On the contrary,

151

they are useful, for the mutation rate induced can be compared with the spontaneous mutation *rate* and the carcinogenicity estimated (taking the mutation spectrum and other caveats into account). By this means, it should be possible to conclude with a satisfactory degree of confidence, that a given exposure level is inconsequential. No assays in species other than man can do better than this with the current level of knowledge of human and animal biology.

8.8 Future assays

Much of this chapter has dealt with gene mutation assays, especially the transgenic mouse assays. It is likely that they represent a transient phase in genetic toxicology. PCR-based assays for mutated oncogenes and tumour suppressor genes are now being widely used, and other methods for the detection and quantification of mutations are being studied. In the not-too-distant future we can expect cheaper methods that will detect mutations of neutral host loci in the more traditional toxicological animals (rat, rabbit, dog and monkey) and in man, possibly even the homologous loci in all species. Detection of the relevant mutations will be possible and the methods will reveal the mutation spectrum, or at least the part of interest, automatically. The lessons we are learning now will influence the design and interpretation of future studies.

Acknowledgements

It is a pleasure to acknowledge the hospitality of the members of the Biology and Biotechnology Program at the Lawrence Livermore National Laboratory, especially J.D. Tucker and J. Felton, where this chapter was written. I have drawn heavily upon the work of the members of my research group at York University to whom I am greatly in debt, especially K.S. Tao, C. Urlando and P. Shaver-Walker. The work in the laboratory has been supported by the National Science and Engineering Research Council, the National Cancer Institute of Canada, and the Ontario Ministry of the Environment, whose support was gratefully received.

References

Albertini RJ, Nicklas JA, O'Neill JP, Robison SH. (1990) *In vivo* somatic mutations in humans: measurement and analysis. *Annu. Rev. Genet.* **24:** 305–326.

Bishop JM. (1983) Cancer genes come of age. *Cell* **32:** 1018.

Breneman JW, Ramsey MJ, Lee DA, Eveleth GG, Minkler JL, Tucker JD. (1993) The development of chromosome-specific composite DNA probes for the mouse and their application to chromosome painting. *Chromosoma* **102:** 591–598.

Cavanee W, Koufos A, Hansen M. (1986) Recessive mutant genes predisposing to human cancer. *Mutat. Res.* **168:** 3–14.

Cho KR, Vogelstein B. (1992) Genetic alterations in the adenoma–carcinoma sequence. *Cancer* **70:** 1727–1731.

Fearon ER, Jones PA. (1992) Progressing toward a molecular description of colorectal cancer development. *FASEB J.* **6:** 2783–2790.

Fearon ER, Vogelstein B. (1990) A genetic model for colorectal tumorigenesis. *Cell* **61:** 759–767.

Fialkow PJ. (1977) Clonal origin and stem cell evolution of human tumors. In: *Genetics of Human Cancer* (eds Mulvihill, Miller and Fraumeni). Raven Press, New York, pp. 439–453.

Gossen JA, de Leeuw WJF, Tan CHT, Lohman PHM, Berends F, Knook DL, Zwarthoff EC, Vijg J. (1989) Efficient rescue of integrated shuttle vectors from transgenic mice: a model for studying mutations *in vivo*. *Proc. Natl Acad. Sci. USA* **86:** 7981–7985.

Goyette MC, Cho K, Fasching CL, Levy DB, Kinzler KW, Paraskeva C, Vogelstein B, Stanbridge EJ. (1992) Progression of colorectal-cancer is associated with multiple tumor suppressor gene defects but inhibition of tumorigenicity is accomplished by correction of any single defect via chromosome transfer. *Mol. Cell. Biol.* **12:** 1387–1395.

Grawé J, Zetterberg G, Amneus H. (1993) Effects of extended low dose rate exposure to Cs^{137} detected by flow cytometric enumeration of micronucleated erythrocytes in mouse peripheral blood. *Int. J. Rad. Biol.* **63:** 339–347.

Halberstadt J, Heddle JA. (1995) Promotion: a new hypothesis. *Mutat. Res.,* in press.

Heddle JA. (1988) Prediction of chemical carcinogenicity from in vitro genetic toxicity: how valid is the validation. *Mutagenesis* **3:** 287–291.

Heddle JA. (1991) Implications for genetic toxicology of the chromosomal breakage syndromes. *Mutat. Res.* **247:** 221–229.

Heddle JA, Carrano AV. (1977) The DNA content of micronuclei induced in mouse bone marrow by X-irradiation: evidence that micronuclei arise from acentric chromosomal fragments. *Mutat. Res.* **44:** 63–69.

Heddle JA, Krepinsky AB, Marshall RR. (1983) Cellular sensitivity to mutagens and carcinogens in the chromosome breakage and other cancer prone syndromes. In: *Chromosome Mutation and Neoplasia* (ed. J. German). Alan R Liss, New York, pp. 203–234.

Heddle JA, Gingerich JD, Urlando C, Pagura M, Shepson P, Khan MK. (1992) Detection of somatic mutations *in vivo* using the concurrent assay. I. Spontaneous frequencies in Chinese hamsters and F344 rats. *Mutat. Res.* **272:** 195–203.

Kohler SW, Provost GS, Fieck A, Kretz PL, Bullock WO, Putman DL, Sorge JA, Short JM. (1991) Spectra of spontaneous and induced mutations using a lambda ZAP®*lacI* shuttle vector in transgenic mice. *Proc. Natl Acad. Sci. USA* **88:** 7958–7962.

Lasko D, Cavanee W. (1991) Loss of constitutional heterozygosity in human cancer. *Annu. Rev. Genet.* **25:** 281–314.

Lee AT, DeSimoone C, Cerami A, Bucala R. (1994) Comparative analysis of DNA mutations in *lacI* transgenic mice with age. *FASEB J.* **8:** 545–550.

MacGregor JT, Wehr CM, Gould DH. (1980) Clastogen-induced micronuclei in peripheral blood erythrocytes: the basis of an improved micronucleus test. *Environ. Mutagen.* **2:** 509–514.

Pinkel D, Landegent J, Collins C, Fuscoe J, Seagraves R, Lucas J, Gray J. (1988) Fluorescence *in situ* hybridization with human chromosome-specific libraries: detection of trisomy 21 and translocations of chromosome 4. *Proc. Natl Acad. Sci. USA* **85:** 9138–9142.

Russell LB. (1983) The mouse spot test as a predictor of heritable genetic damage and other endpoints. In: *Chemical Mutagens* (ed. FJ de Serres). Plenum Press, New York, pp. 95–110.

Samson L, Cairns J. (1977) A new pathway for DNA repair. *Nature* **267:** 281–283.

Shephard SE, Lutz WK, Schlatter C. (1994) The *lacI* transgenic mouse mutagenicity assay: quantitative evaluation in comparison to tests for carcinogenicity and cytogenetic damage *in vivo*. *Mutat. Res.* **306:** 119–128.

Stanbridge EJ. (1990) Human tumor suppressor genes. *Annu. Rev. Genet.* **24:** 615–657.

Stephenson DA, Searle AG. (1986) The effects of X-rays on the induction of somatic mutations and growth in the retinal pigmented epithelium during development of the mouse eye. *Mutagenesis* **1:** 135–141.

Tao KS, Heddle JA. (1994) The accumulation and persistence of somatic mutations. *Mutagenesis* **9**: 187–191.

Tao KS, Urlando C, Heddle JA. (1993a) Comparison of somatic mutation in a transgenic versus host locus. *Proc. Natl Acad. Sci. USA* **90**: 10681–10685.

Tao KS, Urlando C, Heddle JA. (1993b) Mutagenicity of methyl methanesulphonate (MMS) *in vivo* in the *Dlb-1* native locus and the *lacI* transgene. *Environ. Mol. Mut.* **22**: 293–296.

Tennant RW, Margolin BH, Shelby MD, Zeiger E, Haseman JK, Spalding J, Caspary W, Resnick M, Stasiewicz S, Anderson B, Minor R. (1987) Prediction of chemical carcinogenicity in rodents from *in vitro* genetic toxicity assays. *Science* **236**: 933–941.

Weinberg RA. (1989) Oncogenes, antioncogenes, and the molecular bases of multistep carcinogenesis. *Cancer Res.* **49**: 3713–3721.

Winton DJ, Blount MA, Ponder BAJ. (1988) A clonal marker induced by mutation in mouse intestinal epithelium. *Nature* **333**: 463–466.

Wolff S. (1992) Is radiation all bad - the search for adaptation. *Rad. Res.* **131**: 117–123.

Worton RG, Thompson MW. (1988) Genetics of Duchenne muscular dystrophy. *Annu. Rev. Genet.* **22**: 601–629.

Zetterberg G, Grawé J. (1993) Flow cytometric analysis of micronucleus induction in mouse erythrocytes by gamma-irradiation at very low dose rates. *Int. J. Rad. Biol.* **64**: 555–564.

Zhu X, Dunn JM, Goddard AD, Squire JA, Becker A, Phillips RA, Gallie BL. (1992) Mechanism of loss of heterozygosity in retinoblastoma. *Cytogenet. Cell Genet.* **59**: 248–252.

In vivo germ cell mutagenesis assays

Jack B. Bishop and Kristine L. Witt

9.1 Introduction

Germ cell mutagenesis assays detect mutants resulting from mutations in germ cells and serve at least two purposes. First, the frequency with which mutants occur among the progeny of exposed individuals provides a means of assessing the impact of exposures to various environmental agents. In the field of environmental mutagenesis, this function is the one most often emphasized. Secondly, mutants identified by these assays are a valuable resource in the study of genetic diseases, and in understanding basic genome organization and gene structure–function relationships. The production of such mutants has been, and remains, an important function of germ cell mutagenesis assays (Russell, 1994).

There has been a rapid increase in our understanding of the genetic basis of human disease, resulting in part from information from the human genome projects. A substantial number of human diseases have been characterized at the molecular (DNA sequence) as well as the genetic level, with a collection of genetic alterations now associated with specific tumours and other diseases. This has led to a greater awareness of the inherited nature of a variety of conditions ranging from birth defects to cancer susceptibility and other late-onset syndromes.

Mutants recovered in germ cell mutagenesis experiments, including those with specific gene alterations as well as those with chromosomal rearrangements, are playing increasingly important roles in the identification and understanding of human diseases (Moyer *et al.*, 1994). For example, in addition to the obvious utility of disease models based upon the mutant phenotypes recovered in one of the dominant or recessive gene mutation tests to be discussed (Shedlovsky *et al.*, 1993), translocation mutants that are recovered in the heritable translocation assay (also discussed), and which give rise to malformed progeny, are useful in identifying the genes responsible for such malformations (Bultman *et al.*, 1992). Positional cloning techniques

155

similar to those used in the identification of human mutations (Collins *et al.*, 1989) are now being applied to such translocations to locate and isolate the gene(s) responsible for the malformation (Stubbs, 1992). Such mutants serve as valuable models in the study of human genetic disease and can provide the molecular probes needed for locating and cloning the affected human gene.

The more well-publicized purpose of germ cell mutation assays is their use in the identification, evaluation and characterization of environmental agents that may be potential human health hazards (Bishop and Shelby, 1990). Evaluation of effects on germ cells normally implies an *in vivo* evaluation in mammalian germ cells. While initial identification may involve *in vitro* (as well as *in vivo*) tests in somatic cells, there are presently no purely *in vitro* assays with cultured germ cells (Vogel, 1993). The only germ cell assay that might be considered *in vitro* is the recently developed assessment of micronuclei in short-term suborgan cultures of testicular tubules (Sjöblom *et al.*, 1994; Toppari *et al.*, 1986). Likewise, although there are a few non-mammalian *in vivo* germ cell assays such as the sex-linked recessive lethal and reciprocal translocation tests in *Drosophila*, most evaluations of germ cell mutagenicity are conducted in rodents, predominantly the mouse.

The assays described here are limited to *in vivo* mammalian germ cell assays and are listed in *Table 9.1*. Detailed descriptions are provided for the traditional assays for heritable damage: the dominant lethal, heritable translocation and morphological specific locus tests. Cytogenetic assays for chromosome damage, various dominant gene mutation tests, and the electrophoretic specific locus test are also included, as well as the new transgenic mouse assays and some of the other molecular assays currently under development. There is also brief mention of several assays for primary DNA damage in germ cells. A number of seldomly used tests, such as an immunological specific locus test (Bailey and Kohn, 1965), a sex-chromosome loss test (Russell, 1976), a test for recessive lethals at non-specific loci (Roderick, 1983), and an X-linked recessive lethal test (Lüning and Eiche, 1975), are not included in this chapter but have been previously described by Russell and Shelby (1985) or Adler (1983).

9.2 Gametogenesis

Germ cells represent a unique population of target cells and some knowledge of mammalian gametogenesis is essential to understanding and interpreting results from germ cell mutagenesis tests. The cell cycles of male and female germ cells are distinctly different (*Figure 9.1*). There is no oogonial stem cell population in the adult female. All of the gametes that an individual female possesses are present at birth and are arrested just before meiosis metaphase I as late primary oocytes. Their subsequent progression consists primarily of maturation processes involving various follicular cell developments and growth, and oocyte potentiation, but few noticeable gametogenic changes occur until ovulation stimulates an oocyte to progress through meiosis I and into the second meiotic division. Chromatin in the arrested primary oocyte is

Table 9.1. Mammalian germ cell tests with treated males[a]

Germ cell assay	Primary genetic lesion induced	Spermatogenic stages exposed	Cell type/phenotype evaluated	References[b]
Cytogenetic tests				
Spermatogonial metaphase analysis	Breakage, rearrangement	Spermatogonia	Spermatogonia	3,19
Meiosis metaphase II analysis	Mis-segregation (hyperhaploidy)	1° spermatocytes	2° spermatocytes	16,17,18,26
First-cleavage embryo analysis	Breakage, rearrangement	Spermatids through spermatozoa	One-celled embryo	3,4,31
Sperm aneuploidy test (FISH)	Mis-segregation	Spermatocytes	Testicular spermatids or epididymal sperm	20,33
Spermatid micronucleus assay	Breakage, rearrangement, mis-segregation	Spermatocytes	Spermatids	15,24,25,32
F_1 assays				
Dominant lethal assay	Breakage, rearrangement, mis-segregation	Spermatogonia through spermatozoa	Decidua, embryos and fetuses *in utero*	6,7,12,13
Heritable translocation test	Breakage, rearrangement (reciprocal translocations)	Spermatids through spermatozoa	F_1 fertility or 1° spermatocytes of F_1	1,2,5,10,11,27
Specific locus tests (a) morphological (b) biochemical	Point mutations, small and large deletions	Spermatogonia through spermatozoa	(a) Visible markers, or (b) Electrophoretic protein variants in F_1	(a) 21,22,23,28 (b) 14
Dominant cataract test	Point mutations, small and large deletions	Spermatogonia through spermatozoa	Cataracts in the F_1	8,9
Skeletal abnormalities	Point mutations, small and large deletions, breakage, rearrangements	Spermatogonia through spermatozoa	Skeletal abnormalities in F_1	29,30

[a] Many of these tests can also be performed with females as the treated sex. All the F_1 tests may be performed with treated females, as well as the first-cleavage embryo analysis.

[b] Numbers indicate the following references: 1, Adler (1978); 2, Adler (1980); 3, Adler (1984); 4, Adler and Brewen (1982); 5, Bishop and Kodell (1980); 6, Bishop *et al.* (1983); 7, Ehling *et al.* (1978); 8, Ehling (1983); 9, Ehling *et al.* (1982); 10, Generoso *et al.* (1980); 11, Generoso *et al.* (1981); 12, Generoso and Piegorsch (1993); 13, Green *et al.* (1985); 14, Johnson and Lewis (1981); 15, Kallio and Lahdetie (1993); 16, Leopardi *et al.* (1993); 17, Miller and Adler (1992); 18, Pacchierotti *et al.* (1983); 19, Preston *et al.* (1981); 20, Robbins *et al.* (1993); 21, Russell (1951); 22, Russell, 1954; 23, Russell and Russell (1959); 24, Russo and Levis, (1991); 25, Russo and Levis (1992) 26, Russo *et al.* (1984); 27, Searle (1974); 28, Searle (1975); 29, Selby (1983); 30, Selby and Selby (1977); 31, Tanaka (1981); 32, Tates (1992); 33, Wyrobek *et al.* (1994).

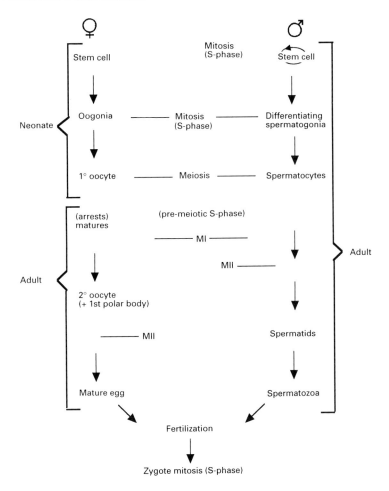

Figure 9.1. The periods of cell division and DNA replication in oogenesis and spermatogenesis. Adapted from Preston (1994).

diffuse and there is no period of DNA replication (other than repair synthesis) in the long interval between the formation of this primary oocyte and its fertilization. In contrast, male germ cells undergo numerous, rapid gametogenic changes and represent a dynamic target population for mutagens with respect to both repair capacities and chromatin structure.

Male germ cells develop through a series of stages which are similar in all mammals, including man. Stem cell spermatogonia provide a pool of slowly dividing cells which generate differentiating spermatogonia. In the mouse (*Figure 9.2*), the cell cycle time for the stem cells is approximately 6–8 days (Oakberg, 1956, 1960, 1975). The differentiating spermatogonia divide more rapidly, with a 26–31 hour cycling time (Monesi, 1962), undergoing six to eight mitotic divisions to produce the primary spermatocyte population. The subsequent meiotic prophase of primary spermatocytes lasts about 2 weeks and

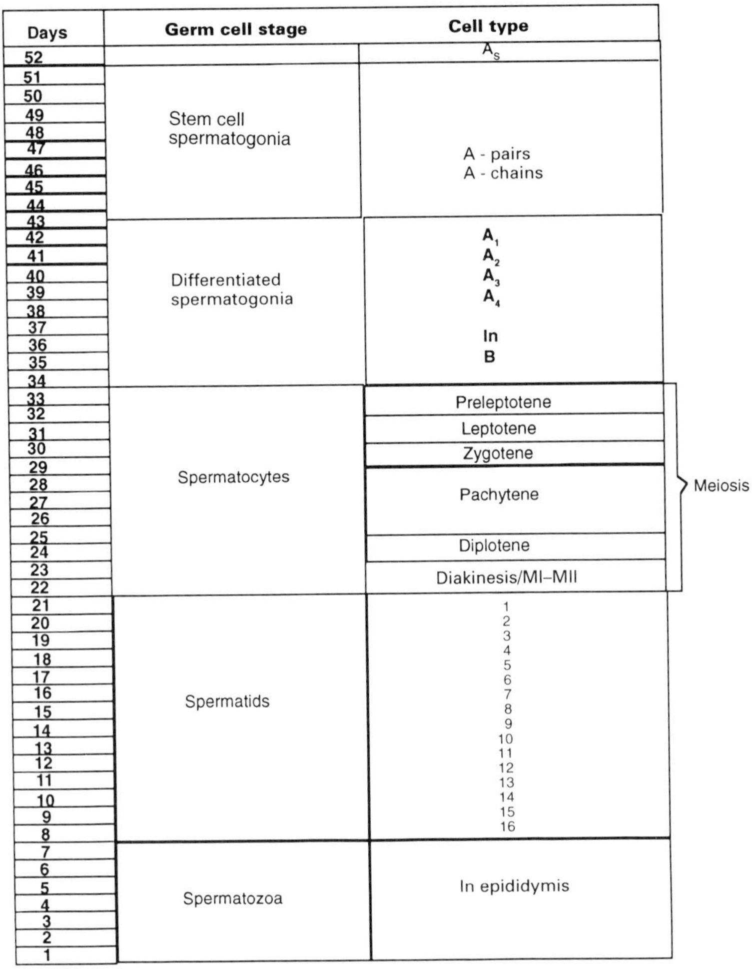

Days	Germ cell stage	Cell type	
52		A_s	
51	Stem cell spermatogonia		
50			
49			
48		A - pairs	
47		A - chains	
46			
45			
44			
43			
42	Differentiated spermatogonia	A_1	
41		A_2	
40		A_3	
39		A_4	
38			
37		In	
36		B	
35			
34			
33	Spermatocytes	Preleptotene	
32			
31		Leptotene	
30		Zygotene	
29			Meiosis
28		Pachytene	
27			
26			
25		Diplotene	
24			
23		Diakinesis/MI–MII	
22			
21	Spermatids	1	
20		2	
19		3	
18		4	
17		5	
16		6	
15		7	
14		8	
13		9	
12		10	
11		11	
10		12	
9		13	
8		14	
7		15	
6		16	
5	Spermatozoa	In epididymis	
4			
3			
2			
1			

Figure 9.2. Timetable for spermatogenesis in mice. Adapted from Preston (1994).

is a time where the DNA of each cell doubles, homologous chromosomes pair and recombination occurs, and the chromatin condenses in preparation for segregation of the homologues. First and second meiotic divisions occur within a relatively short span of less than 24 hours, creating haploid spermatids that, within the next 2 weeks, undergo major nuclear condensation and cytoplasmic reduction. During this latter period, repair capacity stops and chromosomal histones are replaced by protamines (Fawcett *et al.*, 1971; Handel, 1987). Spermatozoa are released from the seminiferous tubules into the epididymis where they mature for another week. According to Oakberg (1983), approximately 49 days are required for development of stem-cell spermatogonia into sperm in the ejaculate of mice. The duration of spermatogenesis in mice, rats and humans is 43, 65 and 90 days, respectively (Oakberg, 1983).

159

With regard to potential effects of chemical exposures, it is important to recognize specialized features of male reproductive biology such as the adluminal compartment of tubules, where exposure of germ cells (other than spermatogonia and spermatocytes) to blood borne substances is restricted by tight Sertoli cell junctions (Dym and Fawcett, 1970; Poorman-Allen *et al.*, 1990; Russell, 1990b) and the fact that the late-step spermatids and sperm lack cytoplasm and its associated capacities for metabolism and DNA repair. Chemical mutagens are characterized by their ability to induce mutations in specific germ cell stages (Ehling *et al.*, 1972) and knowledge of this stage specificity is essential to the evaluation of genetic hazards (Albanese, 1987). Female germ cells represent a similarly specialized target with the loosely packaged chromatin of the maturing oocyte being distinctly different from that of either somatic cells or spermatogenic cells. Because the various germ cells possess different features, it is important to know the type and features of the cell that is being treated as well as the cell that is being sampled. For a thorough review of mammalian gametogenesis, readers are referred to reviews by Sharpe (1994) on spermatogenesis, and Espey and Lipner (1994) and Crisp (1992) on oogenesis.

9.3 Cytogenetic assays

Cytogenetic assays provide the most direct and quickest means of identifying agents which induce chromosome aberrations in germ cells. In these assays, changes in chromosome structure or number are evaluated in either the germ cell exposed or its immediate descendants.

Spermatogonial metaphase cells, first and second meiotic metaphase cells, and first-cleavage metaphase embryonic cells have each been used in cytogenetic evaluations of germ cell damage. Spermatogonial metaphase tests have been used to detect chromosomal damage, particularly chromatid type aberrations, induced in differentiating spermatogonia and detected during the first post-treatment division (Adler, 1984). In this test, animals exposed to the test substance are treated with a spindle inhibitor before they are killed to accumulate cells in the metaphase stage of cell division. Cell preparations are treated with hypotonic fluid, dropped on to slides, air-dried, stained and then microscopically analysed.

Radiation-induced chromosome damage in spermatogonia has also been detected through analysis of the descendant cell population at diakinesis–metaphase I (Leonard, 1973). However, the aberrations observed at this stage are generally restricted to non-lethal reciprocal translocations because chromatid or chromosome deletions are cell-lethal events that will not survive the numerous cell divisions that occur during progression from spermatogonia to primary spermatocytes. Therefore, diakinesis–metaphase I analysis does not reflect the total damage induced in spermatogonia and it has not proved particularly useful in the evaluation of most chemical mutagens (Leonard and Linden, 1972). Analysis of first or second meiotic division is useful for detecting cytogenetic damage induced in spermatocytes, but exposures must involve either preleptotene cells or chemicals whose clastogenic action is independent of DNA synthesis (Russo and Levis, 1992). Evaluation of chromosome aberrations induced in post-meiotic spermatid and spermatozoa, as well as in earlier stages, can also be accomplished by analysis of embryos at first-cleavage metaphase division (Brewen *et al.*, 1975; Pacchierotti *et al.*, 1994).

Examination of synaptonemal complexes (SC) of chromosomes at diplotene–diakinesis constitutes a special meiotic analysis examining damage in primary spermatocytes (Allen *et al.*, 1987; Moses *et al.*, 1985; Poorman-Allen *et al.*, 1990). Theoretically, SC abnormalities could play a mechanistic role in aneuploidy and represent potentially transmissible structural and numerical chromosome defects, but the biological significance of SC damage in terms of heritable genetic effects has yet to be determined (Backer *et al.*, 1988).

Aneuploidy in male and female germ cells has been studied by analysis of second meiotic divisions of oocytes and spermatocytes as well as analysis of first-cleavage (or subsequent) mitotic metaphases of zygotes (or developing embryos). Second meiotic metaphase analyses usually employ special centromeric staining procedures to aid in identifying numbers of chromosomes (Leopardi *et al.*, 1993; Russo *et al.*, 1984). Second meiotic and first-cleavage metaphase analyses are particularly useful in identifying agents which induce aneuploidy in female germ cells (Aardema *et al.*, 1992; Generoso *et al.*, 1989; Mailhes and Marchetti, 1994; Mailhes *et al.*, 1988, 1990; Yuan and Mailhes, 1987).

Analysis of round spermatids for micronuclei (MN) is another means of assessing chromosome damage induced in premeiotic stages of spermatogenesis (Lahdetie, 1983; Russo and Levis, 1992; Tates, 1992). The presence of MN is thought to represent whole chromosomes or fragments of chromosome and therefore indicate numerical or structural chromosome damage. However, the exact relationship of the MN observed in spermatids to that of numerical and/or structural chromosome aberrations observed in meiotic or first-cleavage metaphase analyses is not yet clear in terms of aetiology or health impact.

9.4 F₁ assays

The three *in vivo* mammalian germ cell mutagenesis assays most often used are the dominant lethal assay, the heritable translocation assay and the morphological specific locus test. Each of these assays is discussed in terms of the phenotypes evaluated, the spectrum of mutant genotypes which may be assessed, general parameters for a valid test, and some of the major advantages and benefits, as well as limitations and disadvantages, in its use.

9.4.1 Dominant lethal mutation assay

The rodent dominant lethal assay, at least for male germ cells, is perhaps the simplest and least expensive of the *in vivo* mammalian germ cell mutagenesis assays to perform. Thus, it is the assay most often used in identifying hazardous agents and in determining the most sensitive germ cell stage(s). More chemicals have been tested for their ability to induce dominant lethality than any other germ cell effect. All of the chemicals to date which have been shown to be positive in the dominant lethal test are effective in post-meiotic stages. Thus, it has been suggested that it is only necessary to examine post-meiotic stages when conducting dominant lethal assays (Zeiger, 1994). However, in addition to providing information on the genetic toxicity of agents to the various germ cell stages, a well-conducted dominant lethal test that involves examination of all stages of spermatogenesis can provide a valuable assessment of overall reproductive effects (Generoso, 1994).

Most rodent dominant lethal assays have been conducted with mice but rats, hamsters and rabbits have also been used. The preference for mice probably arises more from economic factors associated with their maintenance than to their greater sensitivity in detecting chemical mutagens. In fact, there are examples of chemicals which induce dominant lethals in male rats and not in male mice, but no examples of the converse (Generoso et al., 1985; Teramoto and Shirasu, 1989).

In the male dominant lethal assay, treated males are caged with untreated females at various intervals after treatment. Each successive mating interval samples sperm that were exposed at progressively earlier stages of spermatogenesis (*Figure 9.3*). Mated females are killed at mid- to late-gestation (Gd 14–18) and their uterine contents evaluated for number of implantation sites, living fetuses, resorption moles, early embryonic deaths, and early- and late-fetal deaths (Lockhart et al., 1991). Resorption moles are deciduamata with no, or very little, visible embryonic tissue that result from early peri-implantation death of Gd 4–8 embryos. Early embryonic death around Gd 10–12 presents as a necrotic tissue mass sometimes having visible limb buds but no eye formation, while early fetal death around Gd 14–16 presents as whitish, fully formed fetuses having visible eyes and limbs with digits. Embryos which die before implantation, when they are between the two-cell and blastocyst stages, may fail to implant and are represented in *in utero* analyses only as reduced numbers of implantation sites. In addition to embryonic and fetal wastage, other informative parameters assessed in the dominant lethal assay include number of pregnant females, and sometimes, the incidence of malformed fetuses.

Statistical analyses can be performed on any of the above parameters. The most frequently used dominant lethal test statistic is mean percentage resorption per female (Generoso and Piegorsch, 1993; Lockhart et al., 1992) where the number of implants is the denominator and, therefore, the critical unit for establishing sample size for a valid test. A minimum of 300 implantations (i.e. about 20–30 females) should be evaluated per data point. Another commonly used measure of lethality is the dominant lethal index, which most often compares mean numbers of living fetuses per female and simply expresses treatment effect as a proportion of the control values (Ehling et al., 1972).

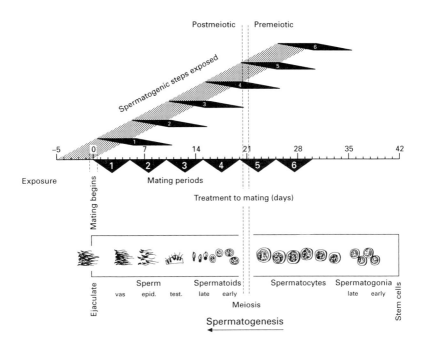

Figure 9.3. Illustration of the stages of spermatogenesis sampled in a male dominant lethal test involving 5-day exposures and six 5-day mating intervals. Mating period 1 samples sperm exposed as mature vas sperm back to late-step testicular spermatids, whereas mating period 6 samples sperm exposed as mid- to early-step spermatocytes and perhaps back to late spermatogonia. Additional mating periods would have sampled back into the spermatogonial stem cell compartment.

Although this proportion (which is generally converted to a percentage) incorporates a measure of both pre- and post-implantation loss, it cannot be subjected to any form of statistical analysis and obscures variations between individual animals. Numbers of females with one, two, three or more resorptions is another, less frequently used measure for expressing dominant lethal effects.

The incidence of resorptions following treatment of males with methyl methanesulphonate was found to correlate with that of chromatid aberrations observed cytologically in first-cleavage metaphase divisions (Brewen *et al.*, 1975). Thus, an increase in the incidence of resorptions in a male dominant lethal assay is generally assumed to indicate that the agent being tested has induced genetic damage in the germ cells that results in breakage of chromosomes. However, the correlation between resorptions and cytogenetic damage has not been established with a wide variety of chemicals. The need to establish this correlation is particularly critical for female dominant lethal studies where resorptions may result from various maternal toxicities instead of from induced genetic damage.

163

The dominant lethal assay is generally not as applicable to the study of induced genetic damage in female germ cells. Logistically, dominant lethal studies with females are more difficult and expensive compared with those with males because each treated female has relatively few germ cells to contribute to the requisite number of implantations per dose and time point. More importantly, however, very labour-intensive first-cleavage metaphase cytogenetic analyses or egg transfer experiments may be required to rule out confounding maternal toxicities. Furthermore, female germ cells were found to be more resistant than male germ cells to the dominant lethal effects of ionizing radiation and ethylnitrosourea (ENU; Generoso and Russell, 1969), and this response pattern has been assumed to hold for chemicals in general. Chemical reactions are, however, more varied than those of radiation, and a number of chemicals have now been identified that induce dominant lethal mutations in female germ cells but not in male germ cells (Katoh et al., 1990; Sudman et al., 1992).

9.4.2 Heritable translocation test

Translocations are chromosomal interchanges involving exchanges of non-homologous segments. If such an exchange occurs in a parental germ cell and is a balanced, reciprocal exchange, any progeny resulting from the union of this gamete with a normal gamete from the opposite-sex parent will be a translocation heterozygote. Translocation heterozygotes are generally of normal appearance but have reduced fertility because they produce duplication-deficient gametes as a result of aberrant multivalent meiotic chromosome pairing and segregation (*Figure 9.4*). This reduced fertility provides the basis for the detection of translocations (Adler, 1978, 1980; Bishop and Kodell, 1980; Generoso et al., 1980).

In mice, permanent semisterility is highly indicative of translocation heterozygosity, and translocation heterozygotes almost always have reduced fertility. In a study where 397 male mice, whose fathers were treated with an extremely clastogenic chemical, triethylenemelamine, were examined independently by both cytogenetic and fertility test methods, more than 95% of those with cytogenetically confirmed translocations had reduced fertility (Morris et al., 1988). Conversely, Adler (1978) reported that cytogenetic misclassification of a true translocation heterozygote should be negligible because fewer than 1% of heterozygotes have a multivalent frequency of 10% or lower (i.e. less than three multivalents in the 25 meiotic metaphases normally scored).

Methodology for a mammalian translocation assay using the mouse was first described by Snell (1935). Translocation heterozygotes were detected among the F_1 male progeny of parental males exposed to X-rays through breeding tests which identified individuals with reduced fertility. The first translocation test with a chemical mutagen, nitrogen mustard, gave inconclusive results (Auerbach and Falconer, 1949), and it was not until almost 10 years later that

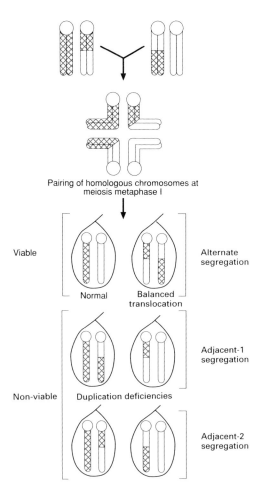

Pairing of homologous chromosomes at meiosis metaphase I

Viable

Normal — Balanced translocation

Alternate segregation

Adjacent-1 segregation

Non-viable — Duplication deficiencies

Adjacent-2 segregation

Figure 9.4. Production of viable and non-viable gametes by translocation heterozygotes. Reproduced from Bishop and Kodell (1980). © Wiley-Liss Inc. Reprinted by permission of Wiley-Liss Inc., a subsidiary of John Wiley & Sons Inc.

a chemical, triethylenemelamine, was shown to induce translocations (Cattanach, 1957, 1959). Subsequently, more than 50 chemicals have been shown to induce translocations in rodent germ cells (Adler 1980; Bishop and Kodell, 1980; Bishop and Shelby, 1990; Generoso *et al.*, 1980).

In a heritable translocation test, the F_1 male progeny from treated parents are screened for the presence of translocation heterozygotes. This is most often accomplished by a fertility test based either upon examination of death *in utero* and/or counting of live F_2 offspring (Generoso *et al.*, 1978). Males who have reduced fertility are presumed to be heterozygous translocation carriers. The presence of a translocation can be confirmed cytogenetically. There is also a purely cytogenetic test method (Adler, 1978) in which testing for fertility is by-passed and all F_1 male progeny are analysed cytogenetically. While this method may be more labour intensive and require more highly skilled personnel than the fertility test method, it requires fewer animals and may be preferred by laboratories with limited space for housing animals.

Completely sterile males are also detected in translocation tests, often comprising as much as one-half to one-third of the presumed translocation carriers. Sterility associated with translocations induced by X-irradiation and by ethylmethane sulphonate (EMS) has been reported to be related to the position of the breakpoint, with a disruption in meiosis more likely to occur when there is a break in the proximal or distal heterochromatic region of one of the pairs of chromosomes involved in the exchange (Cacheiro et al., 1974). The presence of multiple translocations might also produce disruptions of meiotic pairing and thus lead to blockages in meiosis that result in sterility (Bishop, unpublished results). Complete sterility, however, may also occur because of physiological factors and may not be associated with translocations. Thus, extra effort should be made to confirm cytogenetically all completely sterile males from a translocation test. Such confirmation may be complicated by the fact that the testes of sterile males are usually small and thus more difficult to process, and often lack meiotic stages for analysis. In such cases, it may be necessary to resort to analysis of mitotic metaphases for confirmation of a translocation.

The background frequency of de novo translocations is assumed to be very small (<0.1%). Required sample sizes are dependent upon the type of test protocol used, assumed spontaneous rate and desired power of the test (Bishop and Kodell, 1980) but confirmation of two or more translocation carriers among 300 progeny would indicate a significant effect (Generoso et al., 1980). A need for sample sizes substantially greater than 300 would make the translocation test too expensive for practical screening. In fact, the high cost involved, and the associated scarcity of laboratories with historical control data and adequate technical expertise to conduct translocation assays, are the greatest deterrents to its more wide-scale use.

Translocations do constitute a type of genetic damage that is known to contribute to the human disease burden. Thus data from translocation tests have been used in estimating genetic risk (Rhomberg et al., 1990). Furthermore, the test is not restrictive in terms of stocks, strains or sex of animals that can be used. Although the vast majority of translocation tests have been performed with male mice, Armenian hamsters (Lavappa, 1974) and rats (Medina et al., 1976) have also been used.

9.4.3 Morphological specific locus test

The morphological specific locus test was developed in the 1950s (Russell 1951, 1954; Russell and Russell, 1959; Searle 1975). It detects mutations at any of several specific loci (seven in Russell's system) for which there are known recessive genes affecting externally visible phenotypes (coat and eye colour, pigment distribution, external ear morphology). The treated animals, which are homozygous for the wild-type or dominant, non-mutant alleles at each of the specific loci, are mated to tester stock animals which are homozygous for the recessive mutant alleles at those same loci. Thus, the offspring produced from these matings will be heterozygous at each of the specific loci except when a mutation has been induced at one or more of the loci in a germ cell of the treated animal. An F_1 individual carrying the newly mutated gene opposite the recessive allele will express the homozygous recessive phenotype for that locus. The specific locus test can be used to test for induction of mutations in either male or female germ cells but is most often conducted with treated males.

The specific locus mutants detected in the morphological specific locus test have been shown to include genetic changes ranging from alteration of only a few bases to large megabase deletions and rearrangements (Rinchik *et al.*, 1990; Russell, 1983; Russell *et al.*, 1990). The type of lesion formed varies with the germ cell stage in which it is induced (Russell, 1990b; 1991). Mutations induced in post-meiotic male germ cells are primarily large lesions such as deletions that affect target genes and flanking regions whereas those recovered from spermatogonial stem cells tend to result from small lesions, primarily intragenic changes (Russell *et al.*, 1990). It has been shown recently, however, that X-ray pretreatment of spermatogonia followed by a second dose of X-rays can result in a greater proportion of large lesions among the mutants produced (Russell and Rinchik, 1993).

A cumulative 'historical control' is generally used for statistical analysis of morphological specific locus test data. For a detailed discussion of the sample sizes required for a specific locus test, see Russell *et al.* (1981).

9.4.4 Biochemical specific locus test

In this test, F_1 progeny from a cross between two inbred strains of mice (C57BL/6 and DBA) are compared to their parents with respect to the one-dimensional electrophoretic pattern for 32 proteins in tissue homogenates and blood. Differences from normal F_1 patterns may consist of changes in the position or number of bands. The electrophoretic analysis is capable of detecting genetic changes both in protein bands for which the parental strains differ and in bands for which they are alike. The tissues used are obtained by biopsy so that any mutant animals identified can be bred for further genetic analysis (Johnson and Lewis, 1981). A number of mouse models for human diseases have been found with this assay. For example, a mutant deficient in carbonic anhydrase-2 that was detected in an experiment with ENU (Lewis *et al.*, 1988) provides a model for carbonic anhydrase deficiency syndrome in humans. A second mouse model, which produces a haemoglobin disorder known in humans as β-thalassaemia, was detected among the control population (Skow *et al.*, 1983). Recently, this assay has been included in a 'combined endpoint assay' for detecting, in addition to electrophoretic mutations, morphological specific locus mutations, dominant visible mutations and cataract mutations (Lewis, 1991). The use of such a multiple-endpoint approach reduces the number of animals required by increasing the number of loci studied.

9.4.5 Other biochemical tests

There have been at least two other methods developed to detect mutants by measuring biochemical endpoints. Assays to detect altered enzyme activities were developed by Bishop and Feuers (1982) as well as by Charles and Pretsch (1983) but only a single chemical has been tested in each. In these assays, tissue homogenates are analysed using automated enzyme analysers. These tests have the potential to assess mutational effects upon a large number and wide variety of genes, including regulatory genes, but there are some difficulties associated with characterization and maintenance of mutants exhibiting relative, quantitative changes. An assay which uses two-dimensional gel electrophoresis to detect mutants with alterations in protein charge and size is also under development (Giometti *et al.*, 1994). This assay, like the one-dimensional assay of Johnson and Lewis (1981) detects both qualitative and quantitative protein changes, but the analysis is not confined to a limited, specific set of proteins. Therefore, like the altered enzyme activity assays (Bishop and Feuers 1982; Charles and Pretsch, 1983), it has the potential to detect a broader spectrum of mutational events.

9.4.6 Dominant mutation tests

There are several tests which serve to measure dominant gene mutations in mice based upon assessments of a variety of morphological parameters. These parameters include gross changes in external morphology [e.g. dominant visible mutation test (Lyon and Morris, 1969; Searle and Beechy, 1986), cataracts (Ehling, 1983; Ehling *et al.*, 1982), and skeletal changes (Selby 1983; Selby and Selby, 1977)]. These tests have direct relevance to human ill-health because they provide a measure of the frequency of phenotypic effects expressed immediately in the offspring of exposed individuals.

Assessments for dominant visibles includes a variety of miscellaneous phenotypes. Searle and Beechy (1986), in addition to observing pigmentation variants, some of which were determined to be mutants of the *Steel* and *Splotch* loci, also assessed growth retardation recorded as small phenotypes based upon visual inspection and weight (less than four-fifths normal weight looks smaller) at weaning, and impaired locomotion. Litter size reduction has also been suggested as a potential indicator of dominant damage (Selby, 1990). Many dominant mutations cause malformations that are indistinguishable from normal variation and probably cause no handicap, but they may be used to demonstrate a treatment effect.

A cataract is an opacity of the lens causing a reduction of visual function. Approximately 30 loci code for dominant cataract mutations compared with seven in the morphological specific locus test. The dominant cataract test, however, has exhibited greater experimental variability and, therefore, requires a larger investment of experimental resources to achieve the same degree of accuracy in determining germ cell mutation rate as the specific locus test (Favor *et al.*, 1988).

Skeletal mutations may cause changes in any of several parts of the skeleton and are visualized through alizarin staining of cleared skeletons. Many of the observed phenotypes closely resemble known human genetic disorders. Not all skeletal changes are expressed in each generation and, in this way, their transmission resembles that of so-called irregularly inherited disorders in man. Initial methods of identifying skeletal mutations utilized breeding of F_1s to test for heritability (Selby and Selby, 1977). Subsequently, non-breeding test methods, based upon comparing the frequency in treatment versus control of specific malformations having a high probability of resulting from mutations, have been developed for quicker scoring (Selby, 1983).

These dominant mutation tests are ideal for use in a multi-endpoint assay (Ehling, 1986; Lewis, 1991; Selby, 1990). When used in combination with the morphological specific locus test they substantially increase the number of loci evaluated and allow a systematic comparison of mutagenic results for different classes of mutations in the same animals.

9.5 Primary DNA damage

There are a number of germ cell tests which measure primary DNA damage in the form of adduct formation (Sega, 1974a, 1990), unscheduled DNA synthesis (UDS; Sega, 1974b; Sega and Sotomayor, 1982), DNA strand breakage detected by alkaline elution (Sega and Generoso, 1988), and sister-chromatid exchanges (SCE; Allen and Latt, 1976). Of these, DNA adduct appears to be the most critical relative to assessments of germ cell risks (Sobels, 1982). In these studies, the alkylating portion of the chemical used to treat the animals was labelled with ^{14}C or tritium. DNA is isolated from mature sperm obtained from the epididymis and/or specific spermatogenic cells isolated from the testis. The amount of radiolabel adducted to the total DNA and/or specific nucleotide components is then determined. Unfortunately, the number of chemicals for which such information exists is very limited, primarily because one generally needs radiolabelled compounds and knowledge of their metabolism to conduct such studies. However, a wider range of chemicals could, in principle, be used if the adducts formed by non-radioactive compounds are detected by a post-labelling method (see Chapter 18).

9.6 New molecular assays

9.6.1 Mutation tests with transgenic mice

A number of transgenic mouse models have recently been developed for measuring mutations *in vivo* in an attempt to fill the gap, relative to cost-effectiveness and complexity, between the identification of mutagens with short-term test systems and *in vivo* studies of heritable mutations using endogenous genes (Lewis, 1994). Transgenic animal models are based on the insertion into the genome of specific target genes that can be recovered from selected tissues and analysed for mutations (see Chapter 15). There are currently three transgenic systems in mice which have been used to examine induction of mutations in male germ cells. One uses the bacteriophage φX174 shuttle vector and detects reverse mutations at a nonsense codon in the phage *am-e* gene (Burkhart *et al.*, 1993; Burkhart and Malling, 1993; Malling and Burkhart, 1989, 1992). The other two utilize the bacteriophage λ shuttle vector carrying an *E. coli lac*I (Kohler *et al.*, 1990) or *lac*Z (Gossen *et al.*, 1989) target gene and detect forward mutations in those bacterial genes. In all three systems, the transgene is exposed to a mutagen while integrated into the animal's genome. Genomic DNA is then isolated from the tissue(s) of interest (testis or sperm, in the case of a germ cell assay), the shuttle vector is recovered and mutations in the target gene evaluated in *E. coli*.

Each of the above transgenic systems has been used to examine induction of mutations in male germ cells (Douglas *et al.*, 1993; Gossen *et al.*, 1989; Kohler *et al.*, 1991). However, most of these studies have been performed on mixtures of cells recovered from whole testis preparations or isolated seminiferous tubules rather than specific stages of spermatogenesis (Gossen *et al.*, 1989; Kohler *et al.*, 1991). There are serious questions regarding the ability of transgenic shuttle-vector systems to detect mutations at stages of spermatogenesis other than spermatogonia (Lewis, 1994) because most chemical mutagens have their greatest effect in post-meiotic stages where the lesions may not be fixed until they are in the egg and the large lesions most frequently produced in post-meiotic stages would not be detected in the transgenic assays. An international collaborative project is currently underway that will address some of these concerns regarding transgenic mouse assays with male germ cells (Ashby, 1995). The transgenic systems as they presently exist are not suitable for the study of mutations in exposed female germ cells because of the markedly fewer germ cells present for analysis.

9.6.2 Fluorescence in situ hybridization

Recently, a number of assays for chromosome damage have been developed that incorporate the use of various molecular biomarkers, such as chromosome-specific paints or repetitive DNA fluorescence probes, to enhance the versatility or efficiency of cytogenetic analyses (see also Chapter 14).

Methods have been developed for detecting aneuploidy in both human and mouse sperm using fluorescence *in situ* hybridization (FISH) with DNA probes specific for repeated genomic sequences on individual chromosomes (Robbins *et al.*, 1993; Wyrobek *et al.*, 1994, 1995a). A similar assay for rat sperm is under development (Wyrobek *et al.*, 1995b). These assays provide bridging biomarkers that are both phenotypically and genotypically similar, allowing more direct cross-species comparison of germ cell mutagenesis events.

These sperm FISH assays for aneuploidy are relatively easy to perform. In the case of humans, sperm is obtained by ejaculation. Mouse or rat sperm can be obtained from the epididymis or from the testis. When mature sperm from an ejaculate or the epididymis are used, slides containing air-dried smears of the sperm cells must first be swollen before they are hybridized. This swelling is accomplished with dithiothreitol and lithium salts. Hybridizations with digoxigenin and biotinylated probes are visualized immunologically with antidigoxigenin antibody conjugated with rhodamine or avidin D conjugated with FITC. Many of the chromosome-specific probes are now commercially available with label directly incorporated into the DNA, thus by-passing the need for immunological staining and, generally, giving much brighter images.

At present, validation of these sperm FISH assays for aneuploidy is still incomplete. However, in addition to providing bridging biomarkers across multiple species, they hold promise for supplementation and/or replacement of the relatively laborious hamster egg–human sperm assay and mouse meiotic metaphase II and first-cleavage metaphase analyses with an assay that evaluates 10 000 cells per exposed individual as opposed to a few hundred cells. As more probes are developed, and with the promise of future automation, these assays may help fill gaps in our knowledge of the aetiology of aneuploidy. Unfortunately, this particular assay is not applicable to female germ cells.

The use of chromosome painting probes, however, should have application to studies of chromosome damage induced in both female and male germ cells. Methods are currently being developed for application of FISH to the identification of translocations, dicentrics and aneuploidies in one-cell zygotes, as well as later stage embryos (Lowe *et al.*, 1995). These techniques promise to be particularly valuable for characterizing the induction of aneuploidy in female and male germ cells, and its persistence during early development. In addition, comparative genomic hybridization procedures (Kallioniemi *et al.*, 1992), in which differentially labelled test DNA (in this case, from early embryos) and normal reference DNA is hybridized simultaneously to normal chromosome spreads, are being evaluated for their ability to detect duplications and deficiencies in early embryos. These and other molecular techniques may improve our ability to assess induction of germ cell aneuploidy and other chromosomal changes.

Acknowledgements

The authors would like to thank the staffs of the NIEHS Library and Environmental Mutagen Information Center for providing reference materials for this chapter. We would also like to thank the staff of Image Associates for graphics support and our secretary, Ms Angie Holland, for reviewing the references and assistance in preparation of figures.

References

Aardema MJ, Mailhes JB, Preston RJ. (1992) Status of the mouse metaphase II oocyte assay for detecting chemicals which induce aneuploidy in germ cells. *Environ. Mol. Mutagen.* **19:** 1.

Adler I-D. (1978) The cytogenetic heritable translocation test. *Biolog. Zeuralblatt.* **97:** 441–451.

Adler I-D. (1980) New approaches to mutagenicity studies in animals for carcinogenic and mutagenic agents. I. Modification of the heritable translocation test. *Teratogen. Mutagen. Carcinogen.* **1:** 75–86.

Adler I-D. (1983) Comparison of types of chemically induced genetic changes in mammals: ICPEMC Working Paper 4/3. *Mutat. Res.* **115:** 293–321.

Adler I-D. (1984) Cytogenetic tests in mammals. In: *Mutagenicity Testing: a Practical Approach* (eds S Venitt, JM Parry). IRL Press, Oxford, pp. 275–306.

Adler I-D and Brewen JG. (1982) Effects of chemicals on chromosome-aberration production in male and female germ cells. In: *Chemical Mutagens: Principles and Methods for their Detection,* Vol. 7 (eds FJ de Serres, A Hollaender). Plenum Press, New York, pp. 1–35.

Albanese R. (1987) Mammalian male germ cell cytogenetics. *Mutagenesis* 2: 79–85.

Allen JW, Latt SA. (1976) *In vivo* BrdU-33258 Hoechst analysis of DNA replication kinetics and sister chromatid exchange formation in mouse somatic and meiotic cells. *Chromosoma* **58:** 325–340.

Allen JW, deWeese GK, Gibson JB, Poorman P, Moses MJ. (1987) Synaptonemal complex damage as a measure of chemical mutagen effects on mammalian germ cells. *Mutat. Res.* **190:** 19–24.

Ashby J. (1995) Transgenic germ cell mutation assays: a small collaborative study. *Environ. Mol. Mutagen.* **25:** 1–3.

Auerbach C, Falconer DS. (1949) A new mutant in the progeny of mice treated with nitrogen mustard. *Nature* **163:** 687–679.

Backer LC, Gibson JB, Moses MJ, Allen JW. (1988) Asynaptonemal complex damage in relation to meiotic chromosome aberrations after exposure of male mice to cyclophosphamide. *Mutat. Res.* **203:** 317–330.

Bailey DW, Kohn HI. (1965) Inherited histocompatibility changes in progeny of irradiated and unirradiated inbred mice. *Genet. Res.* **6:** 330–340.

Bishop JB, Feuers RJ. (1982) Development of a new biochemical mutation test in mice based upon measurement of enzyme activities. II. Test results with ethyl methanesulfonate (EMS). *Mutat. Res.* **95:** 273–285.

Bishop JB, Kodell RL. (1980) The heritable translocation assay: its relationship to assessment of genetic risks for future generations. *Teratogen. Carcinogen. Mutagen.* **1:** 305–332.

Bishop JB, Shelby MD. (1990) Mammalian heritable effects research in the National Toxicology Program. *Banbury Report 34: Biology of Mammalian Germ Cell Mutagenesis.* Cold Spring Harbor Laboratory Press, Cold Spring Harbor, NY, pp. 425–435.

Bishop JB, Kodell RL, Whorton EB, Domon OE. (1983) Dominant lethal test response with IMS and TEM using different combinations of male and female stocks of mice. *Mutat. Res.* **121:** 273–280.

Brewen JG, Payne HS, Jones KP, Preston RJ. (1975) Studies on chemically induced dominant lethality. I. The cytogenetic basis of MMS-induced dominant lethality in post-meiotic male germ cells. *Mutat. Res.* **33:** 239–250.

Bultman SJ, Michaud EJ, Woychik RP. (1992) Molecular characterization of the mouse agouti locus. *Cell* **71:** 1195–1204.

Burkhart JG, Malling HV. (1993) Mutagenesis and transgenic systems: perspective from the mutagen *N*-ethyl-*N*-nitrosourea. *Environ. Mol. Mutagen.* **22:** 1–6.

Burkhart JG, Burkhart BA, Sampson KS, Malling HV. (1993) ENU-induced mutagenesis at a single A:T base pair in transgenic mice containing fX174. *Mutat. Res.* **292:** 69–81.

Cacheiro NLA, Russell LB, Swartout MS. (1974) Translocations, the predominant cause of total sterility in sons of mice treated with mutagens. *Genetics* **76:** 73–91.

Cattanach BM. (1957) Induction of translocations in mice by triethylenemelamine. *Nature* **180:** 1364–1365.

Cattanach BM. (1959) The sensitivity of the mouse testis to the mutagenic action of triethylenemelamine. *Zeit. fur Vererbiol.* **90:** 1–6.

Charles DJ, Pretsch W. (1983) Detection of mouse mutants with altered enzyme activity. *Genet. Res.* **41**: 304.

Collins FS, O'Connell P, Ponder BAJ, Seizinger BR. (1989) Progress towards identifying the neurofibromatosis (NF1) gene. *Trends Genet.* **5**: 217–221.

Crisp TM. (1992) Organization of the ovarian follicle and events in its biology: oogenesis, ovulation or atresia. *Mutat. Res.* **296**: 89–106.

Douglas GR, Gingerich JD, Gossen JA. (1993) Time course and sequence spectra of ethylnitrosourea-induced lacZ mutations in transgenic mouse somatic and germ cells. *Environ. Mol. Mutagen.* **21**(suppl. 22): 18.

Dym M, Fawcett DW. (1970) The blood–testis barrier in the rat and the physiological compartmentation of the seminiferous epithelium. *Biol. Reprod.* **3**: 300–326.

Ehling UH. (1983) Cataracts — indicators for dominant mutations in mice and man. In: *Utilization of Mammalian Specific Locus Studies in Hazard Evaluation and Estimation of Genetic Risk* (eds FJ deSerres, W Sheridan). Plenum Press, New York, pp. 169–190.

Ehling UH. (1986) Induction of gene mutations in mice: the multiple endpoint approach. *Prog. Clin. Biol. Res.* **209B**: 501–510.

Ehling UH, Doherty DG, Malling H. (1972) Differential spermatogenic response of mice to the induction of dominant lethal mutations by *n*-propyl methanesulfonate and isopropyl methanesulfonate. *Mutat. Res.* **15**: 175–184.

Ehling UH, Machemer L, Buselmaier W, Dycka J, Frohberg H, Kratochvilova J, Lang R, Lorke D, Muller D, Peh J, Röhrborn G, Roll R, Schulze-Schencking M, Wiemann W. (1978) Standard protocol for the dominant lethal test on male mice set up by the Working Group Dominant Lethal Mutations of the *ad hoc* Committee Chemogenetics. *Arch. Toxicol.* **39**: 173–185.

Ehling UH, Favor J, Kratochvilova J, Neuhauser-Klaus A. (1982) Dominant cataract mutations and specific locus mutations in mice induced by radiation or ethylnitrosourea. *Mutat. Res.* **92**: 181–192

Espey LL , Lipner H. (1994) Ovulation. In: *The Physiology of Reproduction*, 2nd Edn (eds E Knobil, JD Neill). Raven Press, New York, pp. 725–780.

Favor J, Neuhauser-Klaus A, Ehling UH. (1988) The effect of dose fractionation on the frequency of ethylnitrosourea-induced dominant cataract and recessive specific locus mutations in germ cells of the mouse. *Mutat. Res.* **198**: 269–275.

Fawcett DW, Anderson WA, Phillips DM. (1971) Morphogenetic factors influencing the shape of the sperm head. *Devel. Biol.* **26**: 220–251.

Generoso WM. (1994) Letter to the Editor: The rodent dominant-lethal assay. *Environ. Mol. Mutagen.* **24**: 332–333.

Generoso WM, Piegorsch WW. (1993) Dominant lethal tests in male and female mice. In: *Methods in Toxicology* (eds. JJ Heindel, R Chapin). Academic Press, New York, pp. 124–142.

Generoso WM, Russell WL. (1969) Strain and sex variations in the sensitivity of mice to dominant-lethal induction with ethyl methanesulfonate. *Mutat. Res.* **8**: 589–598.

Generoso WM, Cain KT, Huff SW, Goslee DG. (1978) Heritable translocation test in mice. In: *Chemical Mutagens*, Vol. 5 (eds A Hollaender, FJ deSerres). Plenum Press, New York, pp. 55–77.

Generoso WM, Bishop JB, Goslee DG, Newell GW, Sheu CJ, vonHalle E. (1980) Heritable translocation test in mice. *Mutat. Res.* **76**: 191–215.

Generoso WM, Krishna M, Cain KT, Sheu CW. (1981) Comparison of two methods for detecting translocation heterozygotes in mice. *Mutat. Res.* **81**: 177–186.

Generoso WM, Cain KT, Hughes LA. (1985) Tests for dominant-lethal effects of 1,2-dibromo-3-chloropropane (DBCP) in male and female mice. *Mutat. Res.* **156**: 103–108.

Generoso WM, Katoh M, Cain KT, Hughes LA, Foxworth LB, Mitchell TJ, Bishop JB. (1989) Chromosome malsegregation and embryonic lethality induced by treatment of normally ovulated mouse oocytes with nocodazole. *Mutat. Res.* **210**: 313–322.

Giometti C, Tollaksen SL, Grahn D. (1994) Altered protein expression detected in the F_1 offspring of male mice exposed to fission neutrons. *Mutat. Res.* **320:** 75–85.

Gossen JA, DeLeeuw WJF, Tan CHT, Zwarthoff EC, Berends F, Lohman PHM, Knook DL, Vijg J. (1989) Efficient rescue of integrated shuttle vectors from transgenic mice: a model for studying mutations *in vivo. Proc. Natl Acad. Sci. USA* **86:** 7971–7975.

Green S, Auletta A, Fabricant J, Kapp R, Manandhar M, Sheu C, Springer J, Whitfield B. (1985) Current status of bioassays in genetic toxicology – the dominant lethal assay: a report of the US Environmental Protection Agency Gene-Tox Program. *Mutat. Res.* **154:** 49–67.

Handel MA. (1987) Genetic control of spermatogenesis in mice. In: *Spermatogenesis, Genetic Aspects, Results and Problems in Cell Differentiation*, Vol. 15 (ed. W Henning). Springer, Berlin, pp. 1–62.

Johnson FM, Lewis SE. (1981) Electrophoretically detected germinal mutations induced in the mouse by ethylnitrosourea. *Proc. Natl Acad. Sci. USA* **78:** 3138–3141.

Kallio M, Lähdetie J. (1993) Analysis of micronuclei induced in mouse early spermatids by mitomycin C, vinblastine sulfate or etoposide using fluorescence *in situ* hybridization. *Mutagenesis* **8:** 561–567.

Kallioniemi A, Kallioniemi O-P, Sudar D, Rutovitz D, Gray JW, Waldman F, Pinkel D. (1992) Comparative genomic hybridization for molecular cytogenetic analysis of solid tumours. *Science* **258:** 818–821.

Katoh MA, Cain KT, Hughes LA, Foxworth LB, Bishop JB, Generoso WM. (1990) Female-specific dominant lethal effects in mice. *Mutat. Res.* **230:** 205–217.

Kohler SW, Provost GS, Kretz PL, Fieck A, Sorge JA, Short JM. (1990) The use of transgenic mice for short-term, *in vivo* mutagenicity testing. *GATA* **7:** 212–218.

Kohler SW, Provost GS, Fieck A, Kretz PL, Bullock WO, Sorge JA, Putman DL, Short JM. (1991) Spectra of spontaneous and mu mutations in the *lacI* gene in transgenic mice. *Proc. Natl Acad. Sci. USA* **88:** 9758–9762.

Lahdetie J. (1983) Meiotic micronuclei induced by adriamycin in male rats. *Mutat. Res.* **119:** 79–82.

Lavappa KS. (1974) Induction of reciprocal translocations in the Armenian hamster. *Lab. Anim. Sci.* **24:** 62–65.

Leonard A. (1973) Observation on meiotic chromosomes of male mouse as a test of the potential mutagenicity of chemicals in mammals. In: *Chemical Mutagens: Principles and Methods for their Detection*, Vol. 3 (ed. A Hollaender). Plenum Press, New York, pp. 21–56.

Leonard A, Linden G. (1972) Observations of dividing spermatocytes for chromosome aberrations induced in mouse spermatogonia by chemical mutagens. *Mutat. Res.* **16:** 297–300.

Leopardi P, Zijno A, Bassani B, Pacchierotti F. (1993) *In vivo* studies on chemical induced aneuploidy in mouse somatic and germinal cells. *Mutat. Res.* **287:** 119–130.

Lewis SE. (1991) The biochemical specific locus test and a new multiple-endpoint mutation detection system: considerations for genetic risk assessment. *Environ. Mol. Mutagen.* **18:** 303–306.

Lewis SE. (1994) A consideration of the advantages and potential difficulties of the use of transgenic mice for the study of germinal mutations. *Mutat. Res.* **307:** 509–515.

Lewis SE, Erickson RP, Barnett LB, Venta P, Tashian R. (1988) Ethylnitrosourea induced null mutation at the mouse Car-2 locus: an animal model for human carbonic anhydrase II deficiency syndrome. *Proc. Natl Acad. Sci USA* **85:** 1962–1966.

Lockhart A-M, Bishop JB, Piegorsch WW. (1991) Issues regarding data acquisition and analysis in the dominant lethal assay. *Proc. Biopharm. Section Am. Stat. Assoc.,* pp. 234–237.

Lockhart A-M, Piegorsch WW, Bishop JB. (1992) Assessing overdispersion and dose-response in the male dominant lethal assay. *Mutat. Res.* **272:** 35–58.

Lowe X, Marchetti F, Ahlborn T, Bishop J, Titenko-Holland N, Smith M, Wyrobek AJ. (1995) Chromosomal abnormalities detected in zygotes and 4-day embryos of mice using multi-color FISH. Proc. 2nd Internat. Conf. Environmental Mutagens in Human Populations. *Environ. Hlth Perspec.,* in press.

Lüning KG, Eiche A. (1975) X-ray-induced recessive lethal mutations in the mouse. *Mutat. Res.* **34**: 163–174.

Lyon MF, Morris T. (1969) Gene and chromosome mutation after large fractionated or unfractionated radiation doses to mouse spermatogonia. *Mutat. Res.* **8**: 191–198.

Malling HV, Burkhart JG. (1989) Use of fX174 as a shuttle vector for the study of *in vivo* mammalian mutagenesis. *Mutat. Res.* **212**: 11–21.

Malling HV, Burkhart JG. (1992) Comparison of mutation frequencies obtained using transgenes and specific-locus mutation systems in male mouse germ cells. *Mutat. Res.* **279**: 149–151.

Mailhes JB, Marchetti F. (1994) Chemically induced aneuploidy in mammalian oocytes. *Mutat. Res.* **320**: 87–111.

Mailhes JB, Preston RJ, Yuan ZP, Payne HS. (1988) Analysis of mouse metaphase II oocytes as an assay for chemically induced aneuploidy. *Mutat. Res.* **198**: 145–152.

Mailhes JB, Yuan ZP, Aardema MJ. (1990) Cytogenetic analysis of mouse oocytes and one-cell zygotes as a potential assay for heritable germ cell aneuploidy. *Mutat. Res.* **242**: 89–100.

Medina FIS, Leonard A, Leonard ED, DeKnudt GH. (1976) Heritable chromosome rearrangements induced in rat spermatids by X-irradiation. *Strahlentherapie* **152**: 482–486.

Miller BM, Adler I-D. (1992) Aneuploidy induction in mouse spermatocytes. *Mutagenesis* **7**: 69–76.

Monesi V. (1962) Autoradiographic study of DNA synthesis and the cell cycle in spermatogonia and spermatocytes of mouse testis using tritiated thymidine. *J. Cell Biol.* **14**: 1–18.

Morris SM, Kodell RL, Domon OE, Bishop JB. (1988) Detection of TEM-induced reciprocal translocations in F_1 sons of CD-1 male mice: comparison of sequential fertility evaluation and cytogenetic analysis. *Environ. Mol. Mutagen.* **11**: 215–223.

Moses MJ, Poorman PA, Dresser ME, DeWeese GK, Gibson JB. (1985) The synaptonemal complex in meiosis: significance of induced perturbations, In: *Aneuploidy – Etiology and Mechanisms* (eds VL Dellarco *et al.*). Plenum Press, New York, pp.337–352.

Moyer J, Lee-Teschler M, Kwon H-Y, Schrick JJ, Avner ED, Sweeney WE, Godfrey VL, Cacheiro NLA, Wilkinson JE, Woychik RP. (1994) Candidate gene associated with a mutation causing recessive polycystic kidney disease in mice. *Science* **264**: 1329–1333.

Oakberg EF. (1956) Duration of spermatogenesis in the mouse and timing of stages of the cycle of the seminiferous epithelium. *Am. J. Anat.* **99**: 507–516.

Oakberg EF. (1960) Irradiation damage to animals and its effect on their reproductive capacity. *J. Dairy Sci.* **43**: 54–67.

Oakberg EF. (1975) Effects of radiation on the testis. In: *Handbook of Physiology*, Section 7, Endocrinology 5. William Wilkins Co., Baltimore, pp. 233-243.

Oakberg EF. (1983) Germ cell toxicity: significance in genetic and fertility effects of radiation and chemicals. In: *Mutation, Cancer and Malformation* (eds EHY Chu, WM Generoso). Plenum Press, New York, pp. 549–590.

Pacchierotti F, Bellincampi D, Civitareale D. (1983) Cytogenetic observations in mouse secondary spermatocytes on numerical and structural chromosome aberrations induced by cyclophosphamide in various stages of spermatogenesis. *Mutat. Res.* **119**: 177–183.

Pacchierotti F, Tiveron C, D'Archivio M, Bassani B, Cordelli E, Leter G, Spano M. (1994) Acrylamide-induced chromosomal damage in male mouse germ cells detected by cytogenetic analysis of one-cell zygotes. *Mutat. Res.* **309**: 273–284.

Poorman-Allen P, Backer LC, Adler ID, Westbrook-Collins B, Moses MJ, Allen JW. (1990) Bleomycin effects on mouse meiotic chromosomes. *Mutagenesis* **5**: 573–581.

Preston RJ. (1994) *Genetic Risk Assessment: Parallelograms and Pragmatism.* Chemical Industry Institute of Toxicology (CIIT) Activities, Vol. 14, pp. 1-7.

Preston RJ, Au W, Bender MA, Brewen JG, Carrano AV, Heddle JA, McFee AF, Wolff S, Wassom JS. (1981) Mammalian *in vivo* and *in vitro* cytogenetic assays: a report of the U.S. EPA Gene-Tox Program. *Mutat. Res.* **87**: 143–188.

Rhomberg L, Dellarco VL, Siegel-Scott C, Dearfield KL, Jacobson-Kram D. (1990) Quantitative estimation of the genetic risk associated with the induction of heritable translocations at low-dose exposures: ethylene oxide as an example. *Environ. Mol. Mutagen.* **16:** 104–125.

Rinchik EM, Bangham JW, Hunsicker PR, Cacheiro NLA, Kwon BS, Jackson IJ, Russell LB. (1990) Genetic and molecular analysis of chlorambucil-induced germ line mutations in the mouse. *Proc. Natl Acad. Sci. USA* **87:** 1416–1420.

Robbins WA, Segraves R, Pinkel D, Wyrobek AJ. (1993) Detection of aneuploid human sperm by fluorescence *in situ* hybridization: evidence for a donor difference in frequency of sperm disomic for chromosomes 1 and Y. *Am. J. Hum. Genet.* **52:** 799–807.

Roderick TH. (1983) Using inversions to detect and study recessive lethals and detrimentals in mice. In: *Utilization of Mammalian Specific Locus Studies in Hazard Evaluation and Estimation of Genetic Risk* (eds FJ deSerres, W Sheridan). Plenum Press, New York, pp. 135–167.

Russell LB. (1976) Numerical sex-chromosome anomalies in mammals: their spontaneous occurrence and use in mutagenesis studies. In: *Chemical Mutagens – Principles and Methods for their Detection,* Vol 4 (ed. A Hollaender). Plenum Press, New York, pp. 55–91.

Russell LB. (1983) Qualitative analysis of mouse specific-locus mutations: information on genetic organization, gene expression, and the chromosomal nature of induced lesions. In: *Utilization of Mammalian Specific Locus Studies in Hazard Evaluation and Estimation of Genetic Risk* (eds FJ deSerres, W Sheridan). Plenum Press, New York, pp. 241–258.

Russell LD. (1990a) Barriers to entry of substances into seminiferous tubules: compatibility of morphological and physiological evidence. *Banbury Report 34: Biology of Mammalian Germ Cell Mutagenesis.* Cold Spring Harbor Laboratory Press, Cold Spring Harbor, NY, pp. 3–17.

Russell LB. (1990b) Patterns of mutational sensitivity to chemicals in poststem-cell stages of mouse spermatogenesis. In: *Mutation and the Environment,* Part C; *Somatic and Heritable Mutations, Adduction and Epidemiology* (eds M Mendelsohn, R. Albertini). Alan R Liss, New York, pp. 101–113.

Russell LB. (1991) Factors that affect the molecular nature of germ-line mutations recovered in the mouse specific locus test. *Environ. Mol. Mutagen.* **18:** 298–302.

Russell LB. (1994) Role of mouse germ-cell mutagenesis in understanding genetic risk and in generating mutations that are prime tools for studies in modern biology. *Environ. Mol. Mutagen.* **23** (suppl. 24): 23–29.

Russell LB, Rinchik E. (1993) Structural differences between specific-locus mutations induced by different exposure regimes in mouse spermatogonial stem cells. *Mutat. Res.* **288:** 187–195.

Russell LB, Shelby MD. (1985) Tests for heritable genetic damage and for evidence of gonadal exposure in mammals. *Mutat. Res.* **154:** 69–84.

Russell LB, Selby PB, vonHalle E, Sheridan W, Valcovic L. (1981) The mouse specific locus test with agents other than radiation: interpretation of data and recommendations for future work. *Mutat. Res.* **86:** 329–354.

Russell LB, Russell WL, Rinchik EM, Hunsicker PR. (1990) Factors affecting the nature of inducing mutations. *Banbury Report 34: Biology of Mammalian Germ Cell Mutagenesis.* Cold Spring Harbor Laboratory Press, Cold Spring Harbor, NY, pp. 271–289.

Russell WL. (1951) X-ray-induced mutations in mice. *Cold Spring Harbor Symp. Quant. Biol.* **16:** 327–336.

Russell WL. (1954) Genetic effects of radiation in mammals. In: *Radiation Biology,* Vol. 1 (ed. A Hollaender). McGraw-Hill, New York, pp. 825–859.

Russell WL, Russell LB. (1959) Radiation-induced genetic damage in mice. In: *Progress in Nuclear Energy,* Series VI, Vol. 2, *Biological Sciences.* Pergamon Press, London, pp. 179–188.

Russo A, Levis AG. (1991) The contribution of the mouse spermatid micronucleus assay to the detection of germinal mutagens. *Prog. Clin. Biol. Res.* **372:** 513–520.

Russo A, Levis AG. (1992) Detection of aneuploidy in male germ cells of mice by means of a meiotic micronucleus assay. *Mutat. Res.* **281:** 187–191.

Russo A, Pacchierotti F, Metalli P. (1984) Nondisjunction induced in mouse spermatogenesis by chloral hydrate, a metabolite of trichloroethylene. *Environ. Mutagen.* **6:** 695–703.

Searle AG. (1974) Nature and consequences of induced chromosome damage in mammals. *Genetics* **78:** 173–186.

Searle AG. (1975) The specific locus test in the mouse. *Mutat. Res.* **31:** 277–290.

Searle AG, Beechey C. (1986) The role of dominant visibles in mutagenicity testing. In: *Genetic Toxicology of Environmental Chemicals*, Part B, *Genetic Effects and Applied Mutagenesis. Progress in Clinical Biological Research*, Vol. 209B, pp. 511–518.

Sega GA. (1974a) Dosimetry studies on the ethylation of the mouse sperm DNA after *in vivo* exposure to 3H-ethylmethanesulfonate. *Mutat. Res.* **24:** 317–333.

Sega GA. (1974b) Unscheduled DNA synthesis in the germ cells of male mice exposed *in vivo* to the chemical mutagen ethyl methanesulfonate. *Proc. Natl Acad. Sci. USA* **71:** 4955–4959.

Sega GA. (1990) Molecular targets, DNA breakage, and DNA repair: their roles in mutation induction in mammalian germ cells. In: *Banbury Report 34: Biology of Mammalian Germ Cell Mutagenesis* (eds JW Allen, BA Bridges, MF Lyon, MJ Moses, LB Russell). Cold Spring Harbor Laboratory Press, Cold Spring Harbor, NY, pp.79–92.

Sega GA, Generoso EE. (1988) Measurement of DNA breakage in spermiogenic germ cell stages of mice exposed to ethylene oxide using an alkaline elution procedure. *Mutat. Res.* **197:** 93–99.

Sega GA, Sotomayor RE. (1982) Unscheduled DNA synthesis in mammalian germ cells – its potential use in mutagenicity testing. In: *Chemical Mutagens: Principles and Methods for Their Detection*, Vol 7 (eds FJ deSerres, A Hollaender). Plenum Press, New York, p. 421.

Selby PB. (1983) Applications in genetic risk estimation of data on the induction of dominant skeletal mutations in mice. In: *Utilization of Mammalian Specific Locus Studies in Hazard Evaluation and Estimation of Genetic Risk* (eds FJ de Serres, W Sheridan). Plenum Press, New York, pp. 191–210.

Selby PB. (1990) The importance of the direct method of genetic risk estimation and ways to improve it. In: *Banbury Report 34: Biology of Mammalian Germ Cell Mutagenesis* (eds JW Allen, BA Bridges, MF Lyon, MJ Moses and LB Russell). Cold Spring Harbor Laboratory Press, Cold Spring Harbor, NY, pp. 437–449.

Selby PB, Selby PR. (1977) Gamma-ray induced dominant mutations that cause skeletal abnormalities in mice, I. Plan, summary of results and discussion. *Mutat. Res.* **43:** 357–375.

Sharpe RM. (1994) Regulation of spermatogenesis. In: *The Physiology of Reproduction*, 2nd Edn, Vol. 1 (eds E Knobil, JD Neill). Raven Press, New York, pp. 1363–1434.

Shedlovsky A, McDonald JD, Symula D, Dove WF. (1993) Mouse models of human phenylketonuria. *Genetics* **134:**1205–1210.

Sjöblom T, Parvinen M, Lähdetie J. (1994) Germ-cell mutagenicity of etoposide: induction of meiotic micronuclei in cultured rat seminiferous tubules. *Mutat. Res.* **323:** 41–45.

Skow LC, Burkhart BA, Johnson FM, Popp RA, Popp DM, Goldberg SZ, Anderson WF, Barnett LB, Lewis SE. (1983) A mouse model for beta-thalassemia. *Cell* **34:** 1043–1052.

Snell GD. (1935) The induction by X-rays of hereditary changes in mice. *Genetics* **20:** 545–567.

Sobels FH. (1982) The parallelogram: an indirect approach for the assessment of genetic risks from chemical mutagens. In: *Progress in Mutation Research*, Vol. 3 (eds KC Bora, GR Douglas, ER Nestman), pp. 323–327.

Stubbs, L. (1992) Long-range walking techniques in positional cloning strategies. *Mamm. Genome* **3:** 127–142.

Sudman PD, Rutledge JC, Bishop JB, Generoso WM. (1992) Bleomycin: female-specific dominant lethal effects in mice. *Mutat. Res.* **296:** 143–156.

Tanaka N. (1981) Studies on chemical induction of chromosomal aberrations in postcopulation germ cells and zygotes of female mice. I. Comparative studies on the frequency of first-cleavage chromosomal aberrations and dominant lethal mutations. *Jap. J. Genet.* **56:** 117–129.

Tates AD. (1992) Validation studies with the micronucleus test for early spermatids of rats: a tool for detecting clastogenicity of chemicals in differentiating spermatogonia and spermatocytes. *Mutagenesis* **7:** 411–419.

Teramoto S, Shirasu Y. (1989) Genetic toxicology of 1,2-dibromo-3-chloropropane (DBCP). *Mutat. Res.* **221:** 1–9.

Toppari J, Lähdetie J, Härkönen P, Eerola E, Parvinen M. (1986) Mutagen effects on rat seminiferous tubles *in vitro:* induction of meiotic micronuclei by adriamycin. *Mutat. Res.* **171:** 149–156.

Vogel R. (1993) *In vitro* approach to fertility research: genotoxicity tests on primordial germ cells and embryonic stem cells. *Reprod. Toxicol.* **7** (suppl. 1): 69–73.

Wyrobek AJ, Robbins WA, Mehraein Y, Pinkel D, Weier HU. (1994) Detection of sex chromosomal aneuploidies X-X, Y-Y, and X-Y in human sperm using multiprobe fluorescence *in situ* hybridization. *Am. J. Med. Genet.* **53:** 1–7.

Wyrobek A, Lowe X, Pinkel D, Bishop J. (1995a) Aneuploidy in late-step spermatids of mice detected by two-chromosome fluorescence *in situ* hybridization. *Mol. Reprod. Dev.* **40:** 259–266.

Wyrobek A, Bishop J, deStoppelaar J, Cassel M, Hoebee B, Lowe X. (1995b) Detection of aneuploidy in rat sperm by FISH: comparisons of sperm aneuploidy frequencies among young rats, mice and humans. *Environ. Mol. Mutagen.* **25** (suppl. 25): 58.

Yuan ZP, Mailhes JB. (1987) Aneuploidy determination in C-banded mouse metaphase II oocytes following cyclophosphamide treatment *in vivo. Mutat. Res.* **179:** 209–214.

Zeiger E. (1994) Commentary: old protocols, as do old habits, die hard. *Environ. Mol. Mutagen.* **24:** 1–2.

Additional sources of information

Adler I-D. (1993a) Overview of mammalian germ cell tests. *International Workshop on Standardisation of Procedures in Genetic Toxicology* (ed. T Sofuni). Scientist Inc., Tokyo, pp. 71-86.

Adler I-D. (1993b) Synopsis of the *in vivo* results obtained with the 10 known or suspected aneugens tested in the CEC collaborative study. *Mutat. Res.* **287:** 131–137.

Brandriff BF, Gordon LA. (1989) Analysis of the first cell cycle in the cross between hamster eggs and human sperm. *Gamete Res.* **23:** 299.

Brewen JG. (1976) Practical evaluation of mutagenicity data in mammals for estimating human risk. *Mutat. Res.* **41:** 15–24.

Cattanach BM, Papworth D, Kirk M. (1984) Genetic tests for autosomal non-disjunction and chromosome loss in mice. *Mutat. Res.* **126:** 189–204.

Ehling UH. (1978) Specific locus mutations in mice. In: *Chemical Mutagens: Principles and Methods for their Detection,* Vol. 5 (eds A Hollaender, FJ deSerres). Plenum Press, New York, pp. 233–256.

Evans EP, Breckon G, Ford CE. (1964) An air-drying method for meiotic preparations from mammalian testes. *Cytogenetics* **3:** 289–294.

Generoso WM, Cain KT, Krishna M, Huff SW. (1979) Genetic lesions induced by chemicals in spermatozoa and spermatids of mice are repaired in the egg. *Proc. Natl. Acad. Sci. USA* **76:** 435–437.

Generoso WM, Rutledge JC, Cain KT, Hughes LA, Braden PW. (1987a) Exposure of female mice to ethylene oxide within hours after mating leads to fetal malformation and death. *Mutat. Res.* **176:** 269–274.

Generoso WM, Rutledge JC, Cain KT, Hughes LA, Downing DJ. (1987b) Mutagen induced fetal anomalies and death following treatment of females within hours after mating. *Mutat. Res.* **199:** 175–181.

Generoso WM, Shourbaji AG, Piegorsch WW, Bishop JB. (1991) Developmental response of zygotes exposed to similar mutagens. *Mutat. Res.* **250:** 439–446.

Hansmann I. (1974) Chromosome aberrations in metaphase II oocytes: stage sensitivity in the mouse oogenesis to amethopterin and cyclophosphamide. *Mutat. Res.* **22:** 175–191.

Lowe X, Collins B, Allen J, Holland N, Breneman J, vanBeek M, Bishop J, Wyrobek A. (1995) Aneuploidies and micronuclei in the germ cells of male mice of advanced age. *Mutat. Res.* (in press).

Martin RH, Rademaker A. (1990) The frequency of aneuploidy among individual chromosomes in 6,821 human sperm chromosome complements. *Cytogen. Cell Genet.* **53:** 103–107.

McDonald J, Shedlovsky A, Dove WF. (1990) Investigating inborn errors of phenylalanine metabolism by efficient mutagenesis of the mouse germ line. In: *Banbury Report 34, Biology of Mammalian Germ Cell Mutagenesis* (eds JW Allen, BA Bridges, MF Lyon, MJ Moses, LB Russell). Cold Spring Harbor Laboratory Press, Cold Spring Harbor, NY, pp. 259–270.

Mirsalis JC, Monforte JA, Winegar RA. (1994) Transgenic animal models for measuring mutations *in vivo*. *Crit. Rev. Toxicol.* **24(3):** 255–280.

Shedlovsky A, Guenet JL, Johnson LL, Dove WF. (1986) Induction of recessive lethal mutations in the T/t-H-2 region of the mouse genome by a point mutagen. *Genet. Res.* **47(2):** 135–142.

Shelby MD, Bishop JB, Mason JM, Tindall KR. (1993) Fertility reproduction and genetic disease: studies on the mutagenic effects of environmental agents on mammalian germ cells. *Environ. Hlth Perspect.* **100:** 283–291.

The mouse spot test

Rudolf Fahrig

10.1 Introduction

Several methods exist for detecting the genotoxicity and especially the mutagenicity of physical or chemical agents. For the extrapolation of experimental results to man, *in vivo* results are more relevant than *in vitro* ones. Risk estimations for man gained from *in vivo* data are more reliable than those from *in vitro* data (see also Chapter 9).

The spot test is the best validated method for detection of genotoxic alterations, namely gene mutations and recombinations, in somatic cells of mice. It has been or is being established in six laboratories in Germany, four in Japan and two each in the USA, Switzerland, UK and Denmark. Like other mutagenicity tests, the spot test not only serves for detecting mutagenic effects but also for detecting carcinogenic potential. An important feature of the spot test is its ability to detect recombinations (Fahrig, 1992a) which can unmask carcinogen-induced mutations (Ames and Gold, 1990), as well as being able to detect gene mutations which play a role during initiation. In recent years, the spot test has gained importance in detecting co-recombinogenic effects, which are regarded as important in tumour promotion (Fahrig, 1984, 1992b).

10.2 Fundamentals

Although colour spots which occur spontaneously in animals have been known for a long time, it was not until 1957 that they were induced experimentally by X-rays (Russell and Major, 1957). In these classical experiments embryos which were heterozygous for different coat colour mutations were irradiated *in utero*. Some of these developing young animals showed colour spots.

The spot test is based on this principle (*Figure 10.1*). It has been used as a mutagenicity test since 1975 when colour spots were induced for the first time with chemical mutagens (Davidson and Dawson, 1976; Fahrig, 1975; Lang, 1978; Neuhäuser-Klaus, 1981). The spot test has been further developed to the point where it is now possible to distinguish between induced gene mutations and induced recombinations (Fahrig, 1984; Fahrig and Neuhäuser-Klaus, 1985).

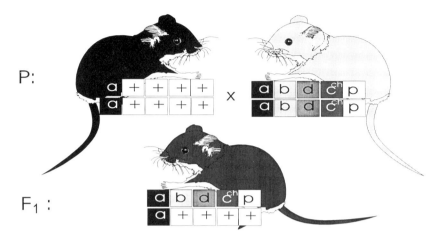

Figure 10.1. Principle of the mouse spot test exemplified with the cross C57BL x T-stock.

10.3 Test principle

Embryos heterozygous for different recessive coat colour mutations are treated *in utero* with a mutagen, preferably between the 9th and the 11th day of fetal development. This is normally performed by intraperitoneal (i.p.) injection or treatment *per os* of the mother. If this treatment leads to the alteration or loss of the wild-type allele of one of the recessive coat colour genes in a pigment precursor cell, a colour spot will develop after several cell divisions.

10.4 Animals

Embryos with the genetic constitution appropriate for the spot test can be obtained by crossing any *a/a* strain (*a/a* produces a black coat and thus, good perceptibility on this background) with the T stock or by the cross T x HT (T, test strain; HT, Harwell test strain). Embryos of the genotype *a/a; b/B; c^{ch} p/C P; d se/D SE; s/S* (dark grey coat, dark eyes) are obtained by crossing C57BL females with T-strain males (*a/a*, 'non-agouti'; *b/b*, 'brown'; c^{ch} *p/c^{ch} p*, 'chinchilla' and 'pink-eyed dilution'; *d se/d se*, 'dilute' and 'short ear'; *s/s*, 'piebald spotting'). Embryos of the genotypes *bp a pa/BP a PA; b/B; c^{ch} p/C P; d se/D SE; s/S; ln fz/LN FZ; pe/PE* (dark grey coat, dark eyes) are obtained by crossing the T strain with the HT strain (*bp pa/bp pa*: 'brachypodism' and 'pallid'; *ln fz/ln fz*: 'leaden' and 'fuzzy'; *pe/pe*: 'pearl')(*see Table 10.1*).

Other suitable crosses can also be used, provided that they are equal or superior to those mentioned with respect to fertility and possess as many as four recessive coat colour mutations, besides non-agouti.

The fertility and sensitivity of a new cross (Hart, 1985) using male PDB (*a/a, b/b, d/d, p/p*) and non-agouti female NMRI (*a/a, c/c*) mice have been assessed. The fertility of this new cross reflects the good breeding performance of the NMRI females, and this, together with the high pregnancy rate, results in substantial economies in the number of treated animals.

Table 10.1. Principal effects of pigment genes expressed in the spot test (Silvers, 1979)

Symbol	Name	General effect on pigment	Probable site of gene action
a	Non-agouti	Black-pigmented hairs except for some yellow hairs around ears, mammae and perineum. The alleles at the agouti locus control a reversible trigger mechanism causing either yellow or black pigment granules	Follicular environment
b	Brown	Brown pigment. The principal effect of substituting brown for black is a qualitative change in pigment from black to brown, with some reduction in pigment volume	Within melanoblast, polymerization of melanin, as related to periodicity of binding sites on protein matrix of melanosome
c^{ch}	Chinchilla	Reduced melanins, particularly phaeomelanin. Hairs have reduced or absent distal pigment with darker bases, giving chinchilla effect	Within melanoblast, alteration of tyrosinase structure
c	Albino	Absence of melanins. The alleles at the albino locus appear to control quantity of pigment only	Within melanoblast, alteration of tyrosinase structure
d	Dilute	Hair and skin pigments appear diluted. The dilute allele causes irregularity of pigment deposition (granular clumping), reduced cortical pigment, and uneven pigment distribution within the cells of the hair	Within melanoblast, alteration of melanocyte morphology
ln	Leaden	Diluted hair and skin pigment; no effect on retinal pigment	Within melanoblast, alteration of melanocyte morphology
p	Pink-eyed dilution	Reduced eye, skin, and hair pigment, particularly eumelanin. Pink-eyed dilution affects size of granules and level of pigmentation. These shred-like granules assume a kind of flocculent clumping, and hairs have greatly reduced pigmentation distally	Within melanoblast, alteration of protein matrix of melanosome itself
pa	Pallid	Reduced eye, skin, and hair pigment, particularly eumelanin	?
pe	Pearl	Reduced eye and hair pigment	?

The PW strain used in Japanese laboratories has the same genotype as the T stock lacking *s/s*. The DBA strain has the genotype *a/a*, *b/b* and *d/d*.

10.5 Genetics of colour spots

In the embryo, the primordial melanoblasts develop in the neural crest at both sides of the back. During their migration to the midline of the abdomen between the 8th and the 12th day of fetal development they continue to divide (Rawles, 1947). Each clone of pigment cells (melanocytes) originating from a primordial melanoblast is responsible for pigmentation of a distinct band of the coat. According to Mintz (1967), 34 pigment clones exist in the mouse. Therefore, no colour spot can be larger than a band of limited width stretching from the midline of the back to the midline of the abdomen. The earlier the mutagen treatment the higher the probability of obtaining a colour spot the size of a whole band. Later treatment will result in smaller spots the size of a half, a quarter, an eighth or even less of a band. Colour spots smaller than an eighth of a band can hardly be detected, if at all. The advantage of later treatment, inducing more spots, is outweighed by the difficulties in detecting these spots.

Figure 10.2 shows the correlation between the time of treatment and the size and frequency of colour spots after treatment with ethyl methanesulphonate (EMS) in (C57BL/6 x T) F_1, with procarbazine in (T x HT) F_1, and with

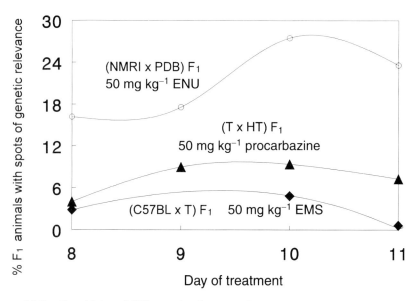

Figure 10.2. Sensitivity of different developmental stages of (T x HT) F_1 embryos to the induction of coat colour spots by 50 mg kg^{-1}-procarbazine (Neuhäuser-Klaus, 1981) and (C57 BL x T) F_1 embryos to the induction of coat colour spots by 50 mg kg^{-1} ethylmethane sulphonate (EMS) (Fahrig, 1978) and with 50 mg kg^{-1} ethylnitrosourea (ENU) in (NMRI x PDB) F_1 (Hart, 1985).

ethylnitrosourea (ENU) in (NMRI x PDB) F_1. Treatment on the 9th or 10th days of fetal development has been shown to be appropriate if the mutagen is given in a single dose.

Spots of genetic relevance can be induced by gene mutations, recombinations, or chromosome aberrations. Of the numerical and structural chromosome aberrations only those that survive the filter of several mitoses cause a colour spot. Chromosome aberrations leading to the death of pigment cells are expressed as genetically non-relevant white ventral spots.

Using the cross C57BL x T, it is possible in two cases to determine which of the different genetic alterations is responsible for the appearance of a colour spot (*Figure 10.3*).

10.5.1 Detection of gene mutations

For the following reason, gene mutations can only be detected as light brown hairs (*Figure 10.4*): c^{ch}/c^{ch} in combination with a/a results in a dark grey spot which does not contrast with the dark grey coat of the F_1 animals. Therefore, the only genetic alteration that can be detected is a mutation (in most cases a small deletion within the *c* locus) to *c*, in combination with c^{ch} showing up as a light brown spot. It is not possible with the naked eye to decide if it is the required spot. Only by microscopical analysis is it possible to achieve a clear picture (Fahrig, 1984).

10.5.2 Detection of reciprocal recombinations

In rare instances, heterozygous animals show a patch of fur for the recessive phenotype, and then the gene has definitely affected the coat. Recombinational processes are among the most potent mechanisms for expression of recessive genes. With reciprocal recombination a single event is sufficient to result in the expression of all recessive mutations of a chromosomal segment, whereas with gene mutations several single events would be needed in order to achieve a similar effect. In spite of that, as possible mechanisms in mammals they have not often been considered because mitotic chromosome pairing has not been observed in mammals.

Nevertheless, 5 years before Stern's (1936) classic work with *Drosophila*, Keeler (1931) considered the possibility of mitotic crossing-over (i.e. reciprocal recombination in mice), but proceeded to give a different interpretation of his data. Therefore, the first case with convincing evidence for mitotic reciprocal recombination in mammals excluding somatic mutation as a possible cause was a mosaic mouse born in 1948 (Carter, 1952).

It is possible to detect reciprocal recombination because the *p* and *c* loci are located on the same chromosome. A genetic alteration leading to $p\ c^{ch}/p\ c^{ch}$ gives rise to near-white coat spots which can be distinguished from white and light grey ones. The most probable reason for appearance of near-white spots is reciprocal recombination due to mitotic crossing-over.

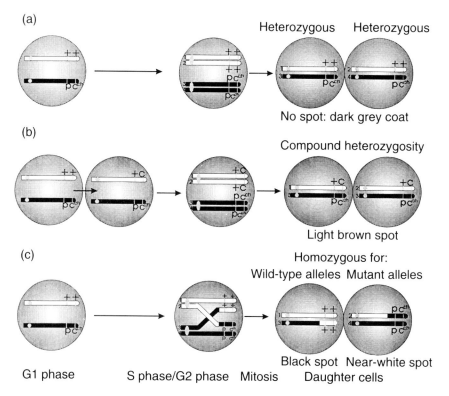

Figure 10.3. Diagrammatic representation of how to distinguish between induced gene mutation and induced reciprocal recombination using the cross C57BL x T. (a) No genetic alteration. Chromosome 7 of the mouse is heterozygous at the *c* and *p* loci. After replication the homologue chromosomes consist of two identical sister chromatids. Mitosis leads to two daughter cells with the parental heterozygous genotype (i.e., will lead to a coat with dark grey colour). (b) Gene mutation: + to *c*. A mutation of the wild-type allele at the *c* locus is (as well as non-reciprocal recombination) possible before replication. After mitosis this mutation leads to daughter cells with the two mutant alleles *c* and *c*ch (compound heterozygosity). This gives rise to a light brown spot. (c) Reciprocal recombination. A crossing-over between centromere and the *p* and *c* loci leads to an exchange of the mutant and the wild-type alleles. Mitosis leads to homozygosity of the wild-type alleles in one of the daughter cells and to homozygosity of the mutant alleles in the other. Homozygous wild-type alleles can be seen as black spots on the dark grey coat, homozygous mutant alleles as near-white spots. If both reciprocal products survive, a twin spot may arise. No genetic alteration other than reciprocal recombination will result in twin spots.

The two reciprocal products are cells of the genotypes *p* c^{ch}/p c^{ch} and *P C/P C*. Cells of the genotype *P C/P C* show up as black spots and can easily be detected, at least on the abdomen. Under the influence of the heterozygous coat colour mutations the level of pigmentation is lower in F_1 animals than in the mother animal. Appearance of both reciprocal products as a twin spot is proof of reciprocal recombination. Experience with yeasts has shown that

Figure 10.4. (C57BL × T) F_1 with a light brown colour spot. Such spots express compound heterozygosity of the two different mutant alleles c and c^{ch}.

reciprocal recombination does not lead in every case to the appearance of both reciprocal products. Therefore, a twin spot is a relatively rare event.

10.6 Microscopical pigment analysis

Using microscopical analysis of hairs from a colour spot it is possible to identify the gene loci affected by a genetic alteration. Microscopical hair analysis of spots of genetic relevance has been described in detail by Fahrig and Neuhäuser-Klaus (1985). Misdifferentiations can be distinguished from spots of genetic relevance by a bright yellow fluorescence, white spots by lack of pigment. The fine structure of mouse hairs is essentially similar for all hair types: a wide central medulla surrounded by a narrow cortex, in turn, surrounded by a thin cuticle.

C57BL mice are homozygous for non-agouti (a/a), their medullary cells being almost filled with black pigment granules. Although (C57BL × T) F_1 mice are also homozygous for non-agouti, they have considerably less pigment granules than their mothers. This phenomenon, presumably caused by the heterozygous recessive coat-colour alleles, allows detection of black spots on the dark grey coat. Such spots can be clearly recognized on the ventral side because it is less pigmented than the back. Black spots are, as already discussed, presumably the result of a reciprocal recombination.

The single gene loci can be identified as follows (*Figure 10.5*):

(i) in most hairs of dilute black (a/a; d/d) mice, several clumps of melanin granules disrupt the otherwise normal pattern of pigment distribution. Alterations in the refraction of light produces a grey colour.

(ii) Non-agouti pink-eyed dilution animals (a/a; p/p) show the same colour of pigment but considerably reduced pigmentation. Moreover, the form of the pigment granules is changed to an irregular, shredded shape with flocculent clumping. The coat colour resulting from these alterations is light grey.

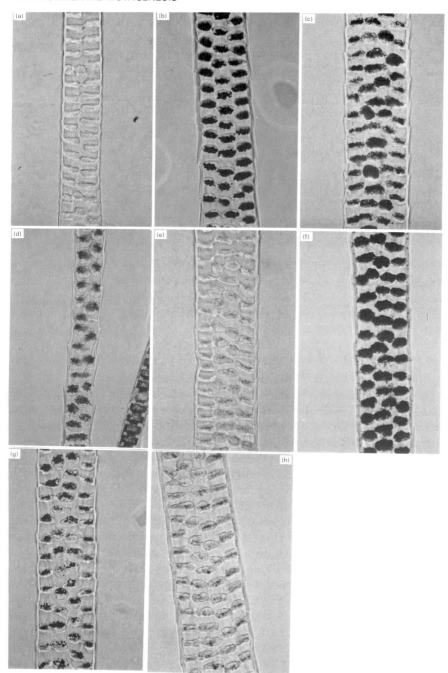

Figure 10.5. Microscopical analysis of hair pigments in (C57BL x T) F₁:
(a) White, (b) dark grey to black of the (C57BL x T) F₁, (c) grey (*d/d*), (d) brown
(*b/b*). (e) Near white (*cᶜʰ p/ cᶜʰ p*), (f) black (*C P/C P*), (g) light grey (*p/p*), (h) light
brown (*c/cᶜʰ*).

(iii) Homozygotes a/a; c^{ch}/c^{ch} and heterozygotes a/a; c^{ch}/c are easily distinguished visually. While a/a; c^{ch}/c^{ch} results in a dull black or sepia colour that is nearly indistinguishable from the (C57BL x T) F_1 black, a/a; c^{ch}/c gives a light brown colour. Spots of such colour can be induced only by gene mutations.

(iv) As b/b mutations show a clear brown and possess round granules instead of oval ones, light brown a/a; c^{ch}/c and brown a/a; b/b coat colours can easily be distinguished from one another.

(v) A drastic reduction in pigmentation can be observed in (a/a; $p\ c^{ch}/p\ c^{ch}$)-mice. Some hairs contain no pigment, others only traces. Such near-white hairs are an indication that reciprocal recombination has occurred.

A deletion covering the c locus as well as the p locus or loss of the whole chromosome would lead as well to black or near-white spots. But it is questionable whether deletions of such size or chromosome loss would lead to viable cells. By way of contrast, recombination always leads to viable cells.

10.7 Treatment

One has to treat enough pregnant females such that for each dose level a sufficient number of offspring is born. At a control frequency of 2% colour spots, for example, 300 F_1 animals per group have to be examined (Miltenburger, 1986). This corresponds to the treatment of about 50 pregnant females. The test substance is given to the pregnant females in such a way that it can be assumed that the test substance or its metabolite will reach the embryonic target cells. Other toxicological experiments, the chemical structure of the substance and its physicochemical properties can be used for determining the appropriate route of exposure (Fahrig *et al.*, 1991).

At least two different doses have to be used, one of which results in signs of toxicity in the mother animals or in reduced litter size. With non-toxic substances, treatment should not exceed 2000 mg kg^{-1}. Treatment is possible on days 8, 9, 10 or 11 of pregnancy, day 1 being the day on which the vaginal plug is detected. Successive treatment on all these days is also possible.

10.8 Categories of colour spots

Three or four weeks after birth the offspring are examined for colour spots. At the same time, the litters should be coded. Three categories of colour spots can be distinguished:

(i) white ventral spots which normally touch the ventral line. It is highly probable that such spots have been induced by pigment cell killing (WMVS, white midventral spots).

(ii) Yellow, agouti-like spots which are in all likelihood the result of misdifferentiations (MDS, misdifferentiated spots).

(iii) Pigmented black, grey, brown or near-white spots randomly distributed over the whole coat, which are the result of genetic alterations (SGR, spots of genetic relevance). By microscopical hair analysis it is possible to assign a single spot to a single coat colour mutation (see Section 10.6). For routine experiments these examinations are unnecessary (Fahrig, 1984; Hart *et al.*, 1982).

10.9 Validation

Since 1975 experimental results with the spot test have been published for 80 substances. Most published data deal with known mutagens; most experiments were aimed at checking the sensitivity and reliability of the spot test for detecting different classes of mutagens. These experiments have shown that the spot test is not only susceptible to standard mutagens (like ethylnitrosourea, methyl methanesulphonate) but also to mutagens that are difficult to detect *in vivo* (e.g. hydrazine and 4-chloro-*o*-toluidine-HC), and specific point mutagens (such as 2-aminopurine and other base analogues, acridine orange, hycanthone and other frame-shift mutagens) (Fahrig, 1978; Fahrig *et al.*, 1982; Lang, 1984; Neuhäuser-Klaus, 1991; Russell *et al.*, 1981; Styles and Penman, 1985). Sometimes the reliability of data originating in the 1970s is restricted by the fact that too few animals were used. Under these circumstances, even with strong mutagens, weak effects will result (Russell *et al.*, 1981). Using high numbers of animals, even with different strains of mice, similar and reproducible results will be gained (Fahrig, 1988; Fahrig and Neuhäuser-Klaus, 1989; Hart and Fahrig, 1985).

In the course of an international collaborative study, pairs of substances have been tested which have similar molecular structures, but only one is a carcinogen and the other is a non-carcinogen [examples: benzo[*a*]pyrene as carcinogen, pyrene as non-carcinogen; 2-acetylaminofluorene as carcinogen, 4-acetylaminofluorene as non-carcinogen]. The spot test was able to distinguish between carcinogens and non-carcinogens (Fahrig, 1988).

The published results for 60 chemicals and X-rays investigated up to 1984 in the mouse spot test were compared with data on the same chemicals tested in the Ames test and lifetime rodent bioassays (Styles and Penman, 1985). The performance of the spot test as an *in vivo* complementary assay to the *in vitro* bacterial mutagenesis test revealed that of 60 agents, 38 were positive in both systems, six were positive only in the spot test. Ten were positive only in the bacterial test and six were negative in both assays. The spot test was also considered as a predictor of carcinogenesis: 45 chemicals were carcinogenic of which 35 were detected as positive by the spot test and three out of six non-carcinogens were correctly identified as negative.

If the results are regarded in sequence (i.e. that a positive result in a bacterial mutagenicity test reveals potential that may or may not be realized *in vivo*) then 48 chemicals were mutagenic in the bacterial mutation assay of which 38 were active in the spot test and 31 were confirmed as carcinogens in bioassays. Twelve chemicals were non-mutagenic to bacteria of which six gave positive responses in the spot test and five were confirmed as carcinogens. These results provide strong evidence that the mouse spot test is an effective complementary test to the bacterial mutagenesis assay for the detection of genotoxic chemicals and as a confirmatory test for the identification of carcinogens. The main deficiency up to 1984 was the paucity of data from the testing of non-carcinogens.

The data published since 1984 are summarized in *Table 10.2*. It is apparent that as well as known mutagens and genotoxic carcinogens, non-carcinogens and non-genotoxic carcinogens/tumour promoters have been examined. A new aspect is the search for co- and anti-mutagens as well as for co- and anti-recombinogens.

10.10 Comments

EC and OECD guidelines exist for the spot test (Guideline 67/548/EWG, 1988; OECD Guideline 484, 1986). Spot tests will be used in particular to determine the relevance of positive *in vitro* data for the *in vivo* situation and for man. The spot test should be performed in those cases where positive results have been gained with cell culture employing gene mutation as the endpoint, and negative results in assays for chromosome aberrations. If it can be assumed that the test substance is systemically distributed and is able to reach the embryonic target cells to a considerable extent, then great importance has to be attached to the spot test.

In *Table 10.3*, 21 compounds are listed for which results were obtained in the specific locus test (Neuhäuser-Klaus and Ehling, 1993) and the spot test. The compounds differ in their germ cell-specific action. Six were mutagenic both in post-spermatogonial germ cell stages and in spermatogonia. Nine compounds were mutagenic only in post-spermatogonia. One compound, urethane, was negative and with four compounds the specific-locus test was only performed for spermatogonia (g), and an effect of the compounds on post-spermatogonial (pg) germ cell stages was not ruled out. The comparison shows a concordance for germ cell mutagens: all germ cell mutagens were also positive in the mouse spot test. Two compounds were negative in both tests, diethylnitrosamine and vincristine. Benzo[*a*]pyrene, hycanthone methane-sulphonate and urethane were positive in the spot test but negative in the germ cell specific-locus test. Four of the five germ cell-negative compounds were incompletely tested for mutagenic effects on post-spermatogonial germ cell stages. The only extensively tested compound with a negative result in the germ cell test and a positive result in the spot test was urethane.

Table 10.2. Results published from 1984

Chemical	Mouse strain	Dose (mg kg⁻¹)	Route	Day	No. of F_1 observed	F_1-with SGR (%)	Result	References
Ethylnitrosourea (ENU)	C57BLxT	30	i.p.	9	325	54 (17 %)	+	Fahrig (1984)
Catechol		3 x 22	i.p.	9,10, 11	216	2 (1 %)	–	
D-limonene		3x215	i.p.	9,10,11	291	4 (1 %)	–	
Catechol + ENU		3 x 22 + 30	i.p.	9,10,11 + 9	469	100 (21 %)	Comutagenic	
D-limonene + ENU		3x215 + 30	i.p.	9,10,11 + 9	797	110 (14 %)	Antimutagen	
Control (buffer)		0	i.p.	9,10,11	288	3 (1 %		
Dimethylbenzanthrazene	C57BLx	12.5	p.o.	10	106	2 (1.9 %)	–	Shibuya and
(DMBA)	PW	25			108	4 (3.7 %)	(±)	Murota (1984)
		50			132	9 (6.8 %)	+	
Control (oil)		0			103	1 (1 %)		
DMBA+ phenobarbital		see *Figure 10. 7*					Antimutagenic	
Dimethylsulphate	C57BLxT	50	i.p.	10	139	1 (0.72 %)	-	Braun *et al.*
Ethylmethanesulphonate		100			56	4 (7.1 %)	+	(1984)
ENU		20			238	13 (5.5 %)	+	
Methylnitrosourea		4			51	1 (2 %)	–	
Trenimon		0.1			75	0	–	
Diethylsulphate		225				3 (3 %)	(±)	
Methylmethanesulphonate		125			125	2 (1.6 %)	–	
Isoniazid		100			190	13 (6.8 %)	+	
Control (saline pooled)		0			1710	9 (0.5 %)		
ENU	NMRIxPDB	50	i.p.	8–11	465	99 (21 %)	+	Hart (1985)
Control(citric acid)		0.1 ml	i.p.	8–11	356	2 (0.6 %)		
Isoniazid		100	i.p.	10	116	3 (2.6 %)	(±)	
		150			243	6 (2.5 %)	+	
		200			93	1 (1.1 %)	–	
Procarbazine		50	i.p.	10	72	7 (9.7 %)	+	
Cyclophosphamide		5	i.p.	10	151	4 (2.6 %)	+	
4-NQO		15	i.p.	10	134	4 (3 %)	+	
Control (saline)		0	i.p.	10	504	4 (0.8 %)		
Dimethylaminoazobenzene	C57BLxT	50	i.p.	9	390	9 (2.3%)	–	Hart and
(DAB)		100			407	13 (3.2%)	–	Fahrig (1985)
	NMRIxPDB	150			307	4 (1.3 %)	–	
		200			272	3 (1.1 %)	–	
4-Cyanodimethylaniline	C57BLxT	100	i.p.	9	278	3 (1.1 %)	–	
(CDA)		200			317	5 (1.6 %)	–	
	NMRIxPDB	150			310	9 (2.9 %)	(±)	
		200			293	4 (1.4 %)	–	
Control (oil)	C57BLxT	0	i.p.	9	224	3 (1.3 %)		
	NMRIxPDB	0			479	3 (0.6 %)		
Ethylnitrosourea	NMRIxPDB	10	i.p.	9	118	8 (6.8 %)	+	Nielsen (1986)
(ENU)		20			130	33 (25.4 %)	+	
		40			128	41 (32 %)	+	
Control (saline)		0	i.p.	9	87	3 (3.5 %)		
Dibromochloropropane	C57BLx	106	i.p.	10	721	21 (2.9 %)	+	Sasaki *et al.*
ENU	PW	50			189	45 (23.8 %)	+	(1986)
Control (oil)		0			643	4 (0.6 %)		
TPA	C57BLx T	2x0.1	i.p	9, 10	478	8 (1.7 %)	–	Fahrig (1987b)
		1x1			72	2 (2.8 %)	–	
Control (buffer)		0			550	10 (1.8 %)		
TPA + ENU		see *Figure 10.7*					Comutagen	
Benzo[a]pyrene	T x HT	0	p.o.	9	559	12 (2.1 %)		Neuhäuser-
(BP)		50			281	14 (5 %)	+	Klaus (1988)
		100			272	11 (4 %)	+	
		150			295	40 (13.6 %)	+	
Pyrene (PYR)	T x HT	0	p.o.	9	559	12 (2.1 %)		
		200			220	4 (1.8 %)	–	
		400			267	6 (2.2 %)	–	
		600			197	4 (2 %)	–	

Table 10.2. Results published from 1984 (continued)

Chemical	Mouse strain	Dose (mg kg⁻¹)	Route	Day	No. of F₁ observed	F₁-with SGR (%)	Result	References
2-Acetylaminofluorene (2AAF)	T x HT	0	i.p.	9	734	12 (1.6 %)		Neuhäuser-
		200			229	5 (2.2 %)	+	Klaus (1988)
		400			268	15 (5.6 %)	+	
		600			238	8 (3.4 %)	+	
		800			178	9 (5.1 %)	+	
	C57BLxT	0	i.p.	9	214	3 (1.4 %)		Fahrig (1988)
		400			175	9 (5 %)	+	
		0	p.o.	11	151	0		Hüttner et al.
		446			195	1 (0.5 %)	−	(1988)
4-Acetylaminofluorene (4AAF)	T x HT	0	i.p.	9	734	12 (1.6 %)		Neuhäuser-
		100			282	5 (1.8 %)	−	Klaus (1988)
		200			310	7 (2.3 %)	−	
		400			261	5 (1.9 %)	−	
		600			157	3 (1.9 %)	−	
	C57BLxT	0	i.p.	9	214	3 (1.4 %)		Fahrig (1988)
		400			420	6 (1.4 %)	−	
		0	p.o.	11	151	0		Hüttner et al.
		446			191	0		(1988)
Caprolactam	T x HT	500	i.p.	9	1018	58 (5.7 %)	+	Neuhäuser-
		700			122	6 (4.9 %)	+	Klaus and
Control (saline)		0			637	22 (3.5 %)		Lehmacher (1989)
Caprolactam	C57BLx T	400	i.p.	9	397	11 (2.8 %)	(±)	Fahrig (1989)
		500			487	11 (2.3 %)	(±)	
		500			490	17 (3.5 %)	+	
Control (buffer)		0			407	6 (1.5 %)		
Acrylamide	T x HT	50	i.p.	12	213	14 (6.6 %)	+	Neuhäuser-
		75		12	211	13 (6.1%)	+	Klaus and
		3x50		10–12	196	26 (13.3 %)	+	Schmahl (1989)
		3x75		10–12	215	21 (9.8 %)	+	
Control (buffer)			i.p.	1–3x	437	11 (2.5 %)		
Hypericum extract	NMRI x	1	p.o.	9	240	3 (1.3 %)	−	Okpanyi et al.
	DBA/2J	5			236	3 (1.3 %)	−	(1990)
		10			285	5 (1.8 %)	−	
Control	0				226	3 (1.1 %)		
Tannic acid	C57BLxPW	500	p.o.	10	266	5 (2.2 %)	−	Sasaki et al.
Control (aqua dest.)		0			566	5 (1 %)		(1990)
Tannic acid+ENU		see Figure 10.7					Antimutagen	
Vanillin	C57BLxPW	3x500	p.o	10	220	1 (0.5 %)	−	Imanishi et al.
Control (DMSO)		0	.		566	0		(1990)
Vanillin + ENU		see Figure 10.7					Antimutagen	
2-Amino-N-hydroxyadenine	T x HT	16	i.p.	9	139	5 (3.6 %)	−	Neuhäuser-
		20			133	11 (8.3 %)	+	Klaus (1991)
		40			103	12 (11.7 %)	+	
Control (aqua dest.)		0			332	13 (3.9 %)		
Cyclophosphamide	C57BLxT	7.5	s.c.	9.5	206	6 (2.9 %)	+	Braun and
		7.5	s.c.	10.5	178	6 (3.4 %)	+	Hüttner (1991)
Control (saline)		0	s.c.	9.5	222	1 (0.5 %)		
		0	s.c.	10.5	238	1 (0.4 %)		
Dioxin mixture	C57BLxT	0.128	i.p.	9	168	1 (0.6 %)	−	Fahrig (1993)
Control (DMSO)		0			208	3 (1.4 %)		
TCDD + ENU		see Figure 10.6					Corecombinogen	
M-Phenylenediamine	C57BLx T	5	i.p	9	488	3 (0.6 %)	−	Umweltbund./
Control (aqua dest.)		15			116	1 (0.6 %)	−	Fahrig (1993)
		0			117	0		
1,2-Dichlorethane	C57BLxT	400	i.p.	9	579	4 (0.7 %)	−	Umweltbund./
Control (oil)		0			298	2 (0.7 %)		Fahrig (1994)
(E)-5-(2-bromovinyl)-2'-deoxyuridine (BVDU)	C57BLxT	150	i.p.	9	498	2 (0.4 %)	−	Berlin-Chemie/
		280			513	1 (0.2 %)	−	Fahrig (1994)
		0			472	4 (0.9 %)		

SGR, spots of genetic relevance.

Table 10.3. Correlation of test results from compounds tested in the specific locus test (germ cells) and the mouse spot test (somatic cells; Neuhäuser-Klaus and Ehling, 1993; and personal communication)

Specific locus test		Compound	Spot test
pg	g		
+		Acrylamide	+
+	+	Ethylnitrosourea	+
+		Methylnitrosourea	+
	+	Mitomycin C	+
+	+	Procarbazine-HCl	+
+	+	Propylnitrosourea	+
+	+	Triethylenemelamine	+
+	–	Busulphan	+
+	–	Chlorambucil	+
+	–	Chlormethine	+
+	–	Cyclophosphamide	+
+	–	Diethylsulphate	+
+	–	Ethyl methanesulphonate	+
+	–	Methyl methanesulphonate	+
+	–	n-Propyl methanesulphonate	+
	–	Benzo[a]pyrene	+
	(–)	Hycanthone methanesulphonate	+
–	–	Urethane	+
	–	Diethylnitrosamine	–
	–	Vincristine	–

10.11 Relationship between mutagenesis and carcinogenesis

The relationship between mutagenesis and carcinogenesis was investigated by Schmähl et al. (1990) in T x HT crossbred mice using diaplacental application of ethylnitrosourea (ENU) at different stages of embryonal development. Mutagenesis was detected by induction of coat colour spots, and the carcinogenic response was investigated in a long-term follow-up study of the F_1 generation. The animals were particularly sensitive to induction of tumours at the central nervous system (CNS)–skull/vertebra interface (30% and 20% in ENU-treated male and female offspring, respectively, compared with <1% in controls). There was a correlation between the appearance of these tumours and the presence of colour spots. This correlation was low but statistically significant in female offspring. Three other types of tumours showed a correlation with the presence of coat colour spots. Liver tumours were significantly increased in colour spot-positive females but unchanged in males. Lung tumours were reduced in colour spot-positive males and appeared earlier in colour spot-positive females. There was a lower incidence of lymphoma/leukaemia in all spot-positive mice. The reduction in tumour incidence below the spontaneous rate in spot-positive animals might be caused by a high cytolethal response to ENU in the relevant organs and tissues.

Another study by Nomura *et al.* (1990) combined X-rays and 12-*O*-tetradecanoyl-phorbol-13-acetate (TPA). Although *in utero* irradiation at early stages induced a high incidence of colour spots in (PT x HT) F_1, it was not carcinogenic by itself. However, *in utero*-irradiated animals did develop skin tumours and hepatomas (but not leukaemias) by the post-natal administration of TPA. The incidence of both tumours and colour spots increased with *in utero* doses of X-rays. Furthermore, a large reduction of tumour incidence, about 80%, was observed by low dose rate irradiation, similar to the 75% reduction in spot frequency. The tumour nodule size was also dramatically reduced by low dose-rate irradiation. Consequently, the induced incidence and size of tumours produced by TPA treatment parallel those observed for coat colour mutations.

10.12 Enhancement or reduction of mutations or recombinations

In contrast to the large number of chemicals which have been tested for their ability to enhance mutations, few chemicals have been tested for corecombinogenic activity. In general, substances that induce mutation can also induce recombination because both effects depend on DNA damage. While in most, but not all, cases gene mutations may arise through a special DNA repair process that introduces errors as part of the correction mechanism, recombination itself is a DNA repair process. Two modes of exchange are involved in recombination, reciprocal exchange (crossing-over) and non-reciprocal exchange (gene conversion). Non-reciprocal recombinations appear unidirectional, as the transfer of genetic information from one gene to its homologue allele leaves the donor gene unchanged. Studies addressing the proportion of products generated by reciprocal and non-reciprocal recombination have generally found a preponderance of gene conversion (Bollag and Liskay, 1988; Rubnitz and Subramani, 1986).

Co-recombinogenic effects were observed in experiments using yeast: tumour promoters were co-recombinogenic and antimutagenic, substances being tumour promoters as well as cocarcinogens were co-recombinogenic as well as comutagenic (Fahrig, 1979, 1984, 1987a). To confirm the results obtained in yeast, spot tests were performed (Fahrig, 1984). In these verification experiments, the co-recombinogen D-limonene and the anti-recombinogen catechol were used. Two substances which had not been tested or were ineffective in yeast were also used: TPA (Fahrig, 1987b) and 2,3,7,8-tetrachloro-dibenzo-*p*-dioxin (TCDD) (Fahrig, 1993). The direct alkylating agent ENU served as the mutagen.

The data are summarized in *Figure 10.6*. As discussed in Section 10.5, light brown spots are the result of induced mutation and black and near-white spots result from recombination. About 20% of all spots induced with ENU belong to these categories. ENU alone induced about 8% of colour spots depending on reciprocal recombination. Under the influence of TPA or D-limonene about 14% of all colour spots are presumed products of reciprocal recombination.

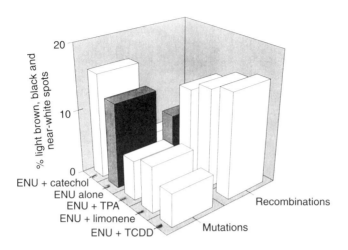

Figure 10.6. Co-recombinogenic and antimutagenic effects. Light brown spots are the result of induced mutation. See text for details.

The frequency for TCDD is 16%. The co-recombinogenic effects are very strong when the low frequency of twin spots is also considered: 0.2% after ENU treatment alone, the enhancement of up to 1.3% for TPA, 7% for D-limonene and 3.7% for TCDD. It is perhaps worth mentioning that TCDD was active at a dose of only 3 µg kg^{-1} and TPA at 0.2–0.33 mg kg^{-1}. These are comparable to doses used in tumour promotion experiments.

Just as co-recombinogenic effects could be observed, the substances showed antimutagenic effects: while ENU alone induced about 13% light brown colour spots depending exclusively on gene mutations (control = 11%), under the influence of TPA the frequency was only 6%, D-limonene 7%, and TCDD 5%. In contrast to this, catechol gave a frequency of light brown spots of 16%, and a frequency of black or near-white spots of only 4%. Twin spots were not observed. This result can be interpreted as comutagenicity and anti-recombinogenicity. Thus, the *in vitro* results with yeast were confirmed by *in vivo* results with mice.

Antimutagenic effects of vanillin, phenobarbital and tannic acid in (C57BL x PW) F_1 mice, and comutagenic effects of TPA in (C57BL x T) F_1 mice can be seen in *Figure 10.7*. TPA was comutagenic/recombinogenic when administered intra peritoneally twice after ENU treatment on days 9 and 10 (Fahrig, 1987b). Shibuya and Murota (1984) investigated 7,12-dimethylbenz[a]anthracene (DMBA) with and without phenobarbital pretreatment. Phenobarbital (80 mg kg^{-1}) was antimutagenic when administered daily (p.o.) for 3 days before to DMBA treatment. Sasaki *et al.* (1990) observed antimutagenic effects of tannic acid in combination with 50 mg kg^{-1} ENU; 200–500 mg kg^{-1} tannic acic were given orally 6 h before intraperitoneal injection of ENU. Imanishi *et al.* (1990) observed antimutagenic effects of vanillin. Pregnant females were given three successive

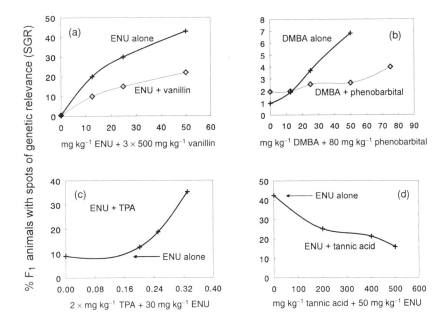

Figure 10.7. Antimutagenic effects of vanillin, phenobarbital and tannic acid in (C57BL x PW) F_1-mice, comutagenic effects of TPA in (C57BL x T) F_1.

(a) Antimutagenic effects of vanillin: pregnant females were given three successive oral administrations of 125–500 mg kg^{-1} vanillin 0, 4 and 24 h after intraperitoneal injection of 50 mg kg^{-1} ENU (Imanishi *et al.*, 1990).

(b) 7, 12-Dimethylbenz[*a*]anthracene (DMBA) with and without phenobarbital pretreatment. Phenobarbital (80 mg kg^{-1}) was antimutagenic when administered daily (p.o.) for 3 days prior to DMBA treatment (Shibuya and Murota, 1984).

(c) 12-*O*-tetradecanoyl-phorbol-13-acetate (TPA) was comutagenic/recombinogenic when administered i.p. twice after ENU treatment on days 9 and 10 (Fahrig, 1987b).

(d) Tannic acid in combination with 50 mg kg^{-1} ENU: 200–500 mg kg^{-1} tannic acid were given orally 6 h before intraperitoneal injection of ENU (Sasaki *et al.*, 1990).

oral administrations of 125–500 mg kg^{-1} vanillin 0.4 and 24 h after intraperitoneal injection of 50 mg kg^{-1} ENU. In experiments on yeast, co-recombinogenic and antimutagenic effects of tannic acid and phenobarbital could be observed (Fahrig, in press). Therefore, it would be interesting to know if the three antimutagens phenobarbital, vanillin and tannic acid are effective as co-recombinogens. This would require microscopical hair pigment analysis.

References

Ames B, Gold LS. (1990) Carcinogenesis debate. *Science* **250:** 1498–1499.
Berlin-Chemie/Fahrig. (1994) Spot test **210/211.**

Bollag RJ, Liskay, RM. (1988) Conservative intrachromosomal recombination between inverted repeats in mouse cells: association between reciprocal exchange and gene conversion. *Genetics* **119:** 161–169.

Braun R, Hüttner E, and Schöneich J. (1984) Transplacental genetic and cytogenetic effects of alkylating agents in the mouse. I: Induction of somatic coat color mutations. *Teratogen. Carcinogen. Mutagen.* **4:** 449–457.

Braun R, Hüttner E. (1991) Influence of ethanol on genetic activity of cyclophosphamide in the mammalian spot test. *Biol. Zbl.* **110:** 284–289.

Carter TC. (1952) A mosaic mouse with an anomalous segregation ratio. *J. Genet.* **51:** 1–6.

Davidson GE, Dawson GWP. (1976) Chemically induced presumed somatic mutations in the mouse. *Mutat. Res.* **38:** 151–154.

Fahrig R. (1975) A mammalian spot test: induction of genetic alterations in pigment cells of mouse embryos with X-rays and chemical mutagens. *Mol. Gen. Genet.* **138:** 309–314.

Fahrig R. (1978) The mammalian spot test, a sensitive *in vivo* method for the detection of genetic alterations in somatic cells of mice. In: *Chemical Mutagens: Principles and Methods for their Detection*, Vol. 5 (eds A Hollaender, FJ de Serres). Plenum Press, New York, pp. 151–176.

Fahrig R. (1979) Evidence that induction and suppression of mutations and recombinations by chemical mutagens in *S. cerevisiae* during mitosis are jointly correlated. *Mol. Gen. Genet.* **169:** 125–139.

Fahrig R. (1984) Genetic mode of action of cocarcinogens and tumour promoters in yeast and mice. *Mol. Gen. Genet.* **194:** 7-14

Fahrig R. (1987a) Effects of bile acids on the mutagenicity and recombinogenicity of triethylene melamine in yeast strain MP1 and D61M. *Arch. Toxicol.* **60:** 192–197.

Fahrig R. (1987b) Enhancement of carcinogen-induced mutations or recombinations by 12-*O*-tetradecanoyl-phorbol-13-acetate in the mammalian spot test. *J. Cancer Res. Clin. Oncol.* **113:** 61–66.

Fahrig R. (1988) Summary Report on the performance of the mammalian spot test. In: *Evaluation of Short-term In Vivo Tests for Carcinogens.* Report of The International Programme on Chemical Safety's Collaborative Study on *in vivo* Assays II (eds J Ashby, FJ de Serres, MD Shelby, BH Margolin, M Ishidate, GC Becking). Cambridge University Press, Cambridge, UK, pp.151–158.

Fahrig R. (1989) Possible recombinogenic effect of Caprolactam in the mammalian spot test. *Mutat. Res.* **224:** 373–375.

Fahrig R. (1992a) Tests for recombinagens in mammals *in vivo*. *Mutat. Res.* **284:** 177–183.

Fahrig R. (1992b) Co-recombinagenic effects. *Mutat. Res.* **284:** 185–193.

Fahrig R. (1993) Genetic effects of dioxins in the spot test with mice. *Environ. Hlth Perspect.* **101:** 257–261.

Fahrig R. (1995) Anti-mutagenic agents are also co-recombinogenic and can be converted into co-mutagens. *Mutation Res.,* in press.

Fahrig R, Neuhäuser-Klaus A. (1985) Similar pigmentation characteristics in the specific locus and the mammalian spot test: a way to distinguish between induced mutation and recombination. *J. Hered.* **76:** 421–426.

Fahrig R, Neuhäuser-Klaus A. (1989) Positive response of Caprolactam in the mammalian spot test. *Mutat. Res.* **224:** 377–378.

Fahrig R, Lang R, Madle S. (1991) General strategy for the assessment of genotoxicity. *Mutat. Res.* **252:** 161–163.

Fahrig R, Dawson GWP, Russell LB. (1982) Mutagenicity of selected chemicals in the mammalian spot test. In: *Comparative Chemical Mutagenesis* (eds FJ de Serres and MD Shelby). Plenum Press, New York, pp. 713–732.

Guideline 67/548/EWG. (1988) In vivo-*Säuger-Fellfleckentest der Maus.* Amtsblatt der EG, no. L 133, May 30, 1988, 82–84.

Hart JW. (1985) The mouse spot test: results with a new cross. *Arch. Toxicol.* **58:** 1–4.

Hart JW, Fahrig R. (1985) Effects of 4-dimethylamino-azobenzene (DAB) and 4-cyanodimethylaniline (CDA) in the mammalian spot test. In: *Comparative Genetic Toxicology* (eds JM Parry, CF Arlett). Macmillan Press, London, pp. 507–513.

Hart JW, Jensen NJ, Larsen S. (1982) Fluorescence microscopy as an aid in differentiation of coat-colour mosaics in the mammalian spot test. *Mutat. Res.* **105:** 357–361.

Hüttner E, Braun R, Schöneich J. (1988) Mammalian spot test with the mouse for detection of transplacental genetic effects induced by 2-acetylaminofluorene and 4-acetylaminofluorene. In: *Evaluation of Short-term* In Vivo *Tests for Carcinogens*. Report of The International Programme on Chemical Safety's Collaborative Study on *in vivo* Assays II. (eds J Ashby, FJ de Serres, MD Shelby, BH Margolin, M Ishidate, GC Becking). Cambridge University Press, Cambridge, UK, pp. 164–167.

Imanishi H, Sasaki YF, Matsumoto K, Watanabe M, Ohta T, Shirasu Y, Tutikawa K. (1990) Suppression of 6-TG-resistant mutations in V79 cells and recessive spot formations in mice by vanillin. *Mutat. Res.* **243:** 151–158.

Keeler CE (1931) A probable new mutation to white-belly in the house mouse, *Mus musculus*. *Proc. Natl Acad. Sci. USA* **17:** 700–703.

Lang R. (1978) Mammalian spot test with moxnidazole and 5-nitroimidazole. *Experientia* **34:** 500–501.

Lang R. (1984) The mammalian spot test and its use for testing of mutagenic and carcinogenic potential: experience with the pesticide chlordimeform, its principal metabolites and the drug lisuride hydrogen maleate. *Mutat. Res.* **135:** 219–224.

Miltenburger HG. (1986) Der Fellfleckentest - ein Routine-Verfahren zum Nachweis somatischer Mutationen *in vivo*. In: *Carcinogenese, Mutagenese, Teratogenese*. Der Bundesminister für Forschung und Technologie, Bonn, pp. 64–76.

Mintz B. (1967) Gene control of mammalian pigmentary differentiation. I. Clonal origin of melanocytes. *Proc. Natl Acad. Sci. USA* **58:** 344–351.

Neuhäuser-Klaus A. (1981) An approach towards the standardization of the mammalian spot test. *Arch. Toxicol.* **48:** 229–243.

Neuhäuser-Klaus A. (1988) Evaluation of somatic mutations in (T x HT) F1. In: *Evaluation of Short-term* In Vivo *Tests for Carcinogens*. Report of the International Programme on Chemical Safety's Collaborative Study on *in vivo* Assays II (eds J Ashby, FJ. de Serres, MD Shelby, BH Margolin, M Ishidate, GC Becking). Cambridge University Press, Cambridge, pp. 168–172.

Neuhäuser-Klaus A. (1991) Mutagenic activity of 2-amino-N^6-hydroxyadenine in the mouse spot test. *Mutat. Res.* **253:** 109–114.

Neuhäuser-Klaus A, Schmahl W. (1989) Mutagenic and teratogenic effects of acrylamide in the mammalian spot test. *Mutat. Res.* **226:** 157–162.

Neuhäuser-Klaus A, Lehmacher W. (1989) The mutagenic effect of caprolactam in the spot test with (T x HT) F1 mouse embryos. *Mutat. Res.* **224:** 369–371.

Neuhäuser-Klaus A, Ehling UH. (1993) Der spezifische Locustest mit Mäusen. In: *Mutationsforschung und Genetische Toxikologie* (ed. R Fahrig). Wissenschaftliche Buchgesellschaft, Darmstadt, pp. 309–320.

Nielsen IM. (1986) Induction of coat colour spots in non-agouti NMRI x PDB, F1 dose response relationship with ethylnitrosurea. *Acta Pharmacol. Toxicol.* **48:** 159–160.

Nomura T, Nakajima H, Hatanaka T, Kinuta M, Hongyo T. (1990) Embryonic mutation as a possible cause of *in utero* carcinogenesis in mice revealed by postnatal treatment with 12-O-tetradecanoylphorbol-13-acetate. *Cancer Res.* **50:** 2135–2139.

OECD. (1986) OECD Guideline for Testing of Chemicals, no. 484, *Genetic Toxicology: Mouse Spot Test*. Paris, Oct. 23, 1986.

Okpanyi SN, Lidzba H, Scholl BC, Miltenburger HG. (1990) Genotoxizität eines standardisierten Hypericum-Extraktes. *Arzneimittelforsch.* **40:** 851–855.

Rawles ME. (1947) Origin of pigment cells from the neural crest in the mouse embryo. *Physiol. Zool.* **20:** 248–266.

Rubnitz J, Subramani S. (1986) Extrachromosomal and chromosomal gene conversion in mammalian cells. *Mol. Cell. Biol.* **6:** 1608–1614.

Russell LB, Major MH. (1957) Radiation-induced presumed somatic mutations in the house mouse. *Genetics* **42:** 161–175.

Russell LB, Selby PB, von Halle E, Sheridan W, Valcovic L. (1981) Use of the mouse spot test in chemical mutagenesis: interpretation of past data and recommendations for future work. *Mutat. Res.* **86:** 355–379.

Sasaki YF, Imanishi H, Watanabe M, Sekiguchi A, Moriya M, Shirasu, Y, Tutikawa K. (1986) Mutagenicity of 1,2-dibromo-3-chloropropane (DBCP) in the mouse spot test. *Mutat. Res.* **174:** 145–147.

Sasaki YF, Matsumoto K, Imanishi H, Watanabe M, Ohta T, Shirasu Y, Tutikawa K. (1990) *In vivo* anticlastogenic and antimutagenic effects of tannic acid in mice. *Mutat. Res.* **244:** 43–47.

Schmahl W, Neuhäuser-Klaus A, Leierseder-Bauer M, Luz A. (1990) Simultaneous induction of mutagenic and cancerogenic effects in T x HT mice with transplacental ethylnitrosourea treatment. *Teratogen. Carcinogen. Mutagen.* **10:** 307–320.

Shibuya T Murota T. (1984) Mouse spot tests with dimethylbenz[a]anthracene with and without phenobarbital pretreatment. *Mutat. Res.* **141:** 105–108.

Silvers WK. (1979) *The Coat Colours of Mice.* Springer-Verlag, New York.

Stern C. (1936) Somatic crossing-over and segregation in *Drosophila melanogaster. Genetics* **21:** 625–730.

Styles JA, Penman MG. (1985) The mouse spot test. Evaluation of its performance in identifying chemical mutagens and carcinogens. *Mutat. Res.* **154:** 183–204.

Umweltbundesamt/Fahrig (1993) Spot test with m-phenylenediamine. *UBA project* 116 06 092/01.

Umweltbundesamt/Fahrig (1994) Spot test with1,2-dichlorethane. *UBA project* 116 06 087.

Mammalian cell mutation assays

Donald Clive

11.1 Introduction

The principal assays for detecting genotoxic activity of test agents are the *Salmonella* reverse mutation assay ('Ames' assay; see Chapter 6) and *in vitro* cytogenetics in various mammalian cells (see Chapter 7). The former responds to several types of viable point mutations involving up to a few base pairs of DNA in a simple bacterial genome; the latter detects microscopically visible structural alterations to mammalian chromosomes that are of unknown viability and which involve at least 10 million base pairs (10 Mb). Viable genetic damage that involves more than 10 bp but less than 10 Mb is known to occur in human disease but has no representation in this minimal test battery. That deficiency is the current rationale for testing in mammalian cell mutation assays.

The diploid nature of eukaryotic cells has long been known to permit multigenic alterations such as gene conversion, mitotic recombination and chromosome rearrangements, in addition to the point mutations so well defined in bacterial systems. Despite this knowledge, the guiding principle behind the development of *in vitro* mammalian cell mutagenicity assays for most of the past 25 years has been to detect bacterial-type point mutations in a mammalian cell environment. The earliest of these assays utilized reverse mutations with nutritional markers which involved multistep biochemical pathways. These were soon displaced by forward mutations at the genetically simpler hypoxanthine–guanine phosphoribosyl transferase (*hprt*) locus, first in Chinese hamster ovary (CHO) cells, and later in Chinese hamster lung (V79), human and mouse cells. Forward mutation to ouabain resistance – involving the dominant Na^+/K^+-*ATPase* gene – was found to be useful for the detection of base-pair substitution mutations in mammalian cells.

Fortunately and serendipitously, these best intentions at developing more expensive and time-consuming versions of bacterial assays were occasionally thwarted. This happened particularly with the heterozygous thymidine kinase locus ($tk^{+/-}{\rightarrow}tk^{-/-}$) in L5178Y mouse lymphoma cells (the mouse lymphoma assay or MLA). This approach recurred in CHO cells with the adenine-guanine phosphoribosyl transferase locus ($aprt^{+/-}{\rightarrow}aprt^{-/-}$). Later still, the heterozygous tk locus was redeveloped in human cells (TK-6 assay).

The history and details of the development and experiences with these assays can be found in two excellent Banbury Conference volumes (Hsie et al., 1979; Moore et al., 1987) and will not be repeated here.

Over the past decade, cytogenetic and molecular studies of tumours and inherited genetic diseases of mice and men have provided a set of mammalian cell mutations that is not biased by the design of our test systems (Ali et al., 1987; Emanuel, 1988; Fearon and Vogelstein, 1990; Ngo et al., 1988; Russell et al., 1989; Sager, 1989; Shapiro et al., 1989; Turner et al., 1985, 1988; Weinberg, 1991). This led to an awareness of the limitations of bacterial and gross chromosomal models of genotoxicity and a debate over the theoretical and practical sufficiency of testing new compounds in only *Salmonella* and *in vitro* cytogenetics (Ashby, 1986a,b; Clive, 1985; DeMarini et al., 1989; Garner and Kirkland, 1986; Gatehouse and Tweats, 1986; Tweats and Gatehouse, 1988; Waters and Stack, 1988). This chapter focuses on the different types of genetic damage that can affect a mammalian gene and on the assay that best detects them. In order to compensate for preconceptions associated with bacterial assays, this discussion will be developed in the context of the organization and functioning of the mammalian genome.

11.2 The mammalian genome: scale, organization and function

The quantitative and qualitative extents to which the mammalian genome differs from that of bacteria are well illustrated in *Table 11.1* and *Figures 11.1* and *11.2*. This model of the mammalian chromosome (Marsden and Laemmli, 1979; Pienta and Coffey, 1984) has been helpful in understanding the variety of mutations detected in the MLA (Applegate et al., 1990; Clive et al., 1990; Liechty et al., 1994; P. Glover and D. Clive, unpublished).

Table 11.1. Amounts of DNA in various subunits of the mammalian genome

Level of structure	Amount of DNA	
	No. of bp	Size
Diameter of ds DNA	1	2.0 nm
Base pair	1	0.34 nm
Mouse *tk* gene	1.5×10^4	3.7 μm
Mouse *hprt* gene	4×10^4	14 μm
Single replicon	1×10^5	30 μm
Single radial loop	1×10^5	30 μm
High-resolution band	2×10^6	560 μm
Single chromatid	1.5×10^8	50 mm
Diploid genome	6×10^9	2.2 m

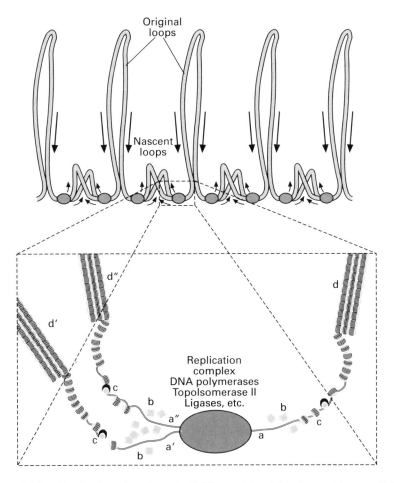

Figure 11.1. Replication of replicons/radial loops (after Marsden and Laemmli, 1979; Pardoll *et al.*, 1980; Vogelstein *et al.*, 1980). *Top:* five loops/replicons are shown, together with four interspersed pairs of daughter loops growing at their expense. *Bottom:* details of bidirectional replication process according to this model. The base end of an extant loop (d) relaxes from its 30 nm diameter solenoid form. Individual nucleosomes (c) dissociate into histone octamers (b) and DNA (a), which is fed into the fixed replication complex. Replicated DNA emerges (a' and a") and the dissociation process is reversed (b, c) to form growing parts of sister chromatids (d' and d").

The mammalian diploid genome contains about 6×10^9 bp (6000 Mb) of DNA compared with only about 4 Mb for a typical bacterial genome. Unlike the essentially naked DNA found in bacteria, DNA in nuclei of mammalian cells is organized into a hierarchy of structures (*Table 11.1, Figures 11.1* and *11.2*) to facilitate both its replication during an S phase of approximately 7 hours, and the accurate distribution of duplicate copies of DNA into two daughter cells. The machinery for accomplishing these feats in a relatively error-free manner is diagrammed in *Figure 11.1*. It involves

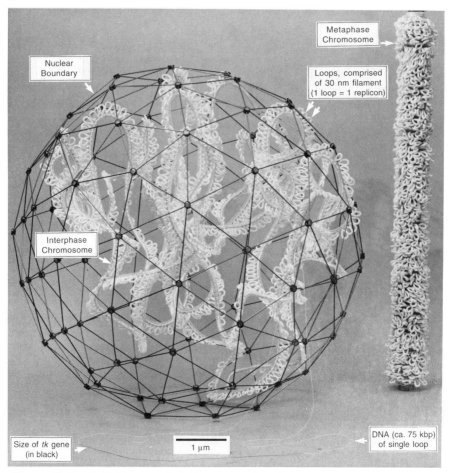

Figure 11.2. Radial loop model of a mammalian chromosome (constructed by the author based on work of Marsden and Laemmli, 1979; Pardoll *et al.*, 1980; Vogelstein *et al.*, 1980). A single average-length interphase chromosome, consisting of approximately 1000 replicons/loops each of 50–200 kb of DNA, is pictured within its nucleus (× 5000). One loop is 'denatured' into naked DNA (foreground) with a 15 kb *tk* gene shown in black. On the right is a tentative model of this same chromosome in metaphase. Bar = 1.0 μm.

serial levels of packaging of the DNA, first into nucleosomes, then these into a 30 nm solenoid filament. Higher-level organization involving this last structure is less well defined, but probably assumes the form of radial loops (Marsden and Laemmli, 1979; Pardoll *et al.*, 1980; Pienta *et al.*, 1989; Vogelstein *et al.*, 1980; *Figure 11.1*, top; *Figure 11.2*). This organization allows for the parallel processing of smaller lengths of DNA (replicons – Huberman and Riggs, 1968; Huberman and Tsai, 1973) by passage through a fixed replication complex (*Figure 11.1*, bottom). On the basis of size and bidirectional replication, radial loops and replicons appear to be equivalent.

Figure 11.2 illustrates a single average-length chromatid (mouse chromosome 11 or human chromosome 15) within a mammalian cell nucleus (about 5000 ×) constructed by the author according to this radial loop model. The following features are of relevance here:

(i)   The basic material is a 30 nm filament comprised of DNA wrapped about histone octamers to form nucleosomes; these nucleosomes are wound into the 30 nm diameter solenoid or filament.

(ii) This filament is organized into loops (replicons) radiating from a central axis that contains the sites of active DNA replication (labelled as 'replication complex' in *Figure 11.1*) and that extends the length of the chromosome. One of these loops is shown in denatured form to illustrate the relative dimensions of its DNA and a relatively small (*tk*; 15 kb) gene, shown in black.

(iii) There are about 1000 loops in this average chromosome and about 50–200 kb per loop; thus, about 50–200 Mb per average chromosome.

(iv) For clarity, the remaining 39 mouse (45 human) chromosomes in this same nucleus are not shown.

11.3 The mammalian genome: 'gene' mutations

Until recently, studies in mammalian cell 'gene' (the reason for the quotation marks will become apparent shortly) mutation assays have probably hindered rather than advanced our knowledge of the nature and mechanisms of mammalian cell mutagenicity. Such a paradox arose because the design of most of these tests was based on what was known at the time about bacterial gene structure and mechanisms of mutagenesis. In contrast, studies of unbiased mutational events occurring in somatic and germ cells of rodents and humans have revealed mechanisms and scales of damage that simply cannot occur in bacteria. Some of these will be discussed in the next section.

Let us pause to consider what is meant by a 'gene' mutation in a mammalian cell. One standard answer to this seemingly trivial question is:

Any qualitative or quantitative alteration to the DNA sequence of a particular gene.

This sounds straightforward. It includes transitions, transversions, frameshifts, small deletions, small insertions – all of the events detected in bacteria. But what if the entire gene in question – let us assume it to be a tumour suppressor gene – is simply deleted from the cell? Has *its* DNA sequence been qualitatively or quantitatively altered? Obviously not, since the entire sequence has simply been moved from inside the altered cell to elsewhere, either to a sibling cell during a previous mitosis, or to the extracellular milieu. By this definition, then, such a loss is not a mutation because the target gene can, in principle at least, be recovered from its new location, sequenced, and be found unchanged.

So we seem to have a cell that is missing a tumour suppressor gene without any gene mutation having occurred, a mutant cell without a mutated gene. To accommodate the cell's version of what happens, a better definition of a 'gene' mutation might be:

Any qualitative or quantitative difference in the DNA sequence of a particular gene between a cell and its parent.

The DNA sequence of our deleted gene has changed from normal in the parental cell to non-existent in the daughter cell, thereby qualifying as a gene mutation by this definition.

But this definition may include too much. For instance, a cell could have undergone homologous recombination over a large part of the chromosome that includes our particular heterozygous tumour suppressor gene. According to this second definition, this would still be a 'gene' mutation, even though tens or hundreds of other linked genes may also have been altered in this loss of heterozygosity (LOH). Most geneticists would balk at such leniency and would probably prefer something like the following – and final – definition of a gene mutation:

Any qualitative or quantitative difference in DNA sequence between a cell and its parent that is essentially restricted to a single gene.

The qualifier 'essentially' permits some leeway in extending the definition to include non-functional DNA that might reside between genes.

This digression into semantics emphasizes a specific point concerning the determination of gene mutations: just because a molecular study of tumour cells or a mutagenesis assay monitors alterations to a single gene does not mean that it is measuring gene mutations in this third sense. This qualification is most relevant for the mouse lymphoma assay. Ongoing studies that involve monitoring the loss of heterozygosity of chromosome 11 heteromorphic microsatellites in $tk^{-/-}$ mutants have demonstrated concomitant losses of up to 150 Mb – almost the entire chromosome (11_b) bearing the initially functional tk gene – in both large and small colony mutants (Liechty et al., 1994; P. Glover and D. Clive, unpublished results). Similar findings hold for molecular studies of tumours showing loss of tumour suppressor genes by LOH (Ali et al., 1987; Sager, 1989; Fearon and Vogelstein, 1990; Weinberg, 1991). To avoid confusion, we shall simply refer to 'mutation' or 'mutation assays' and rely on context to convey the sense intended.

11.4 The scale of mutations in mammalian cells

Molecular and cytogenetic studies of viable mutations in both germ and somatic cells of mammals suggest a preponderance of large-scale alterations. These range from hundreds to hundreds of millions ($10^2–10^8$) of base pairs of DNA and include visible alterations to a significant fraction of a chromosome (Ali et al., 1987; Emanuel, 1988; Fearon and Vogelstein, 1990; Fuscoe et al., 1994; Ngo et al., 1988; Russell et al., 1989; Shapiro et al., 1989; Solomon et al., 1991; Turner et al., 1985, 1988; Weinberg, 1991; Yunis, 1985). Similar large-scale damage has also been seen in mammalian cell mutation assays (Applegate and Hozier, 1987; Applegate et al., 1990; Blazak et al., 1986, 1989; Bradley et al., 1988; Clive et al., 1990; DeMarini et al., 1987, 1989; Evans et al., 1986; Liber et al., 1989; Liechty et al., 1994; Yandell et al., 1986, 1990). One example of this variety of mutations can be seen in the model of the development of

Stage	Location	Type of mutational event

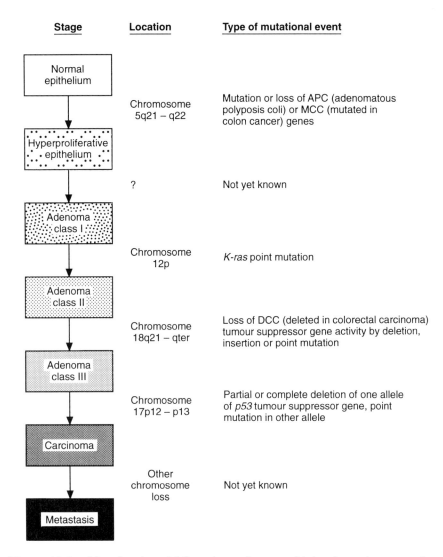

Figure 11.3. Mutational model for colorectal cancer. Molecular and cytogenetic alterations responsible for the indicated transitional steps in colorectal carcinogenesis are indicated (after Fearon and Vogelstein, 1990).

colorectal cancer worked out by Vogelstein's group (Fearon and Vogelstein, 1990; Stanbridge, 1990; *Figure 11.3*). The scale of these alterations, ranging from a single base pair up to a significant portion of a chromosome, does not readily yield to explanation by a direct attack on DNA. If we arrange these somatic and germ cell mutations in order of increasing size (*Figure 11.4* columns 1, 2) we can address the question of how well these scales of known genetic damage would likely be detected in our current genotoxicity tests.

207

Type of mutational event	Example	Detected by: Salmonella	MLA	Chromosome abnormality
Point mutation	*ras* activation	YES	YES	No
Oligonucleotide insertions, deletions	Loss of tumour suppressor gene; reduction in heterozygosity	YES	YES	No
Allele loss	Loss of tumour suppressor gene; reduction in heterozygosity	No	YES	No
Multi-locus mutations	Contiguous gene syndromes	No	YES	Sometimes, with banding
Small chromosome alterations	Loss of tumour suppressor gene, contiguous gene syndromes; many heritable traits	No	YES	Not without banding
Large chromosome alteration	Ph chromosome; partial trisomy, balanced translocation syndromes	No	YES	YES
Aneuploidy	Down's, Turner's, Klinefelter's syndromes	No	Possibly	Not with standard protocols

Figure 11.4. Ability of three short-term tests to detect various scales of genetic damage. Scales of damage (column 1) are arranged in order of increasing size. Column 2 lists human tumour or germline examples of each corresponding scale of damage. Columns 3–5 indicate the ability of *Salmonella*, the mouse lymphoma assay (MLA) or chromosome aberration assays to detect that scale of damage. Because of their genetic design, *Salmonella* strains are inherently incapable of detecting any but the smallest scales of damage. Owing to the limitations of light microscopy, cytogenetic assays are restricted to only the largest scales of damage. Molecular and cytogenetic studies on the tk$^{-/-}$ mutants (see test) have demonstrated the capability of MLA to detect damage ranging from less than 50 bp to as much as 150 Mb.

This question is answered for the two most widely used *in vitro* tests: the Ames *Salmonella* reverse mutation assay, and *in vitro* cytogenetics. For comparison, the responses of the less frequently used L5178Y/*tk$^{+/-}$→tk$^{-/-}$* mouse lymphoma assay (MLA) are also shown. It is clear that the scales of most types of mutations seen in colorectal tumours are too large (greater than a few base pairs) to be detected by *Salmonella* (column 3), too small (less than about 10 Mb) to be detected by *in vitro* cytogenetics (column 5), but can be detected in the mouse lymphoma assay (column 4; Clive *et al.*, 1990; Liechty *et al.*, 1994).

There are two other reasons for preferring mammalian cell mutation assays for mutagenicity testing: viability of the mutant cells and relevance of the genome. Although *Salmonella* revertants are viable, they result from only the smallest-scale damage to a genome that is far smaller and less complex than the one we are trying to protect. On the other hand, classical chromosome aberrations, while they occur in a relevant genome, are of unknown viability.

The rest of this chapter sets out to demonstrate that the mouse lymphoma assay – and, to a lesser extent, certain other mammalian cell mutation assays – can detect, in a relevant genome, the wide range of viable and heritable genetic damage indicated in column 4 of *Figure 11.4*.

11.5 Mechanisms of genetic damage: non-DNA targets

The central tenet of bacterial mutagenesis is that mutations arise solely from the chemical attack by genotoxins or their metabolites on DNA. The mechanisms of such direct attack on DNA have been well explored over the past three decades, and have led to our current understanding of electrophilicity, DNA adduct formation, structure–mutagenic activity relationships, and to the *Salmonella* assay (Ashby and Tennant, 1988). Needless to say, this first mechanism of genetic damage also occurs in mammalian cells but, it turns out, as a minor component.

A second set of mechanisms of mutagenesis is implied by the existence of mutant strains of bacteria and other organisms which exhibit elevated mutant frequencies in the absence of any exogenous mutagen, and by the existence of enzyme poisons. Many different classes of these 'mutator' genotypes have been characterized in both prokaryotic and eukaryotic cells over the past 30 or so years. Typically, these cells possess a heritable alteration to one or another of the dozen or so enzymes of DNA metabolism (e.g. polymerases, ligases, topoisomerases, repair enzymes). Although not yet well explored, it is quite likely that mutations to any of these can lead to various 'mutator' phenotypes, each characterized by elevated frequencies of 'spontaneous' genetic damage.

A specific poison to one of these enzymes of DNA metabolism would result in a 'mutator' phenocopy – a genetically normal cell with a temporarily impaired capacity to replicate DNA accurately. The enzymes which can be so poisoned to produce a 'mutator' phenocopy are non-DNA targets. The poisons themselves are not likely to possess familiar structural alerts. And, at the experimental level, increasing concentrations of these poisons could induce increasing proportions of 'mutator' phenocopies among the treated population; each of these could possess an increased probability of misreplication; the resulting dose-related increase in mutant frequency would identify our enzyme poison as a genotoxin.

Of these two mechanisms for increasing genetic damage in mammalian cells, the first – direct attack on DNA by an exogenous agent – has been the subject of the vast majority of mutation studies over the past quarter century, and needs no further comment here.

Heritable defects in the machinery of DNA metabolism are recognized as being causal in a number of cancer predisposition syndromes in humans (see Chapter 4) and in experimental animals. As already mentioned, they are usually associated with high levels of 'spontaneous' genetic damage and are often associated with mutations to DNA repair enzymes. However, they are not really part of genetic toxicology and were mentioned only as a heuristic lead-in to the concept of mutator phenocopies, our second mechanism by which genetic damage can be increased by exogenous agents and the main concept of this chapter.

If we wish to detect agents that induce genetic damage through function-impairing interaction with one or more of the enzymes of DNA metabolism (i.e. non-DNA targets) we need: evidence that such agents exist; knowledge of the types of genetic damage that such impairment might induce; and assays that are sensitive to this variety of damage.

Numerous studies with topoisomerase inhibitors convincingly demonstrate the existence of genotoxins that attack this enzyme of DNA metabolism rather than DNA itself (Ashby *et al.*, 1994; DeMarini *et al.*, 1987; Gaulden, 1987; Marshall and Ralph, 1985). Some of these inhibitors either do not interact with *Salmonella* DNA at all or induce mutations that are too large to show up in that assay (Ashby *et al.*, 1994); however, their genetic damage is detectable as chromosome aberrations in a variety of cells. Most importantly from the viewpoint of this chapter, these topoisomerase inhibitors induce high frequencies of multigenic mutations – usually associated with a chromosome aberration – in the mouse lymphoma assay (Ashby *et al.*, 1994). From this it follows that appropriately designed mammalian cell mutation assays should be used to detect the spectrum of mutations known to occur in human and rodent diseases. Serendipity appears to have given rise to one or more adequate assays for the detection of such damage.

11.6 Mammalian cell mutagenesis

In light of the multifaceted nature of DNA metabolism, the following characteristics play a significant role in mammalian cell mutagenesis:

(i) Direct interaction with DNA occurs relatively infrequently and seems to induce the same types of small-scale mutations as are seen in bacteria. Most mutations in mammalian cells result from impairment of one or more enzymes of DNA metabolism, presumably located within the replication complex (Earnshaw *et al.*, 1985; Earnshaw and Heck, 1985; Gasser and Laemmli, 1986; Gasser *et al.*, 1986; Nelson *et al.*, 1986; Pardoll *et al.*, 1980; Pienta and Coffey, 1984).

(ii) The basic unit of some of these mutations (e.g. those resulting from inhibition of topoisomerase II) is a large integral number of radial loops/replicons.

(iii) The consequences of impairment of other enzymes of DNA metabolism are less well known in mammalian cells, but probably include point mutations and chromosome aberrations from impaired DNA polymerases and ligases, respectively.

For topoisomerase II, the best-explored non-DNA target for mutagens, the basic mechanism of mutation is impaired rejoining of normally transient double-strand breaks in DNA. These transient breaks are introduced by the topoisomerase enzyme for the purpose of unwinding DNA in preparation for its replication. Impaired rejoining by a suitable topoisomerase poison can lead to more or less permanent double-strand breaks which can be resolved by either recombination between homologous or non-homologous chromosomes, loss of one or more loops of DNA, or a chromosome break. The mutations that result are LOH, loss of one or more genes, or chromosome aberrations. The scale of such mutations can be from hundreds to over 100 000 kb.

The mutational consequences of inhibiting other enzymes of DNA metabolism are relatively unexplored.

11.7 Mammalian cell mutation assays

We are finally in a position to discuss the various mammalian cell mutation assays that are in common use. This we will do both briefly (see Hsie *et al.*, 1979, and Moore *et al.*, 1987, for more detailed reviews) and primarily in the context of the radial loop model of the mammalian chromosome and the scales of mutations known to occur in genetic disease.

To a very good first approximation, cell type is usually less important than the gene that is the target for mutagenesis; consequently, cell type will be mentioned only in passing.

11.7.1 Target genes and selection of mutant phenotypes

Of the estimated 50 000 genes in a mammalian cell, only about 10 have been used in any systematic manner to study mammalian cell mutagenesis. Over 90% of such studies have been performed in assays based on only four of these genes (Na^+/K^+-$ATPase$; *hprt*; *aprt*, used as the $aprt^{+/-}$ heterozygote; and *tk*, used as the $tk^{+/-}$ heterozygote). The AS-52 cell line, which measures mutations at the bacterial *gpt* (xanthine–guanine phosphoribosyl transferase) gene, is the equivalent of a heterozygous *hprt* gene (Stankowski and Tindall, 1987). It was engineered by randomly inserting a single copy of the *gpt* gene into *hprt*-deficient hamster cells in the expectation that this would duplicate the capabilities of the human and mouse native *tk* heterozygotes. If we list this assay by target gene, it is not novel because it is a functional heterozygote; however, it could be classified as a fifth, potentially major endpoint. This relative paucity of test-system constructs is due largely to the difficulties in designing selective conditions that can readily and clearly distinguish between mutant and non-mutant cells, and to the logistics of quantitating both cell types.

All of the most commonly used mammalian cell mutation systems use the principle of mass selection to permit colonial growth of mutant cells while inhibiting the growth of normal cells. Following an appropriate expression period (during which functional wild-type enzyme is replaced by mutant enzyme), a treated population of up to several million cells is plated out in the presence of a selective agent. This agent is ouabain to select for ouabain-resistant $Na^+/K^+-ATPase$ (OUA^{res}) mutants; 6-thioguanine (TG) for $hprt^-$ or gpt^- mutants; 2,6-diaminopurine (DAP) for $aprt^{-/-}$ mutants; and trifluorothymidine (TFT) for $tk^{-/-}$ mutants. Cloning can be done either two-dimensionally (on the solid surface of a petri dish, for instance), or three-dimensionally in soft agar cloning medium.

Mutations to ouabain-resistance are usually studied in CHO and lung (V79) cells, and in human cells. They are studied less frequently in the $tk^{+/-}$ heterozygote of L5178Y mouse lymphoma cells (henceforth referred to as mouse lymphoma cells). Mutations to the $hprt$ gene have been studied in the same four systems, but most commonly in CHO, V79 and human cells. Heterozygous $aprt$ cell lines have been developed only from CHO cells, while tk heterozygotes exist for mouse lymphoma cells and for human cells (TK-6 cells and derivatives).

Let us briefly compare the types of mutations that are known to occur in mammalian cells, and how well these systems detect and distinguish them.

11.7.2 Point (oligonucleotide) mutations

These can be detected as gene mutations at the ouabain-resistance (OUA^{res}) marker ($Na^+/K^+-ATPase$ gene). This ATPase activity is essential to cell survival. In a homozygous normal cell ouabain, a glucocorticoid, binds to this enzyme, thereby inhibiting Na^+/K^+-ATPase activity and cell viability. Fortuitously, the ouabain-binding site and the ATPase active site are independent. Thus, mutations that result in appropriate subtle alterations to the ouabain-binding site of the enzyme can occur without affecting its ATPase activity, leading to a ouabain-resistant Na^+/K^+-ATPase-competent variant. Being dominant, this single altered allele will be sufficient to maintain adequate Na^+/K^+-ATPase activity, and hence viability, of the cell in the presence of ouabain, even though enzyme from the other allele is inhibited.

Because of the need to retain ATPase activity, anything but the smallest-scale mutations to this gene, while certainly capable of eliminating the ouabain-binding site, will also affect the ATP-binding site, resulting in a Na^+/K^+-ATPase-deficient (and hence dead) cell. Thus, only agents capable of inducing certain specific point mutations will be detected as mutagens in this assay.

Point mutations can also inactivate the $hprt$, gpt, $aprt$ or tk genes in their respective assays, if such mutations affect critical sites of the normal enzyme.

However, assays based on these genes also detect larger-scale alterations which are phenotypically indistinguishable from point mutations (Applegate and Hozier, 1987; Applegate *et al.*, 1990; Blazak *et al.*, 1986, 1989; Bradley *et al.*, 1988; Clive *et al.*, 1990; DeMarini *et al.*, 1987, 1989; Evans *et al.*, 1986; Liber *et al.*, 1989; Liechty *et al.*, 1994; Yandell *et al.*, 1986, 1990). Various molecular analyses (Applegate *et al.*, 1990; Liechty *et al.*, 1994) are readily available to overcome this limitation.

11.7.3 Single gene mutations

By single gene mutations we mean any mutation that is greater than a few nucleotides in extent, but which does not extend to other genes. This scale of mutations is not detected in ouabain-resistance assays, but is at the *hprt, gpt, aprt* or *tk* loci; however, without appropriate supplementary molecular analyses, such mutants are indistinguishable from even larger scale events. For instance, *hprt⁻* mutants that involve the loss of as much as 2 Mb of DNA (nearly 50 times the 44 kb length of the *hprt* gene itself) have been recovered from human cells (Fuscoe *et al.*, 1994).

11.7.4 Larger (multilocus) mutations

Although mutations involving the equivalent of more than a single gene have been isolated at the *hprt* and *aprt* loci, they have been best defined at the *tk* locus of human TK-6 (Liber *et al.*, 1989; Yandell *et al.*, 1986, 1990) and L5178Y mouse lymphoma (Applegate *et al.*, 1987, 1990; Clive *et al.*, 1990; DeMarini *et al.*, 1989; Evans *et al.*, 1986; Liechty *et al.*, 1994) cells. The most recent studies involve the analysis of several heteromorphic microsatellite sequences that span the entire chromosome 11 between the centromere and the distally located *tk* gene. In approximately 80% of large or small $tk^{-/-}$ mutants induced by nine mutagens, the tk_b gene (i.e. the recently mutated *tk* gene located on chromosome 11_b) and one or more of these microsatellite sequences is missing from chromosome 11_b (Liechty *et al.*, 1994; P. Glover and D. Clive, unpublished), indicating that the DNA containing the missing microsatellite(s) and the distally located tk_b gene, inclusive, has been lost.

11.8 The spectrum of L5178Y/$tk^{+/-} \rightarrow tk^{-/-}$ mutations

From the preceding discussion, it appears that approximately 80% of all mutations detected at the heterozygous *tk* locus of the mouse lymphoma assay involve the loss of between 6.5 kb and 150 Mb of DNA from the chromosome that bears the target *tk* gene (Applegate *et al.*, 1990; Clive *et al.*, 1990; Liechty *et al.*, 1994; P. Glover and D. Clive, unpublished). The maximum observed losses represent over 95% of the affected chromosome and occur at least as frequently in large-colony mutants as in small-colony mutants. Such losses might be responsible for the impaired growth of small-colony mutants.

However, most of these small-colony mutants also bear chromosome 11_b-specific aberrations (Hozier *et al.*, 1981, 1982, 1983; Moore *et al.*, 1985; Blazak *et al.*, 1986, 1989), which could also cause this impaired growth by gene dosage effects.

However, the cytogenetically normal chromosomes of large-colony mutants that have lost up to 150 Mb of 11_b DNA pose an enigma. First, all large-colony mutants that have been analysed cytogenetically have two normally banded chromosomes 11; because all have the two distinguishable 11_a and 11_b centromeres, we can rule out the occurrence of non-disjunction (Hozier *et al.*, 1982). Second, small-colony mutants that show no measurable loss of contiguous DNA (multiple-locus point mutations) can be induced by potent point mutagens (Clive *et al.*, 1991; Moore *et al.*, 1991). Thus, the enigma: heritably impaired growth (i.e. some small-colony mutants) can result from essentially point mutations, yet losses of 150 Mb (i.e. some large-colony mutants) can be without effect.

We are currently proceeding with the working hypothesis that: (a) the two chromosomes 11 are mostly homologous, based on these cells' origin from highly inbred mice, and (b) the large-scale event that is being measured in large-colony mutants is not simple loss of chromosome 11_b DNA (since a normal banded 11_b is present in such mutants), but replacement of 11_b DNA by genetically equivalent 11_a DNA (i.e. homologous mitotic recombination). Earlier studies imply that the short region of chromosome 11_b distal to the *tk* gene does not share in this homology, but includes one or more genes that are necessary for normal growth and/or survival (Clive *et al.*, 1983).

If this hypothesis is correct, the preponderance of this class of mutants would suggest that the mouse lymphoma assay is measuring primarily loss of heterozygosity (LOH) in large-colony mutants, with a few point mutations thrown in, and primarily chromosome aberrations (with or without LOH) in small-colony mutants. These possibilities are currently under investigation.

11.9 Conclusions: relevance to human risk

There is no reason to believe that these findings are unique to the L5178Y/*tk*[+/-]-3.7.2C mouse lymphoma cell line. This cell line is karyotypically stable (Hozier *et al.*, 1981, 1982). Spontaneous mutant frequencies of (20–80) x 10^{-6} are high relative to those at the *hprt* locus (about 10^{-6}), but are consistent with the larger variety of genetic damage detected by *tk*, particularly the viable chromosome aberrations seen in the majority of small-colony mutants. (In standard cytogenetic analyses, these typically occur at about 1% per genome, or about 200 x 10^{-6} per chromosome, and are of unknown viability.) However, because this cell line is from a highly inbred animal with correspondingly low heterozygosity among homologous chromosomes, the consequences of LOH are exceptionally benign. In marked contrast, LOH in outbred species has significant biological consequences, the most important of which is loss of tumour suppressor genes.

Finally, the reasons for a number of properties of the mouse lymphoma assay are readily explicable in light of the models presented of the mammalian cell genome and types of mutations it can endure. The relatively high spontaneous mutant frequencies have already been explained and the same argument holds for induced frequencies. A second property is the presence of a bimodal distribution of colony sizes: that can be explained by the variety of genetic damage detected by this system, as has been discussed in the previous sections. Third is this assay's high sensitivity (proportion of carcinogens detected as mouse lymphoma assay mutagens) and low specificity (proportion of non-carcinogens that are non-mutagenic in the mouse lymphoma assay). In light of *Figure 11.4*, the high sensitivity is not surprising. The low specificity (or high incidence of 'false positives') is likely to be a consequence of the combination of the variety of genetic endpoints detected and the fact that many non-DNA targets that lead to such damage may involve thresholds which are readily exceeded *in vitro* but which would not be in the whole animal without compromising its survival. Since many of these non-DNA targets also lead to chromosome aberrations, it would be expected that properly performed *in vitro* cytogenetic assays would share this low specificity, and that is the case.

It is surprising that the most commonly used assays to estimate human genetic risk are so inappropriate due to either phylogenetic distance (bacterial assays), an endpoint of unknown viability (cytogenetic assays), or limited scale of damage (bacterial and cytogenetic assays). There is a large void between these assays that can be filled with the mouse lymphoma assay and possibly other appropriate mammalian cell mutation assays which represent the best genetic model of the types of damage that are of the greatest significance to our species.

References

Ali IU, Lidereau R, Theillet C, Callahan R. (1987) Reduction to homozygosity of genes on chromosome 11 in human breast neoplasia. *Science* 238: 185–188.

Applegate ML, Hozier JC. (1987) On the complexity of mutagenic events at the mouse lymphoma *tk* locus. In: *Progress in Mammalian Cell Mutagenesis* (eds MM Moore *et al.*). Banbury Report 28, Cold Spring Harbor Laboratory Press, New York, pp. 213–224.

Applegate ML, Moore MM, Broder CB, Burrell A, Juhn G, Kasweck KL, Lin P-F, Wadhams A, Hozier JC. (1990) Molecular dissection of mutations at the heterozygous thymidine kinase locus in mouse lymphoma cells. *Proc. Natl Acad. Sci. USA* 87: 51–55.

Ashby J. (1986a) The prospects for a simplified and internationally harmonized approach to the detection of possible human carcinogens and mutagens. *Mutagenesis* 1: 3–16.

Ashby J. (1986b) Letter to the editor. *Mutagenesis* 1: 309–317.

Ashby J, Tennant RW. (1988) Chemical structure, Salmonella mutagenicity and extent of carcinogenicity as indicators of genotoxic carcinogenesis among 222 chemicals tested in rodents by the U.S. NCI/NTP. *Mutat. Res.* 204: 17–115.

Ashby J, Tinwell H, Glover P, Poorman-Allen P, Krehl R, Callander RD, Clive D. (1994) Potent mutagenicity of the human carcinogen etoposide to mouse bone marrow cells and mouse lymphoma L5178Y cells. *Mutagenesis* 555: 345–456.

Blazak WF, Stewart BE, Galperin I, Allen KL, Rudd CJ, Mitchell AD, Caspary WJ. (1986) Chromosome analysis of trifluorothymidine-resistant L5178Y mouse lymphoma cell colonies. *Environ. Mutagen.* 8: 229–240.

Blazak WF, Los FJ, Rudd CJ, Caspary WJ. (1989) Chromosome analysis of small and large L5178Y mouse lymphoma cell colonies: comparison of trifluorothymidine-resistant and unselected cell colonies from mutagen-treated and control cultures. *Mutat. Res.* **224:** 197–208.

Bradley WEC, Belouchi A, Messing K. (1988) The *aprt* heterozygote/hemizygote system for screening mutagenic agents allows detection of large deletions. *Mutat. Res.* **199:** 131–138.

Clive D. (1985) Mutagenicity in drug development: interpretation and significance of test results. *Regul. Toxicol. Pharmacol.* **5:** 79–100.

Clive D, Glover P, Applegate M, Hozier J. (1990) Molecular aspects of chemical mutagenesis in L5178Y/*tk*+/- mouse lymphoma cells. *Mutagenesis* **5:** 191–197.

Clive D, Glover P, Krehl R, Poorman-Allen P. (1991) Mutagenicity of 2-amino-N^6-hydroxyadenine (AHA) at three loci in L5178Y/*tk*+/- mouse lymphoma cells: molecular and preliminary cytogenetic characterizations of AHA-induced *tk*-/- mutants. *Mutat. Res.* **253:** 73–82.

Clive D, Hozier J, Moore MM. (1983) 'Single-gene' and viable chromosome mutations affecting the TK locus in L5178Y mouse lymphoma cells. *Ann. N.Y. Acad. Sci.* **407:** 420–422.

DeMarini DM, Doerr CL, Meyer MK, Brock KH, Hozier J, Moore MM. (1987) Mutagenicity of *m*-AMSA and *o*-AMSA in mammalian cells due to clastogenic mechanism: possible role of topoisomerase. *Mutagenesis* **2:** 349–355.

DeMarini DM, Brockman HE, de Serres FJ, Evans HH, Stankowski LF Jr, Hsie AW. (1989) Specific-locus mutations induced in eukaryotes (especially mammalian cells) by radiation and chemicals: a perspective. *Mutat. Res.* **220:** 11–29.

Earnshaw WC, Heck MMS. (1985) Localization of topoisomerase II in mitotic chromosomes. *J. Cell Biol.* **100:** 1716–1725.

Earnshaw WC, Halligan B, Cooke CA, Heck MMS, Liu LF. (1985) Topoisomerase II is a structural component of mitotic chromosome scaffolds. *J. Cell Biol.* **100:** 1706–1715.

Emanuel BS. (1988) Molecular cytogenetics: toward dissection of the contiguous gene syndromes. *Am. J. Hum. Genet.* **43:** 575–578.

Evans HH, Mencl J, Horng M-F, Ricanati M, Sanchez C, Hozier J. (1986) Locus specificity in the mutability of L5178Y mouse lymphoma cells: the role of multilocus lesions. *Proc. Natl Acad. Sci. USA* **83:** 4379–4383.

Fearon ER, Vogelstein B. (1990) A genetic model for colorectal tumorigenesis. *Cell* **61:** 759–767.

Fuscoe JC, Nelsen AJ, Pilia G. (1994) Detection of deletion mutations extending beyond the HPRT gene by multiplex PCR analysis. *Som. Cell Mol. Gen.* **20:** 39–46.

Garner RC, Kirkland DJ. (1986) Letter to the editor (Reply to Ashby). *Mutagenesis* **1:** 233–235.

Gasser SM, Laemmli UK. (1986) The organization of chromatin loops: characterization of a scaffold attachment site. *EMBO J.* **5:** 511–518.

Gasser SM, Laroche T, Falquet J, Boy de la Tour E, Laemmli UK. (1986) Metaphase chromosome structure: involvement of topoisomerase II. *J. Mol. Biol.* **188:** 613–629.

Gatehouse DG, Tweats DJ. (1986) Letter to the editor (reply to Ashby). *Mutagenesis* **1:** 307–308.

Gaulden ME. (1987) Hypothesis: some mutagens directly alter specific chromosomal proteins (DNA topoisomerase II and peripheral proteins) to produce chromosome stickiness, which causes chromosome aberrations. *Mutagenesis* **2:** 357–365.

Hozier J, Sawyer J, Moore M, Howard B, Clive D. (1981) Cytogenetic analysis of the L5178Y/TK+/-→TK-/- mouse lymphoma mutagenesis assay system. *Mutat. Res.* **84:** 169–181.

Hozier J, Sawyer J, Clive D, Moore M. (1982) Cytogenetic distinction between the TK+ and TK- chromosomes in the L5178Y TK+/- 3.7.2C mouse lymphoma cell line. *Mutat. Res.* **105:** 451–456.

Hozier J, Sawyer J, Clive D, Moore M. (1983) Cytogenetic analysis of small-colony L5178Y TK-/- mutants early in their clonal history. *Ann. N.Y. Acad. Sci.* **407:** 423–425.

Hsie AW, O'Neill JP, McElheny VK. (1979) (eds) *Mammalian Cell Mutagenesis: the Maturation of Test Systems.* Banbury Report 2. Cold Spring Harbor Laboratory Press, Cold Spring Harbor, NY.

Huberman JA, Riggs AD. (1968) On the mechanism of DNA replication in mammalian chromosomes. *J. Mol. Biol.* **32:** 327–341.

Huberman JA, Tsai A. (1973) Direction of DNA replication in mammalian cells. *J. Mol. Biol.* **75:** 5–12.

Liber HL, Yandell DW, Little JB. (1989) A comparison of mutation induction at the *tk* and *hprt* loci in human lymphoblastoid cells: quantitative differences are due to an additional class of mutations at the autosomal *tk* locus. *Mutat. Res.* **216:** 9–17.

Liechty MC, Hassanpour Z, Hozier JC, Clive D. (1994) Use of microsatellite DNA polymorphisms on mouse chromosome 11 for *in vitro* analysis of thymidine kinase gene mutations. *Mutagenesis* **9:** 423–427.

Marsden MPF, Laemmli UK. (1979) Metaphase chromosome structure: evidence for a radial loop model. *Cell* **17:** 849–858.

Marshall B, Ralph RK. (1985) The mechanism of action of mAMSA. *Adv. Cancer Res.* **44:** 267–293.

Moore MM, Clive D, Hozier JC, Howard BE, Batson AG, Turner NT, Sawyer J. (1985) Analysis of trifluorothymidine-resistant (TFT^r) mutants of L5178Y/TK$^{+/-}$ mouse lymphoma cells. *Mutat. Res.* **151:** 161–174.

Moore MM, DeMarini DM, de Serres FJ, Tindall KR. (1987) (eds) *Mammalian Cell Mutagenesis.* Banbury Report 28. Cold Spring Harbor Laboratory Press, Cold Spring Harbor, NY.

Moore MM, Harrington-Brock K, Parker L, Doerr CL, Hozier JC. (1991) Genotoxicity of 2-amino-N^6-hydroxyadenine (AHA) to mouse lymphoma and CHO cells. *Mutat. Res.* **253:** 63–71.

Nelson WG, Pienta KJ, Barrack ER, Coffey DS. (1986) The role of the nuclear matrix in the organization and function of DNA. *Annu. Rev. Biophys. Biophys. Chem.* **15:** 457–475.

Ngo KY, Glotz VT, Koziol JA, Lynch DC, Gitschier J, Ranieri P, Ciavarella N, Ruggeri ZM, Zimmerman TS. (1988) Homozygous and heterozygous deletions of the von Willebrand factor gene in patients and carriers of severe von Willebrand disease. *Proc. Natl Acad. Sci. USA* **85:** 2753–2757.

Pardoll DM, Vogelstein B, Coffey DS. (1980) A fixed site of DNA replication in eucaryotic cells. *Cell* **19:** 527–536.

Pienta KJ, Coffey DS. (1984) A structural analysis of the role of the nuclear matrix and DNA loops in the organization of the nucleus and chromosome. *J. Cell Sci.* (suppl. 1): 123–135.

Pienta KJ, Partin AW, Coffey DS. (1989) Cancer as a disease of DNA organization and dynamic cell structure. *Cancer Res.* **49:** 2525–2532.

Russell LB, Hunsicker PR, Cacheiro NLA, Bangham JW, Russell WL, Shelby MD. (1989) Chlorambucil effectively induces deletion mutations in mouse germ cells. *Proc. Natl Acad. Sci. USA* **86:** 3704–3708.

Sager R. (1989) Tumour suppressor genes: the puzzle and the promise. *Science* **246:** 1406–1412.

Shapiro LJ, Yen P, Pomerantz D, Martin E, Rolewic L, Mohandas T. (1989) Molecular studies of deletions at the human steroid sulfatase locus. *Proc. Natl Acad. Sci. USA* **86:** 8477–8481.

Solomon E, Borrow J, Goddard AD. (1991) Chromosome aberrations and cancer. *Science* **254:** 1153–1160.

Stanbridge EJ. (1990) Identifying tumour suppressor genes in human colorectal cancer. *Science* **247:** 12–13.

Stankowski LF Jr, Tindall KR. (1987) Characterization of the AS52 cell line for use in mammalian cell mutagenesis studies. In: *Mammalian Cell Mutagenesis* (eds MM Moore *et al.*). Banbury Report 28, Cold Spring Harbor Laboratory Press, Cold Spring Harbor, NY, pp. 71–79.

Turner DR, Morley AA, Haliandros M, Kutlaca R, Sanderson BJ. (1985) *In vivo* somatic mutations in human lymphocytes frequently result from major gene alterations. *Nature* **315:** 343–345.

Turner DR, Grist SA, Janatipour M, Morley AA. (1988) Mutations in human lymphocytes commonly involve gene duplication and resemble those seen in cancer cells. *Proc. Natl Acad. Sci. USA* **85:** 3189–3192.

Tweats DJ, Gatehouse DG. (1988) Further debate of testing strategies (reply to Ashby). *Mutagenesis* **3:** 95-102.

Vogelstein B, Pardoll DM, Coffey DS. (1980) Supercoiled loops and eucaryotic DNA replication. *Cell* **22:** 79–85.

Waters MD, Stack HF. (1988) Letter to the editor (reply to Ashby). *Mutagenesis* **3:** 89–94.

Weinberg RA. (1991) Tumour suppressor genes. *Science* **254:** 1138–1146.

Yandell DW, Dryja TP, Little JB. (1986) Somatic mutations at a heterozygous autosomal locus in human cells occur more frequently by allele loss than by intragenic structural alterations. *Somat. Cell Mol. Genet.* **12:** 255–263.

Yandell DW, Dryja TP, Little JB. (1990) Molecular genetic analysis of recessive mutations at a heterozygous autosomal locus in human cells. *Mutat. Res.* **229:** 89–102.

Yunis JJ. (1985) Genes and chromosomes in human cancer. *Prog. Med. Virol.* **32:** 58–71.

Assays for unscheduled DNA synthesis *in vivo*

Jon C. Mirsalis

12.1 Unscheduled DNA synthesis assays

The repair of damage to DNA involves the excision of DNA adducts and resynthesis of the excised regions (Djordevic and Tolmach, 1967). It was originally identified by Rasmussen and Painter (1964), who observed that radiation exposure of cells in culture produced an increase in the uptake of [^{3}H]thymidine. This DNA synthesis occurred at times in the cell cycle other than the 'scheduled' S phase and is therefore often referred to as 'unscheduled' DNA synthesis (UDS).

Assays for the measurement of UDS have been devised in a variety of cell lines (Rasmussen and Painter, 1964; San and Stich, 1975; Trosko and Yager, 1974). However, these assays lacked sophisticated metabolic activation systems and were generally insensitive to many electrophiles. A major advance in the field was the development of an assay that measures UDS in primary cultures of rat hepatocytes (Williams, 1977; Williams *et al.*, 1982). This assay offered many advantages over other *in vitro* genetic toxicology endpoints:

(i) Primary hepatocytes more closely mimic *in vivo* metabolism than microsomal preparations because both phase I and phase II metabolism reactions are generally preserved, whereas phase II reactions are largely absent in microsomes.
(ii) Hepatocytes serve as both the activator and the target, thereby increasing the sensitivity for detection of highly reactive electrophilic intermediates.
(iii) The assay is relatively inexpensive in comparison with other, more labour intensive assays such as *in vitro* cytogenetics (see Chapter 7) or the mouse lymphoma mutagenesis assay (see Chapter 11). The rat hepatocyte assay became one of the most popular short-term genetic toxicology screening tests of the 1980s, and a large database of test results has been established by several laboratories (Kornbrust and Barfknecht, 1985; McCarroll *et al.*, 1984; Williams *et al.*, 1982).

Despite the advantages of the *in vitro* hepatocyte DNA repair assay, several limitations were quickly identified. This assay is generally unresponsive to many hepatocarcinogenic nitro- and azo-compounds (Bermudez *et al.*, 1979; Doolittle *et al.*, 1983; Kornbrust and Barfknecht, 1985). Compounds such as 2,6-dinitrotoluene, 2-nitrotoluene, 1,2-dimethylhydrazine and cycasin are all potent genotoxic liver carcinogens that are not detected in this assay; likewise, several non-genotoxic agents (2,6-diaminotoluene, HC Blue 2, 4-acetyl-aminofluorene) produced positive results.

Investigators in our laboratory became interested in the development of an *in vivo* model for the measurement of genetic damage in the liver. An *in vivo* assay would take into account the complex uptake, distribution, metabolic activation, detoxification and elimination of chemicals that occurs in whole animals. Several earlier studies had evaluated the feasibility of measuring UDS in various tissues following *in vivo* treatment (Brambilla *et al.*, 1978; Craddock and Henderson, 1978; Kaufmann *et al.*, 1979; Lee and Suzuki, 1979), but all of these approaches had technical difficulties and were generally insensitive, largely because they relied on administration of [^{3}H]thymidine to animals, which severely limited the amount of radiolabel available for incorporation into DNA during UDS.

We adopted the novel approach of treating animals *in vivo*, isolating hepatocytes and incubating these cultures with [^{3}H]thymidine to maximize the amount of radiolabel available to the cells during periods of DNA repair. Following treatment of rats or mice with test chemicals, hepatocytes are isolated by liver perfusion and incubated with [^{3}H]thymidine. The rationale behind the assay is that genetic damage occurs *in vivo*, and *in vitro* incubation with [^{3}H]thymidine allows incorporation of a radiolabelled nucleotide into DNA in regions where excision repair occurs. This *in vivo–in vitro* approach was highly sensitive to electrophiles that required metabolic activation as well as to direct-acting mutagens (Mirsalis and Butterworth, 1980).

12.2 Measurement of UDS *in vivo*

The *in vivo–in vitro* hepatocyte DNA repair assay, often referred to simply as the *in vivo* UDS assay, might have been dismissed as yet another variation of an established assay if not for dinitrotoluene (DNT). Technical-grade DNT is a major industrial chemical that is a potent hepatocarcinogen in rats, yet it failed to produce positive results in nearly all genetic toxicology tests, including the *in vitro* rat hepatocyte UDS assay (Bermudez *et al.*, 1979). When DNT was evaluated in the *in vivo* UDS assay, it was an extremely potent inducer of UDS (Mirsalis and Butterworth, 1982). In addition, we demonstrated that the discrepancy between the *in vivo* and *in vitro* results was due to the requirement for metabolism of DNT by intestinal microflora (*Figure 12.1*; Mirsalis *et al.*, 1982b), a requirement that is difficult to mimic in a tissue culture dish. An interesting secondary observation from this study was that a few animals contaminated with low levels of *Streptococcus faecalis* demonstrated slight

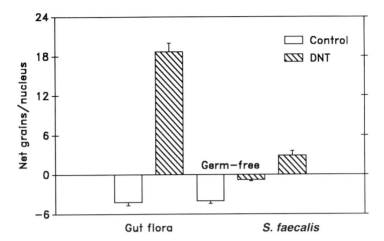

Figure 12.1. Induction of unscheduled DNA synthesis (UDS) by a single dose of technical-grade dinitrotoluene (DNT; 100 mg kg^{-1}) in germ-free rats and rats with normal gut flora. DNT produced a significant increase in UDS in the presence of gut flora, but no increase was produced in the absence of gut flora. Two rats were contaminated with low levels of *Streptococcus faecalis*, and these animals demonstrated a slight increase in UDS, suggesting that they were carrying out limited metabolism of DNT. Adapted from Mirsalis *et al.* (1982b).

increases in UDS, indicating that changes in gut flora can significantly alter the genotoxicity of chemicals. The *in vivo* UDS assay also predicted that 2,6-DNT was the most potent carcinogen of the major DNT isomers, and this prediction was confirmed when purified isomers were evaluated in bioassays. Thus, the evaluation of a single compound underscored the severe limitations of *in vitro* assays for the detection of compounds that are activated by complex (*in vivo*) metabolic pathways.

Our laboratory and others identified several other compounds that were generally negative in *in vitro* genetic toxicology assays but produced positive results in the *in vivo* UDS assay. These included 2-nitrotoluene (Doolittle *et al.*, 1983), several azo dyes (Kornbrust and Barfknecht, 1985), and 1,2-dimethylhydrazine (Mirsalis *et al.*, 1982a). More recently, some novel pharmaceuticals have been identified that are negative in most genetic toxicology batteries but positive in the *in vivo* UDS assay (Short *et al.*, 1994; Suzuki *et al.*, 1992).

Of equal importance to the detection of genotoxic agents is the ability of *in vivo* UDS to discriminate accurately non-genotoxic chemicals. For example, 2,4- and 2,6-diaminotoluene (DAT) are both potent *in vitro* genotoxic agents, but only 2,4-DAT induces UDS *in vivo*; these results are consistent with the potent hepatocarcinogenicity of 2,4-DAT and the non-carcinogenicity of 2,6-DAT (Mirsalis *et al.*, 1982a). Methapyrilene is a rat hepatocarcinogen that has been extensively studied in short-term tests (Mirsalis, 1987a), and although it

221

produces positive responses in selected tests such as the mouse lymphoma L5178Y assay, it does not bind to DNA in rat liver or produce other genotoxic effects in liver. The *in vivo* UDS assay accurately identified methapyrilene as a non-genotoxic carcinogen (Steinmetz *et al.*, 1988a). The hair dyes HC Blue 1 and 2 have also been shown to be potent genotoxic agents *in vitro*, but genotoxicity is not observed *in vivo* in the organs where tumours are produced. This finding was first reported from studies using the *in vivo* UDS assay (Mirsalis, 1987b).

In the 15 years since its development, the *in vivo* UDS assay has become the second most widely used *in vivo* genetic toxicology assay, after the mouse bone marrow micronucleus assay (Heddle *et al.*, 1983; see also Chapter 8). Following the development of the *in vivo* UDS assay in the rat, our laboratory developed methods for evaluation of UDS in the B6C3F1 mouse liver (Mirsalis *et al.*, 1985), and Doolittle *et al.* (1987a,b) used essentially identical methods for evaluation in the CD-1 mouse. The assay has been shown to be quite sensitive to genotoxic liver carcinogens, and can detect UDS in cells containing approximately 13 DNA adducts per 10^8 nucleotides (Gallagher *et al.*, 1991).

Several studies, evaluating a large number of chemicals in this system, have been published. Our laboratory published the results of evaluating 20 compounds in this assay (Mirsalis *et al.*, 1982a). This study demonstrated that genotoxic liver carcinogens consistently induce UDS in hepatocytes following *in vivo* treatment. In contrast, potent genotoxic agents that are not hepatocarcinogens [e.g. benzo[a]pyrene] generally fail to produce positive responses in the liver. This finding underscored the tissue-specific nature of this assay. A second major collaborative study evaluated induction of UDS in rats and mice treated with 2- or 4-acetylaminofluorene (AAF), benzo[a]pyrene, or pyrene. This study, sponsored by the World Health Organization's International Program on Chemical Safety (IPCS), compared interlaboratory results from eight different laboratories using this assay. The excellent agreement between laboratories further supports the reliability of this assay (Mirsalis, 1988).

The single largest study correlating results of *in vivo* UDS assays with rodent carcinogenicity was sponsored by the US National Toxicology Program (NTP). In this programme, 51 compounds were evaluated for their ability to induce UDS and/or S-phase synthesis in rat or mouse liver. Rodent carcinogenicity data were available for many of the compounds tested. Several conclusions resulted from this study. Most significant was the finding that a large number of compounds which induced liver tumours failed to induce genotoxic effects in liver. This was particularly evident for the mouse, in which many chlorinated solvents were potent hepatocarcinogens, yet failed to induce genotoxicity in most short-term tests (Tennant *et al.*, 1987). The correlation between S-phase induction and cancer was, in fact, better than the correlations between any genotoxic endpoint and cancer (Spalding and Mirsalis, 1986). Carcinogenicity, UDS and S phase results for 24 of the compounds in this study have previously been published (Mirsalis *et al.*, 1989), and these results indicate that the *in vivo* UDS assay is useful for discriminating between genotoxic and non-genotoxic liver carcinogens.

12.3 Methods for conduct of *in vivo* UDS assays

The methods for the conduct of the *in vivo* UDS assay have changed surprisingly little since the first methods were reported (Mirsalis and Butterworth, 1980). Under the auspices of the American Society of Testing and Materials (ASTM), a set of guidelines for the conduct of this assay was proposed (Butterworth *et al.*, 1987). We had previously published detailed methods for the conduct of this assay (Mitchell and Mirsalis, 1984), and later we published revised methods where some of the experimental procedures were optimized (Hamilton and Mirsalis, 1987). Other guidelines and recommendations have also been published (Ashby *et al.*, 1985; Madle *et al.*, 1994). The recommended methods for the conduct of the assay are more or less the same in all these published reports. The principal methods for the assay as conducted in our laboratory are summarized next.

12.3.1 Animals and treatment

The assay has most commonly been conducted using male Fischer-344 rats, but female rats and several other strains have also been used, including CD-1 and B6C3F1 mice, Sprague–Dawley and Alderley Park rats, and Syrian golden hamsters. Animals are generally used at 8–12 weeks of age. Using younger animals can complicate scoring of slides because of the higher incidence of S-phase cells in younger animals. The test chemical should be administered by whatever is the most appropriate route of exposure. Most compounds are administered by gavage, but inhalation exposure (Doolittle *et al.*, 1984) and intravenous exposure (for drugs intended for i.v. administration) have also been used. It is recommended that an absolute minimum of three animals per dose be evaluated. There have been no examples of compounds that induce UDS in females but not in males; therefore, using males only is considered an acceptable study design. However, if there is reason to believe that there might be a sex difference in metabolism of the compound in question, using both sexes is a conservative and prudent approach.

12.3.2 Treatment times

Most investigators evaluate UDS at two time-points: an early time-point (usually 1–4 h) and a later time point (usually 12–18 h). This strategy allows for detection of rapidly metabolized chemicals that produce peak responses early (e.g. methylmethane sulphonate and dimethylnitrosamine produce peak responses within minutes of treatment and fall off rapidly with time), as well as compounds that require more time for activation and binding to DNA (e.g. DNT induces no

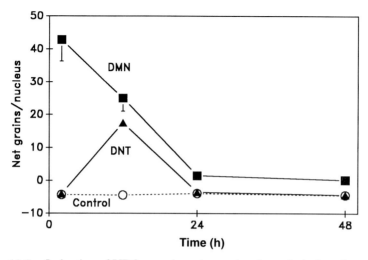

Figure 12.2. Induction of UDS at various time-points by technical-grade dinitrotoluene (DNT) and dimethylnitrosamine (DMN). DMN is rapidly metabolized and produces DNA damage within minutes of administration. Conversely, DNT produces no increase in UDS 2 h after administration, induces a peak response at 12 h, and by 24 h returns to control levels.

UDS response at 2 h and a very strong response at 12 h; *Figure 12.2*). Some studies have evaluated the induction of UDS at steady-state following repeated doses of the test compound (Doolittle *et al.*, 1987a; Mirsalis *et al.*, 1993; Steinmetz *et al.*, 1990). These reports generally show that sensitivity of the UDS assay decreases considerably when multiple (lower) doses are administered. Therefore, UDS should optimally be conducted as a single-dose assay, where animals are given a single large dose of the test chemical and evaluated within the first 24 h after treatment. A promising variation of this approach is the administration of two doses of chemical, one dose at 16 h before killing and a second dose at 2 h before killing (Ian Mitchell, personal communication).

12.3.3 Establishment of hepatocyte cultures

Methods for isolation of hepatocytes are fairly standard and have been discussed in detail elsewhere (Butterworth *et al.*, 1987; Mitchell and Mirsalis, 1984; Williams, 1977). We recommend *in situ* perfusion of livers with EGTA in Hank's buffer, followed by type I collagenase in Williams Medium E (WME). Cells are then combed out, pelleted and resuspended in WME without serum. Approximately 5×10^5 viable hepatocytes are seeded into six-well culture dishes containing WME with 10% fetal bovine serum and plastic coverslips. Cells are allowed to attach to coverslips for 1.5–2 h, washed once with WME, then reseeded with serum-free WME containing 10 μCi ml^{-1} [^{3}H]thymidine. Cultures are incubated for 4 h, washed with WME, then incubated overnight in

WME containing unlabelled thymidine. This 'pulse–chase' approach is particularly effective for reducing the cytoplasmic background (Hamilton and Mirsalis, 1987). Longer incubation times with [^{3}H]thymidine result in unacceptably high cytoplasmic backgrounds (Hamilton and Mirsalis, 1987).

12.3.4 Autoradiography

Following overnight incubation in unlabelled thymidine, cultures are washed, nuclei are swelled with a sodium citrate solution (to increase nuclear area and to 'flatten' the nuclei on the coverslip), and the cells are fixed in acetic acid:ethanol. Coverslips are then dried and glued to microscope slides with Permount, after which autoradiograms are prepared. Slides are dipped in Kodak NTB-2 emulsion, prediluted 1:1 with water, followed by exposure at –20°C for 7 days. Longer exposure times increase the cytoplasmic background but do not improve the sensitivity of the assay, and they are therefore not recommended (Hamilton and Mirsalis, 1987). Slides are developed in Kodak D-19 developer for 3 min at 15°C, fixed in Kodak Fixer for 8 min and washed in running water for 25 min. Slides are stained with methyl green pyronin Y, dried and coverslipped before scoring.

12.3.5 Scoring of slides

Individual silver grains may be enumerated manually, but this method is so tedious and inaccurate that it makes the assay completely impractical. The most efficient means for scoring is to use an external video camera mounted on a light microscope and connected to a colony counter (e.g. ARTEK model 880 or 980 counter). Data may be captured directly on to a computer using an RS232 interface. Our laboratory has developed a software package for collection and summary of data on IBM-compatible PCs, but commercial software packages are also available for UDS data collection.

All UDS slides should be scored blind (i.e. random identification codes should be assigned to all slides by an individual who will not score the experiment). Individuals who score UDS slides need to be aware of several potential artefacts that can affect results. Pyknotic nuclei, nuclei without a surrounding cytoplasm, cells with staining artefacts that are 'flagged' by the automatic counter as grains and cells with obvious morphological defects should not be scored. In other words, only 'normal' healthy cells should be evaluated for UDS. A minimum of 75–100 cells should be scored for each animal.

There are three general measures of UDS that may be calculated: the net grains per nucleus (NG; the nuclear count minus the cytoplasmic background), the percentage of cells in repair (%IR; the percentage of cells in a population that fall above a predetermined upper range) and the net grains of cells in repair (NGIR; the average net grain count of the cells in repair). Calculation of the %IR should not be based on an arbitrary cut-off published in the literature (e.g. 5 NG), but calculated based on historical data in each laboratory. The cut-off should generally be 2 standard deviations above the mean of the individual cell counts, and this value is generally around 5 NG (Hamilton and Mirsalis, 1987) if the methods described in this chapter are followed.

Several methods for scoring have been reported (Harbach *et al.*, 1991; Hill *et al.*, 1989; Mitchell and Mirsalis, 1984), and these are all based on the same principle. UDS is quantitated by determining the amount of nuclear incorporation of radiolabelled thymidine. This is accomplished by counting the number of exposed silver grains over the nucleus; however, hepatocytes show significant incorporation of thymidine into the cytoplasm (possibly owing to mitochondrial DNA synthesis), and the cytoplasm lying over the nucleus can bias the results. Compounds that inhibit cytoplasmic backgrounds will decrease the cytoplasmic background, whereas compounds that induce mitochondrial DNA synthesis will increase cytoplasmic backgrounds. Some methods subtract the highest of several cytoplasmic counts (the most conservative approach), whereas others subtract an average cytoplasmic count taken from predetermined coordinates relative to the nucleus. These different methods may shift the NG values up or down by 1–3 grains, but the choice of scoring method makes little difference in the qualitative response of the assay.

12.3.6 Analysis and interpretation of results

Interpretation of UDS results has been a controversial topic since the introduction of the *in vitro* UDS assay. There are no widely accepted statistical methods for evaluation of UDS assays, although some authors have proposed specific methods (Thakur, 1987). The most sensible approach is to transform data (e.g. log transformation) and then apply standard parametric statistics, such as Student's *t*-test.

Regardless of the statistical significance of the results, the following general guidelines should be followed when evaluating data.

(i) A positive response requires both an increase in the NG and an increase in the absolute nuclear counts. *Figure 12.3* shows a hypothetical case where cytoplasmic backgrounds are decreased (because of toxicity, inhibition of mitochondrial DNA synthesis, or other effects), the result being a perfect 'dose–response' for the test chemical that will be highly statistically significant. In fact, there is no increase in nuclear counts and the response for this compound should be reported as negative.

(ii) A positive response requires an NG count that is greater than zero. As in the case above (*Figure 12.3*), decreasing cytoplasmic counts can cause the NG value to approach zero asymptotically, but a true positive response will always produce NG values that are greater than zero.

(iii) A positive result need not be dose related. Benzidine produces a significant increase in UDS, but as the dose increases, the UDS response drops to control levels (*Figure 12.4*). This effect has been observed both *in vitro* (Steinmetz *et al.*, 1988c) and *in vivo* (Mirsalis *et al.*, 1989) and may be related to inhibition of DNA repair at higher doses of benzidine.

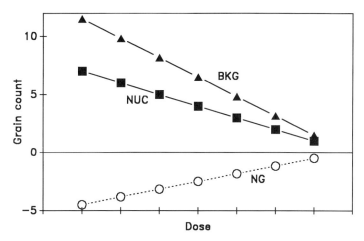

Figure 12.3. Hypothetical data showing induction of a (false) positive response due to inhibition of cytoplasmic incorporation of [³H]thymidine. Cytoplasmic background (BKG) grain counts decline, nuclear counts (NUC) decline slightly, and the result is an increase in net grains/nucleus (NG) that would incorrectly be interpreted as a positive UDS response. The NG count will asymptotically approach zero but should not exceed it. A true UDS response requires an increase in the absolute number of NUC counts in addition to an increase in NG.

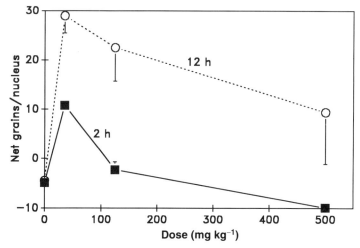

Figure 12.4. Induction of UDS in F-344 rat liver by benzidine. Inhibition of DNA repair by benzidine results in a decrease in UDS at higher doses. Despite the absence of a dose response, benzidine is clearly positive in this assay at both 2 and 12 hours.

12.4 Regulatory applications

In the early 1990s, most regulatory agencies began to gravitate toward a minimal battery of genetic toxicology tests. Although there are disagreements between various countries (and even between agencies within countries) over the ideal battery of tests to use, the two most widely used *in vivo* assays for detection of genetic damage are the bone marrow micronucleus assay and the *in vivo–in vitro* hepatocyte DNA repair assay. The micronucleus assay is useful for detecting chromosomal damage in rapidly proliferating tissues (for example, bone marrow, colonic epithelium; see also Chapter 8) but is not useful for detecting effects in tissues with low rates of cell division (for example, liver, kidney). Therefore, both the micronucleus assay and the *in vivo–in vitro* hepatocyte DNA repair assay have been advocated for evaluating the genotoxic risk of chemicals in specific target tissues (Canadian Minister of National Health and Welfare, 1986; Gatehouse, 1990). The *in vivo* UDS assay is frequently required by regulatory agencies in a variety of situations. These include: (i) when extrahepatic metabolism (gut flora, other tissues) of a chemical is suspected, (ii) when there is evidence of an *in vivo* effect on the liver (hepatotoxicity, liver enlargement), (iii) when a structural alert suggests the liver as a probable target site, or (iv) when positive results are obtained in *in vitro* systems in the presence of liver microsomes. The US Food and Drug Administration 'Red Book' lists the *in vivo* UDS assay as a useful secondary assay for the evaluation of some classes of chemicals (US FDA, 1993).

The micronucleus and UDS assays are quite complementary and provide information on a wide variety of genotoxic chemicals (*Table 12.1*). Several liver carcinogens that are detected in the UDS assay are missed in the micronucleus assay. Conversely, several non-liver carcinogens are readily detected in the micronucleus assay but fail to induce UDS in the liver. Both tests have fewer false-positive results than corresponding *in vitro* assays.

Table 12.1. Comparison of results of selected chemicals in the *in vivo* unscheduled DNA synthesis (UDS), bone marrow micronucleus and Ames assays

Compound	Ames	UDS	Micronucleus	Cancer
Dimethylnitrosamine	+	+	−	+
2,6-Dinitrotoluene	±	+	−	+
1,2-Dimethylhydrazine	−	+	−	+
Urethane	−	−	+	+
Cyclophosphamide	−	−	+	+
Benzene	−	−	+	+
Pyrene	+	−	−	−
2,6-Diaminotoluene	+	−	−	−

12.5 *In vivo* UDS assays in other tissues

Subsequent to the development of a UDS assay in liver, our laboratory and others have reported methods for measurement of UDS in other tissues, including kidney (Tyson and Mirsalis, 1986), pancreas (Steinmetz and Mirsalis, 1984), spermatocytes (Working and Butterworth, 1984), trachea

(Doolittle and Butterworth, 1984) and glandular stomach (Bakke and Mirsalis, 1988; Furihata and Matsushima, 1982). Data for these tissues are far less extensive than data for liver. The best-characterized assay is kidney, where approximately 20 compounds have been evaluated (Loury *et al.*, 1987; Steinmetz *et al.*, 1988b; Tyson and Mirsalis, 1986). These assays have shown similar results to those observed in the liver; that is, tissue-specific genotoxicity correlates with tissue-specific carcinogenicity. For example, the liver carcinogens 2-AAF and aflatoxin B_1 are potent inducers of UDS in rat liver (Mirsalis *et al.*, 1982a) but fail to elicit a UDS response in rat kidney (Loury *et al.*, 1987).

A limitation of UDS assays in tissues other than the liver is the apparent insensitivity of other tissues. This may simply be related to their inability to induce high levels of excision repair. From an evolutionary standpoint, one would expect an organ like the liver, which has considerable exposure to electrophilic metabolites, to evolve an efficient DNA repair mechanism, whereas the kidney or pancreas may support only much lower levels of excision repair.

The spermatocyte UDS assay (Bentley and Working, 1988; Working and Butterworth, 1984) has been adopted by the US Environmental Protection Agency as an assay for identification of germ-cell mutagens.

12.6 Conclusions

The *in vivo* UDS assay has been adopted throughout the world as a useful endpoint for monitoring genotoxic effects in the liver. When combined with results from the bone marrow micronucleus assay, it provides highly useful information on the genotoxic potential of chemicals *in vivo*.

Negative *in vivo* UDS data are often contrasted with positive data from *in vitro* tests as support for the 'insensitivity' of this assay. This argument loses some validity when it is recognized that non-carcinogens such as 2,6-diaminotoluene and HC Blue 2 consistently induce *in vitro* positive responses in many assays, including *in vitro* UDS (Tennant *et al.*, 1987), yet are negative in *in vivo* systems such as UDS.

There are few, if any, examples of hepatocarcinogens that induce genotoxicity in the liver (as measured by DNA adducts, covalent binding, etc.) but fail to induce UDS. Unless several such examples are found, measurement of UDS will remain a valuable tool for the assessment of genotoxic damage to the liver.

References

Ashby J, Lefevre PA, Burlinson B, Penman MG. (1985) An assessment of the *in vivo* rat hepatocyte DNA repair assay. *Mutat. Res.* **156:** 1–18.

Bakke JP, Mirsalis JC. (1988) Measurement of unscheduled DNA synthesis (UDS) in rat stomach following *in vivo* exposure. *Environ. Mutagen.* **11** (suppl. 11): 8.

Bentley KS, Working PK. (1988) Use of seminiferous tubule segments to study stage specificity of unscheduled DNA synthesis in rat spermatogenic cells. *Environ. Mol. Mutagen.* **12:** 285–297.

Bermudez E, Tillery D, Butterworth BE. (1979) The effect of 2,4-dinitrotoluene and isomers of dinitrotoluene on unscheduled DNA synthesis in primary rat hepatocytes. *Environ. Mutagen.* **1:** 391–398.

Brambilla G, Cavanna M, Carlo P, Finollo R, Parodi S. (1978) DNA repair synthesis in primary cultures of kidneys from BALB/c and C3H mice treated with dimethylnitrosamine. *Cancer Lett.* **5:** 153–159.

Butterworth BE, Ashby J, Bermudez E, Casciano D, Mirsalis J, Probst G, Williams G. (1987) A protocol and guide for the *in vivo* rat hepatocyte DNA repair assay. *Mutat. Res.* **189:** 123–133.

Canadian Minister of National Health and Welfare. (1986) Guidelines on the use of mutagenicity tests in the toxicological evaluation of chemicals. Ottawa, Canada, pp. 61–64.

Craddock VM, Henderson AR. (1978) *De novo* and repair replication of DNA in liver of carcinogen-treated animals. *Cancer Res.* **38:** 2135–2143.

Djordevic B, Tolmach LJ. (1967) Responses of synchronous populations of HeLa cells to ultraviolet irradiation at selected stages of the generation cycle. *Radiat. Res.* **32:** 327–346.

Doolittle DJ, Butterworth BE. (1984) Assessment of chemical-induced DNA repair in rat tracheal epithelial cells. *Carcinogenesis* **5:** 773–779.

Doolittle DJ, Sherrill JM, Butterworth BE. (1983) Influence of intestinal bacteria, sex of the animal, and position of the nitro group on the hepatic genotoxicity of nitrotoluene. *Cancer Res.* **43:** 2836–2842.

Doolittle DJ, Bermudez E, Working PK, Butterworth BE. (1984) Measurement of genotoxic activity in multiple tissues following inhalation exposure to dimethylnitrosamine. *Mutat. Res.* **141:** 123–127.

Doolittle DJ, Muller G, Scribner HE. (1987a) A comparative study of hepatic DNA repair, DNA replication and hepatotoxicity in the CD-1 mouse following multiple administrations of dimethylnitrosamine. *Mutat. Res.* **188:** 141–147.

Doolittle DJ, Muller G, Scribner HE. (1987b) The *in vivo–in vitro* hepatocyte assay for assessing DNA repair and DNA replication: studies in the CD-1 mouse. *Food Chem. Toxicol.* **25:** 399–405.

Furihata C, Matsushima T. (1982) Unscheduled DNA synthesis in rat stomach – Short-term assay of potential stomach carcinogens. In: *Banbury Report 13: Indicators of Genotoxic Exposure* (eds BA Bridges, BE Butterworth, IB Weinstein). Cold Spring Harbor Laboratory Press, Cold Spring Harbor, NY, pp. 123–135.

Gallagher JE, Shank T, Lewtas J, Lefevre PA, Ashby J. (1991) Relative sensitivity of ^{32}P-postlabelling of DNA and the autoradiographic UDS assay in the liver of rats exposed to 2-acetylaminofluorene (2AAF). *Mutat. Res.* **252:** 247–257.

Gatehouse DG. (1990) An industrial and UK perspective on short-term testing. In: *Mutation and the Environment, Part D* (ed. M Mendelsohn). Wiley-Liss, New York, pp. 249–259.

Hamilton CM, Mirsalis JC. (1987) Factors that affect the sensitivity of the *in vivo–in vitro* hepatocyte DNA repair assay in male rats. *Mutat. Res.* **189:** 341–347.

Harbach PR, Rostami HJ, Aaron CS, Wiser SK, Grzegorczyk CR. (1991) Evaluation of four methods for scoring cytoplasmic grains in the *in vitro* unscheduled DNA synthesis (UDS) assay. *Mutat. Res.* **252:** 139–148.

Heddle JA, Hite M, Kirkhart B, Mavournin K, MacGregor JT, Newell GW, Salamone MF. (1983) The induction of micronuclei as a measure of genotoxicity. A report of the U.S. Environmental Protection Agency Gene-Tox Program. *Mutat. Res.* **123:** 61–118.

Hill LE, Yount DJ, Garriott ML, Tamura RN, Probst GS. (1989) Quantification of unscheduled DNA synthesis by a whole cell counting method. *Mutat. Res.* **224:** 447–451.

Kaufmann WK, Kaufman DG, Grisham JW. (1979) Unscheduled DNA synthesis in isolated hepatic nuclei after treatment of rats with methylnitrosourea *in vivo*. *Biochem. Biophys. Res. Commun.* **91:** 297–302.

Kornbrust D, Barfknecht T. (1985) Testing of 24 food, drug, cosmetic, and fabric dyes in the *in vitro* and the *in vivo/in vitro* hepatocyte primary culture/DNA repair assay. *Environ. Mutagen.* **7:** 101–120.

Lee IP, Suzuki K. (1979) Induction of unscheduled DNA synthesis in mouse germ cells following 1,2-dibromo-3-chloropropane (DBCP) exposure. *Mutat. Res.* **68:** 169–173.

Loury DJ, Smith-Oliver T, Butterworth BE. (1987) Assessment of unscheduled and replicative DNA synthesis in rat kidney cells exposed *in vitro or in vivo* to unleaded gasoline. *Toxicol. Appl. Pharmacol.* **87:** 127–140.

Madle S, Dean SW, Andrae U, Brambilla G, Burlinson B, Doolittle DJ, Furihata C, Hertner T, McQueen CA, Mori H. (1994) Recommendations for the performance of UDS test *in vitro* and *in vivo*. *Mutat. Res.* **312:** 263–286.

McCarroll NE, Farrow MG, McCarthy KL, Scribner HE. (1984) A survey of genetic toxicology testing in industry, contract laboratories and government. *J. Appl. Toxicol.* **4**(2): 66–74.

Mirsalis JC. (1987a) Genotoxicity, toxicity, and carcinogenicity of the antihistamine methapyrilene. *Mutat. Res.* **185:** 309–317.

Mirsalis JC. (1987b) *In vivo* measurement of unscheduled DNA synthesis and hepatic cell proliferation as an indicator of hepatocarcinogenesis in rodents. *Cell Biol. Genet. Toxicol.* **3:** 165-173.

Mirsalis JC. (1988) Summary report on the performance of the *in vivo* DNA repair assays. In: *Report of the International Program on Chemical Safety's Collaborative Study on* in vivo *Assays* (eds J Ashby, FJ de Serres, MD Shelby, BH Margolin, M Ishidate, GC Becking). Cambridge University Press, Cambridge, pp. 345–351.

Mirsalis JC, Butterworth BE. (1980) Detection of unscheduled DNA synthesis in hepatocytes isolated from rats treated with genotoxic agents: an *in vivo–in vitro* assay for potential carcinogens and mutagens. *Carcinogenesis* **1:** 621–625.

Mirsalis JC, Butterworth BE. (1982) Induction of unscheduled DNA synthesis in rat hepatocytes following *in vivo* treatment with dinitrotoluene. *Carcinogenesis* **3:** 241–245.

Mirsalis JC, Tyson CK, Butterworth BE. (1982a) Detection of genotoxic carcinogens in the *in vivo–in vitro* hepatocyte DNA repair assay. *Environ. Mutagen.* **4:** 553–562.

Mirsalis JC, Hamm TE, Sherrill JM, Butterworth BE. (1982b) Role of gut flora in the genotoxicity of dinitrotoluene. *Nature* **295:** 322–323.

Mirsalis JC, Tyson CK, Loh EN, Steinmetz KL, Bakke JP, Hamilton CM, Spak DK, Spalding JW. (1985) Induction of hepatic cell proliferation and unscheduled DNA synthesis in mouse hepatocytes following *in vivo* treatment. *Carcinogenesis* **6:** 1521–1524.

Mirsalis JC, Tyson CK, Steinmetz KL, Loh EN, Hamilton CM, Bakke JP, Spalding JW. (1989) Measurement of unscheduled DNA synthesis and S-phase synthesis in rodent hepatocytes following *in vivo* treatment: testing of 24 compounds. *Environ. Mutagen.* **14:** 155–164.

Mirsalis JC, Steinmetz KL, Blazak WF, Spalding JW. (1993) Evaluation of the potential of riddelliine to induce unscheduled DNA synthesis, S-phase synthesis or micronuclei following *in vivo* treatment with multiple doses. *Environ. Mol. Mutagen.* **21:** 265–271.

Mitchell AD, Mirsalis JC. (1984) Unscheduled DNA synthesis as an indicator of genotoxic exposure. In: *Single-Cell Mutation Monitoring Systems* (eds AA Ansari, FJ de Serres). Plenum Press, New York, pp. 165–216.

Rasmussen RE, Painter RB. (1964) Evidence for repair of ultraviolet damaged deoxyribonucleic acid in cultured mammalian cells. *Nature* **203:** 1360–1362.

San RHC, Stich HF. (1975) DNA repair synthesis of cultured human cells as a rapid bioassay for chemical carcinogens. *Int. J. Cancer* **16:** 284–291.

Short B, Hamilton C, Zimmerman D, Hart R, Wier P, Mirsalis J, Schwartz L. (1994) Effect of SK&F 105809 on hepatocellular proliferation, foci, and DNA repair in the rat. *Toxicologist* **14:** 252.

Spalding JW, Mirsalis JC. (1986) The high correspondence of S-phase induction in rodent hepatocytes with chemical carcinogenicity *in vivo*. *In Vitro* **22:** 15.

231

Steinmetz KL, Mirsalis JC. (1984) Induction of unscheduled DNA synthesis in primary cultures of rat pancreatic cells following *in vivo* and *in vitro* treatment with genotoxic agents. *Environ. Mutagen.* **6:** 321–330.

Steinmetz KL, Tyson CK, Meierhenry EF, Spalding JW, Mirsalis JC. (1988a) Examination of genotoxicity and morphologic alteration in hepatocytes following *in vivo* or *in vitro* exposure to methapyrilene. *Carcinogenesis* **9:** 959–963.

Steinmetz KL, Stack CR, Bakke JP, Hamilton CM, Pardo K, Ramsey M, Mirsalis JC. (1988b) Evaluation of unscheduled DNA synthesis and S-phase synthesis following treatment with *p*-dichlorobenzene. *Toxicologist* **8:** 160.

Steinmetz KL, Green CE, Bakke JP, Spak DK, Mirsalis JC. (1988c) Induction of unscheduled DNA synthesis in primary cultures of rat, mouse, hamster, monkey and human hepatocytes. *Mutat. Res.* **206:** 91–102.

Steinmetz KL, Garin KE, Hamilton CM, Bakke JP, MacGregor JT, Mirsalis JC. (1990) Multiple endpoint *in vivo* genetic toxicology assay: evaluation of hepatic unscheduled DNA synthesis, S-phase synthesis and peripheral blood micronuclei following repeated dosing of male B6C3F1 mice. *Environ. Mol. Mutagen.* **15** (suppl. 17): 58.

Suzuki H, Ohsawa K, Watanabe C, Yamashiro M, Watanabe H, Nakane S, Mirsalis JC, Hamilton CM, Steinmetz KL. (1992) Mutagenicity of a new chemical which has a carbazole structure. *Proc. Environ. Mutagen Soc. Jpn,* November 1992, p. 54 (in Japanese).

Tennant RW, Spalding J, Stasiewicz S, Caspary W, Mason JM, Resnick MA. (1987) Comparative evaluation of genetic toxicity patterns of carcinogens and noncarcinogens: strategies for predictive use of short-term assays. *Environ. Hlth Perspect.* **75:** 87–95.

Thakur, AK. (1987) Unscheduled DNA synthesis: some statistical thoughts. *Cell Biol. Toxicol.* **3:** 175–192.

Trosko JE, Yager JD. (1974) A sensitive method to measure physical and chemical carcinogen-induced 'unscheduled DNA synthesis' in rapidly dividing eukaryotic cells. *Exp. Cell Res.* **88:** 47–55.

Tyson CK, Mirsalis JC. (1986) Induction of unscheduled DNA synthesis in isolated kidney cells following *in vivo* exposure to genotoxic agents. *Environ. Mutagen.* **7:** 889–899.

U. S. Food and Drug Administration. (1993) *Toxicological Principles for the Safety Assessment of Direct Food Additives and Color Additives Used in Food.* Center for Food Safety and Applied Nutrition, Washington, DC, 1993.

Williams GM. (1977) Detection of chemical carcinogens by unscheduled DNA synthesis in rat liver primary cell cultures. *Cancer Res.* **37:** 1845–1851.

Williams GM, Laspia MF, Dunkel VC. (1982) Reliability of the hepatocyte primary culture/DNA repair test in testing of coded carcinogens and non-carcinogens. *Mutat. Res.* **97:** 359–370.

Working P, Butterworth BE. (1984) An assay to detect chemically induced DNA repair in rat spermatocytes. *Environ. Mutagen.* **6:** 273–286.

Use of genetically engineered cells for genetic toxicology testing

Charles L. Crespi

13.1 Introduction

Most environmental mutagens require metabolic conversion in order to generate electrophilic species (Miller and Miller, 1971). However, many target cell lines used for testing environmental chemicals for mutagenic activity are markedly deficient in expression of the enzymes that carry out this conversion. Two approaches have been taken to overcome this limitation; firstly, addition of an exogenous metabolic activation system (usually fractions derived from rodent liver; Krahn and Heidelberger, 1977; Malling, 1971); and secondly, the development of metabolically competent target cell lines. Two approaches have been used to attain metabolic competence in cell lines. The first is the identification of cell lines with endogenous activity and subsequent development of the genotoxicity assays (Crespi and Thilly, 1984; Loquet and Wiebel, 1982; Tong et al., 1983). The development of assays using these cell lines has been hampered by the limited array of promutagen-activating enzymes which are expressed in stable cell culture and by the often poor growth properties of cell lines which possess native metabolic competence. The second approach is the introduction, by gene transfer, of promutagen-activating enzymes into pre-existing target cells for genotoxicity assays (Davies et al., 1989; Doehmer et al., 1988). The resulting cell lines typically retain their original usefulness for genotoxicity testing and have augmented metabolic competence. However, while this approach has been successful with some target cells, it is currently unclear whether all target cells for genotoxicity will support efficient expression of promutagen-activating enzymes.

The focus of this chapter is recent progress in the incorporation of xenobiotic metabolizing enzymes into target cells for genotoxicity assays via gene transfer. This has been made possible by two developments: (i) the isolation and characterization of cDNAs encoding xenobiotic metabolizing enzymes (Gonzalez, 1988), and (ii) the development of methods for the introduction and efficient expression of cDNAs in target cells via DNA-mediated gene transfer or retroviral infection.

This chapter is not a comprehensive review of all studies involving introduction of xenobiotic metabolizing enzymes into target cells. Instead, the differences between metabolically competent cells and the traditional (extracellular) approach to promutagen activation are explored; the toxicological significance of the individual enzymes which have been expressed is discussed; and considerations for the validation of the systems are discussed. In addition, several specific examples of applications are provided.

Metabolic competence can be constitutive to the target cell or introduced either by a mutation/selection process or by cDNA expression. The means of obtaining the target cell are not as important as the properties of the cell once it is obtained. For some cases, the metabolic competence of the resulting cell line is the product of both endogenous and cDNA-expressed enzymes. In what follows, the term 'metabolically competent' is used instead of the more restrictive terms of 'cDNA-expressing' or 'genetically engineered'.

The term 'xenobiotic' will be used instead of the more restrictive terms 'promutagen' or 'procarcinogen', since it allows use of a broader literature on human enzyme-specific biotransformations, specifically in the area of drug metabolism. Importantly, it is ethical to administer drugs to humans. This allows the development in humans of data on rates of biotransformation by specific enzymes. Such data are important to validate systems, since promutagens usually cannot be administered to humans.

A variety of mammalian target cells have been used as hosts for cDNA expression and a wide variety of enzymes from rodents, human and other species have been expressed directly in the target cell. Typical target cells which have been used for cDNA expression studies are V79 cells (Doehmer *et al.*, 1988), CHO cells (Trinidad *et al.*, 1991), CHL cells (Sawada *et al.*, 1993) and human lymphoblasts (Crespi *et al.*, 1993a).

A strength of this approach is the availability of isogenic, control cells without the introduced cDNA and thus, enzyme-specific activation can be analysed readily. Non-mutagenic, specific inhibitors have also been identified for many enzymes. Through such metabolic inhibition, the cDNA-expressing cell can serve as its own metabolically deficient 'control'. Mutagenic activity can, therefore, be attributed to activation by a specific enzyme and not merely to the presence of an extracellular mixture of enzymes/cofactors. In addition, the stability of these systems should allow standardization of testing methods between independent laboratories.

Using the cDNA expression approach, the induction of genotoxicity in the target cell is modulated by the intrinsic properties of the specific xenobiotic metabolizing enzyme. The data thus obtained can be related to that on enzyme expression *in vivo* (tissue-specific expression, polymorphism, induction, inhibition, etc.). Where human enzymes are studied, a clearer picture of potential human risk should be possible. Such insight cannot be gained by studying rodent enzymes. cDNA expression is useful for determining the relative efficiency of different enzymes for activation of promutagens. Extension of these data to absolute rates of mutagenesis *in vivo* is not appropriate. Further extension to absolute risks of a multistep process such as carcinogenesis is even more tenuous.

It is also important to consider that in addition to the level of a cDNA-expressed enzyme within the cell, the relationship of that enzyme to other, accessory enzymes may also modulate the genotoxic response. Examples of this effect are discussed later in this chapter. The dependence of genotoxicity on metabolic capacity is also a feature of extracellular activation systems (S-9, for example). However, with extracellular activating systems, the relationship of rates of activation by individual enzymes to metabolic activation is not clear owing to the multiplicity of enzymes present in these crude preparations.

The development of metabolically competent target cells for genotoxicity assays represents an important step in the transition from correlative *in vitro* genetic toxicology to mechanism-based analyses. Specific activation steps can be linked to specific human or rodent enzymes. The affinity and capacity of individual enzymes can be compared to determine which enzyme is likely to be most active at low, environmental exposure levels. The tissue-specificity of enzyme expression can be used to develop hypotheses as to which tissue may be most at risk for genetic damage.

The reader is warned that cDNA expression of human promutagen-activating enzymes in target cells is not a surrogate for predicting human carcinogenesis. However, these systems, used thoughtfully, can contribute significantly to our understanding of human promutagen/procarcinogen activation. Moreover, escape from the over-dependence of *in vitro* genotoxicology on rodent liver homogenates for promutagen activation should be regarded as a significant step forward.

13.2 Xenobiotic-metabolizing enzymes

Discussions of xenobiotic-metabolizing enzymes centre on the cytochrome P450 (CYP) superfamily of mixed-function oxygenases. CYP enzymes are membrane-bound haemoproteins present in the endoplasmic reticulum. Catalytic activity of CYP enzymes requires electron transfer from NADPH to the haem through the action of NADPH cytochrome P450 oxidoreductase (OR). Stimulation of the catalytic activity of some enzymes also occurs by electron transfer from NADH via cytochrome b_5 (b_5). When CYP enzymes are introduced by cDNA expression, the level of catalytic activity will be influenced by the levels of OR and b_5.

A unified nomenclature for the CYP enzymes, based on sequence similarities, has been developed (Nelson *et al.*, 1993) and will be used throughout this chapter. It uses the CYP prefix followed by a number designating the family, a letter designating the subfamily and a number designating the individual form (see Chapter 5). For subfamilies where enzyme number and function are well conserved across species, the same number for the individual forms is used. For subfamilies where evolutionary relationship is uncertain, all individual forms are given different numbers. While this nomenclature can be confusing to those not active in the field (this confusion has been compounded by the renaming of some forms as the nomenclature was updated), it does represent a great improvement over the previous use of trivial names.

Two functionally distinct classes of CYP enzymes exist in mammals: the steroidogenic enzymes and the xenobiotic-metabolizing enzymes. The latter class of enzymes is expressed at highest levels in liver and has been the focus of promutagen activation studies. CYP enzymes play a crucial role in the metabolism of many drugs, pollutants and other xenobiotics. The extent of metabolism and the specific pathways of metabolism can influence the safety and/or efficacy of a drug or other xenobiotic. Certain cytochrome P450 enzymes and other xenobiotic metabolizing enzymes are known to be polymorphic in human populations with different allele frequencies for different ethnic populations (Caporaso et al., 1991; see also Chapter 5). The levels of other cytochrome P450 enzymes are known to be modulated in humans by treatment with some drugs, the consumption of alcohol, cigarette smoking or disease (Murray, 1992). Therefore, information on the specific enzymes involved in activation is helpful for predicting whether certain subpopulations may be more or less sensitive to a xenobiotic.

The composition of cytochrome P450 enzymes in rodent species is markedly different from that of humans. Rats and mice sometimes exhibit sexually dimorphic expression of these enzymes (Gonzalez et al., 1991; Smith, 1991). Therefore, there is no *a priori* reason to expect rodent xenobiotic metabolism to be similar to that in humans.

Because of the central role of cytochrome P450s in drug metabolism and carcinogenesis, considerable effort has been devoted by many laboratories to the development of model systems to study human P450 metabolism (Gonzalez, 1988; Guengerich, 1988; Wrighton and Stevens, 1992; Wrighton et al., 1993a). A relatively good understanding of the relative levels, and substrate specificity, of different cytochrome P450 enzymes in human liver has emerged.

A brief summary of the significance of different human cytochrome P450 forms is presented next. This summary will focus on what is known about the capacity of the major human forms to activate promutagens/procarcinogens and will provide the reader with some salient properties of the different enzymes. A more comprehensive discussion of the individual human enzymes can be found in Wrighton and Stevens (1992). Some probe substrates are listed. A probe substrate is a compound that is metabolized specifically (exclusively by a single enzyme) or selectively (nearly exclusively by a single enzyme). With probe substrates, the catalytic activity of the desired CYP form can be assayed without substantial interference from other CYP forms which may be present in a preparation.

13.2.1 CYP1A1

CYP1A1 is the major polycyclic aromatic hydrocarbon (PAH)-metabolizing enzyme in rodents. However, while human CYP1A1 shares specificity for PAH, it appears to be expressed at very low levels in human liver and is usually not detectable in extrahepatic tissues of non-smokers (Murray et al., 1993; Shimada et al., 1992). Therefore, the role of this enzyme in human hepatic xenobiotic metabolism may be relatively minor. CYP1A1 is the most common enzyme found to be expressed in stable cell lines where it is inducible upon exposure to PAH (Crespi et al., 1985; Diamond et al., 1980).

13.2.2 CYP1A2

In rodents and humans, this enzyme appears to be expressed exclusively in the liver (Kimura et al., 1986; deWazier et al., 1990). CYP1A2 is the principal enzyme responsible for the metabolism of aflatoxin B_1, and of aromatic and heterocyclic amines (Aoyama et al., 1989a, 1990a; Crespi et al., 1990a; Shimada et al., 1989a). It also has some capacity to activate nitrosamines (Crespi et al., 1991b; Smith et al., 1992). A diagnostic biotransformation for CYP1A2 in vivo is caffeine 3-demethylation (Butler et al., 1989).

13.2.3 CYP2A6

CYP2A6 metabolizes some relatively low-molecular-weight promutagens including nitrosamines (Crespi et al., 1990b, 1991b; Yamazaki et al., 1992). It is a relatively low-affinity form for the activation of aflatoxin B_1. Coumarin 7-hydroxylase is a probe biotransformation for this enzyme in humans (Yamano et al., 1990).

13.2.4 CYP2B6

CYP2B6 is variably expressed in human liver and the substrate specificity of this enzyme is poorly understood (Yamano et al., 1989). CYP2B6 has been reported to activate cyclophosphamide and ifosphamide. However, it appears to be one of several low-affinity enzymes in human liver (Chang et al., 1993).

13.2.5 CYP2C

There are multiple CYP2C enzymes, including CYP2C8, CYP2C9, CYP2C18 and CYP2C19. Their substrate specificity often overlaps and the identity of the specific form that performs a particular biotransformation is often uncertain (except in cDNA expression systems). Therefore, the final digit in the name is often deleted in order to convey this uncertainty. None of these enzymes has yet been found to play a prominent role in the activation of promutagens. However, there is some evidence for CYP2C-mediated metabolism of benzo[a]pyrene in human liver microsomes (Yun et al., 1992). CYP2C protein(s) appear to be high-affinity forms for the activation of cyclophosphamide (Chang *et al.*, 1993). CYP2C19 is polymorphic in humans with about 5–10% of Caucasians and 20% of Asians deficient in this enzyme (Wrighton *et al.*, 1993b).

13.2.6 CYP2D6

CYP2D6 plays a principal role in the metabolism of many drugs with basic amine functionalities. It is polymorphic in humans, 5–10% of Caucasians being deficient in this enzyme (Eichelbaum and Gross, 1990). It has not been found to play an important role in the activation of mutagens, although it has been found to activate the tobacco smoke nitrosamine 4-(methylnitrosamino)-1-(3-pyridyl)-1-butanone (NNK; Crespi *et al.*, 1991b; Penman *et al.*, 1993). In addition, some epidemiological studies have linked the CYP2D6 polymorphism to risk of smoking-induced lung cancer (Caporaso *et al.*, 1991). The antimalarial drug, quinidine, is a potent, selective inhibitor of CYP2D6.

13.2.7 CYP2E1

CYP2E1 plays a prominent role in the metabolic activation of many low-molecular-weight promutagens/procarcinogens (Guengerich *et al.*, 1991). *N*-nitrosodimethylamine, a promutagen which is often poorly activated by S-9-based activation systems, is activated at high affinity by CYP2E1. CYP2E1 also activates other nitrosamines, urethane, vinyl halides, butadiene and a variety of low-molecular-weight halogenated compounds. The catalytic activity of CYP2E1 is strongly inhibited by many organic solvents which may be used to add promutagens to cell cultures (Yoo *et al.*, 1987). Chlorzoxazone is a useful probe substrate for CYP2E1 catalytic activity.

13.2.8 CYP3A

There are two principal CYP3A proteins expressed in the human adult liver. CYP3A4 is the most abundant P450 in many humans and CYP3A5 is present in about 20% of human liver samples (Aoyama *et al.*, 1989b). CYP3A4 metabolizes a wide variety of structurally diverse xenobiotics (Wrighton *et al.*, 1993a) and can metabolize compounds of up to 1000 mol. wt (for example, cyclosporine). CYP3A enzymes play a role in the metabolic activation of aflatoxins and some PAH-dihydrodiols (Shimada *et al.*, 1989b).

Within the human liver, which is the best-characterized tissue for P450 expression, CYP3A appears to be the most abundant protein, comprising about 30% of total P450. CYP2C9, CYP1A2 and CYP2E1 are of intermediate abundance, while the other P450 forms are of lower abundance (Shimada *et al.*, 1989a). This contrasts sharply with rodent (rat) liver, where CYP2C enzymes are the most abundant in uninduced animals and these enzymes exhibit sexually dimorphic expression.

When substrate specificities between P450 orthologues from different species are compared, CYP1A1, CYP1A2, CYP2E1 and CYP3A appear to have relatively well-conserved substrate preferences for metabolizing PAH, aromatic/heterocyclic amines, low-molecular-weight compounds and high-molecular-weight compounds, respectively. For the other P450 forms, substantial divergence in substrate specificity has been observed.

In addition to the cytochrome P450 systems, other enzymes can be important for promutagen activation. These include epoxide hydrolases, which can influence the genotoxicity of PAH, and *N*-acetyltransferases (NAT) which can influence aromatic amine genotoxicity. These enzymes are briefly discussed in the last section of this chapter.

13.3 Approaches to cDNA expression

A wide variety of model expression systems has been developed utilizing bacteria, yeast, rodent cells, human cells and, more recently, intact organisms (Komori *et al.*, 1993; Waterman and Johnson, 1991). Many of the practical aspects of the development of CYP cDNA expression systems can be found in Waterman and Johnson (1991). This chapter focuses exclusively on mammalian target cell expression of promutagen-activating enzymes. An alternative approach is the use of cDNA-expressed enzymes as an extracellular activation system. This has been used successfully by several laboratories using microbial target cells (Aoyama *et al.*, 1989a,b; McManus *et al.*, 1990). The amount of activating enzyme needed and the relative inefficiency of extracellular versus intracellular activation systems will probably make extracellular activation impractical for mammalian target cell systems.

A schematic diagram of an approach to cell line development is shown in *Figure 13.1*. cDNA is prepared from mRNA isolated from tissue. The appropriate cDNA is identified and purified by traditional phage plaque screening or by polymerase chain reaction. The cDNA, if it is a new isolate, should be sequenced to ensure its identity and the lack of base substitution mutations. The cDNA is then subcloned into an expression vector and introduced into the host cell. Cells bearing the vector are selected, purified and the level of cDNA expression analysed, usually at the protein level. Clonal populations expressing high levels of the cDNA are thus identified. The stability of expression is then monitored during long-term growth in culture. In addition to stability, it is important that the levels of cDNA expression are homogeneous in the population. Heterogeneous expression can result in aberrant concentration–response curves owing to differential survival and genotoxicity in the sensitive and resistant populations.

All approaches couple the cDNA to be expressed to a heterologous promoter which functions efficiently in the host cell. Therefore, while the cDNA is expressed efficiently, the transcriptional regulation of the native protein is not retained. The vector containing the cDNA and promoter sequences also contains a gene which confers drug/antibiotic resistance on the

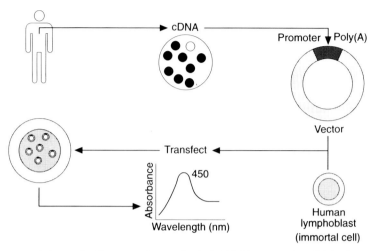

Figure 13.1. Schematic diagram of a method of cDNA expression. This shows the overall approach for cDNA expression in human lymphoblastoid cells using an extrachromosomal vector. Successful, high-level CYP cDNA expression should result in sufficient expression to allow spectrophotometric quantitation of P450 content in cellular membrane fractions.

mammalian cell to, for example, G418, hygromycin B or L-histidinol. Alternatively, DNA bearing the resistance gene can be cotransfected. The resistance gene allows identification and propagation of cells bearing introduced DNA. The best promoter/resistance gene will vary from host cell to host cell. Combinations which have been reported to work successfully include: V79 cells with the SV40 early promoter and G418 resistance (integrating DNA, Doehmer *et al.*, 1988), AHH-1 TK$^{+/-}$ cells with the herpes simplex virus thymidine kinase promoter and either hygromycin B or L-histidinol resistance (extrachromosomal DNA; Crespi *et al.*, 1993a), C3H 10T1/2 cells with a long terminal repeat promoter and G418 resistance (retrovirus; Tiano *et al.*, 1993), NIH 3T3 cells with a long terminal repeat promoter and G418 resistance (retrovirus; Battula, 1989), and CHO cells with SV40 early promoter and G418 resistance (integrating DNA; Trinidad *et al.*, 1991).

The vector DNA is introduced by a variety of methods including calcium phosphate precipitation, electroporation, protoplast fusion or retroviral infection. For the bulk of these techniques, the vector DNA integrates into the cellular genome, and, depending on the integration site, is stably maintained in a subset of transfected cells.

Extrachromosomal vectors have also been used with considerable success (Crespi *et al.*, 1993a). The introduced DNA replicates extrachromosomally with these vectors and stability is achieved by maintaining selection for a resistance marker present on the vector. With both integrating and extrachromosomal vectors, multiple, different DNAs can be introduced using independent means of selection.

From a practical point of view, the cDNA-expressing cells usually have culture properties (growth rate, plating efficiency, etc.) similar to (but not necessarily identical with) unmodified cells. At relatively high cDNA expression levels in the human lymphoblastoid system we have observed a decrease in cellular growth rate (from 18-h doubling time to 24-h doubling time) and in plating efficiency (from 50% to 30%). We have not observed any substantive difference in negative control (background) mutant frequency when conducting mutation assays at the thymidine kinase (*tk*) locus or the hypoxanthine-guanine phosphoribosyl transferase (*hprt*) locus (see Chapter 11). Since the changes in growth characteristics upon cDNA expression are relatively minor, testing protocols usually can be applied to cDNA-expressing cells with few modifications.

Within the cytochrome P450 system, a desirable experimental approach is to have an isogenic panel of cell lines which differ only in the CYP cDNA that is expressed. The experimental approach is shown in *Figure 13.2.* Concentration–response data for the toxicity/genotoxicity of the promutagen are developed in each cell line (shown as a survival curve). It is often desirable to conduct analyses of the metabolites produced by the cell line in parallel with

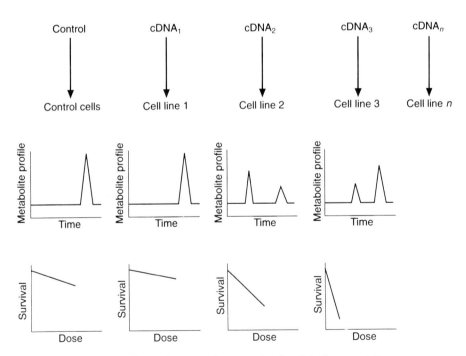

Figure 13.2. Scheme for an integrated approach of toxicity/genotoxicity measurement with concurrent metabolite analysis, which shows the use of a panel of isogenic cell lines each expressing a different cDNA. The production of metabolites with different retention times (by HPLC, for example) is depicted. Cells lines which produce different metabolites have different sensitivities to the toxicity of a protoxin.

the toxicity studies (shown as HPLC chromatograms with peaks of different retention times). Such an analysis can indicate which CYP enzymes are capable of activating the promutagen. Correcting for any differences in CYP expression levels among the cell lines allows determination of which enzyme is the most effective at activating the promutagen.

There are two challenges to developing such a CYP-expressing cell line panel. The first is the multiplicity of human cytochromes P450 expressed *in vivo*. There are over 12 distinct, xenobiotic-metabolizing cytochrome P450 enzymes expressed in human liver. Each enzyme constitutes a separate project. Two systems have been most extensively developed. In V79 cells, human CYP1A1 and CYP1A2 expression has been reported (Wolfel *et al.*, 1992; Schmalix *et al.*, 1993) while CYP2E1-, CYP2A6-and CYP3A4-expressing cells have recently been developed (J. Doehmer, personal communication). For the AHH-1 TK[+/-] cell system, cells expressing CYP1A1 (Penman *et al.*, 1994), CYP1A2 (Penman *et al.*, 1994), CYP2A6 (Crespi *et al.*, 1990b), CYP2B6 (Chang *et al.*, 1993), CYP2C9 (unpublished), CYP2D6 (Crespi *et al.*, 1991b; Penman *et al.*, 1993), CYP2E1 (Crespi *et al.*, 1991b,c) and CYP3A4 (Crespi *et al.*, 1991a) have been reported. With the human lymphoblast system, all of the major human promutagen-activating cytochrome P450s have now been expressed individually and several have been coexpressed in the same host cell (Crespi *et al.*, 1991c).

The second challenge is achieving a stable cell line which contains cytochromes P450 at levels which are adequate for the intended studies. We can offer the following guidelines regarding necessary expression levels for different applications. The expression levels below are based on a routine, 1-day exposure to the promutagen. Naturally, shorter exposure times would require higher enzyme expression. In general, stable expression levels below 10 pmol P450 per mg microsomal protein are consistent with the data reported for most mammalian systems. Only with the human lymphoblast system have higher expression levels, up to 160 pmol P450 per mg, been reported (Penman *et al.*, 1993).

Our experience with cDNA expression of many human and rodent P450 cDNAs indicates that at least 1 pmol P450 mg^{-1} microsomal protein is required in order to observe genotoxic effects. In fact, promutagen activation is one of the most sensitive indicators of successful cDNA expression in mammalian cells, often detecting an effect (for rapidly turned-over substrates) at expression levels below those detectable by enzyme assay or Western blot. However, at these low expression levels the system has very limited applications to studies of xenobiotic metabolism. In order to have a system which is useful for xenobiotic metabolism/activation, at least 10-fold higher expression levels, of about 10 pmol P450 mg^{-1} microsomal protein, are required. At this level, primary metabolites of substrates with high or moderate rates of turnover are easily detected, particularly if radiolabelled material is used. This is also the practical lower limit for spectrophotometric quantitation of P450 content. A system of even greater utility for studying xenobiotic metabolism/activation should contain about 50 pmol P450 mg^{-1} microsomal protein. At this expression level secondary metabolites are readily detected, low turnover substrates can be studied and metabolism can often be measured by loss of the parent compound. The utility of these expression levels assumes adequate, but not necessarily saturating, cytochrome P450 reductase levels. The 50 pmol mg^{-1} microsomal protein level of expression is roughly comparable to levels of individual P450s observed in human liver microsomes. It is likely that only at this highest expression level would significant promutagen deactivation be observed.

At the highest expression levels it is easier to support observations of genotoxicity with analytical studies of biotransformation under the conditions of the assay or under closely related conditions. The principal effect of increasing cDNA expression level on genotoxicity of a promutagen is to shift the concentration–response curve to lower concentrations. Therefore, detectable mutagenic responses are more likely to be observed at promutagen concentrations which are below the K_m of the activating enzyme (Crespi *et al.*, 1993b).

Our laboratory has focused on development of human cells that express human P450 (Crespi *et al.*, 1990a,b,c, 1991a,b,c, 1993a, b; Davies *et al.*, 1989; Penman *et al.*, 1993, 1994). We have chosen to focus on human cells because of their greater relevance to human toxicity than other mammalian cells, simple eukaryotic systems or prokaryotic systems. Our laboratory has used human B lymphoblastoid cells because of the relative ease of culture and scale-up of this anchorage-independent cell type and the availability of a flexible extrachromosomal vector system (Sugden *et al.*, 1985). The particular cell line used in our laboratory, AHH-1 TK$^{+/-}$ (Crespi and Thilly, 1984), is a versatile indicator cell line for a variety of *in vitro* toxicological endpoints (Penman and Crespi, 1987). Therefore, P450-mediated metabolism can be readily related to toxic or genotoxic effects at the cellular level. This cell line has some native CYP1A1 activity which is detectable upon pretreatment with appropriate inducers and does not have detectable microsomal epoxide hydrolase activity (Crespi *et al.*, 1985).

The other commonly used system, V79 cells, is also a versatile indicator cell line for a variety of genotoxicity endpoints. These cells do not appear to have substantial native CYP activities (McGregor *et al.*, 1991). Therefore, the rare competition for metabolic activation by native CYP1A1 in AHH-1 TK$^{+/-}$ cells is avoided. However, some promutagens are mutagenic to native V79 cells (Doehmer *et al.*, 1988; Dogra *et al.*, 1990). The enzymes responsible for this activation have not been identified. In addition, mutations cannot be scored at the *tk* locus and the mutation assay at the *hprt* locus is statistically less precise than in human lymphoblastoid culture due to the smaller number of cells which can be used in practice. This cell line is anchorage dependent which makes it more cumbersome to scale up cultures. For the few cDNAs which have been expressed to date in both AHH-1 TK$^{+/-}$ cells and V79 cells, human lymphoblasts support higher expression.

13.4 Theoretical and technical considerations

13.4.1 Availability of human enzymes

Cytochrome P450 cDNA expression allows the study of human promutagen activation with the same ease as was once only possible for non-human enzymes. Data from such studies can be used to understand better human susceptibility to environmental mutagens and the contribution of genetic and environmental factors to such susceptibility. Several xenobiotic-metabolizing enzymes, most prominently CYP2D6, CYP2C19 and NAT2, are polymorphic in humans and the frequency of these polymorphisms varies in different populations. (A more complete discussion of these polymorphisms can be found in Chapter 5.) In addition, the levels of some enzymes are known to be induced in humans on treatment with certain drugs. Other drugs and components of the diet can inhibit enzymes specifically. Those genetic and environmental factors that are already known to modulate responses to drugs will probably also modulate responses to environmental mutagens.

13.4.2 Promutagen exposure time

The instability of extracellular activation systems presents a limitation on the length of incubation of the target cells with the promutagen. S-9 activation systems are active for a few hours, and whole primary cells typically lose much of their CYP enzymes in a few days. Such time limitations do not apply to metabolically competent cells. Long-term, low-dose treatment regimens are possible. Such protocols allow quantitation of mutagenic activity at low concentrations in order to determine whether mutagenic effects observed at high concentrations also occur at low concentrations where minimal perturbation of cellular homoeostasis occurs (Penman *et al.*, 1983).

13.4.3 Sensitivity to promutagens

Expression of the promutagen-activating enzyme within the target cells (close to the target DNA) results in a substantially more sensitive system for studying genotoxicity. The cellular membrane no longer presents a barrier to the passage of electrophiles into the cell. Therefore, mutagenic activity can often be measured at concentrations similar to those expected in the environment. *Figure 13.3* provides a comparison of the mutagenic activity of *N*-nitrosodimethylamine to TK6 human lymphoblasts with an Aroclor 1254-induced rat-liver S-9 and in an AHH-1 $TK^{+/-}$ cell derivative expressing CYP2E1 cDNA.

13.4.4 Experimental design

Sensitivity of the endpoint to the activated promutagen and the level of cDNA expression present in the target cell are important considerations in

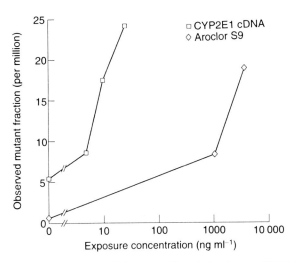

Figure 13.3. Mutagenicity of *N*-nitrosodimethylamine to AHH-1 TK$^{+/-}$ cells expressing human CYP2E1 cDNA and TK6 cells incubated in the presence of rat liver S-9. The inductions of gene locus mutations were measured at the *hprt* and *tk* loci in AHH-1 TK$^{+/-}$ cells and TK6 cells, respectively. Exposure times were 28 h and 3 h for AHH-1 TK$^{+/-}$ cells and TK6 cells, respectively. The final S-9 concentration was 5% (vol./vol.). Values are plotted as the mean of duplicates.

experimental design. Since promutagens can cause different types of genetic change, it is important to use the same genotoxicity endpoint when comparing relative promutagen activation by different cDNA-derived proteins. In this way, intrinsic properties of the activating enzyme are compared. In addition, exposure time should be consistent with all cell lines and the investigator should consider, particularly at low, submicromolar promutagen concentrations, whether non-linearities may be caused by complete metabolism of the promutagen in some but not all cultures. Finally, the investigator should be aware of interactions between the cDNA-expressed enzyme and organic solvents. Most importantly, CYP2E1 is inhibited by most organic solvents (Yoo *et al.*, 1987), but we have found also that the expression levels of some cDNAs (for example, CYP2D6) can be elevated by treatment with the commonly used solvent dimethylsulphoxide (Crespi *et al.*, 1991b).

13.4.5 Qualitative vs. quantitative conclusions

With these systems, concentration–response data can be interpreted per unit cell, per unit enzyme or per unit enzyme with weighting applied for differences in enzyme abundance in a target tissue. Interpretation at the level of the unit cell provides qualitative data as to which enzymes are capable of activating a promutagen. However, because the absolute levels of enzymes typically varies among cell lines expressing different cDNAs, no information as to which enzyme is intrinsically the most efficient is gained. In order to gain insight into quantitative aspects of activation at the unit enzyme level one can take two approaches: (i) develop cells with similar enzyme contents (Crespi *et al.*, 1991a), or (ii) transform the concentration–response data based on relative enzyme content (Crespi *et al.*, 1993b). The principal disadvantage of the first approach is that all cell lines will by definition have expression levels comparable to the most poorly expressed cDNA. This can elevate the concentrations needed to observe a response. Naturally, quantitative approaches require accurate means to measure expression levels in the cell lines (quantitative Western blots or spectrophotometric quantitation). Finally, the quantitative data developed above can be further transformed to adjust for the relative enzyme abundance in the target tissue. In this way insight can be gained into which enzyme, based on its intrinsic activation potential and its abundance in the target tissue, is likely to play the principal role in mutagenic activation *in vivo*.

13.4.6 In vivo *complexity*

When using cDNA expression systems, the investigator should always be aware that while these systems offer the capability to study promutagen activation at the molecular level and can identify specifically which enzymes are involved in the activation of a xenobiotic, the systems may model only a portion of the complete *in vivo* pathways. Other pathways, not present in the target cell, such as specific glutathione-*S*-transferase forms, other conjugation enzymes, or other P450s which deactivate the xenobiotic, may modulate the response. At low concentrations, all enzymatic processes will exhibit first-order kinetics. Therefore, competing, non-activating pathways can reduce, but not eliminate promutagen activation.

There is clearly a need to extend these systems to achieve higher levels of complexity. The only *in vitro* systems available which maintain a high degree of the complexity seen *in vivo* are primary human cells (for example, hepatocytes). However, these cells are of limited availability, highly variable depending on donor and are only amenable to a limited array of genotoxicity endpoints.

13.4.7 Speculation on possible effects of test sensitivity and specificity

The landmark paper of Tennant *et al.* (1987) compared results of genotoxicity tests conducted *in vitro* with results obtained in rodent carcinogenicity bioassays. The authors differentiated the relationship on the basis of sensitivity (the percentage of carcinogens detected as genotoxic) and specificity (the percentage of non-carcinogens detected as non-genotoxic) for four different short-term tests. The mouse lymphoma system and CHO sister chromatid exchange system were sensitive but lacked specificity. The Ames assay and the CHO chromosomal aberration tests were specific but lacked sensitivity. All of these systems shared an Aroclor 1254-induced rat-liver S-9-based activation system. Target cells expressing xenobiotic metabolizing enzymes, by introducing the relevant metabolizing enzymes, have the potential to increase the sensitivity of systems without affecting specificity. It will be interesting to see if this suggestion will be supported by experimental evidence.

13.5 Validation

The traditional approach to validating genetic toxicology assays, correlation with *in vivo* endpoints such as carcinogenicity, is completely inappropriate for systems containing single or even multiple cDNA-expressed enzymes. cDNA expression systems express only a subset of the biotransformation enzymes present *in vivo* and, therefore, will not correlate for xenobiotics which are metabolized by other than the cDNA-expressed enzymes. Only systems which contain a more complete array of xenobiotic metabolizing enzymes allow correlation with genotoxic or carcinogenic effects at the level of the organism.

Systems expressing human cDNA-derived promutagen-activating enzymes have not been developed to be predictive of carcinogenicity in rodents. Our knowledge of the differences in biotransformation by human and rodent enzymes indicates that mechanism-based one-to-one correspondence should not exist. It is possible that systems using cDNA-expressed rodent enzymes may offer a better correlation between mutagenicity and rodent carcinogenicity, but progress in achieving expression of rodent enzymes in cell lines has lagged substantially behind the work on human enzymes.

Validation of single-enzyme, cDNA-expressing target cells must use an approach different from correlation with data derived from studies on whole animals. We propose below a two-pronged approach focusing separately on the endpoint measured and the cDNA-derived enzyme expressed. The enzyme-specific substrates needed for validation have only recently become available, as have the supporting data from the literature. Therefore, many of the cell lines developed to date have not been validated using this approach. However, we believe this approach is sound and should be applied to systems which have already been developed as well as to new systems.

13.5.1 Endpoint

As part of the development of the target cell, the genotoxicity endpoint must be validated. Mutants scored should be true mutants and they must be induced, not selected, by the treatment. Assay parameters such as treatment time, expression time and cell concentrations must be optimized. The expression of a xenobiotic metabolizing enzyme within the target cell should not affect this endpoint validation. In addition, the sensitivity of the target cell to different types of DNA damage (for example that caused by low-molecular-weight alkylating agents, by bulky mutagens or by cross-linking agents) should be verified to assure that if each type of adduct were produced it would lead to a measurable response. Mutagens for the validation of the endpoint can be direct-acting and several different agents should be tested in each class.

13.5.2 Enzyme

Since a principal focus of developing metabolically competent cell lines is the determination of which (human) enzymes are capable of promutagen activation, the catalytic fidelity of the cDNA-expressed enzyme is the principal concern. There is very little published data on catalytic constants for CYP metabolism of promutagens. However, there is an extensive literature on catalytic constants for drugs. These data can be used to validate the catalytic properties of the cDNA-expressed enzyme. However, this approach makes the assumption that kinetic 'fidelity' for drugs will extend to promutagens. However, there is no *a priori* reason why this assumption is not valid. While the interests of investigators may be restricted to certain chemical classes, enzymes carry no such biases.

Therefore, the catalytic properties of the cDNA-expressed enzyme should be measured and compared with those observed for the same enzyme from other sources. For most human cytochrome P450 enzymes, this represents human liver microsomes. At present, human liver microsomes (which are available commercially from several sources) should be regarded as the 'gold standard' for verification of enzyme catalytic properties. However, care must be taken in data analysis because of the multiplicity of P450 enzymes present in human liver microsomes and the variability of levels of specific enzymes in preparations from different donors. Quantitative deviations of two- to three-fold should be expected.

Apparent K_m is probably the most important parameter and is easily measured in human liver microsomes and for cDNA-expressed enzymes. There is no reason why apparent K_m should not be measured and reported for any cDNA expression system. True turnover number in human tissue samples usually has to be inferred because of the mixture of enzymes present in human liver microsomes and the lack of quantitative data on the level of the active enzyme. Measurement of turnover number for cDNA expression systems requires accurate quantitation of the enzyme content in the incubations (preferably by spectrophotometric means). In general, absolute turnover numbers for cDNA-expressed enzyme in mammalian cell systems should be higher than those reported for purified/reconstituted P450s. However, the level of OR and b_5 in the target cell can have a large effect on turnover number.

Probably more important than absolute turnover (which can be affected by levels of OR and b_5) is the relative turnover for multiple, specific substrates. Relative turnover provides a means to establish that the cDNA-expressed enzyme discriminates properly between substrates with respect to rate of metabolism. This a key application of the system. The rank order of rates of metabolism for the different substrates should be the same for the cDNA-expressed enzyme as for the enzyme in human liver microsomes.

For most human cytochrome P450s, a number of enzyme-specific substrates have been identified and apparent K_m and turnover numbers have been reported for human liver microsomes, purified P450s or from cDNA-expressed P450s. Some enzyme/substrate combinations (and literature citations for the methods) which can be used are: CYP1A2, 7-ethoxyresorufin O-deethylation (Burke *et al.*, 1977), caffeine 3-demethylation (Butler *et al.*, 1989); CYP2A6, coumarin 7-hydroxylation (Yamano *et al.*, 1990); CYP2C9, (S)-warfarin 7-hydroxylation (Rettie *et al.*, 1992), diclofenac 4'-hydroxylation (Leeman *et al.*, 1993); CYP2D6, bufuralol 1'-hydroxylation, debrisoquine 4-hydroxylation, dextromethorphan O-demethylation (Kronbach *et al.*, 1987); CYP2E1, chlorzoxazone 6-hydroxylation (Peter *et al.*, 1990); CYP3A4, testosterone 6β-hydroxylation (Waxman *et al.*, 1983).

13.6 Examples of applications

In this section the earliest cDNA expression work in target cells for genotoxicity assays is summarized and more recent applications are reviewed. cDNA expression in target cells for genotoxicity assays has enjoyed rapid progress in both the number of cDNAs expressed in different systems and also the level of expression. Many human enzymes have been successfully expressed at high levels in mammalian target cells. Cell lines expressing multiple enzymes have also been developed. Given the rate of progress, we can look forward to the development of even more robust systems in the near future.

The first demonstration of cytochrome P450 cDNA expression in target cells for mutagenicity studies was by Doehmer *et al.* (1988). Rat CYP2B1 cDNA was introduced into V79 cells under control of a SV40 early promoter in a vector conferring resistance to G418. Aflatoxin B_1 was found to be a modest two- to three-fold more active gene mutagen in cells expressing the CYP2B1 cDNA than in untransfected control cells. CYP2B1 expression was also detected by Western blot and by CYP2B1-selective enzyme assays. These cells have also been found to activate cyclophosphamide and ifosphamide to gene mutagens (Doehmer *et al.*, 1990). The V79 cell system, using the same vector, has been extended to express additional rat and human cytochrome P450 enzymes (CYP1A1 and CYP1A2, both rat and human). This system has also been used in studies of micronucleus induction (Ellard *et al.*, 1991).

The first example of human cell expression of a transfected cytochrome P450 was by Crespi *et al.* (1989). In this study a human *CYP1A1* gene was introduced into AHH-1 TK$^{+/-}$ human lymphoblasts under control of the herpes simplex thymidine kinase gene promoter in an extrachromosomal vector conferring resistance to hygromycin B (this human lymphoblastoid cell line is not sensitive to G418). This cell line has endogenous CYP1A1 activity. A modest, approximate doubling in basal CYP1A1 activity and aflatoxin B_1 mutagenicity was observed. More recently, CYP1A1 cDNA has been expressed at much higher levels in this system (Penman *et al.*, 1994) and the role of CYP1A1 in the activation of aflatoxin B_1 was confirmed. However, CYP1A2 appears to have a much more prominent role in the activation of aflatoxin B_1 as determined by cDNA expression and studies in human liver microsomes (Gallagher *et al.*, 1994; Penman *et al.*, 1994).

Applications can be divided into three broad categories:

(i)   Use of a specific cDNA-expressing cell line to examine aspects of the genotoxicity of chemicals already believed to be activated by the cDNA-encoded enzyme.

(ii) Use of a panel of cDNA-expressing cell lines to determine the enzymes activating a promutagen with an uncharacterized metabolic activation pathway.

(iii) Use of single cell lines expressing multiple cDNAs in order to screen chemicals for mutagenic activity.

Examples of the three types of application are discussed next.

13.6.1 Single cell lines

Rat CYP1A1 has been expressed in V79 cells (Dogra *et al.*, 1990). Two cell lines, designated XEM1 and XEM2 were isolated. These lines differed in the level of cDNA expression (XEM2 had higher expression). The gene-locus mutagenicity of benzo[*a*]pyrene and benzo[*a*]pyrene-7,8-dihydrodiol was examined in both cell lines and also in control cells and cells expressing rat CYP2B1. These promutagens were much more active in the CYP1A1-expressing cell lines, of which the XEM2-cell line was more sensitive that the XEM1 cell line, consistent with the relative levels of cDNA expression. Benzo[*a*]pyrene-7,8-dihydrodiol was more active than benzo[*a*]pyrene, consistent with the single step activation of the former promutagen and the multistep activation of the latter promutagen. These studies were extended to human CYP1A1 (Schmalix *et al.*, 1993) and similar observations were made for the same two promutagens with a single CYP1A1-expressing cell line. In both these cases, endogenous microsomal epoxide hydrolase probably participated in the activation of benzo[*a*]pyrene.

Human CYP1A1 has also been expressed in human lymphoblastoid cells (Penman *et al.*, 1994). CYP1A1-expressing cells were found to activate benzo[*a*]pyrene, cyclopenta(*cd*)pyrene, NNK and aflatoxin B_1. The AHH-1 $TK^{+/-}$ cell line does not have detectable microsomal epoxide hydrolase activity (Davies *et al.*, 1989). Recent data from our laboratory demonstrate the effect of microsomal epoxide hydrolase expression (*Figure 13.4*). Introduction of human microsomal epoxide hydrolase into cells expressing CYP1A1 cDNA resulted in a 25- to 30-fold increase in mutagenicity by benzo[*a*]pyrene at the *tk* locus. Human CYP1A2 and CYP1A1-mediated activation of aflatoxin B_1 was compared with compensation for the spectrophotometric P450 contents of the CYP1A1 and CYP1A2-expressing cell lines. CYP1A2 was found to be more than 40-fold more active than CYP1A1 for the activation of this promutagen. Kinetic constants for a series of model substrates were also reported.

Human CYP1A2 has also been expressed in V79 cells (Wolfel *et al.*, 1991, 1992). For this study, two different V79 cell strains were used. One strain was deficient in NAT and the other was proficient. NAT is an important enzyme in

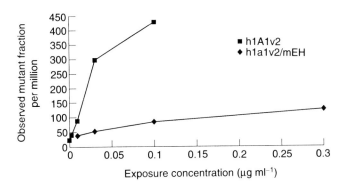

Figure 13.4. Mutagenicity of benzo[a]pyrene to AHH-1 TK$^{+/-}$ cells expressing human CYP1A1 cDNA with and without coexpression of human microsomal epoxide hydrolase (mEH) cDNA. h1A1v2 cells (Penman *et al.*, 1994) and h1A1v2 cells bearing the mEH expression plasmid p167Dtk2 (Davies *et al.*, 1989), were tested for the induction of gene locus mutations at the *tk* locus after a 3-day exposure to benzo[a]pyrene. Values are plotted as the mean of duplicates.

the generation of potent electrophiles from the *N*-hydroxy aromatic amines produced by CYP1A2. The importance of NAT activity was elegantly demonstrated by Yanagawa *et al.* (1994). In this study, CHL cells were transfected with human CYP1A2 and then subsequently transfected with human NAT1 (the monomorphic form) or human NAT2 (the polymorphic form). Activation of 2-amino-3-methylimidazo[4,5-*f*]quinoline (IQ) and 2-amino-3,8-dimethylimidazo[4,5-*f*]quinoline (MeIQx) to gene mutagens was examined in the cell lines. Cells expressing CYP1A2 alone or cells expressing CYP1A2 with NAT1 were no more sensitive to these promutagens than untransfected control cells. Cells expressing CYP1A2 and NAT2 were 100-fold and 370-fold more sensitive to IQ and MeIQx, respectively. This study demonstrates the importance of all the specific enzymes in an activation pathway. The selectivity of human NAT1 and NAT2 suggests that data developed using endogenous rodent NAT may have little relevance to human activation potential.

Activation of heterocyclic amines was also studied in repair-proficient and repair-deficient CHO cells expressing mouse CYP1A2 cDNA (Buonarati *et al.*, 1991). In this study, 2-amino-1-methyl-6-phenylimidazo[4,5-*b*]pyridine (PhIP) was clastogenic to DNA-repair deficient, CYP1A2-expressing cells while non-clastogenic to isogenic cells without the CYP1A2 cDNA. The status of NAT expression in these cell lines was not reported. A similar approach of CYP cDNA expression in DNA repair-proficient and repair-deficient cells has been reported using human fibroblasts (States *et al.*, 1993).

The ability to perform malignant transformation studies in mammalian cells expressing human CYP2A6 has recently been demonstrated (Tiano *et al.*, 1993). In this study, C3H10T1/2 cells were transduced with CYP2A6 and malignant transformation was induced by NNK treatment in CYP2A6-expressing, but not control, cells.

251

13.6.2 Panel of cell lines

The ability to perform analyses of promutagen activation with a panel of cell lines requires a considerable effort in establishing the panel. Appropriate panels of cell lines have become available only recently. Some of the studies reviewed below are early experiments in which a relatively small number of cDNAs had been expressed.

CYP2E1 is known to be a high-affinity enzyme for the metabolic activation of nitrosamines. Early cDNA expression work indicated that human CYP2A6 also had activity for nitrosamine activation (Davies *et al.*, 1989). The ability of CYP2A6 and CYP2E1 to activate *N*-nitrosodimethylamine (NDMA) and *N*-nitrosodiethylamine (NDEA) was examined in isogenic human lymphoblastoid cells individually expressing these cDNAs (Crespi *et al.*, 1990b). CYP2E1 was found to be much more efficient than CYP2A6 for NDMA activation while CYP2A6 and CYP2E1 were roughly comparable in their ability to activate NDEA. This finding was subsequently confirmed in studies using human liver microsomes (Yamazaki *et al.*, 1992).

Epidemiological studies have linked the CYP2D6 polymorphism to lung cancer risk in smokers (Caporaso *et al.*, 1991). However, the mechanistic basis for this linkage was unclear as CYP2D6 was not known to activate any promutagens/procarcinogens and most CYP2D6 substrates contain a basic amine group, a functionality not common in procarcinogens. A panel of human lymphoblastoid cell lines expressing human CYP1A2, CYP2A6, CYP2D6 and CYP2E1 was used to examine which human CYP forms were capable of activating the tobacco smoke nitrosamine NNK to a gene mutagen. All of the above enzymes, including the polymorphic CYP2D6, were found to be capable of activating NNK (Crespi *et al.*, 1991b). This observation was confirmed and extended in a subsequent study using a cell line with higher CYP2D6 expression levels. The mutagenicity of NNK was attenuated by the addition of quinidine, a potent CYP2D6 inhibitor and elevated NNK metabolism was demonstrated in extracts for the CYP2D6-expressing cell line. Subsequently, independent studies examining NNK metabolites from human liver microsomes and cDNA-expressed enzyme produced in vaccinia virus revealed a principal role for CYP1A2 and CYP2E1 in NNK activation (Smith *et al.*, 1992). CYP2A6 and CYP2D6 appear to play minor roles in hepatic NNK activation because of the low NNK turnover by these enzymes and also the lower abundance of these enzymes relative to CYP1A2 and CYP2E1. Nevertheless, the study of Crespi *et al.* (1991b) has demonstrated that CYP2D6 can activate promutagens.

Aflatoxin B_1 (AFB) is a fungal toxin to which there is widespread human exposure through contamination of improperly stored grains. The role of specific human cytochrome P450 enzymes in the metabolic activation of AFB has been examined by a number of laboratories using several methods. Shimada and Guengerich (1989) correlated AFB activation to a bacterial mutagen with the levels of CYP3A protein and catalytic activity in a panel of human liver microsomes. This activation could be inhibited by anti-CYP3A

IgG. Because the bacterial mutation assay was relatively insensitive, high AFB concentrations (three to four orders of magnitude above human dietary exposure levels), were needed to detect a response in the bacterial mutagenicity assay. Forrester *et al.* (1990) confirmed these results using essentially the same methods. However, based on inhibition by anti-CYP1A and anti-CYP2A antisera, they suggested an additional role for CYP1A2 and CYP2A6 in the activation of AFB.

Aoyama *et al.* (1990b) examined AFB activation using a panel of cDNA-expressed cytochrome P450s and vaccinia virus/HepG2 cells. Five forms, CYP1A2, CYP2A6, CYP2B6, CYP3A3 and CYP3A4 were found to be capable of activating AFB to a species which bound to cellular DNA. Given the multiplicity of enzymes capable of activating AFB, it is important to know which P450 forms are most active at low AFB concentrations.

Using a panel of human B-lymphoblastoid cell lines expressing CYP1A2, CYP2A6 and CYP3A4 cDNAs and specifically developed to contain approximately equal cytochrome P450 contents (verified by quantitative Western blot), Crespi *et al.* (1991a) demonstrated that cells expressing CYP1A2 were substantially more sensitive to AFB mutagenicity than cells expressing CYP3A4 which were in turn markedly more sensitive to AFB than cells expressing CYP2A6. This rank order was supported by parallel studies of binding of AFB to cellular DNA. This study suggested that CYP1A2 has a higher affinity for AFB activation than CYP3A4. However, this study compared AFB activation on a unit P450 enzyme basis. The relative contribution of CYP1A2 and CYP3A4 to AFB activation in human liver is also dependent on the relative abundance of the two enzymes.

The relative roles of CYP1A2 and CYP3A4 in the activation of AFB by cDNA-expressed enzymes and human liver microsomes were also examined by Gallagher *et al.* (1994). In this study, the apparent K_m for the formation of AFB-8,9-oxide (the electrophilic species) was determined for CYP1A2 and CYP3A4 expressed from cDNAs in the human lymphoblast system. The relative contribution of CYP1A2 and CYP3A4 for human liver microsomal production of AFB-8,9-oxide was examined by the use of the CYP1A2-selective inhibitor furafylline and the CYP3A-selective inhibitor troleandomycin. These studies revealed that, at low AFB concentrations, activation is mediated primarily by CYP1A2, and that CYP3A4 has a lower affinity for AFB activation.

Styles *et al.* (1994) have examined the genotoxicity of tamoxifen using a panel of human B-lymphoblastoid cells expressing individual and multiple cytochrome P450 cDNAs. Tamoxifen is beneficial to women with breast cancer and it is currently undergoing trials to assess its effectiveness as a prophylactic in healthy women considered to be at high risk of breast cancer. Tamoxifen has been demonstrated to induce micronuclei in MCL-5 cells which express five human P450 cDNAs (White *et al.*, 1992).

The study by Styles *et al.* (1994) examined whether the genotoxicity of tamoxifen was P450-dependent and which particular P450(s) were capable of metabolic activation. Human lymphoblastoid cells expressing human CYP1A1, CYP1A2, CYP2D6, CYP2E1 and CYP3A4 cDNAs, as well as control cells with vector only, were used as target cells for tamoxifen-induced formation of micronuclei. Cells expressing CYP2E1, CYP3A4 and CYP2D6 were found to be capable of activating tamoxifen. Significantly higher tamoxifen concentrations were needed to observe genotoxicity in CYP2D6-expressing cells relative to the other cell lines which produced positive responses.

This study clearly demonstrates that human enzymes are capable of activating tamoxifen to genotoxic species. The observation of human enzyme-mediated genotoxicity in human cells raises concerns regarding treatment with tamoxifen of women who do not have breast cancer. It is important to note, however, that this study does not prove that tamoxifen has substantial carcinogenic potential in humans. Other phase II enzymes may serve to protect cells *in vivo* from the genotoxic effects of tamoxifen. The observation of activation by CYP2E1 and CYP3A4 suggests that coadministration of specific inhibitors (disulfaram and ketoconazole for example) might redirect metabolism to pathways which do not result in genotoxic metabolites.

13.6.3 Cell lines with multiple cDNAs

Given the multiplicity of human CYP enzymes involved in the activation of human promutagens, there is clearly a need for cell lines expressing multiple cDNAs for use as a screening tool for promutagen activation. Studies with panels of cell lines are very informative but also very labour intensive. It is not economically feasible to conduct 12 mutation assays for every unknown compound.

Crespi *et al.* (1991c) used two independent vectors each containing three cDNA-containing expression units to develop a derivative of AHH-1 TK$^{+/-}$ cells which expresses CYP1A1 (endogenously), CYP1A2, CYP2A6, CYP2E1 and CYP3A4 as well as microsomal epoxide hydrolase. These enzymes were chosen because studies with human liver microsomes and cDNA-expressed enzymes have indicated their principal role in promutagen activation. The resulting cell line was designated MCL-5. Expression of the cDNAs was verified by specific catalytic assays and the level of expression determined by quantitative Western blotting. The expression levels for the individual CYP enzymes ranged from 1.4 to 3 pmol mg^{-1} microsomal protein. The stability of expression of the specific CYP and epoxide hydrolase enzymes was measured individually. cDNA expression was found to be stable for 2 months of continuous cell culture. MCL-5 cells were found to activate benzo[*a*]pyrene, 3-methylcholanthrene, AFB, NDMA and NDEA to gene mutagens from 1000-fold to 100 000-fold more efficiently than the unmodified AHH-1 TK$^{+/-}$ cell parent line. A modest increase in sensitivity to 2-acetylaminofluorene was observed and no difference in benzidine mutagenicity was observed.

Crofton-Sleigh *et al.* (1993) evaluated MCL-5 cells for their ability to detect micronucleus induction by a variety of chemicals in two different laboratories with slightly different testing protocols. AFB, sterigmatocystin, MeIQx, NDMA, NDEA, benzidine, 2-aminofluorene, cyclophosphamide, benzene, benzo[a]pyrene, dibenz(a,h)anthracene, 3-methylcholanthrene, tamoxifen, 2-acetylaminofluorene and omeprazole were found to be clastogenic. Anthracene, phenanthrene and pyrene were non-clastogenic. There was good agreement between independent determinations in the same laboratory and between laboratories. For several chemicals, micronucleus induction was also measured in untransfected control cells. In all cases, except benzo[a]pyrene, the promutagen was substantially more mutagenic in MCL-5 cells than in control cells. This observation points to a cDNA-derived enzyme being responsible for promutagen activation.

For several promutagens, the lowest effective concentration for micronucleus induction (Crofton-Sleigh *et al.* 1993) can be compared with the lowest effective concentration for gene locus mutation induction (Crespi *et al.*, 1991c). The two endpoints show similar sensitivities to NMDA, NDEA, AFB, and 2-acetylaminofluorene. Benzo[a]pyrene and 3-methylcholanthrene appear to be more potent gene mutagens. Benzidine appears to be more potent for micronucleus induction.

13.7 Conclusions

Considerable progress has been made in the development of metabolically competent target cells for genotoxicity testing. Cell lines expressing most of the major human CYP enzymes and all of the major human promutagen-activating CYP have been developed. Cell lines expressing multiple human CYP enzymes show considerable promise as a useful screening system for human promutagen activation. Metabolically competent target cells, especially those expressing human enzymes, will provide useful insights into human promutagen activation and the factors which may affect this activation *in vivo*.

References

Aoyama T, Gonzalez FJ, Gelboin HV. (1989a) Mutagen activation of cDNA-expressed P₃450, and P450a. *Mol. Carcinogen.* **1:** 253–259.

Aoyama T, Yamano S, Waxman DJ, Lapenson DP, Meyer UA, Fischer V, Tyndale R, Inaba T, Kalow W, Gelboin HV, Gonzalez FJ. (1989b) Cytochrome P-450 hPCN3, a novel cytochrome P-450 IIIA gene product that is differently expressed in adult human liver. *J. Biol. Chem.* **264:** 10388–10395.

Aoyama T, Gelboin HV, Gonzalez FJ. (1990a) Mutagenic activation of 2-amino-3-methylimidazo[4,5-f]quinoline by complementary DNA-expressed human liver P-450. *Cancer Res.* **50:** 2060–2063.

Aoyama T, Yamano S, Guzelian PS, Gelboin HV, Gonzales FJ. (1990b) Five of twelve forms of vaccinia virus-expressed human hepatic cytochrome P450 metabolically activate aflatoxin B₁. *Proc. Natl Acad. Sci. USA* **87:** 4790–4793.

Battula N. (1989) Transduction of cytochrome P3-450 by retroviruses: constitutive expression of enzymatically active microsomal hemoprotein in animal cells. *J. Biol. Chem.* **264:** 2991–2996.

Buonarati MH, Tucker JD, Minkler JL, Wu RW, Thompson LH, Felton JS. (1991) Metabolic activation and cytogenetic effects of 2-amino-1-methyl-6-phenylimidazo[4,5-*b*]pyridine (PhIP) in Chinese hamster ovary cells expressing murine cytochrome P450 IA2. *Mutagenesis* **6:** 253–259.

Burke MD, Prough RA, Mayer RT. (1977) Characteristics of a microsomal cytochrome P-448-mediated reaction ethoxyresorufin *O*-de-ethylation. *Drug Metab. Disp.* **5:** 1-8.

Butler MA, Iwasaki M, Guengerich FP, Kadlubar FF. (1989) Human cytochrome P-450$_{pa}$ (P-450IA2), the phenacetin *O*-deethylase, is primarily responsible for the hepatic 3-demethylation of caffeine and *N*-oxidation of carcinogenic arylamines. *Proc. Natl Acad. Sci USA* **86:** 7696–7700.

Caporaso N, Landi MT, Vineis P. (1991) Relevance of metabolic polymorphisms to human carcinogenesis: evaluation of epidemiologic evidence. *Pharmacogenetics* **1:** 4–19.

Chang TKH, Weber GF, Crespi CL, Waxman DJ. (1993) Differential activation of cyclophosphamide and ifosphamide by cytochromes P-450 2B and 3A in human liver microsomes. *Cancer Res.* **53:** 5629–5637.

Crespi CL, Thilly WG. (1984) Assay for mutation in a human lymphoblastoid line, AHH-1, competent for xenobiotic metabolism. *Mutat. Res.* **128:** 221–230.

Crespi CL, Altman JD, Marletta MA. (1985). Xenobiotic metabolism in a human lymphoblastoid cell line. *Chem.–Biol. Interact.* **53:** 257–272.

Crespi CL, Langenbach R, Rudo K, Chen Y-T, Davies RL. (1989). Expression of a transfected human cytochrome P450 gene in a human lymphoblastoid cell line. *Carcinogenesis* **10:** 295–301

Crespi CL, Steimel DT, Aoyoma T, Gelboin HV, Gonzalez FJ. (1990a) Stable expression of human cytochrome P450IA2 cDNA in a human lymphoblastoid cell line: role of the enzyme in the metabolic activation of aflatoxin B$_1$. *Mol. Carcinogen.* **3:** 5–8.

Crespi CL, Penman BW, Leakey JAE, Arlotto MP, Start A, Turner T, Steimel D, Rudo K, Davies RL, Langenbach R. (1990b) Human cytochrome P450IIA3: cDNA sequence, role of the enzyme in the metabolic activation of promutagens, comparison to nitrosamine activation by human cytochrome P450IIE1. *Carcinogenesis* **8:** 1293–1300.

Crespi CL, Langenbach R, Penman BW (1990c) The development of a panel of human cell lines expressing specific human cytochrome P450 cDNAs. In: *Mutation and the Environment* (eds ML Mendelsohn, RL Albertini). Alan R Liss, New York, pp. 97–106.

Crespi CL, Penman BW, Steimel DT, Gelboin HV, Gonzalez FJ. (1991a) The development of a human cell line stably expressing human CYP3A4: role in the metabolic activation of aflatoxin B$_1$ and comparison to CYP1A2 and CYP2A3. *Carcinogenesis* **12:** 355–359.

Crespi CL, Penman BW, Gelboin HV, Gonzalez FJ. (1991b). A tobacco smoke-derived nitrosamine, 4-(methylnitrosamino)-1-(3-pyridyl)-1-butanone, is activated by multiple human cytochrome P450s including the polymorphic human cytochrome P4502D6. *Carcinogenesis* **12:** 1197–1201.

Crespi CL, Gonzalez FJ, Steimel DT, Turner TR, Gelboin HV, Penman BW, Langenbach R. (1991c) A metabolically competent cell line expressing five cDNAs encoding procarcinogen-activating enzymes: application to mutagenicity testing. *Chem. Res. Tox.* **4:** 566–572.

Crespi CL, Langenbach R, Penman BW. (1993a) Human cell lines, derived from AHH-1 TK$^{+/-}$ human lymphoblasts, genetically engineered for expression of cytochromes P450. *Toxicology* **82:** 89–104.

Crespi CL, Penman BW, Gonzalez FJ, Gelboin HV, Galvin M, Langenbach R. (1993b) Genetic toxicology using human cell lines expressing P-450. *Biochem. Soc. Trans.* **21:** 1023–1028.

Crofton-Sleigh C, Doherty A, Ellard S, Parry EM, Venitt S. (1993) Micronucleus assays using cytochalasin-blocked MCL-5 cells, a proprietary human cell line expressing five human cytochromes P-450 and microsomal epoxide hydrolase. *Mutagenesis* **8:** 363–372.

Davies RL, Crespi CL, Rudo K, Turner TR, Langenbach R. (1989) Development of a human cell line by selection and drug-metabolizing gene transfection with increased capacity to activate procarcinogens. *Carcinogenesis* **10:** 885–891.

deWazier I, Cugnenc PH, Yang CS, Leroux J-P, Beaune PH. (1990) Cytochrome P450 isoenzymes, epoxide hydrolase and glutathione transferases in rat and human hepatic and extrahepatic tissues. *J. Pharmacol. Exp. Ther.* **253:** 387–394.

Diamond L, Kruszewski F, Aden DP, Knowles BB, Baird WM. (1980) Metabolic activation of benzo[a]pyrene by a human hepatoma cell line. *Carcinogenesis* **1:** 871–875.

Doehmer J, Dogra S, Friedberg T, Monier S, Adesnik M, Glatt H, Oesch F. (1988) Stable expression of rat cytochrome P-450IIB1 cDNA in Chinese hamster cells (V79) and metabolic activation of aflatoxin B1. *Proc. Natl Acad. Sci. USA* **85:** 5769–5773.

Doehmer J, Seidel A, Oesch F, Glatt HR. (1990) Genetically engineered V79 Chinese hamster cells metabolically activate the cytostatic drugs cyclophosphamide and ifosfamide. *Environ. Hlth Perspect.* **88:** 63–65.

Dogra S, Doehmer J, Glatt H, Molders H, Siegert P, Friedberg T, Seidel A, Oesch F. (1990) Stable expression of rat cytochrome P-4501A1 cDNA in V79 Chinese hamster cells and their use in mutagenicity testing. *Mol. Pharmacol.* **37:** 608–613.

Eichelbaum M, Gross AS. (1990) The genetic polymorphism of debrisoquine/sparteine– clinical aspects. *Pharmacol. Ther.* **46:** 377–394.

Ellard S, Mohammed Y, Dogra S, Wolfel C, Doehmer J, Parry JM. (1991) The use of genetically engineered V79 Chinese hamster cultures expressing rat liver CYP1A1, 1A2 and 2B1 cDNAs in micronucleus assays. *Mutagenesis* **6:** 461–470.

Forrester LM, Neal GE, Judah DJ, Glancey MJ, Wolf CR. (1990) Evidence of involvement of multiple forms of cytochrome P-450 in aflatoxin B_1 metabolism in human liver. *Proc. Natl Acad. Sci. USA* **87:** 8306–8310.

Gallagher EP, Wienkers LC, Stapleton PL, Kunze KL, Eaton DL. (1994) Role of human microsomal and human complementary DNA-expressed cytochromes P4501A2 and P4503A4 in the bioactivation of aflatoxin B_1. *Cancer Res.* **54:** 101–108.

Gonzalez FJ. (1988) The molecular biology of cytochrome P450s. *Pharmacol. Rev.* **40:** 243–288.

Gonzalez FJ, Crespi CL, Gelboin HV. (1991) cDNA-expressed human cytochrome P450s: a new age of molecular toxicology and human risk assessment. *Mutat. Res.* **247:** 113–127.

Guengerich FP. (1988) Roles of cytochrome P-450 enzymes in chemical carcinogenesis and cancer chemotherapy. *Cancer Res.* **48:** 2946–2954.

Guengerich FP, Kim D-H, Iwasaki M. (1991) Role of human cytochrome P-450 IIE1 in the oxidation of many low molecular weight cancer suspects. *Chem. Res. Toxicol.* **4:** 168–179.

Kimura S, Gonzalez FJ, Nebert DW. (1986) Tissue-specific expression of the mouse dioxin-inducible $P_1$450 and $P_3$450 genes: differential transcriptional activation and mRNA stability in liver and extrahepatic tissues. *Mol. Cell. Biol.* **6:** 1471–1477.

Komori M, Kitamura R, Fukuta H, Inoue H, Baba H, Yoshikawa K, Kamataki T. (1993) Transgenic *Drosophila* carrying mammalian cytochrome P-4501A1: an application to toxicology testing. *Carcinogenesis* **14:** 1683–1688.

Krahn DM, Heidelberger C. (1977) Liver homogenate-mediated mutagenesis in Chinese hamster V79 cells by polycyclic aromatic hydrocarbons and aflatoxin. *Mutat. Res.* **46:** 27–44.

Kronbach T, Mathys D, Gut J, Catin T, Meyer UA. (1987) High-performance liquid chromatographic assays for bufuralol 1'-hydroxylase, debrisoquine 4-hydroxylase and dextromethorphan O-demethylase in microsomes and purified cytochrome P450 isozymes of human liver. *Anal. Biochem.* **162:** 24–32.

Leemann T, Transon C, Dayer P. (1993) Cytochrome P450tb (CYP2C): a major monooxygenase catalyzing diclofenac 4'-hydroxylation in human liver. *Life Sci.* **1:** 29–34.

Loquet C, Wiebel FJ. (1982) Geno- and cytotoxicity of nitrosamines, aflatoxin B_1, and benzo[a]pyrene in continuous culture of rat hepatoma cells. *Carcinogenesis* **3:** 1213–1218.

Malling HV. (1971) Dimethylnitrosamine: formation of mutagenic compounds by interaction with mouse liver microsomes. *Mutat. Res.* **13:** 425–429.

257

McGregor DB, Edwards I, Wolf CR, Forrester LM, Caspary WJ. (1991) Endogenous xenobiotic enzyme levels in mammalian cells. *Mutat. Res.* **261:** 29–39.

McManus ME, Burgess WM, Veronese ME, Huggett A, Quattrochi LC, Tukey RH. (1990) Metabolism of 2-acetylaminofluorene and benzo[a]pyrene and activation of food-derived heterocyclic amine mutagens by human cytochromes P-450. *Cancer Res.* **50:** 3367–3376.

Miller EC, Miller JA. (1971) The mutagenicity of chemical carcinogens: correlations, problems and interpretation. In: *Chemical Mutagens: Principles and Methods for their Detection* (ed. A Hollaender). Plenum Press, New York, pp. 83–119.

Murray M. (1992) P450 enzymes, inhibition mechanisms, genetic regulation and effects of liver disease. *Clin. Pharmacokinet.* **23:** 132–146.

Murray BP, Edwards RJ, Murray S, Singleton AM, Davies DS, Boobis AR. (1993) Human hepatic CYP1A1 and CYP1A2 content, determined with specific anti-peptide antibodies, correlates with the mutagenic activation of PhIP. *Carcinogenesis* **14:** 585–592.

Nelson DR, Kamataki T, Waxman DJ, Guengerich FP, Estabrook RW, Feyereisin R, Gonzalez FJ, Coon MJ, Gunsalus IC, Gotoh O, Okuda K, Nebert DW. (1993) The P450 superfamily: update on new sequences, gene mapping, accession numbers, early trivial names of enzymes and nomenclature. *DNA Cell Biol.* **12:** 1–51

Penman BW, Crespi CL. (1987) Analysis of human lymphoblast mutation assays by using historical negative control data bases. *Environ. Mol. Mutagen.* **10:** 35–60.

Penman BW, Crespi CL, Komives EA, Liber HL, Thilly WG. (1983). Mutation in human lymphoblasts exposed to low concentrations of mutagens for long periods of time. *Mutat. Res.* **108:** 417–436.

Penman BW, Reece J, Smith T, Yang CS, Gelboin HV, Gonzalez FJ, Crespi CL. (1993) Characterization of a human cell line expressing high levels of cDNA-derived CYP2D6. *Pharmacogenetics* **3:** 28–39.

Penman BW, Chen L, Gelboin HV, Gonzalez FJ, Crespi CL. (1994) Development of a human lymphoblastoid cell line constitutively expressing human CYP1A1 cDNA: substrate specificity with model substrates and promutagens. *Carcinogenesis* **15:** 1931–1937.

Peter R, Bocker R, Beaune PH, Iwasaki M, Guengerich FP, Yang CS. (1990) Hydroxylation of chlorzoxazone as a specific probe for human liver cytochrome P450IIE1. *Chem. Res. Toxicol.* **3:** 566–573.

Rettie AE, Korzekwa KR, Kunze KL, Lawrence RJ, Eddy AC, Aoyama T, Gelboin HV, Gonzalez FJ, Trager WF. (1992) Hydroxylation of warfarin by human cDNA-expressed cytochrome P-450: a role for P-4502C9 in the etiology of (S)-warfarin-drug interactions. *Chem. Res. Toxicol.* **5:** 54–59.

Sawada M, Kitzmura R, Ohgiya S, Kamataki T. (1993) Stable expression of mouse NADPH-cytochrome P450 reductase and monkey P4501A1 cDNAs in Chinese hamster cells: establishment of cell lines highly sensitive to aflatoxin B_1. *Arch. Biochem. Biophys.* **300:** 164–168.

Schmalix WA, Maser H, Kiefer F, Reen R, Wiebel FJ, Gonzalez F, Seidel A, Glatt H, Greim H, Doehmer J. (1993) Stable expression of human cytochrome P450 1A1 cDNA in V79 Chinese hamster cells and metabolic activation of benzo[a]pyrene. *Eur. J. Pharmacol.* **248:** 251–261.

Shimada T, Guengerich FP. (1989) Evidence for cytochrome P-450nf, the nifedipine oxidase, being the principal enzyme involved in the bioactivation of aflatoxins in human liver. *Proc. Natl Acad. Sci. USA* **86:** 462–465.

Shimada T, Iwasaki M, Martin MV, Guengerich FP. (1989a) Human liver microsomal cytochrome P-450 enzymes involved in the bioactivation of procarcinogens detected by *umu* gene response in *Salmonella typhimurium* TA 1535/pSK1002[1]. *Cancer Res.* **49:** 3218–3228.

Shimada T, Martin MV, Pruess-Schwartz D, Marnett LJ, Guengerich FP. (1989b) Roles of individual human cytochrome P-450 enzymes in the bioactivation of benzo[a]pyrene, 7,8-dihydroxy-7,8-dihydrobenzo[a]pyrene, and other dihydrodiol derivatives of polycyclic aromatic hydrocarbons. *Cancer Res.* **49:** 6304–6312.

Shimada T, Yun C-H, Yamazaki H, Gautier JH-C, Beaune PH, Guengerich FP. (1992) Characterization of human lung microsomal cytochrome P-450 1A1 and its role in the oxidation of chemical carcinogens. *Mol. Pharmacol.* **41**: 856–864.

Smith DA. (1991) Species differences in metabolism and pharmacokinetics: are we close to an understanding? *Drug Metab. Rev.* **23**: 355–373.

Smith TJ, Guo Z, Gonzalez FJ, Guengerich FP, Stones GD, Yang CS. (1992) Metabolism of 4-(methylnitrosamino)-1-(3-pyridyl)-1-butanone in human lung and liver microsomes and cytochromes P-450 expressed in hepatoma cells. *Cancer Res.* **52**: 1757–1763.

States JC, Quan T, Hines RN, Novak RF, Runge-Morris M. (1993) Expression of human cytochrome P450 1A1 in DNA repair deficient and proficient human fibroblasts stably transformed with an inducible expression vector. *Carcinogenesis* **14**: 1643–1649.

Styles JA, Davies A, Lim CK, DeMatteis F, Stanley LA, White INH, Yuan Z-X, Smith LL. (1994) Genotoxicity of tamoxifen, tamoxifen epoxide and toremifene in human lymphoblastoid cells containing human cytochrome P450s. *Carcinogenesis* **15**: 5–9.

Sugden B, Marsh K, Yates J. (1985) A vector that replicates as a plasmid and can be efficiently selected in B-lymphoblasts transformed by Epstein–Barr virus. *Mol. Cell. Biol.* **5**: 410–413.

Tennant RW, Margolin BH, Shelby MD, Zeiger E, Haseman JK, Spalding J, Caspary W, Resnick M, Stasiewicz S, Anderson B. (1987) Prediction of chemical carcinogenicity in rodents from *in vitro* genetic toxicology assays. *Science* **236**: 933–941.

Tiano HF, Hosokawa M, Chulada PC, Smith PB, Wang R-L, Gonzalez FJ, Crespi CL, Langenbach R. (1993) Retroviral mediated expression of human cytochrome P450 2A6 in C3H/10T1/2 cells confers transformability by 4-(methynitrosamino)-1-(3-pyridyl)-1-butanone (NNK). *Carcinogenesis* **14**: 1421–1427.

Tong C, Telang S, Williams GM. (1983) Differences in responses of 4 adult rat-liver epithelial cell lines to a spectrum of chemical mutagens. *Mutat. Res.* **130**: 53–61.

Trinidad AC, Wu RW, Thompson LH, Felton JS. (1991) Expression of mouse cytochrome P4501A1 cDNA in repair-deficient and repair proficient CHO cells. *Mol. Carcinogen.* **4**: 510–518.

Waterman MR, Johnson EF. (1991) *Cytochromes P450. Methods in Enzymology*, Vol. 206. Academic Press, New York.

Waxman DJ, Ko A, Walsh C. (1983) Regioselectivity and stereoselectivity of androgen hydroxylations catalyzed by cytochrome P-450 isozymes purified from phenobarbital-induced rat liver. *J. Biol. Chem.* **258**: 11937–11947.

White INH, DeMatteis F, Davies A, Smith LL, Crofton-Sleigh C, Venitt S, Hewer A, Phillips DH. (1992) Genotoxic potential of tamoxifen and analogues in female Fischer F344/N rats, DBA/2 and C57B$_l$/6 mice and in human MCL-5 cells. *Carcinogenesis* **13**: 2197–2203.

Wolfel C, Platt K-L, Dogra S, Glatt H, Wachter F, Doehmer J. (1991) Stable expression of rat cytochrome P4501A2 cDNA and hydroxylation of 17β-estradiol and 2-aminofluorene in V79 chinese hamster cells. *Mol. Carcinogen.* **4**: 489–498.

Wolfel C, Heinrich-Hirsch B, Schulz-Schlage T, Seidel A, Frank H, Ramp U, Wachter F, Wiebel FJ, Gonzalez F, Greim H, Doehmer J. (1992) Genetically engineered V79 Chinese hamster cells for stable expression of human cytochrome P4501A2. *Eur. J. Pharmacol.* **228**: 95–102.

Wrighton SA, Stevens JC. (1992) The human hepatic cytochromes P450 involved in drug metabolism. *Crit. Rev. Toxicol.* **22**: 1–21.

Wrighton SA, Vandenbranden M, Stevens JC, Shipley LS, Ring FJ. (1993a) *In vitro* methods for assessing human hepatic drug metabolism: their use in drug development. *Drug Metab. Rev.* **25**: 453–484.

Wrighton SA, Stevens JC, Becker GW, Vandenbranden M. (1993b) Isolation and characterization of human liver cytochrome P450 2C19: correlation between 2C19 and *S*-mephenytoin 4′-hydroxylation. *Arch. Biochem. Biophys.* **306**: 240–245.

Yamano S, Nhamburo PT, Aoyama T, Meyer UA, Inaba T, Kalow W, Gelboin HV, McBride OW, Gonzalez FJ. (1989) cDNA cloning and sequence and cDNA directed expression of human P450 IIB1: identification of a normal and two variant cDNAs derived from the *CYP2B* locus on chromosome 19 and differential expression of the IIB mRNAs in human liver. *Biochemistry* **28:** 7340–7348.

Yamano S, Tatsuno J, Gonzalez FJ. (1990) The *CYP2A3* gene product catalyzes coumarin 7-hydroxylation in human liver microsomes. *Biochemistry* **29:** 1322–1329.

Yamazaki H, Inui Y, Yun C-H, Guengerich FP, Shimada T. (1992) Cytochrome P450 2E1 and 2A6 enzymes as major catalysts for metabolic activation of *N*-nitrosodialkylamines and tobacco-related nitrosamines in human liver microsomes. *Carcinogenesis* **13:** 1789–1794.

Yanagawa Y, Sawada M, Deguchi T, Gonzalez FJ, Kamataki T. (1994) Stable expression of human CYP1A2 and *N*-acetyltransferases in Chinese hamster CHL cells: mutagenic activation of 2-amino-3-methylimidazo[4,5-*f*]quinoline and 2-amino-3,8-dimethylimidazo[4,5-*f*]quinoxaline. *Cancer Res.* **54:** 3422–3427.

Yoo J-S, Cheung RJ, Patten CJ, Wade D, Yang CS. (1987) Nature of *N*-nitrosodimethylamine demethylase and its inhibitors. *Cancer Res.* **47:** 3378–3383.

Yun C-H, Shimada T, Guengerich FP. (1992) Roles of human liver cytochrome P4502C and 3A enzymes in the 3-hydroxylation of benzo[*a*]pyrene. *Cancer Res.* **52:** 1868–1874.

Fluorescence *in situ* hybridization: application to environmental mutagenesis

David A. Eastmond and D.S. Rupa

14.1 Introduction

Over the past several decades, a variety of cytochemical and immunological techniques have been developed to detect the presence and location of specific molecules and organelles within cells. *In situ* hybridization, developed independently by several groups of investigators (Buongiorno-Nardelli and Amaldi, 1970; John *et al.*, 1969; Pardue and Gall, 1969), is one such technique which allows the presence of specific nucleic acid sequences to be detected and localized in morphologically preserved cells and tissues. *In situ* hybridization is based upon the specific annealing of a labelled or chemically modified nucleic acid probe to complementary DNA or RNA sequences in fixed tissues or cells, followed by visualization of the labelled probe. The basis of *in situ* hybridization using a DNA probe is illustrated in *Figure 14.1*. Initially radioisotopes were the only labels available for DNA, and autoradiography was required to detect the hybridized probe. In recent years, the synthesis of chemically modified nucleotides (or nucleic acid sequences) and the incorporation of these modified nucleotides into sequences that efficiently hybridize to target RNA or DNA has allowed the hybridized sequences to be visualized using fluorescence, enzymatic precipitates or chromogenic detection methods (Lewis and Baldino, 1990; Tucker *et al.*, 1993). Fluorescence detection is currently the most widely used due to its ease of use, speed in obtaining results, enhanced spatial resolution, ability to distinguish multiple fluorescent labels and reduced safety concerns (Hofler, 1990). Through the use of probes to repetitive DNA sequences or probes generated from

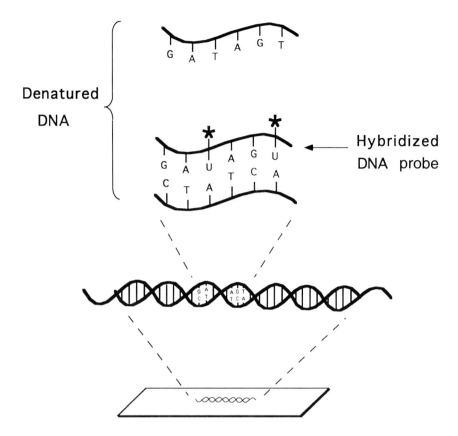

Figure 14.1. An illustration of *in situ* hybridization with a DNA probe. The chemically modified or radiolabelled nucleotides which have been incorporated into the probe are indicated by an asterisk (*).

chromosome libraries, fluorescence *in situ* hybridization (FISH) allows specific chromosomes or chromosomal regions to be recognized easily in both interphase and metaphase cells (Cremer *et al.*, 1986; Pinkel *et al.*, 1986). As a result, this approach can allow the detection and quantitation of structural and numerical aberrations occurring in normal and aberrant cells.

A considerable amount of molecular and cytogenetic evidence indicates that chromosomal alterations are involved in a range of disorders and pathophysiological conditions including congenital anomalies, pregnancy loss, infertility and cancer (Hecht and Hecht, 1987; Hook, 1985; Oshimura and Barrett, 1986). Historically chromosome banding techniques have been relied upon to identify individual chromosomes in metaphase preparations and detect structural and numerical aberrations in normal and affected cells. However, the usefulness of conventional cytogenetic analyses is limited to actively dividing tissues or cell types which divide readily in culture. Studies of cells which grow poorly in culture, such as certain epithelial cells or solid tumours, have been particularly challenging in that these cells typically exhibit a low mitotic index, yield relatively few viable cells, are prone to infiltration by other cell types and display poor chromosome morphology (Sandberg *et al.*, 1988; Teyssier, 1989). In addition, conventional cytogenetic studies are labour intensive, require highly skilled personnel and are prone to other technical problems such as chromosomal loss or poor chromosome spreading during metaphase preparation.

The use of FISH circumvents a number of the problems inherent in conventional cytogenetics by allowing specific cytogenetic information to be obtained rapidly from interphase as well as metaphase cells. Hybridization with fluorescently labelled DNA probes results in a staining of the chromosomal region targeted by the DNA probe, permitting the copy number and location of the chromosome of interest to be rapidly determined in a large number of cells. The application of FISH with RNA and DNA probes has become a valuable complement to conventional cytogenetic studies and has begun to have a significant impact on genetic and biomedical research by facilitating the diagnosis of cancer and congenital anomalies, assisting in the detection of bacterial and viral infections, improving therapeutic strategies and contributing to an understanding of the mechanisms underlying normal cell biology and specific pathological conditions (Boehringer Mannheim, 1992; Horn *et al.*, 1986; Le Beau, 1993; Ward *et al.*, 1993). FISH is also widely used to identify the chromosomal location of specific genes or other DNA sequences and physically to map the relative positions of various loci (Brandriff *et al.*, 1991; Viegas-Pequignot, 1992). Many of these applications and procedures have been reviewed recently (Boehringer Mannheim, 1992; Chesselet, 1990; Gray and Pinkel, 1993; Keller and Manak, 1992; Polak and McGee, 1990; Trask and Pinkel, 1990; Wilkinson, 1992). In addition to the mentioned applications, FISH with chromosome-specific or region-specific DNA probes is being used increasingly to identify chemical and physical agents capable of inducing chromosomal alterations in animal and human cells and to identify human populations exhibiting increased frequencies of chromosome abnormalities. This chapter provides an introduction to the basic procedures of FISH, illustrates various strategies using FISH that have been applied in environmental mutagenesis, reviews observed results and discusses briefly recent developments in FISH technology and their potential application in the study of environmental mutagenesis.

14.2 Hybridization procedures

FISH with DNA and RNA probes consists of four main steps: (i) preparation of the probe; (ii) tissue and cell preparation; (iii) hybridization and associated washes; and (iv) visualization of the hybridized probe. Each step is described briefly next. These descriptions are intended as an introduction to the procedures used in performing FISH. For additional details on methods and procedures, the reader is recommended to refer to the following: Boehringer Mannheim (1992), Keller and Manak (1993), Oncor (1992), Trask and Pinkel (1990), Viegas-Pequignot (1992).

14.2.1 Probe preparation

A variety of different types of DNA probes has been used for FISH. Double-stranded DNA probes have been most commonly used but other types of probes such as single-stranded DNA, RNA and synthetic oligonucleotide probes have been used in specific applications (Lewis and Baldino, 1990; Tagarro *et al.*, 1994). Because virtually all FISH studies in environmental mutagenesis to date have used DNA probes, the following discussion is largely restricted to these types of probes. Genomic DNA probes consist of total genomic DNA which has been extracted from cells and labelled, allowing chromosomes or chromosomal regions from one species to be distinguished from those of another species (Goodwin *et al.*, 1989; Pinkel *et al.*, 1986). Repetitive sequence probes recognizing *alpha* and classical satellite sequences in humans have been used to label specifically the centromeric and pericentric regions of one or all human chromosomes (Meyne *et al.*, 1989; Mitchell *et al.*, 1985; Moyzis *et al.*, 1987; Willard and Waye, 1987). Similar probes have been used to label telomeric sequences (Moyzis *et al.*, 1988). Hybridization with fluorescently labelled repetitive DNA probes results in a compact and bright staining of the chromosomal region targeted by the DNA probe in both metaphase and interphase cells (*Figure 14.2a,b*). Probes to smaller regions, such as cosmid, plasmid or yeast artificial chromosome (YAC) probes, have been used successfully to localize genes or gene families (Brandriff *et al.*, 1991; Le Beau, 1993) but have had limited application in studies of mutagenesis.

Alternatively, composite probes generated from human chromosome libraries (whole chromosome or 'painting' probes) have been used to label specifically the length of one or more chromosomes (Cremer *et al.*, 1988; Pinkel *et al.*, 1988). Hybridization with these probes results in a bright speckled appearance covering most of the chromosome as a result of hybridization by large numbers of small unique sequence or short repetitive sequence probes (*Figure 14.3a*). Examples of various DNA probes which have been used in genotoxicity studies or currently available for FISH analyses are shown in *Table 14.1*. Region-specific or chromosome-specific probes are currently available for most human chromosomes through commercial vendors (for example, Boehringer Mannheim, Mannheim, Germany; Oncor, Gaithersburg, MD, USA; Vysis, Framingham, MA, USA). Cloned sequences for some chromosomal regions which are suitable for FISH can also be obtained from DNA repositories or directly from the investigators responsible for their isolation and cloning (for example, American Type Culture Collection, 1993). In addition, using polymerase chain reaction (PCR)-based techniques, probes can be generated directly from genomic DNA, cell hybrids or *in situ* (Dunham *et al.*, 1992; Goodwin *et al.*, 1989; Hindkjaer *et al.*, 1994). As indicated in *Table 14.1*, similar probes have been or are currently being developed for other commonly used laboratory animal species such as the mouse and rat.

Table 14.1. Examples of DNA probes for FISH

DNA probes	Reference
Human	
Probes for highly repetitive sequences	
Pancentromeric *alpha* satellite probes	Mitchell *et al.* (1985)
Chromosome-specific *alpha* satellite probes	Willard and Waye (1987)
Chromosome-specific classical	Cooke and Hindley (1979),
satellite probes	Moyzis *et al.* (1987)
Telomeric probes	Moyzis *et al.* (1988), Lucas *et al.* (1989)
Chromosome-specific library probes	Pinkel *et al.* (1988), Cremer *et al.* (1988)
Genomic DNA	Pinkel *et al.* (1986), Goodwin *et al.* (1989)
Mouse	
Probes for highly repetitive sequences	
Pancentromeric[a] major satellite probe	Weier *et al.* (1991a), Miller *et al.* (1991)
Pancentromeric[a] minor satellite probe	Pietras *et al.* (1983)
Chromosome-specific satellite probes	Vourc'h *et al.* (1993), Lowe *et al.* (1994)
Telomeric probes	Moyzis *et al.* (1988),
	Miller and Nüsse (1993)
Chromosome-specific library probes	Breneman *et al.* (1993), Boei *et al.* (1994)
Rat	
Probes for highly repetitive sequences	
Centromeric satellite probes	de Stoppelaar *et al.* (1994)
Chromosome-specific satellite probes	de Stoppelaar *et al.* (1994)
Chromosome-specific library probes	Breneman *et al.* (1994),
	Hoebee *et al.* (1994),
	de Stoppelaar *et al.* (1994)

[a] The mouse major and minor probes recognize 39 of the 40 mouse chromosomes. The Y chromosome is not labelled by these probes.

Labelling of the probes is generally accomplished through enzymatic incorporation of a labelled nucleotide or through a chemical reaction (Chan and McGee, 1990; Lewis and Baldino, 1990). Using nick translation, random priming or PCR-based procedures, the modified nucleotide is incorporated into newly synthesized DNA by replacing the native nucleotide in the reaction mixture with the modified derivative. Modified nucleotides can also be incorporated into short DNA sequences during oligonucleotide synthesis. Modified oligomers complementary to highly repetitive sequences such as satellite DNA can be used directly as probes, or modified oligomers without a complementary sequence can be attached to an unlabelled complementary DNA sequence by end-labelling with a terminal transferase to create a labelled probe (Chan and McGee, 1990; Tagarro *et al.*, 1994). DNA sequences can also be labelled by chemical modification after polymerization using a hapten or recognizable moiety attached to a DNA-reactive group. Photobiotin, biotin hydrazide, biotin ester and *N*-acetoxy-2-acetylaminofluorene have been successfully used to attach a biotin or an acetylaminofluorene moiety to DNA sequences for use as probes (Chan and McGee, 1990).

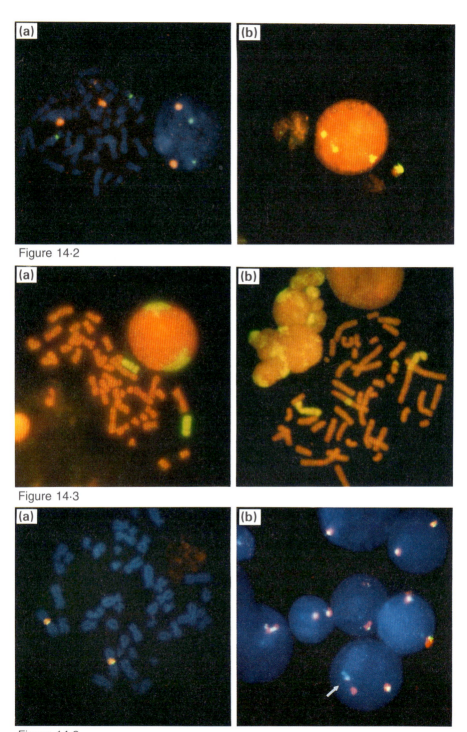

Figure 14·2

Figure 14·3

Figure 14·6

Figure 14.2. FISH of human cultured lymphocytes using three chromosome-specific repetitive DNA sequence probes. (a) A normal metaphase and an interphase cell are shown following hybridization with a classical satellite probe labelled with Texas Red® for chromosome 1 and an *alpha* satellite probe labelled with fluorescein for chromosome 7. Two hybridization regions corresponding to the centromeric/pericentric region of each chromosome are visible in each cell using a triple-band-pass filter. The DNA-specific stain DAPI has been used to counterstain the chromosomes and nucleus. (b) An abnormal interphase nucleus and a micronucleus following hybridization with a fluorescein-labelled classical satellite probe for chromosome 9. Three hybridization domains are visible in the interphase nucleus and one in the micronucleus using a blue filter. The DNA has been counterstained with propidium iodide.

Figure 14.3. FISH of human cultured lymphocytes using a whole chromosome painting probe for chromosome 4. (a) A normal metaphase and an interphase cell are shown exhibiting two copies of the fluorescein-labelled chromosome. The DNA has been counterstained with propidium iodide. (b) A radiation-treated metaphase exhibiting several chromosome exchanges.

Figure 14.6. Multicolour fluorescence *in situ* hybridization with two adjacent centromeric/pericentromeric DNA probes for chromosome 1. (a) A normal metaphase lymphocyte following hybridization with a fluorescein-labelled *alpha* satellite probe and an adjacent Texas Red®-labelled classical satellite probe. A shift in the emission spectrum from green to yellow occurs when a green fluorescein signal is adjacent to a Texas Red® signal. (b) Several irradiated interphase nuclei following hybridization with the tandem labelled probes. The wide separation between the Texas Red® signal and the fluorescein signal (marked by the arrow) indicates that a break has occurred within the chromosomal region between these two probes.

Currently two types of FISH procedures are used: a direct method and an indirect method. The direct method uses a fluorescent dye covalently attached to the nucleic acid probe so that the hybridized probe can be visualized microscopically immediately following hybridization. For example, fluorescein-labelled uridine can replace thymidine in the probe. Following hybridization, the hybridized complex can be detected by microscopy using the appropriate excitation and emission wavelengths to detect the green fluorescein dye. Recently, a series of direct fluorochrome-labelled nucleotides became available. In the indirect method, the chemically modified probe must be detected by a fluorescently labelled antibody or other affinity reagent (for example, avidin). The most common indirect nucleotide modifications are with biotin or digoxigenin although the detection of other attached moieties such as acetylaminofluorene, mercury, sulphone and dinitrophenol has also been performed successfully. The direct method has the advantage that hybridizations require less time to perform as fewer detection steps are necessary. However, the signals often tend not to be as strong as amplified signals obtained using the indirect method. Through the use of antibody combinations, amplification of probes labelled by the indirect methods can be performed easily. The resulting amplified signals are typically stronger and more intense than those obtained with the direct-labelled probes.

14.2.2 Tissue and cell preparation

In situ hybridization techniques have been used on a variety of biological specimens including whole cells, tissue sections, isolated chromosomes and nuclei (Boehringer Mannheim, 1992; Trask and Pinkel, 1990). Tissues must be fixed before hybridization to maintain cellular and nuclear morphology. The most common fixatives are methanol:acetic acid or paraformaldehyde. Formalin- or glutaraldehyde-fixed tissues can also be used. In most cases, cells are attached to glass slides, although hybridization can also be performed on nuclei or cells in suspension (Trask *et al.*, 1985; Trask and Pinkel, 1990). Before hybridization, cells are often pretreated with detergents, proteases, RNase, hydrochloric acid or alcohol solutions in order to facilitate probe penetration and to increase the specificity of binding.

For *in situ* hybridization to cellular DNA, the target DNA (and probe DNA, if double stranded) must be denatured. This is commonly performed by thermal denaturation of the DNA in the presence of formamide (70%) and salts (2 x SSC) for several minutes, which stabilizes the single-stranded DNA and generates a clearly defined melting temperature while preserving chromosome morphology.

14.2.3 Hybridization procedure

Hybridization is performed by incubating the single-stranded modified probe DNA with the denatured cellular DNA in the presence of formamide, salts, dextran sulphate, carrier DNA and water for 2–72 h (typically overnight). As most probes will bind to a limited extent to sequences exhibiting less than perfect homology, carrier DNA, such as herring testes DNA for chromosome-specific repetitive sequences, is used routinely to block these non-specific interactions. However, for probes targeting an entire chromosome or other large genomic region that would contain repetitive sequences located at many locations throughout the genome, unlabelled sonicated genomic DNA is preincubated with the probe. This allows the repetitive sequences in the probe to anneal with the complementary carrier DNA sequences in solution and thereby increases the specificity of the probe for the targeted chromosome or region of the cellular DNA.

In addition to the use of carrier DNA, washing of the hybridized cells is performed to decrease the non-specific binding of the probe. Through the use of elevated temperatures and varying concentrations of formamide and salts in the wash solution, conditions can be optimized to remove probe bound to partially homologous non-target sequences while retaining hybridization of the probe to the desired target sequences.

14.2.4 Visualization of the hybridized probe

For probes directly labelled with a fluorochrome, the hybridized probe can be visualized directly at this point using the appropriate filters on a microscope equipped for fluorescence detection. For probes containing nucleotides labelled with a hapten or other moiety such as biotin, the hybridized probe must be incubated with a fluorochrome-conjugated antibody or other affinity reagent (for example, fluorescein-conjugated avidin). Fluorescein-, rhodamine- and coumarin-based fluorescent labels have been widely used because they are relatively stable and have wide spectral separation. However, in recent years a series of new fluorescent dyes such as BODIPY®, Texas Red® and Cascade Blue® (Molecular Probes, Eugene, OR, USA) have been increasingly used. Counter-staining of the genomic DNA to allow chromosome location and morphology to be observed is generally performed using the red fluorescent dye propidium iodide or 4,6-diamidino-2-phenylindole (DAPI), a blue fluorescent dye. For most applications, an antifading reagent, such as *p*-phenylenediamine, diazobicyclooctane or *n*-propylgallate is incorporated in the mounting medium. A series of new fluorescent filters have been developed recently which enable fluorochromes that excite and emit at different wavelengths to be visualized simultaneously (Omega, Brattleboro, VT, USA; Chroma, Brattleboro, VT, USA). Using these filters, multicolour FISH can be performed which allows cytogenetic alterations affecting more than one region to be detected at the same time, on the same slide.

14.3 Hybridization strategies in studies using FISH for environmental mutagenesis

A variety of probes has been used in FISH studies for environmental mutagenesis (*Table 14.1*). Although the detection and localization of sequences smaller than 1 kb have been reported using FISH (Viegas-Pequignot, 1992), almost all studies on the genotoxic effects of various chemical and physical agents have used probes targeting much larger chromosomal regions. As the size of the targeted sequence decreases, the ease of visibility and the efficiency of hybridization tend to decrease, increasing the difficulty of scoring. Most probes used to date have targeted regions larger than 500 kb. In addition to the use of various types of probes, a number of different strategies have been used in the application of FISH to genotoxicity testing and human biomonitoring. The most common strategies and the results obtained are described in the following sections.

14.3.1 Use of probes to highly repeated sequences

Chromosome-specific DNA probes: in vitro *studies.* One of the first applications of FISH to genotoxicity studies was to use probes to the highly repeated *alpha* satellite or classical satellite sequences located on specific chromosomes to detect alterations in chromosome number in metaphase and interphase cells. Hybridization to these highly repetitive sequences results in a large but compact signal at the targeted region. The number of copies of the chromosome of interest is determined by counting the number of hybridization domains in the nucleus. Several examples of human lymphocyte nuclei following FISH with several chromosome probes are shown in *Figure 14.2a,b*. *Figure 14.2a* shows a normal metaphase and interphase nucleus following hybridization with a fluorescein-labelled *alpha* satellite probe for chromosome 7 and a Texas Red-labelled classical satellite probe for chromosome 1. Two hybridization regions are seen in each nucleus for each probe. An abnormal nucleus containing three hybridization regions (as well as a probe-labelled micronucleus) that was analysed by FISH with a fluorescein-labelled classical satellite probe for human chromosome 9 is shown in *Figure 14.2b*.

Using this approach, information on aneuploidy can be rapidly obtained for a large number of cells. Initial studies by Eastmond and Pinkel (1990) determined the feasibility of using FISH with chromosome-specific DNA probes to detect aneuploidy and aneuploidy-inducing agents in interphase human lymphocytes. Probes specific for chromosomes 1, 7, 9, 17, X and Y were used to determine the number of hybridization regions in 72-h phytohaemagglutinin (PHA)-stimulated lymphocytes. The combined frequencies of nuclei containing zero, one, two, three or four hybridization regions for these autosomes were 0.004, 0.084, 0.909, 0.003 and 0.001, respectively. These results indicated that this approach should be fairly sensitive for the detection of hyperdiploidy but should be relatively insensitive for the detection of hypodiploidy owing to a relatively high frequency of nuclei exhibiting only one hybridization region. Additional experiments reported by Eastmond and Pinkel (1990) indicated that the relatively high frequency of nuclei exhibiting one hybridization region was largely related to an apparent fusion of the fluorescent signal of the probes when the targeted regions lie either on top of one another or immediately adjacent. Subsequent observations have also indicated that the frequency of nuclei with one hybridization region can be strongly influenced by poor probe penetration as well as by associations of chromosomal regions within the interphase nucleus (Eastmond *et al.*, 1994; Ferguson and Ward, 1992; Lewis *et al.*, 1993). Zijno *et al.* (1994) have shown that, by using cytochalasin B to block cytokinesis during cell culture, and by scoring hyperdiploidy and hypodiploidy in the resulting binucleated cells, FISH can be effectively used to detect hypodiploidy in cultured interphase human lymphocytes.

Eastmond and Pinkel (1990) demonstrated, using a classical satellite probe specific for chromosome 9, that a significant increase in the frequency of hyperdiploid cells could be observed when PHA-stimulated human lymphocytes were exposed during cell culture to the aneuploidy-inducing chemicals colchicine, vincristine sulphate and diethylstilboestrol. Similar studies published the same year by De Sario *et al.* (1990) also showed that FISH with a probe specific for the Y chromosome could be used to detect increases in hyperploid nuclei induced by diethylstilboestrol. Subsequent studies have used various chromosome-specific DNA probes to detect hyperdiploidy induced in cultured lymphocytes or HL-60 cells by the benzene metabolites hydroquinone and 1,2,4-trihydroxybenzene, respectively (Eastmond *et al.*, 1994; Zhang *et al.*, 1994).

Dutrillaux and colleagues have used FISH with probes specific to the classical and *alpha* satellite regions on human lymphocytes treated with the hypomethylating agent 5-azacytidine to determine the extent of decondensation in the heterochromatic regions of chromosomes 1 and 16 as well as to identify somatic associations and breakage affecting the decondensed regions (Kokalj-Vokac *et al.*, 1993). Similar alterations affecting the heterochromatic regions of these chromosomes were seen in hypomethylated lymphoblastoid cells obtained from a Fanconi anaemia patient and a patient with ataxia telangiectasia (Almeida *et al.*, 1993).

Two groups of investigators (van Dekken and Bauman, 1988; Lucas *et al.*, 1989) used FISH with a combination of a centromeric and a telomeric probe for chromosome 1, a classical satellite probe targeting the heterochromatin (1q12) and a second probe to the 1p36 region, to detect translocations in human metaphase cells. *Figure 14.4* illustrates the application of this strategy for metaphase analysis. As can be seen in the illustration, this approach has limited applicability for interphase analyses. Lucas and colleagues (Lucas *et al.*, 1989) evaluated approximately 32 000 metaphase cells for dicentrics and translocations following the *in vitro* exposure of human lymphocytes to ionizing radiation from both a caesium and a cobalt source. Significant increases of both dicentrics and translocations were observed, indicating the application of this type of FISH analysis to quantify these classes of structural aberrations in metaphase cells.

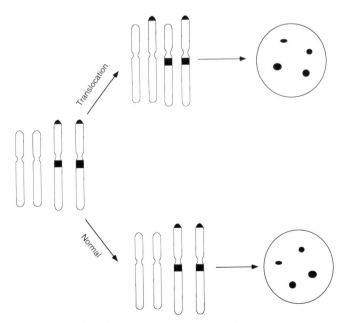

Figure 14.4. An illustration showing the use of repetitive sequence probes for a centromeric (1q12) and a telomeric (1p36) region to detect translocations in metaphase cells.

In the aneuploidy studies of Eastmond and Pinkel (1990), a highly elevated frequency of nuclei containing multiple hybridization regions was observed using classical satellite probes for chromosomes 1 and 9 following treatment of human lymphocytes with ionizing radiation, a potent clastogenic agent. Since previous studies using conventional cytogenetics had indicated that the heterochromatin regions targeted by these probes were prone to breakage (Brogger, 1977; Meyne *et al.*, 1979), these results suggested that breakage occurring within labelled heterochromatin might have been falsely interpreted as a hyperdiploid nucleus. Additional studies from our laboratory (Eastmond *et al.*, 1994) using these same probes plus an *alpha* satellite probe for chromosome 7 demonstrated that the *in vitro* exposure of lymphocytes to the benzene metabolite hydroquinone resulted in an increased frequency of nuclei with multiple hybridization regions, presumably hyperdiploid nuclei. However, a comparison of the frequencies for the three probes revealed that the frequencies of 'hyperdiploid' nuclei were significantly higher for the classical satellite probes than for the *alpha* satellite probe.

To determine whether these differences were the result of a non-random involvement of chromosomes 1 and 9 in hydroquinone-induced hyperdiploidy or breakage occurring within the heterochromatin regions targeted by the classical satellite probes, a new multicoloured FISH assay was developed (Eastmond *et al.*, 1994). An illustration of this approach is shown in *Figure 14.5* and examples of a metaphase and several interphase cells are shown in *Figure 14.6a,b*. Multicolour FISH was used to label two adjacent regions on chromosome 1. The pericentric heterochromatic region which is large and prone to breakage by chemical and physical agents, was labelled with a Texas Red classical satellite probe. An adjacent centromeric region, which is somewhat smaller and much less prone to breakage by genotoxic agents, was labelled by a green fluorescein-labelled *alpha* satellite probe. The presence of an interphase nucleus containing three *alpha* satellite probes adjacent to three classical satellite probes indicates a nucleus which has three copies of chromosome 1. However, if a similar interphase nucleus contains only two *alpha* satellite probes next to two of the three classical satellite probes, this indicates that a breakage event has occurred within the chromosomal region targeted by one of the classical satellite probes. Alternatively, a wide separation between the regions labelled by the *alpha* and classical satellite probes can also indicate that chromosomal breakage has occurred. An example of an interphase cell exhibiting a wide separation between the two probes is shown in *Figure 14.6b*.

Using this new strategy, which we have called tandem labelling, we demonstrated that *in vitro* exposure to hydroquinone had indeed induced hyperdiploidy in the lymphocytes of this donor (Eastmond *et al.*, 1994). However, the observed difference in frequencies between the chromosome 7 *alpha* satellite probe and the chromosome 1 classical satellite probe was most likely the result of breakage within the 1q12 region targeted by the classical satellite probe. These results indicated that this tandem label assay could be used as an assay to detect chromosome breakage in interphase cells. Additional studies

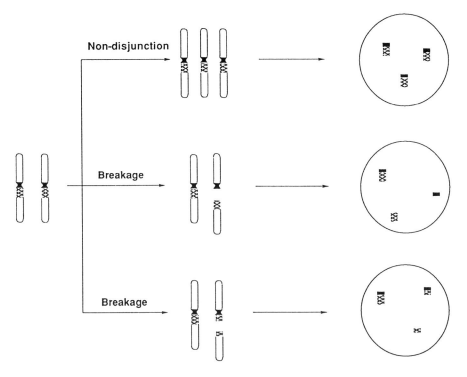

Figure 14.5. Schematic illustration of the use of two adjacent DNA probes to detect and distinguish chromosome breakage and hyperdiploidy in metaphase and interphase cells.

from our laboratory (Rupa *et al.*, 1995) have used the tandem labelled probes for chromosome 1 to compare the frequency of radiation-induced breakage within or next to the 1q12 region in cultured human lymphocytes and compared them with untreated control lymphocytes. A highly significant increase in breakage affecting these regions was seen in the radiation-treated cells. The frequencies of breakage observed in both the metaphase and interphase cells were very similar for both the untreated and the treated cells. Interestingly, metaphase analyses indicated that approximately 50% of the breakage events observed under the treatment conditions used were the result of translocations, inversions and complex exchanges, potentially stable aberrations which might be transferred to the cell's progeny and persist for relatively long periods of time. Many of these alterations involved relatively small regions of centromeric DNA which may have been missed in routine cytogenetic analysis. Additional studies (Rupa, Hasegawa and Eastmond, unpublished) have indicated that this approach can be used with other chromosomes and to detect structural aberrations occurring in granulocytes and buccal mucosa cells, cell types which previously have not been amenable to cytogenetic analyses. These results indicate the utility and potential of this multicolour FISH assay to detect hyperdiploidy and structural aberrations in interphase cells following exposure to genotoxic agents.

Chromosome-specific DNA probes: population studies. Within the past several years, a series of studies has been initiated to determine the feasibility of using chromosome-specific repetitive DNA probes for monitoring chromosomal alterations occurring in human populations exposed to carcinogenic or reproductive toxicants. Most of the studies to date have attempted to establish baseline frequencies of hyperploidy in individuals without known exposure to genotoxic agents. These studies have focused on the use of FISH for analysis of lymphocytes, exfoliated oral mucosa and bladder epithelial cells as well as spermatocytes. Owing to the difficulties of obtaining cytogenetic information in human germ cells, FISH with chromosome-specific DNA probes appears to be an especially promising technique for detecting chromosome alterations in human sperm.

Initial studies by Pieters *et al.* (1990) and Wyrobek *et al.* (1990) focused on determining the frequency of the Y chromosome in sperm from human donors. As expected, following FISH with a Y chromosome probe, approximately 50% of the sperm exhibited one hybridization signal. The study by Pieters *et al.* (1990) as well as one published the following year by Coonen *et al.* (1991) used a classical satellite probe for chromosome 1 to determine the frequency of sperm disomic for this chromosome. In these studies, the frequency of sperm containing two hybridization domains was significantly higher than would be expected based upon the results of the human sperm/hamster egg fusion technique or from live births (Robbins *et al.*, 1993). Since the publication of these results, an increasing number of papers have been published reporting the disomy frequencies for most of the human chromosomes (Bischoff *et al.*, 1994; Han *et al.*, 1993; Holmes and Martin, 1993; Lu *et al.*, 1994; Robbins *et al.*, 1993; Williams *et al.*, 1993). In these studies, a wide range of baseline frequencies have been reported. For example, the frequency of sperm disomic for chromosome 1 was reported to be 0.8% by Pieters *et al.* (1990) and 0.06% by Holmes and Martin (1993), a 13-fold difference. The frequency of YY sperm was reported to be 0.21% by Han *et al.* (1993) and 0.056% by Robbins *et al.* (1993), a four-fold difference. Some of this variability is undoubtedly due to different hybridization procedures and scoring criteria used in different laboratories as well as frequencies based upon a relatively small number of evaluated samples. In particular, procedures used to decondense the sperm chromatin before hybridization, the presence of somatic cells in the ejaculate and the use of different probes, in addition to individual-to-individual variability, are all likely to contribute to the highly variable frequencies observed in different laboratories. The standardization of hybridization and scoring procedures should help establish more accurate frequencies of aneuploid sperm. In addition, the application of multicolour FISH to sperm analyses should improve the accuracy of the analyses by allowing several different chromosomes to be scored simultaneously in the same sperm nucleus (Bischoff *et al.*, 1994; Williams *et al.*, 1993).

Two studies have recently been conducted using FISH analyses to detect abnormal numbers of chromosomes in sperm from infertile males and cancer patients exposed to chemotherapeutic agents. A report by Miharu and coworkers compared the frequencies of disomy for chromosomes 1, 16, 17, 18, X and Y in sperm from nine fertile and 12 infertile males. Similar frequencies of disomy were seen in the two groups leading the investigators to conclude that aneuploidy in sperm was unlikely to be a major contributor for unexplained infertility (Miharu *et al.*, 1994). In addition, initial studies by Robbins *et al.* (1994) using multicolour FISH with probes for chromosomes 8, X and Y on the sperm of seven patients with Hodgkin's disease have indicated that the frequency of hyperhaploid sperm increased following the administration of a chemotherapeutic regimen containing the known aneuploidy-inducing agents, vinblastine sulphate and vincristine sulphate (Robbins *et al.*, 1994).

Moore and associates (1993b) used a probe to the repetitive classical satellite region on chromosome 9 to demonstrate the feasibility and optimize conditions for the use of FISH to detect numerical aberrations in the exfoliated buccal mucosal cells and bladder epithelial cells of 10 non-smoking males and females. The frequencies of nuclei exhibiting three and four hybridization domains in these exfoliated epithelial cells, 1.6 and 1.3% for the buccal and urothelial cells, respectively, were higher than those reported for lymphocytes by Eastmond and Pinkel (1990) and Rupa and associates (Rupa, Hasegawa, Thompson and Eastmond, unpublished) but similar to those reported by Kibbelaar *et al.* (1993). In spite of these differences, these relatively low frequencies of nuclei with multiple hybridization regions indicate that FISH is a useful technique for detecting aneuploidy in exfoliated cells and should allow researchers to detect numerical alterations affecting the oral mucosa and the urinary bladder in human populations occupationally or environmentally exposed to genotoxic agents.

Probes specific for chromosomes X and Y have been used by a number of investigators to determine the contribution of the sex chromosomes to the formation of micronuclei in lymphocytes from unexposed human donors. Previous studies have indicated that the micronucleus frequency in cultured lymphocytes increases with age and that this increase is pronounced in older women (Fenech, 1993). Hando *et al.* (1994) obtained blood samples from eight newborn females and 38 adult females ranging in age from 19 to 77 years and cultured isolated lymphocytes from each donor in the presence of cytochalasin B to block cytokinesis in the dividing cells. Two thousand binucleated cells per individual were scored for the presence of micronuclei. The presence of the X chromosome in each of the identified micronuclei was then determined using an *alpha* satellite probe. Of the micronuclei, 72% were labelled with the X chromosome-specific probe, indicating the presence of the X chromosome in the micronucleus. In addition, the frequency of X-chromosome micronuclei increased with age of donor. An increased involvement of the X chromosome in mononucleated lymphocytes was also observed by Richard *et al.* (1994). Studies from our laboratory (Rupa *et al.*, 1994) have examined the contribution of the X and Y chromosomes to the origin of micronuclei in cultured binucleated lymphocytes obtained from 21 males and 21 females, aged 21–53 years. In these studies, multicolour FISH was used allowing the presence of both the X chromosome and chromosome 7, a representative autosome, to be identified simultaneously in the female cells and the presence of chromosomes X, Y and 7 to be identified in the male cells. For both the males and females, the frequencies of the sex chromosomes in the micronuclei were significantly higher than that observed for chromosome 7. A significant age effect on the loss of the sex chromosomes was observed in the lymphocytes obtained from females but not males within the age range covered in this study. These results indicate that loss of the sex chromosomes contributes significantly to the elevated frequencies of micronuclei observed in lymphocytes isolated from women and older individuals.

Several recent studies have used FISH to detect chromosomal alterations in human populations exposed to genotoxic agents. Studies from our laboratory used FISH with a classical satellite probe for human chromosome 1 to investigate the frequency of hyperdiploidy and hypodiploidy in cultured lymphocytes obtained from 99 smokers and 57 non-smokers ranging in age from 20 to 60 years (Eastmond, 1993; Rupa, Thompson, Hasegawa and Eastmond, unpublished results). The frequency of nuclei containing both three and four hybridization regions was significantly higher in the smokers (mean 0.0055) than in the non-smokers (mean 0.0035). An increase in the number of cells containing only one hybridization region was also observed in the smokers using rigorous scoring criteria to minimize problems with overlap. Smith and associates have used FISH with probes specific to chromosomes 7, 8, and 9 to detect aneuploidy in cultured lymphocytes obtained from 43 benzene-exposed Chinese workers and 44 age- and sex-matched controls (Smith et al., 1995). Significant increases in both hyperdiploidy and hypodiploidy were observed using each of the probes in the cells of workers with the highest benzene exposure (>31 p.p.m.). Using the chromosome 7 probe, chromosomal alterations were also observed in the lymphocytes of workers exposed at lower exposure levels suggesting a more frequent involvement of this chromosome in benzene-induced aneuploidy. This observation is particularly interesting as monosomy of chromosome 7 is one of the most common genetic alterations observed in individuals with chemically induced leukaemias. Additional studies from our laboratory (Rupa et al., 1995) used multicoloured FISH with the tandem-labelled probes to detect hyperdiploidy and chromosomal breakage affecting the heterochromatin of chromosome 1 in cultured blood obtained from workers involved in mixing and spraying pesticides on cotton fields in India. Samples were obtained from 26 exposed and 19 non-exposed males and indicated that workers exposed to the pesticides had significantly higher frequencies of both hyperdiploidy and breakage than non-exposed controls. These results illustrate the feasibility of using this approach to detect structural as well as numerical chromosomal alterations in chemically exposed human populations.

Pancentromeric and pantelomeric DNA probes. Probes to DNA sequences located in the centromeric or pericentric region of all or nearly all of the chromosomes have been developed for both humans and mice. These probes have had two main applications: (1) to assist in identifying the origin of micronuclei formed in human or mouse cells following exposure to genotoxic agents; and (2) to facilitate the identification of centromeres in translocation analysis using composite library probes (described in the following section). The basis for the use of pancentromeric probes in the micronucleus assay is illustrated in *Figure 14.7.* Micronuclei are formed when either an entire chromosome or a chromosome fragment fails to segregate properly during mitosis. The resulting small nucleus or micronucleus will typically remain in the cytoplasm of the cell throughout interphase and therefore can serve as an indicator that large-scale chromosomal alterations have occurred. To distinguish between the two types of micronuclei, the presence or absence of centromeric DNA sequences (or centromeric proteins) in a micronucleus has been used to distinguish a micronucleus formed as a result of chromosome loss (centromere-containing) from a micronucleus formed as a result of chromosome breakage (centromere-lacking).

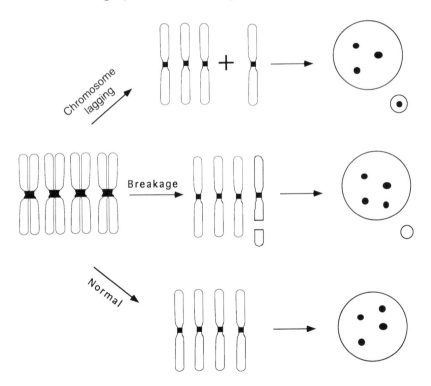

Figure 14.7. A schematic illustration of the use of a pancentromeric DNA probe to distinguish micronuclei formed as a result of chromosome loss from micronuclei formed through chromosome breakage.

The initial experiments using this approach in human cells used an *alpha* satellite probe (p82H) to identify centromeric DNA in micronuclei induced *in vitro* by colcemid, an aneuploidy-inducing agent, and bleomycin, a clastogen (Becker *et al.*, 1990). As expected, micronuclei induced by colcemid contained a higher proportion of centromeric signals (78% of the classifiable micronuclei) than micronuclei induced by bleomycin (31%). Similar experiments were published the next year using FISH following the *in vivo* administration of various clastogenic and aneuploidy-inducing agents to mice (Miller *et al.*, 1991). The presence of pericentromeric DNA sequences in micronucleated bone marrow erythrocytes was determined using a mouse major (*gamma*) satellite probe, which hybridizes to 39 of the 40 mouse chromosomes. This study indicated that this approach was effective for detecting chromosome loss induced by aneuploidy-inducing agents. However, following the administration of mitomycin C, a clastogenic agent, the frequency of micronuclei labelling with the centromeric major satellite probe was greater than that observed using the CREST antibody, which recognizes a centromeric protein located within the kinetochore region. A number of studies using pancentromeric DNA probes have been conducted in recent years to investigate the origin of micronuclei arising spontaneously, or induced by chemical and physical agents, in primary human or mouse cells such as erythrocytes, lymphocytes, spermatids, buccal cells and urothelial cells, as well as in immortalized and malignant cells (Afshari *et al.*, 1994; Chen *et al.*, 1994a,b; Eastmond *et al.*, 1993; Farooqi *et al.*, 1993; Grawé *et al.*, 1994; Hayashi *et al.*, 1994; Kallio and Lahdetie, 1993; Kirchner *et al.*, 1993; Migliore *et al.*, 1993; Miller and Nüsse, 1993; Moore *et al.*, 1993a; Norppa *et al.*, 1993; Salassidis *et al.*, 1992; Stopper *et al.*, 1993; Titenko-Holland *et al.*, 1994). In some of these subsequent investigations, discrepancies similar to that seen by Miller *et al.* (1991) in the frequency of micronuclei labelling with the CREST antibody and the major satellite probe were observed following treatment with benzene, mitomycin C, 2-aminoanthraquinone and hydroquinone (Afshari *et al.*, 1994; Chen *et al.*, 1994b; Eastmond *et al.*, 1993; Hayashi *et al.*, 1994). This observed difference was proposed to be due either to a chemically induced detachment of the kinetochores or to a high frequency of breakage within the mouse heterochromatin which would result in acentric fragments containing major satellite sequences.

To address this issue, a tandem multicolour FISH strategy similar to that described earlier for human cells was developed by our laboratory which requires the presence of both the mouse major and minor satellite probes in a micronucleus for a classification of chromosomal loss (Eastmond *et al.*, 1993). The minor probe targets a centromeric region physically linked to the short arm of mouse chromosomes, whereas the major probe hybridizes to the centromeric heterochromatin next to the long arm. Using this tandem label approach, approximately 30% of the micronuclei induced by the benzene metabolite hydroquinone hybridized with both the major and minor satellite probes, indicating chromosome loss; an additional 35–40% labelled with only the major satellite probe indicating that breakage had occurred within the centromeric heterochromatin (Chen *et al.*, 1994b). This multicolour hybridization approach shows promise as an accurate technique for distinguishing micronuclei arising from chromosome loss from those originating from chromosome breakage, particularly when breakage occurs within the mouse heterochromatin, a breakage-prone region targeted by the major satellite probe.

Similar studies by Adler and coworkers used the mouse major and minor satellite probes as well as a telomeric probe to investigate the possible contribution of breakage within the centromeric heterochromatin by mitomycin C to the discrepant results observed using the CREST antibody and FISH with the major satellite probe. Breakage within the centromeric heterochromatin targeted by the major satellite probe was determined to be the major cause (Schriever-Schwemmer and Adler, 1994). These investigators recommended the use of the minor satellite probe for future mouse studies of this nature as the regions targeted by the minor satellite probe seem to be more resistant to chromosome breakage than those targeted by the major satellite probe. Pantelomeric probes have also been used in metaphase preparations by a number of investigators to study chromosomal instability and non-random breakage affecting these repeated regions (Alvarez *et al.*, 1993; Balajee *et al.*, 1994; Day *et al.*, 1993).

14.3.2 Use of chromosome library or painting probes

Chromosome-specific library or painting probes, originally developed by Pinkel *et al.* (1988) and Cremer *et al.* (1988), are comprised of many different unique or small repeat elements which are distributed along the length of an entire chromosome (or large chromosomal region). Following hybridization, staining covers the length of the chromosome, giving a 'painted' appearance. By labelling the entire chromosome, these chromosomes can provide information on the effect of a genotoxic agent along the length of a chromosome rather than alterations affecting only one specific region. As illustrated in *Figure 14.8*, the primary application of these probes in studies of environmental mutagenesis has been in the analysis of structural chromosomal aberrations, particularly translocations. *Figure 14.3a* shows a normal human metaphase and interphase lymphocyte following FISH with a painting probe for chromosome 4. In *Figure 14.3b* an irradiated lymphocyte metaphase is shown following FISH with the same probe. Numerous chromosomal exchanges can be seen. Chromosome library probes can be used to enumerate the number of copies of a chromosome in an interphase nucleus (*Figure 14.3a*). However, these estimates are accurate only for detecting the modal number of labelled chromosomes for most of the cells due to a high probability of domain overlap and apparent fragmentation resulting from the large domain size and dispersed interphase localization (Gray and Pinkel, 1992; Kuo *et al.*, 1991). As a consequence, these types of probes have limited usefulness for interphase genotoxicity studies of induced aneuploidy.

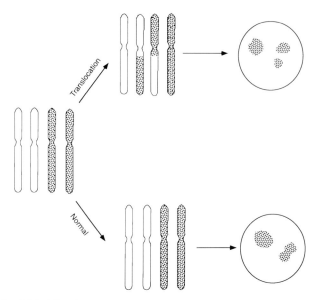

Figure 14.8. Hybridization strategy using a whole chromosome painting probe to detect translocations in metaphase cells.

In vitro *studies.* FISH with chromosome-specific composite probes is being increasingly used for translocation analysis in human cells. Initial studies by Cremer *et al.* (1990) used painting probes for chromosomes 1 and 7 to detect structural chromosomal alterations induced in cultured human lymphocytes by ionizing radiation. Significant increases in translocations, deletions and dicentrics were seen at doses of 4 and 8 Gy. Follow-up studies by a number of investigators have shown that with these types of probes FISH can detect efficiently both dicentrics and translocations resulting from damage induced by radiation exposure or restriction endonucleases *in vitro* (Abella Columna *et al.*, 1993; Bauchinger *et al.*, 1993; Matsuoka *et al.*, 1994; Natarajan *et al.*, 1992; Tucker *et al.*, 1993). Although these FISH analyses can often be performed more quickly and with increased sensitivity than analyses using chromosome banding, only alterations affecting one chromosome pair can be detected. As a result many more metaphases must be scored to obtain the same degree of sensitivity as conventional scoring. By using probes for several chromosomes simultaneously, scoring efficiency can be enhanced (Lucas *et al.*, 1992; Tucker *et al.*, 1993). Studies comparing FISH analyses and standard cytogenetic analyses have shown a generally good agreement after the FISH data have been corrected for the proportion of the genome that is labelled (Pandita *et al.*, 1994; Tucker *et al.*, 1993). However, more detailed evaluations of the FISH results have shown that, in contrast to expectation, the frequency of translocations detected using FISH is somewhat higher than the frequency of dicentrics observed in the same samples (Bauchinger *et al.*, 1993; Lucas *et al.*, 1989; Matsuoka *et al.*, 1994; Natarajan *et al.*, 1992; Tucker *et al.*, 1993). Multicolour

FISH studies combining the use of painting probes and a pancentromeric probe have indicated that part of this discrepancy may be due to a difficulty in seeing the centromeres in the dicentric chromosomes when scoring chromosomes labelled with only the painting probes (Straume and Lucas, 1993; Weier *et al.* 1991b). However, this offers only a partial explanation as excess translocation frequencies continue to be seen in studies employing both painting and pancentromeric probes (Bauchinger *et al.*, 1993; Tucker *et al.*, 1993, 1995).

Premature chromosome condensation (PCC) is a technique in which an interphase cell is fused with a mitotic cell causing the chromosomes of the interphase cell to condense. This allows cytogenetic analyses to be performed in cells at various stages of the cell cycle. Scoring chromosomes in PCC cells is problematic due to the difficulty in distinguishing fragments from intact chromosomes and the relatively poor quality of chromosome banding, which makes exchange aberrations difficult to detect (Evans *et al.*, 1991). Applying chromosome painting to PCC cells significantly facilitates cytogenetic analyses. A number of investigators have used chromosome painting combined with PCC to investigate DNA damage and repair occurring at various times after radiation exposure (Brown *et al.*, 1992; Evans *et al.*, 1991; Kovacs *et al.*, 1994). Evidence from these studies has indicated that chromosomal damage and repair occurred randomly among the various chromosomes but that a loss of cells or a delay in cell division occurred in cells with unjoined breaks (Kovacs *et al.*, 1994).

Chromosome painting has also been used to study chromosomal instability in hamster–human cell hybrids exposed to X-irradiation or human lymphoblastoid mutants selected in 2,6-diaminopurine (Marder and Morgan, 1993; Smith and Grosovsky, 1993). In both these studies FISH with painting probes was found to be an efficient way to detect and follow karyotypic alterations which occurred in the exposed cells. Painting probes have also been used to provide insights into the basic mechanisms responsible for chromosomal exchanges in irradiated cells (Brown *et al.*, 1993; Lucas and Sachs, 1993).

Population studies. In addition to *in vitro* studies, chromosome painting probes have been used to detect stable chromosomal aberrations occurring in the cultured lymphocytes of normal untreated individuals or in human populations previously exposed to genotoxic agents. Lucas and associates used whole-chromosome probes to measure the frequency of translocations in the peripheral blood of 20 individuals exposed to radiation at Hiroshima and four workers exposed during the Y-12 accident (Lucas *et al.*, 1992). The frequencies of translocations observed using FISH (after correction for the proportion of the genome that was labelled) were approximately the same as those obtained by G-banding. In addition, the studies also showed that the dose–response curves for translocation frequency estimated using FISH in the A-bomb survivors was similar to the frequency of translocations observed in first-division metaphases irradiated *in vitro*. This supports the work of previous researchers which indicated that translocation frequencies measured in peripheral blood lymphocytes appeared to be generally stable over time (Buckton *et al.*, 1978). This work also suggests that translocation frequencies measured by FISH can be used as a biological dosimeter to determine approximate levels of radiation exposure in situations where the true exposure is unknown or is thought to be in error. Based on this research, Straume *et al.* (1992) used measurements of chromosome translocations determined by FISH to estimate the extent of radiation exposure in a 59-year-old radiation worker who believed that his actual exposure was significantly underestimated by the dosimetry records maintained by his employer. Although the translocation frequency observed in the worker was higher than in controls, the frequency was within the range expected based upon his workplace dosimetry.

The stability of translocations which allows them to serve as a biodosimeter for radiation exposure may also allow them to quantify genotoxic effects resulting from chronic exposure to environmental agents. To determine the influence of age and lifestyle on the frequency of stable aberrations, Tucker *et al.* (1994b) used FISH with whole-chromosome probes to measure the frequency of stable aberrations in metaphase lymphocytes obtained from the peripheral blood of 47 healthy adults with ages ranging from 19 to 77 years, and from the umbilical cord blood of eight healthy newborns. A strong association between stable aberrations and square of age was observed. A less impressive, albeit statistically significant, association between age and dicentric frequency was also seen. These results support the hypothesis that stable aberrations accumulate with time and may therefore provide an index of chronic exposure to clastogenic agents. As a result, age must be accounted for in all population studies, particularly those attempting to assign exposure doses retrospectively such as the one described earlier. However, it is also conceivable that the observed increase in aberrations is related to an age-related deficiency in DNA repair or increased chromosome instability. Studies in animal models will most likely be required to address these issues adequately. These studies will soon be feasible as whole-chromosome painting probes are becoming available for the mouse and rat (Boei *et al.*, 1994; Breneman *et al.*, 1993, 1994; Hoebee *et al.*, 1994; de Stoppelaar *et al.*, 1994).

14.4 Summary and future prospects

A number of recent technical developments in the field are likely to contribute to future studies of environmental mutagenesis. The application of flow cytometry and image analysis should dramatically increase the speed of scoring and the ability of FISH to detect routinely very small probes. Initial studies have shown the feasibility of this approach but further refinements will be necessary for their routine application (Miller and Nüsse, 1993; Trask *et al.*, 1985). As additional probes become available, it will also be possible to study mutagen-induced changes within regions specifically involved in carcinogenesis and other genetic diseases. For example, probes have been developed to detect the *bcr–abl* fusion in chronic myelogenous leukaemia or the loss of the 17p region which contains the *p53* gene (Matsumura *et al.*, 1992; Tkachuk *et al.*, 1990). The combination of multicolour FISH with computerized image analysis will allow alterations occurring at numerous loci to be detected simultaneously. Recent studies have shown that up to 16 different probes can be used on a single slide (Weber-Matthiesen *et al.*, 1993). Furthermore, the recent and ongoing development of probes for common laboratory species will allow new studies to be designed, or ongoing toxicity studies to be more efficiently used, to compare chromosomal alterations occurring in different species and in different tissues (MacGregor *et al.*, 1995). In addition, a new FISH technique called comparative genomic hybridization has been developed which uses competitive hybridization between labelled genomic DNA obtained from an aberrant cell such as a tumour cell and labelled normal genomic DNA to identify chromosomal alterations affecting all chromosomes in the aberrant cell simultaneously (Kallioniemi *et al.*, 1992, 1994). This approach has been shown to effectively detect multiple alterations such as deletions, duplications and amplifications in solid tumours. The application of this technology to environmental mutagenesis should allow genetic alterations occurring at each stage of neoplastic development to be identified throughout the genome in cells isolated at various stages of a cell transformation assay such as the Syrian hamster embryo (SHE) cell assay. The resulting information should provide significant insights into the nature and temporal sequence of the genetic alterations induced by different classes of environmental carcinogens.

In summary, FISH has been shown to be a valuable complement to traditional cytogenetic analyses. The application of FISH techniques to the field of environmental mutagenesis has begun to make a significant impact on our understanding of the frequency and mechanisms of chromosomal alterations occurring either spontaneously or as a result of exposure to genotoxic agents. The use of chromosome-specific repetitive probes is facilitating the detection of aneuploidy and chromosome breakage in cells by speeding cytogenetic analyses, increasing sample sizes and allowing analyses to be performed on interphase cells. As a result, this approach is allowing cytogenetic information to be obtained rapidly from cells such as sperm and epithelial cells which have previously been refractory to cytogenetic analyses. The use of chromosome library probes has also been shown to be a rapid and efficient method for detecting and quantifying structural chromosomal aberrations in metaphase preparations. The stability of translocations combined with their ease of detection by FISH indicates that they may be useful as a biodosimeter to detect and quantify chronic exposure to clastogenic agents such as ionizing radiation. Future developments and novel applications of existing FISH techniques promise to enhance significantly our ability to detect environmental mutagens and understand the mechanisms underlying the induced genetic lesions.

Acknowledgements

The authors wish to thank Dr Martyn T. Smith and Dr James D. Tucker for communicating their data prior to publication and Dr Maik Schuler for his assistance in editing the manuscript. Financial support from the US Environmental Protection Agency (R 820994-01-1) is gratefully acknowledged.

References

Abella Columna E, Giaccia AJ, Evans JW, Yates BL, Morgan WF. (1993) Analysis of restriction enzyme-induced chromosomal aberrations by fluorescence *in situ* hybridization. *Environ. Mol. Mutagen.* **22:** 26–33.

Afshari AJ, McGregor PW, Allen JW, Fuscoe JC. (1994) Centromere analysis of micronuclei induced by 2-aminoanthraquinone in cultured mouse splenocytes using both a *gamma*-satellite DNA probe and anti-kinetochore antibody. *Environ. Mol. Mutagen.* **24:** 96–102.

Almeida A, Kokalj-Vokac N, Lefrancois D, Viegas-Pequignot E, Jeanpierre M, Dutrillaux B, Malfoy B. (1993) Hypomethylation of classical satellite DNA and chromosome instability in lymphoblastoid cell lines. *Hum. Genet.* **91:** 538–546.

Alvarez L, Evans JW, Wilks R, Lucas JN, Brown JM, Giaccia AJ. (1993) Chromosomal radiosensitivity at intrachromosomal telomeric sites. *Genes Chrom. Cancer* **8:** 8–14.

American Type Culture Collection (1993) *ATTC/NIH Repository Catalogue of Human and Mouse DNA Probes and Libraries,* 7th edn. American Type Culture Collection, Rockville, MD.

Balajee AS, Oh HJ, Natarajan AT. (1994) Analysis of restriction enzyme-induced chromosome aberrations in the interstitial telomeric repeat sequences of CHO and CHE cells by FISH. *Mutat. Res.* **307:** 307–313.

Bauchinger M, Schmid E, Zitzelsberger H, Braselmann H, Nahrstedt U. (1993) Radiation-induced chromosome aberrations analysed by two-colour fluorescence *in situ* hybridization with composite whole chromosome-specific DNA probes and a pancentromeric DNA probe. *Int. J. Radiat. Biol.* **64:** 179–184.

Becker P, Scherthan H, Zankl H. (1990) Use of a centromere-specific DNA probe (p82H) in nonisotopic *in situ* hybridization for classification of micronuclei. *Genes Chrom. Cancer* **2:** 59–62.

Bischoff FZ, Nguyen DD, Burt KJ, Shaffer LG. (1994) Estimates of aneuploidy using multicolour fluorescence *in situ* hybridization on human sperm. *Cytogenet. Cell Genet.* **66:** 237–243.

Boehringer Mannheim (1992) *Nonradioactive* in situ *Hybridization Application Manual.* Boehringer Mannheim GmbH, Mannheim, Germany.

Boei JJ, Balajee AS, de Boer P, Rens W, Aten JA, Mullenders LH, Natarajan AT. (1994) Construction of mouse chromosome-specific DNA libraries and their use for the detection of X-ray-induced aberrations. *Int. J. Radiat. Biol.* **65:** 583–590.

Brandriff BF, Gordon LA, Trask BJ. (1991) DNA sequence mapping by fluorescence *in situ* hybridization. *Environ. Mol. Mutagen.* **18:** 259–262.

Breneman JW, Ramsey MJ, Lee DA, Eveleth GG, Minkler JL, Tucker JD. (1993) The development of chromosome-specific composite DNA probes for the mouse and their application to chromosome painting. *Chromosoma* **102:** 591–598.

Breneman JW, Swiger RR, Ramsey MJ, Minkler JL, Eveleth GG, Langlois R, Brooks AL, Tucker JD. (1994) The development of dual color and multiple chromosome painting probes for the mouse and rat. *Environ. Mol. Mutagen.* **23** (suppl. 23): 5.

Brogger A. (1977) Non-random localization of chromosome damage in human cells and targets for clastogenic action. *Chrom. Today* **6:** 297–305.

Brown MJ, Evans J, Kovacs MS. (1992) The prediction of human tumor radiosensitivity *in situ:* an approach using chromosome aberrations detected by fluorescence *in situ* hybridization. *Int. J. Radiat. Oncol. Biol. Phys.* **24:** 279–286.

Brown MJ, Evans J, Kovacs MS. (1993) Mechanisms of chromosome exchange formation in human fibroblasts: insights from 'chromosome painting'. *Environ. Mol. Mutagen.* **22:** 218–224.

Buckton KE, Hamilton GE, Paton L, Langlands AG. (1978) Chromosome aberrations in irradiated ankylosing spondylitis patients. In: *Mutagen-Induced Chromosome Damage in Man* (eds H Evans, D Lloyd). Edinburgh University Press, Edinburgh, pp. 142–150.

Buongiorno-Nardelli M, Amaldi F. (1970) Autoradiographic detection of molecular hybrids between rRNA and DNA in tissue sections. *Nature* **225:** 946–947.

Chan VT-W, McGee JO'D. (1990) Non-radioactive probes: preparation, characterization, and detection. In: *In Situ Hybridization: Principles and Practice* (eds JM Pollack, JO'D McGee). Oxford University Press, Oxford, pp. 59–70.

Chen HW, Rupa DS, Tomar R, Eastmond DA. (1994a) Chromosomal loss and breakage in mouse bone marrow and spleen cells exposed to benzene *in vivo. Cancer Res.* **54:** 3533–3539.

Chen HW, Tomar R, Eastmond DA. (1994b) Detection of hydroquinone-induced nonrandom breakage in the centromeric heterochromatin of mouse bone marrow cells using multicolour fluorescence *in situ* hybridization with the mouse major and minor satellite probes. *Mutagenesis* **9:** 563–569.

Chesselet M-F. (1990) *In Situ Hybridization Histochemistry.* CRC Press, Boca Raton, FL.

Cooke HJ, Hindley J. (1979) Cloning of human satellite III DNA: different components are on different chromosomes. *Nucleic Acids Res.* **6:** 3177–3197.

Coonen E, Pieters MHEC, Dumoulin JCM, Meyer H, Evers JLH, Ramaekers FCS, Geraedts JPM. (1991) Nonisotopic *in situ* hybridization as a method for nondisjunction studies in human spermatozoa. *Mol. Reprod. Dev.* **28:** 18–22.

Cremer T, Landegent J, Bruckner A, Scholl HP, Schardin M, Hager HD, Devilee P, Pearson P, van der Ploeg M. (1986) Detection of chromosome aberrations in the human interphase nucleus by visualization of specific target DNAs with radioactive and non-radioactive *in situ* hybridization techniques: diagnosis of trisomy 18 with probe L1.84. *Hum. Genet.* **74:** 346–352.

Cremer T, Lichter P, Borden J, Ward DC, Manuelidis L. (1988) Detection of chromosome aberrations in metaphase and interphase tumor cells by *in situ* hybridization using chromosome-specific library probes. *Hum. Genet.* **80:** 235–246.

Cremer T, Popp S, Emmerich P, Lichter P, Cremer C. (1990) Rapid metaphase and interphase detection of radiation-induced chromosome aberrations in human lymphocytes by chromosomal suppression *in situ* hybridization. *Cytometry* **11:** 110–118.

Day JP, Marder BA, Morgan WF. (1993) Telomeres and their possible role in chromosome stabilization. *Environ. Mol. Mutagen.* **22:** 245–249.

De Sario A, Vagnarelli P, De Carli L. (1990) Aneuploidy assay on diethylstilbestrol by means of *in situ* hybridization of radioactive and biotinylated DNA probes on interphase nuclei. *Mutat. Res.* **243:** 127–131.

de Stoppelaar JM, Essers J, Mohn GR, Hoebee B. (1994) Detection of chromosome aberrations in the rat by fluorescence *in situ* hybridization. *Environ. Mol. Mutagen.* **23** (suppl. 23): 13.

Dunham I, Lengauer C, Cremer T, Featherstone T. (1992) Rapid generation of chromosome-specific alphoid DNA probes using the polymerase chain reaction. *Hum. Genet.* **88:** 457–462.

Eastmond DA. (1993) Tobacco smoke, benzene and mechanisms of leukemogenesis. *First Scientific Conference, University of California Tobacco-Related Disease Research Program, Book of Abstracts.* p. 35.

Eastmond DA, Pinkel D. (1990) Detection of aneuploidy and aneuploidy-inducing agents in human lymphocytes using fluorescence *in situ* hybridization with chromosome-specific DNA probes. *Mutat. Res.* **234:** 303–318.

Eastmond DA, Rupa DS, Chen HW, Hasegawa L. (1993). Multicolour fluorescence *in situ* hybridization with centromeric DNA probes as a new approach to distinguish chromosome breakage from aneuploidy in interphase cells and micronuclei. In: *Chromosome Segregation and Aneuploidy* (ed. BK Vig). Springer-Verlag, Berlin, pp. 377–390.

Eastmond DA, Rupa DS, Hasegawa LS. (1994) Detection of hyperdiploidy and chromosome breakage in interphase human lymphocytes following exposure to the benzene metabolite hydroquinone using multicolour fluorescence *in situ* hybridization with DNA probes. *Mutat. Res.* **322:** 9–20.

Evans JW, Chang JA, Giaccia AJ, Pinkel D, Brown JM. (1991) The use of fluorescence *in situ* hybridisation combined with premature chromosome condensation for the identification of chromosome damage. *Br. J. Cancer* **63:** 517–521.

Farooqi Z, Darroudi F, Natarajan AT. (1993) The use of fluorescence *in situ* hybridization for the detection of aneugens in cytokinesis-blocked mouse splenocytes. *Mutagenesis* **8:** 329–334.

Fenech M. (1993) The cytokinesis-block micronucleus technique: a detailed description of the method and its application to genotoxicity studies in human populations. *Mutat. Res.* **285:** 35–44.

Ferguson M, Ward DC. (1992) Cell cycle dependent chromosomal movement in premitotic human T-lymphocyte nuclei. *Chromosoma* **101:** 557–565.

Goodwin E, Blakely E, Ivery G, Tobias C. (1989) Repair and misrepair of heavy-ion-induced chromosomal damage. *Adv. Space Res.* **9:** 83–89.

Grawé J, Abramsson-Zetterberg L, Eriksson L, Zetterberg G. (1994) The relationship between DNA content and centromere content in micronucleated mouse bone marrow erythrocytes analysed by flow cytometry and fluorescent *in situ* hybridization. *Mutagenesis* **9:** 31–38.

Gray J, Pinkel D. (1992) Molecular cytogenetics in human cancer diagnosis. *Cancer* (suppl.) **69:** 1536–1542.

Han TL, Ford JH, Webb GC, Flaherty SP, Correll A, Matthews CD. (1993) Simultaneous detection of X- and Y-bearing human sperm by double fluorescence *in situ* hybridization. *Mol. Reprod. Dev.* **34:** 308–313.

Hando JC, Nath J, Tucker JD. (1994) Sex chromosomes, micronuclei and aging in women. *Chromosoma* **103:** 186–192.

Hayashi M, Maki-Paakkanen J, Tanabe H, Honma M, Suzuki T, Matsuoka A, Mizusawa H, Sofuni T. (1994) Isolation of micronuclei from mouse blood and fluorescence *in situ* hybridization with a mouse centromeric DNA probe. *Mutat. Res.* **307:** 245–251.

Hecht F, Hecht BK. (1987) Aneuploidy in humans: dimensions, demography, and dangers of abnormal numbers of chromosomes. In: *Aneuploidy, Part A: Incidence and Etiology* (eds BK Vig, AA Sandberg). Alan R. Liss, New York, pp. 9–49.

Hindkjaer J, Koch J, Terkelsen C, Brandt CA, Kolvraa S, Bolund L. (1994) Fast, sensitive multicoloured detection of nucleic acids *in situ* by PRimed *IN Situ* labelling (PRINS). *Cytogenet. Cell Genet.* **66:** 152–154.

Hoebee B, de Stoppelaar JM, Suijkerbuijk RF, Monard S. (1994) Isolation of rat chromosome-specific paint probes by bivariate flow sorting followed by degenerate oligonucleotide primed-PCR. *Cytogenet. Cell Genet.* **66:** 277–282.

Hofler H. (1990) Principles of *in situ* hybridization. In: In situ *Hybridization: Principles and Practice* (eds JM Pollack, JO'D McGee). Oxford University Press, Oxford, pp. 15–29.

Holmes JM, Martin RH. (1993) Aneuploidy detection in human sperm nuclei using fluorescence *in situ* hybridization. *Human Genet.* **91:** 20–24.

Horn JE, Quinn T, Hammer M, Palmer L, Falkow S. (1986) Use of nucleic acid probes for the detection of sexually transmitted infectious agents. *Diag. Microbiol. Infect. Dis.* **4:** 101S–109S.

Hook EB. (1985) The impact of aneuploidy upon public health: mortality and morbidity associated with human chromosome abnormalities. In: *Aneuploidy, Etiology and Mechanisms* (eds VL Dellarco, PE Voytek, A Hollaender). Plenum Press, New York, pp. 7–33.

John H, Birnstiel M, Jones K. (1969) RNA:DNA hybrids at the cytological level. *Nature* **223:** 582–585.

Kallio M, Lahdetie J. (1993) Analysis of micronuclei induced in mouse early spermatids by mitomycin C, vinblastine sulphate or etoposide using fluorescence *in situ* hybridization. *Mutagenesis* **8:** 561–567.

Kallioniemi A, Kallioniemi O, Piper J, Tanner M, Stokke T, Chen L, Smith HS, Pinkel D, Gray JW, Waldman FM. (1992) Detection and mapping of amplified DNA sequences in breast cancer by comparative genomic hybridization. *Proc. Natl Acad. Sci. USA* **91:** 2156–2160.

Kallioniemi A, Kallioniemi O, Piper J, Tanner M, Stokke T, Chen L, Smith HS, Pinkel D, Gray JW, Waldman FM. (1994) Comparative genomic hybridization for molecular cytogenetic analysis of solid tumors. *Science* **258:** 818–821.

Keller GH, Manak MM. (1993) *DNA Probes: Background, Applications, Procedures*, 2nd Edn. Stockton Press, New York.

Kibbelaar RE, Kok F, Dreef EJ, Kleiverda JK, Cornelisse CJ, Raap AK, Kluin PhM. (1993) Statistical methods in interphase cytogenetics: an experimental approach. *Cytometry* **14:** 716–724.

Kirchner S, Stopper H, Papp T, Eckert I, Yoo HJ, Vig BK, Schiffmann D. (1993) Cytogenetic changes in primary, immortalized and malignant mammalian cells. *Toxicol. Lett.* **67:** 283–295.

Kokalj-Vokac N, Almeida A, Viegas-Pequignot E, Jeanpierre M, Malfoy B, Dutrillaux B. (1993) Specific induction of uncoiling and recombination by azacytidine in classical satellite-containing constitutive heterochromatin. *Cytogenet. Cell Genet.* **63:** 11–15.

Kovacs MS, Evans JW, Johnstone IM, Brown JM. (1994) Radiation-induced damage, repair and exchange formation in different chromosomes of human fibroblasts determined by fluorescence *in situ* hybridization. *Radiat. Res.* **137:** 34–43.

Kuo W-L, Tenjin H, Segraves R, Pinkel D, Golbus MS, Gray J. (1991) Detection of aneuploidy involving chromosomes 13, 18, or 21, by fluorescence *in situ* hybridization (FISH) to interphase and metaphase amniocytes. *Am. J. Hum. Genet.* **49:** 112–119.

Le Beau M. (1993) Fluorescence *in situ* hybridization in cancer diagnosis. In: *Important Advances in Oncology 1993* (eds VT DeVita, S Hellman, SA Rosenberg). J.B. Lippincott Co., Philadelphia, pp. 29–45.

Lewis ME, Baldino F Jr (1990) Probes for *in situ* hybridization histochemistry. In: In situ *Hybridization Histochemistry* (ed. M-F Chesselet). CRC Press, Boca Raton, FL, pp. 1–21.

Lewis JP, Tanke HJ, Raap AK, Beverstock GC, Kluin-Nelemans HC. (1993) Somatic pairing of centromeres and short arms of chromosome 15 in the hematopoietic and lymphoid system. *Hum. Genet.* **92:** 577–582.

Lowe X, Baulch J, Quintana L, Collins B, Allen J, Ramsey M, Breneman J, Tucker J, Holland N, Wyrobek A. (1994) Molecular detection of chromosomal abnormalities in germ and somatic cells of aged male mice. *Environ. Mol. Mutagen.* **23** (suppl. 23): 40.

Lu PY, Hammitt DG, Zinsmeister AR, Dewald GW. (1994) Dual color fluorescence *in situ* hybridization to investigate aneuploidy in sperm from 33 normal males and a man with a t(2;4;8)(q23;q27;p21). *Fertility Sterility* **62:** 394–399.

Lucas JN, Sachs RK. (1993) Using three color chromosome painting to test chromosome aberration models. *Proc. Natl Acad. Sci. USA* **90:** 1484–1487.

Lucas JN, Tenjin T, Straume T, Pinkel D, Moore D, Litt M, Gray JW. (1989) Rapid human chromosome aberration analysis using fluorescence *in situ* hybridization. *Int. J. Radiat. Biol.* **56:** 35–44. [Erratum *Int. J. Radiat. Biol.* **56:** 201.]

Lucas JN, Awa A, Straume T, Poggensee M, Kodama Y, Nakano M, Ohtaki K, Weier H-U, Pinkel D, Gray J. (1992) Rapid translocation frequency analysis in humans decades after exposure to ionizing radiation. *Int. J. Radiat. Biol.* **62:** 53–63.

MacGregor JT, Tucker JD, Eastmond DA, Wyrobek AJ. (1995) Integration of cytogenetic assays with toxicology studies. *Environ. Mol. Mutagen.,* in press.

Marder BA, Morgan WF. (1993) Delayed chromosomal instability induced by DNA damage. *Mol. Cell. Biol.* **13:** 6667–6677.

Matsumura K, Kallioniemi A, Kallioniemi O, Chen L, Smith HS, Pinkel D, Gray J, Waldman FM. (1992) Deletion of chromosome 17p loci in breast cancer cells detected by fluorescence *in situ* hybridization. *Cancer Res.* **52:** 3474–3477.

Matsuoka A, Tucker JD, Hayashi M, Yamazaki N, Sofuni T. (1994) Chromosome painting analysis of X-ray-induced aberrations in human lymphocytes *in vitro*. *Mutagenesis* **9:** 151–155.

Meyne J, Lockhart LH, Arrigh FE. (1979) Nonrandom distribution of chromosomal aberrations induced by three chemicals. *Mutat. Res.* **63:** 201–209.

Meyne J, Littlefield LG, Moyzis RK. (1989) Labelling of human centromeres using an alphoid DNA consensus sequence: application to the scoring of chromosome aberrations. *Mutat. Res.* **226:** 75–79.

Migliore L, Bocciardi R, Macri C, Lo Jacono F. (1993) Cytogenetic damage induced in human lymphocytes by four vanadium compounds and micronucleus analysis by fluorescence *in situ* hybridization with a centromeric probe. *Mutat. Res.* **319:** 205–213.

Miharu N, Best RG, Young SR. (1994) Numerical chromosome abnormalities in spermatozoa of fertile and infertile men detected by fluorescence *in situ* hybridization. *Hum. Genet.* **93:** 502–506.

Miller BM, Nüsse M. (1993) Analysis of micronuclei induced by 2-chlorobenzylidene malonitrile (CS) using fluorescence *in situ* hybridization with telomeric and centromeric DNA probes, and flow cytometry. *Mutagenesis* **8:** 35–41.

Miller BM, Zitzelsberger HF, Weier H-U, Adler I-D. (1991) Classification of micronuclei in murine erythrocytes: immunofluorescent staining using CREST antibodies compared to *in situ* hybridization with biotinylated *gamma* satellite DNA. *Mutagenesis* **6:** 297–302.

Mitchell AR, Gosden JR, Miller DA. (1985) A cloned sequence, p82H, of the alphoid repeated DNA family found at the centromeres of all human chromosomes. *Chromosoma* **92:** 369–377.

Moore LE, Titenko-Holland N, Quintana PJE, Smith MT. (1993a) Novel biomarkers of genetic damage in humans: use of fluorescence *in situ* hybridization to detect aneuploidy and micronuclei in exfoliated cells. *J. Toxicol. Environ. Hlth* **40:** 349–357.

Moore LE, Titenko-Holland N, Smith MT. (1993b) Use of fluorescence *in situ* hybridization to detect chromosome-specific changes in exfoliated human bladder and oral mucosa cells. *Environ. Mol. Mutagen.* **22:** 130-137.

Moyzis RK, Albright KL, Bartholdi MF, Cram LS, Deaven LL, Hildebrand CE, Joste NE, Longmire JL, Meyne J, Schwarzacher-Robinson T. (1987) Human chromosome-specific repetitive DNA sequences: novel markers for genetic analysis. *Chromosoma* **95:** 375–386.

Moyzis RK, Buckingham JM, Cram LS, Dani M, Deaven LL, Jones MD, Meyne J, Ratliff RL, Wu J-R. (1988) A highly conserved repetitive DNA sequence, $(TTAGGG)_n$, present at the telomeres of human chromosomes. *Proc. Natl Acad. Sci. USA* **85:** 6622–6626.

Natarajan AT, Vyas RC, Darroudi F, Vermeulen S. (1992) Frequencies of X-ray-induced chromosome translocations in human peripheral lymphocytes as detected by *in situ* hybridization using chromosome-specific DNA libraries. *Int. J. Radiat. Biol.* **61:** 199–203.

Norppa H, Renzi L, Lindholm C. (1993) Detection of whole chromosomes in micronuclei of cytokinesis-blocked human lymphocytes by antikinetochore staining and *in situ* hybridization. *Mutagenesis* **8:** 519–525.

Oncor. (1992) *Catalog and Source Book.* Oncor, Gaithersburg, MD.

Oshimura M, Barrett JC. (1986) Chemically induced aneuploidy in mammalian cells: mechanisms and biological significance in cancer. *Environ. Mutagen.* **8:** 129–159.

Pandita TK, Gregoire V, Dhingra K, Hittelman WN. (1994) Effect of chromosome size on aberration levels caused by *gamma* radiation as detected by fluorescence *in situ* hybridization. *Cytogenet. Cell Genet.* **67:** 94–101.

Pardue ML, Gall JG. (1969) Molecular hybridization of radioactive DNA to the DNA of cytological preparations. *Proc. Natl Acad. Sci. USA* **64:** 600–604.

Pieters MHEC, Geraedts JPM, Meyer H, Dumoulin JCM, Evers JLH, Jongbloed RJE, Nederlof PM, van der Flier S. (1990) Human gametes and zygotes studied by nonradioactive *in situ* hybridization. *Cytogenet. Cell Genet.* **53:** 15–19.

Pietras DF, Bennet KL, Siracusa LD, Woodworth-Gutai M, Chapman VM, Gross KW, Kane-Hass C,Hastie ND. (1983) Construction of a small *Mus musculus* repetitive DNA library: identification of a new satellite sequence in *Mus musculus*. *Nucleic Acids Res.* **11:** 6965–6983.

Pinkel D, Straume T, Gray JW. (1986) Cytogenetic analysis using quantitative, high-sensitivity, fluorescence hybridization. *Proc. Natl Acad. Sci. USA* **83:** 2934–2938.

Pinkel D, Landegent J, Collins C, Fuscoe J, Segraves R, Lucas J, Gray J. (1988) Fluorescence *in situ* hybridization with human chromosome-specific libraries: detections of trisomy 21 and translocations of chromsosome 4. *Proc. Natl Acad. Sci. USA* **85:** 9138–9142.

Polak JM, McGee JO'D. (1990) In situ *Hybridization: Principles and Practice.* Oxford University Press, Oxford.

Richard R, Muleris M, Dutrillaux B. (1994) The frequency of micronuclei with X chromosome increases with age in human females. *Mutat. Res.* **316:** 1–7.

Robbins WA, Segraves R, Pinkel D, Wyrobek AJ. (1993) Detection of aneuploid human sperm by fluorescence *in situ* hybridization: evidence for a donor difference in frequency of sperm disomic for chromosomes 1 and Y. *Am. J. Hum. Genet.* **52:** 799–807.

Robbins WA, Cassel MJ, Blakey DH, Meistrich ML, Wyrobek AJ. (1994) Induction of aneuploidy in the sperm of Hodgkin's disease patients treated with NOVP chemotherapy (detection by multi-chromosome fluorescence *in situ* hybridization). *Environ. Mol. Mutagen.* **23** (suppl. **23**): 58.

Rupa DS, Hasegawa L, Eastmond DA. (1994) Chromosomal loss and the involvement of chromosomes X, Y and 7 in spontaneous micronuclei (MN) formed in binucleated lymphocytes of non-smoking females and males. *Environ. Mol. Mutagen.* **23** (suppl. 23): 58.

Rupa DS, Hasegawa L, Eastmond DA. (1995) Detection of chromosomal breakage in the 1cen – 1q12 region of interphase human lymphocytes using multicolour fluorescence *in situ* hybridization with tandem DNA probes. *Cancer Res.* **55:** 640–645.

Salassidis K, Huber R, Zitzelsberger H, Bauchinger M. (1992) Centromere detection in vinblastine- and radiation-induced micronuclei of cytokinesis-blocked mouse cells by using *in situ* hybridization with a mouse *gamma* (major) satellite DNA probe. *Environ. Mol. Mutagen.* **19:** 1–6.

Sandberg AA, Turec-Carel C, Germill RM. (1988) Chromosomes in solid tumors and beyond. *Cancer Res.* **48:** 1049–1059.

Schriever-Schwemmer G, Adler I-D. (1994) Differentiation of micronuclei in mouse bone marrow cells: a comparison between CREST staining and fluorescent *in situ* hybridization with centromeric and telomeric DNA probes. *Mutagenesis* **9:** 333–340.

Smith LE, Grosovsky AJ. (1993) Genetic instability on chromosome 16 in a human B lymphoblastoid cell line. *Somat. Cell Mol. Genet.* **19:** 515–527.

Smith MT, Zhang L, Rothman N, Wang Y, Hayes RB, Li G-L, Yin S-N. (1995) Interphase cytogenetics of workers exposed to benzene. *Toxicologist* **15:** 87.

Stopper H, Korber C, Spencer DL, Kirchner S, Caspary WJ, Schiffmann D. (1993) An investigation of micronucleus and mutation induction by oxazepam in mammalian cells. *Mutagenesis* **8:** 449–455.

Straume T, Lucas JN, Tucker JD, Bigbee WL, Langlois RG. (1992) Biodosimetry for a radiation worker using multiple assays. *Hlth Physics* **62:** 122–130.

Straume T, Lucas JN. (1993) A comparison of the yields of translocations and dicentrics measured using fluorescence *in situ* hybridization. *Int. J. Radiat. Biol.* **64:** 185–187.

Tagarro I, Fernandez-Peralta A, Gonzalez-Aguilera JJ. (1994) Chromosomal localization of human satellites 2 and 3 by FISH method using oligonucleotides as probes. *Hum. Genet.* **93:** 383–388.

Teyssier JR. (1989) The chromosomal analysis of human solid tumors: a triple challenge. *Cancer Genet. Cytogenet.* **37:** 103–125.

Titenko-Holland N, Moore LE, Smith MT. (1994) Measurement and characterization of micronuclei in exfoliated human cells by fluorescence *in situ* hybridization with a centromeric probe. *Mutat. Res.* **312:** 39–50.

Tkachuk DC, Westbrook CA, Andreeff M, Donlon TA, Cleary ML, Suryanarayan K, Homge M, Redner A, Gray J, Pinkel D. (1990) Detection of *bcr-abl* fusion in chronic myelogenous leukemia by *in situ* hybridization. *Science* **250**: 559–562.

Trask B, Pinkel D. (1990) Fluorescence *in situ* hybridization with DNA probes. *Methods Cell Biol.* **33**: 383–400.

Trask B, van den Engh G, Landegent J, Jansen in de Wal N, van der Ploeg M. (1985) Detection of DNA sequences in nuclei in suspension by *in situ* hybridization and dual beam flow cytometry. *Science* **230**: 1401–1403.

Tucker JD, Ramsey MJ, Lee DA, Minkler JL. (1993) Validation of chromosome painting in human peripheral lymphocytes following acute exposure to ionizing radiation *in vitro*. *Int. J. Radiat. Biol.* **64**: 27–37.

Tucker JD, Lee DA, Ramsey MJ, Briner J, Olsen L, Moore DH II. (1994) On the frequency of chromosome exchanges in a control population measured by chromosome painting. *Mutat. Res.* **313**: 193–202.

Tucker JD, Lee DA, Minkler JL. (1995) Validation of chromosome painting. II. A detailed analysis of aberrations following high doses of ionizing radiation *in vitro*. *Int. J. Radiat. Biol.* **67**: 19–28.

van Dekken H, Bauman JGJ. (1988) A new application of *in situ* hybridization: detection of numerical and structural chromosome aberrations with a combination centromeric-telomeric DNA probe. *Cytogenet Cell Genet.* **48**: 188–189.

Viegas-Pequignot E. (1992) In situ *Hybridization: a Practical Approach* (ed. DG Wilkinson). IRL Press, Oxford, pp. 136–158.

Vourc'h C, Taruscio D, Boyle AL, Ward DC. (1993) Cell cycle-dependent distribution of telomeres, centromeres, and chromosome-specific subsatellite domains in the interphase nucleus of mouse lymphocytes. *Exp. Cell Res.* **205**: 142–151.

Ward BE, Gersen SL, Carelli MP, McGuire NM, Dackowski WR, Weinstein M, Sandlin C, Warren R, Klinger KW. (1993) Rapid prenatal diagnosis of chromosomal aneuploidies by fluorescence *in situ* hybridization: clinical experience with 4500 specimens. *Am. J. Hum. Genet.* **52**: 854–865.

Weber-Matthiesen K, Deerberg J, Muller-Hermelink A, Winkemann M, Schlegelberger B, Grote W. (1993) Rationalization of *in situ* hybridization: testing up to 16 different probes on a single slide. *Cancer Genet. Cytogenet.* **68**: 91–94.

Weier H-UG, Zitzelsberger HF, Gray JW. (1991a) Non-isotopical labelling of murine heterochromatin *in situ* by hybridization with *in vitro*-synthesized biotinylated *gamma* (major) satellite DNA. *Biotechniques* **10**: 498–505.

Weier H-UG, Lucas J, Poggensee M, Segraves R, Pinkel D, Gray JW. (1991b) Two-color hybridization with high complexity chromosome-specific probes and a degenerate *alpha* satellite probe DNA allows unambiguous discrimination between symmetrical and asymmetrical translocations. *Chromosoma* **100**: 371–376.

Wilkinson DG. (1992) In situ *Hybridization: a Practical Approach*. IRL Press, Oxford.

Willard HF, Waye JS. (1987) Hierarchical order in chromosome-specific human *alpha* satellite DNA. *Trends Genet.* **3**: 192–198.

Williams BJ, Ballenger CA, Malter HE, Bishop F, Tucker M, Zwingman TA, Hassold TJ. (1993) Non-disjunction in human sperm: results of fluorescence *in situ* hybridization studies using two and three probes. *Hum. Mol. Genet.* **2**: 1929–1936.

Wyrobek AJ, Alhborn T, Balhorn R, Stanker L, Pinkel D. (1990) Fluorescence *in situ* hybridization to Y chromosomes in decondensed human sperm nuclei. *Mol. Reprod. Dev.* **27**: 200–208.

Zhang L, Venkatesh P, Creek MLR, Smith MT. (1994) Detection of 1,2,4-benzenetriol induced aneuploidy and microtubule disruption by fluorescence *in situ* hybridization and immunocytochemistry. *Mutat. Res.* **320**: 315–327.

Zijno A, Marcon F, Leopardi P, Crebelli R. (1994) Simultaneous detection of X-chromosome loss and non-disjunction in cytokinesis-blocked human lymphocytes by *in situ* hybridization with a centromeric DNA probe; implications for the human lymphocyte *in vitro* micronucleus assay using cytochalasin B. *Mutagenesis* **9**: 225–232.

Measuring genetic events in transgenic animals

Roy Forster

15.1 Introduction

Transgenic animals have rapidly become familiar objects in the research environment even though it is little more than 15 years since the first demonstrations of the deliberate introduction and successful integration of exogenous genes into mammalian embryos. The development of transgenic technology has enabled us to manipulate the expression of genes *in vivo* and, more recently, to inactivate endogenous genes, as routine activities. Transgenic technology was rapidly adopted in a number of fields (notably developmental biology, pathology and immunology), where it has revolutionized experimental approaches to some fundamental problems. The techniques required to make transgenic animals are now well established (Pinkert, 1994), even if the generation of transgenics remains inefficient. The applications of transgenic animals have been extensively reviewed (First and Haseltine, 1991; Grosveld and Kollias, 1992; Lathe and Mullins, 1993).

15.2 Transgenic animals: current possibilities

A useful definition of transgenic animals is that they 'harbour transmissible new gene combinations'; this description is true of animals bearing exogenous DNA sequences, and also of 'knock-out' and 'gene replacement' animal strains. Most of the work reported in this chapter involves transgenics made by microinjection of DNA into fertilized eggs. The introduced transgene may have any configuration, but in most cases it is composed of regulatory DNA (which determines the pattern of expression), a structural gene and a polyadenylation sequence 'tail' (*Figure 15.1*).

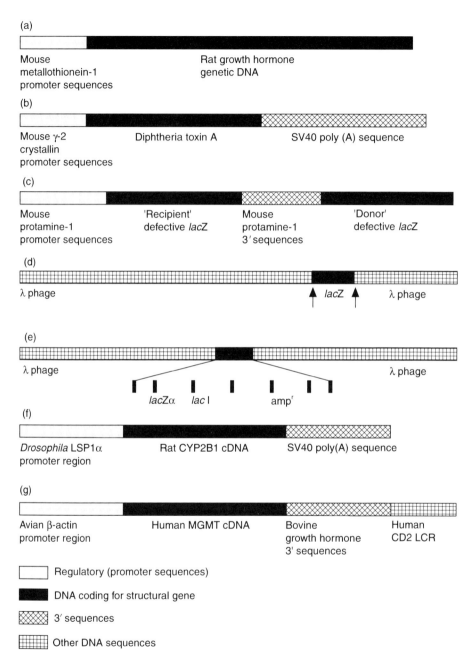

Figure 15.1. Schematic representation of some transgene constructs discussed in the text.(a) Mouse growth hormone (MGH) construct, 5.0 kbp, (b) γ-2 DTA construct, 2.98 kbp, (c) mP1-lacFin-COR construct, 7.5 kbp, (d) λ gt10-lacZ vector, ≈ 50 kbp, (e) lambda LIZα construct, circa 45 kbp, (f) α-CYP2B1 construct (inserted in *Carnegie 30* P-element), (g) MGMT construct, 3.93 kbp. Not drawn to scale.

The introduced transgene constitutes an addition to the genetic material of the host genome, thus allowing 'dominant' or 'gain of function' approaches. Expression of the transgene may result in dramatic phenotypic changes, as illustrated by the giant mice expressing growth hormone transgenes (Palmiter *et al.*, 1982; *Figure 15.1a*). By selection of appropriate regulatory sequences for the transgene construct, the pattern of expression in time and space may be controlled to specific developmental stages or specific tissues, or may be subordinated to an inducible genetic switch.

An example of the 'gain of function' approach is the use of reporter genes (for example, *lacZ*, *CAT*, *luciferase*), which code for easily identified or measured attributes, in order to follow gene expression. Placing the reporter under the control of regulatory DNA from, for example, heat shock genes (Kothary *et al.*, 1989) or cytochrome P450 isozymes (Jones *et al.*, 1991), allows study of the normal (and inducible) expression of these genes. Similar reporter constructs could be imagined to study the expression patterns associated with inducible responses of physiological (e.g. inflammation, proliferation), toxicological (e.g. acute phase response, apoptosis) or genetic relevance (e.g. DNA repair genes, oxidative defences).

'Gain of function' approaches can also be used in more refined ways: the directed expression of potent cytotoxins can be used to achieve the ablation of expressing tissues. Expression of the diphtheria toxin A chain in the crystalline lens of the eye (directed by γ-2-crystallin promoter sequences) in transgenic mice results in microphthalmia (Breitman *et al.*, 1987; *Figure 15.1b*). Similarly, 'gain of function' can be used to achieve dominant *negative* approaches by the introduction of ribozyme or antisense sequences or by the expression of a mutant gene product that interferes with the correct functioning of the endogenous non-mutant gene product (see, for example, Stacey *et al.*, 1988).

Most transgenic work has been performed with the mouse, for which there is extensive background information on genetics and reproductive biology. The techniques can be applied equally well to other species including the rabbit, rat, cow, goat, sheep, pig and some fish. The availability of transgenic rat models would clearly be of particular relevance for toxicology. The rat presents some technical difficulties (the fertilized eggs are reputed to be less amenable to microinjection than mouse eggs) but a number of transgenic rat models have been prepared and we may expect transgenic rats to become increasingly common.

Until recently, it was generally considered that the size of transgenes that could be effectively integrated was limited to about 50 kb. Studies demonstrating the integration of yeast artificial chromosome (YAC) constructs which potentially permit integration of more than 500 kb of DNA (Schedl *et al.*, 1992), and approaches based on the cointegration and recombination of large gene fragments (Strouboulis *et al.*, 1992) suggest that the size of the transgene is not a constraint.

There is currently great interest in techniques for the generation of 'knock-out' mice, in which a selected gene is deliberately inactivated, using homologous recombination to introduce interrupting sequences. In contrast to microinjection transgenics, knock-out techniques allow 'loss of function' approaches, and also permit subsequent 'gene correction' and 'double replacement' approaches. These techniques are based on the genetic manipulation of embryonic stem cell (ES cell) cultures, which are subsequently used to prepare chimaeric embryos, and hence genetically modified mice (Bradley *et al.*, 1992; Doi *et al.*, 1992).

15.3 Role of transgenic assays in mutagenicity testing

Since almost by definition transgenic animals create new experimental approaches in whole animals, this technology would seem exquisitely adapted to the *in vivo* science of toxicology, and even more so to genetic toxicology. By comparison with immunologists or developmental biologists, however, researchers in toxicology and genetic toxicology have been relatively slow to exploit the possibilities offered by this new approach. Nonetheless, transgenic models have been developed for the measurement of mutation induction in whole animals, and this topic has been the subject of review articles (Mirsalis *et al.*, 1994) and monographs (Short, 1994).

Over recent years the view has emerged that *in vivo* mutagenicity assays play a critical role in the assessment of the risk posed by mutagenic agents. While sensitive *in vitro* assays serve to demonstrate a potential mutagenic hazard, it is held that studies in whole animals permit the evaluation of risk. Only in whole animals can the influence of factors such as absorption, distribution, metabolism, elimination and cell turnover be established. Also, in contrast to *in vitro* assays, in whole animals there are physiological limitations on the highest dose levels that can be administered. The International Programme on Chemical Safety (IPCS) *in vivo* collaborative study played an important role in the development of this view, which is now also recognized in some national mutagenicity testing guidelines (COM, 1989).

There is, however, a relatively limited repertoire of routine whole-animal mutagenicity assays (see *Table 15.1*). *In vivo* assays such as the micronucleus test (chromosomal damage) (Chapter 8), alkaline elution, *in vivo* unscheduledDNA synthesis (UDS) (Chapter 12) and ^{32}P-post-labelling (DNA damage/repair)

Table 15.1. Some *in vivo* mutagenicity assays[a]

Genetic endpoint	Somatic tissues	Germ cells
Point mutation	Spot test	Specific-locus test
	HGRPT mutation in lymphocytes	
	Granuloma pouch assay	
Chromosomal aberrations	Micronucleus test	Dominant lethal test
	Bone marrow chromosomal aberrations	
DNA damage, adducts and repair	*In vivo* UDS	Spermatid UDS
	In vivo SCE assay	
	Alkaline elution	
	^{32}P-post-labelling	

[a] See also Chapters 8, 10, 12, 17 and 18.

(Chapter 18) can be performed on a routine basis, but there is a paucity of routine assays for gene mutation (in both somatic and germline cells). In principle, methods such as the granuloma pouch assay, the specific locus test, the spot test, and hypoxanthine–guanine phosphoribosyl transferase (HPRT) mutation in circulating lymphocytes are available to study this endpoint. In reality, because of prohibitive costs, insufficient validation or lack of robustness, none of these assays has become a reference method for the routine detection of gene mutations *in vivo*. A further weakness, which applies to all genetic endpoints, is that for most of these assays only a restricted tissue range is available for study (consisting principally of lymphocytes, erythrocytes, bone marrow and liver).

The promise of novel transgenic mutagenicity assays for all genetic endpoints but in particular for the detection of point mutations, is therefore attractive. In addition, methods that permit the study of a wide range of tissues (both proliferating and non-proliferating) would further enrich the capabilities of genetic toxicology.

15.4 Transgenic mutation assay systems

Although much attention has focused on the two mutation systems which have been made widely available on a commercial basis (see Sections 15.4.3 and 15.4.4), the broad potential of transgene technology to contribute to mutation research can be illustrated by reference also to other initiatives in this area.

15.4.1 Gene conversion

A transgenic model for the detection of intrachromosomal gene conversion events in male germ cells has been described by Murti *et al.* (1992, 1994). The construct is targeted to the male germinal tissue by the protamine-1 promoter, which drives one of two tandemly duplicated defective copies of *lacZ* bearing different mutations (*Figure 15.1c*). Gene conversion can result in the rescue and expression of *lacZ*, and hence the production of β-galactosidase in elongated spermatids (detected by histochemistry or by flow cytometry). It is assumed that expression of *lacZ* in spermatids does not result in any selective disadvantage. Using this system, Murti *et al.* were able to measure the spontaneous gene conversion frequency, and the induced conversion frequencies following treatment with chlorambucil or acrylamide (with a 21 day expression period). Results were confirmed using two transgenic lines, in which both the construct and the site of integration differed.

While much remains to be done in the development of this proposed system, some advantages of the assay emerge. Unlike shuttle-based methods, with this approach there is no ambiguity that mutation fixation occurs in the host animal. The method is rapid and results can be obtained on the day that the animals are killed. In principle, the method could be refined in order to distinguish between meiotic and mitotic gene conversion events.

There has also been much interest in the use of transgenes to study the role of gene conversion, somatic recombination and mutation events in the generation of antibody diversity by somatic hypermutation, and to characterize the *cis*-acting elements involved. These studies generally rely either on the analysis of hybridomas or direct DNA sequencing, and would not readily lend themselves to the measurement of chemically induced genetic events. Matsuoka *et al.* (1991) developed a reporter transgene construct in order to assess the extent of somatic recombination events during the development of the mouse brain. The construct contained an inverted *lac*Z gene, flanked by two recombination signal sequences (RSS), under the control of the chicken β-actin regulatory sequences (which direct ubiquitous expression). Although non-random β-galactosidase activity was detected in the brains of mice from two transgenic lines, the findings were contested by Abeliovich *et al.*(1992), who claimed that the expression of *lac*Z could result from backward transcription of the *lac*Z gene directed by downstream promoter sequences.

15.4.2 Detection of aneuploidy

Natarajan *et al.* (1990) have used transgene sequences as chromosomal markers that can easily be detected with fluorescent probes in order to quantify somatic cell non-disjunction following treatment with X-rays. Two transgenic lines were used: C57BL/6 mice bearing 10 copies of a 47 kb λ virus transgene on chromosome 2, and NMRI mice bearing 50 copies of a c-*myc* transgene on chromosome 8. Probing was performed with isolated nuclei from cell suspensions prepared from spleen, kidney and bone marrow. This permitted demonstration of a dose-related increase in aneuploid cells following X-ray treatment. In this system the transgenes provide a convenient DNA insert which can be detected readily in those tissues amenable to cytological preparations.

15.4.3 Gene mutation assays: lacZ shuttle vector systems

The generation of transgenic mice bearing a bacterial *lac*Z sequence as a mutation target in recoverable (λ) phage vectors offers an approach to study gene mutations *in vivo*. This approach was initially followed by two groups (Gossen *et al.*, 1989; Kohler *et al.*, 1990), although the subsequent development of *lac*Z models has been principally undertaken by researchers in The Netherlands in collaboration with Hazleton Laboratories.

Gossen *et al.* (1989) placed the bacterial *lac*Z sequence in a λgt10 phage vector (*Figure 15.1d*), and generated four transgenic lines in CD2F1 mice (BALB/c × DBA/2 F$_1$): line 20.2 (copy number about 80), line 34.1 (copy number about 8), line 35.5 (copy number about 3) and line 40.6 (copy number about 40). Homozygous animals contain twice as many copies of the transgene target. The *lac*Z target for mutation is 3.1 kb, while the λgt10 construct has a total length of about 47 kb. The phage can be recovered from the DNA of these animals, purified and 'packaged' *in vitro*. Mutation in the *lac*Z sequence is detected by the loss of blue coloration of phage plaques on an *Escherichia coli* lawn in the presence of Xgal.

Strain 40.6 was subsequently commercialized by Hazleton Laboratories as Muta™Mouse, and a number of publications make reference to the strain under this name. Publications from Gossen and others continue to make reference to strain 40.6. These animals were found to be free of rearrangements at the integration site, and the transgene is transmitted in Mendelian fashion. The construct does not contain promoter sequences upstream of *lacZ*, and the gene should not be expressed *in vivo*; no disruption of normal physiology or development has been noted (Myhr, 1991).

Integrated vectors from the liver were completely methylated at CCGG sites (Gossen *et al.*, 1989). *In vivo* methylation of *lacZ* transgenes was found to result in bacterial restriction during recovery and thus to reduce the rescue efficiency of λ. Appropriate restriction-deficient *E. coli* strains were developed to circumvent this problem for *lacZ* (*E. coli* C) and for *lacI* (*E. coli* SCS-8) transgenic mutation systems, and have been subsequently described (Gossen and Vijg, 1988; Kohler *et al.*, 1990).

The measurement of induced mutations in these animals was illustrated by treatment with ethylnitrosourea (ENU). An increased mutation rate was detected in the brain and liver, 7 days after treatment, but not on the first day after treatment. This was taken to suggest that the observed mutations resulted from processing of DNA lesions and mutation fixation in the host mouse and not in the *E. coli* used for recovery of the λ phage. The results from the brain were also taken as a demonstration of mutation in post-mitotic tissue (Gossen *et al.*, 1989). The general utility of the Muta™Mouse in genetic toxicology has been illustrated by data presented by Myhr (1991) and extended by Hoorn *et al.* (1993) with ENU, chlorambucil, cyclophosphamide, procarbazine, acrylamide and 7,12-dimethylbenz[*a*]anthracene.

The original scoring system for this assay involved the counting of (rare) colourless *lacZ⁻* λ plaques against a background of blue *lacZ⁺* plaques (following plating on a lawn of *E. coli* C in the presence of Xgal). Since the scoring of colourless plaques on a background of blue plaques was laborious and difficult, an improved system has been developed to select for *lacZ⁻* λ. This selective system is based on the toxicity of galactose (the product of metabolism by β-galactosidase, and hence produced in the presence of *lacZ⁺* λ) for *E. coli* which bear a deficiency in *gal*E. Using an appropriate strain of *gal*E bacteria, rapid and convenient selection against *lacZ⁺* phage and for *lacZ⁻* phage can be achieved (Dean and Myhr, 1994; Gossen *et al.*, 1993a; Mientjes *et al.*, 1994; Tinwell *et al.*, 1994). This system promises to be widely adopted.

A second *lacZ*-based mutation system has been described, although publication has been fragmentary, based on the recovery of a *lacZ*-containing plasmid which is integrated into mouse DNA (Gossen *et al.*, 1993b, 1994). This system takes advantage of the binding between *lac* repressor protein and the operator sequence upstream of *lacZ*. Repressor protein bound to magnetic beads is used to achieve a high recovery efficiency of the *lacZ*-bearing plasmid, which can then be released and electroporated into host *E. coli* for the determination of mutant numbers. The single-step recovery is also claimed to reduce the time required for the mutation assay.

The use of *lac*Z transgenic mutation models has recently been reviewed (Gossen *et al.*, 1994).

15.4.4 Gene mutation assays: lacI shuttle vector systems

Following the initial development of transgenic *lac*Z mutation systems, Kohler *et al.* (1991a,b) introduced a refined transgenic mutation assay based on the use of *lac*I as the mutational target.

The *lac*I gene codes for a homotetrameric repressor protein which binds to the operator sequence (*lac*O) regulating transcription of the *lac*Z gene. In the absence of a functional *lac*I gene, *lac*Z is expressed producing the enzyme β-galactosidase, which can be detected following incubation with the chromogenic substrate Xgal, by a blue precipitate.

A shuttle vector λ-LIZα has been prepared containing these various elements within the λ phage (Kohler *et al.*, 1991a,b; *Figure 15.1e*). A *lac*Z fragment (α-*lac*Z), consisting of the amino-terminal 675 nucleotides of *lac*Z, produces a truncated protein which can complement the carboxyl-terminal part of β-galactosidase, produced by the *E. coli* host (ω-*lac*Z) to give working enzyme activity. Since a functional *lac*I is present, expression of the genes within the phage vector inside the bacterial hosts results in colourless plaques on Xgal media. When mutations occur either in the *lac*I gene (to produce an inactive protein) or in the *lac*O operator sequence (to prevent effective binding of the *lac*I protein) transcription of *lac*Z will occur, resulting in blue plaques on Xgal media.

This mutation system offers several potential advantages. Mutant plaques are detected as blue plaques against a background of colourless plaques, rendering detection easy. The target sequences for mutation (*lac*I and *lac*O) are small, resulting in easy sequencing of mutations. The construct permits ready extraction of the target sequences by recovery as a plasmid which can be packaged individually in a 'phagemid' (Provost *et al.*, 1993). Furthermore, the mutational spectrum of the *lac*I gene has been extensively studied.

A number of transgenic founder lines were generated with the λ-LIZα construct, using the inbred C57BL/6 mouse strain (Dycaico *et al.*, 1994). The choice of an inbred background reduces variability in offspring and permits the production of hybrid strains for the study of transgene mutation in different genetic backgrounds. Transgenic line 'A1' was selected for further development and is now available commercially as the BigBlue™ transgenic assay system. This transgenic line contains 40 copies of the transgene construct at a single integration site on chromosome 4, probably near the *brown* locus (Dycaico *et al.*, 1994). The transgene is transmitted in Mendelian fashion. BigBlue™ mice are available as C57BL/6 or as hybrid B6C3F1 mice, prepared by crossing transgenic C57BL/6 with CBA3 mice.

The mutation assay is performed by treatment of transgenic tester mice with the test agent, followed by an appropriate expression period (discussed later) before killing. After killing, genomic DNA is recovered from the tissues selected for study. The DNA is mixed with *in vitro* λ phage packaging extract, which packages the phage as single units, recognizing and cleaving the *cos* sites at each end. The phage stock is adsorbed to restriction-deficient *E. coli* SCS-8, mixed with Xgal-containing top agar and plated. The number of blue plaques per total plaque-forming units (pfu) gives a measure of mutation frequency.

Methods have been described to simplify and facilitate further the determination of mutant numbers using genetic selection techniques (Lundberg *et al.*, 1993). Selection is based on the use of a host bacterial strain which expresses the λ *cI*+ protein, hence inducing λ to establish a lysogenic state rather than to enter a lytic cycle. Where the λ contains a mutant *lacI* gene, the *lacZ*-α fragment is expressed, complements the omega portion expressed by the bacterial host, and allows the lysogenic host bacteria to grow on minimal medium containing lactose. Where *lacI* is intact (non-mutant), it represses the expression of *lacZ*-α, resulting in lysogenic bacteria which are unable to grow on minimal medium containing lactose. The *lacI* system would lend itself to the development of several other selective systems (Dycaico *et al.*, 1994).

The range of genetic events which the *lacI* transgenic mutation system can detect, as indicated by the spectrum of spontaneous mutations, includes base-pair substitutions, single base frame-shifts and small deletions. The majority (87%) of spontaneous mutations result from base substitutions (Kohler *et al.*, 1991b). It has been suggested that large deletions will also be detected in this system, although findings with X-rays indicate that the efficiency of detection is low (Tao *et al.*, 1993b). Similarly, Suzuki *et al.* (1994) were unable to detect *lacZ* mutations in bone marrow at dose levels of mitomycin-C that were strongly clastogenic. Nevertheless, Winegar *et al.* (1994) have reported that up to 20% of the recovered mutants contained deletions, following treatment of *lacI* mice with ionizing (γ) radiation.

Spontaneous mutation rates in the A1-derived transgenic strains are in the order of 10^{-6} for germinal tissues and 10^{-5} for somatic tissues (Kohler *et al.*, 1991a,b), regardless of C57BL/6 or B6C3F1 background. These values are comparable to data derived from the *lacZ* transgenic mutation assays. The rather lower spontaneous mutation frequency observed in the ΦX174 system (described later) probably results from the fact that it is a reversion assay with a discrete mutation target. When the lactose agar selection system is used with the *lacI* assay, slightly higher spontaneous mutation frequencies are obtained (Provost *et al.*, 1993) and the range of mutation events efficiently detected appears to be shifted (Dycaico *et al.*, 1994).

The use of the *lacI* transgenic mutation system has recently been reviewed (Provost *et al.*, 1993). A clear and structured review of the genetics and practical conduct of *lacI* mutation assays has been published by Mirsalis *et al.* (1994). Preliminary data have been presented on a *lacI* transgenic rat model, and an F_1 mouse strain bearing the λ-LIZα transgene in a p53-deficient background is under study (Dycaico *et al.*, 1994).

15.4.5 Gene mutation assays: ΦX174 shuttle vector systems

A further interesting shuttle vector system has been described by Malling and coworkers (Malling and Burkhart, 1989; Burkhart *et al.*, 1993). Transgenic C57BL/6 mice have been produced with about 50 stably integrated copies of bacteriophage ΦX174 at a single insertion site. The phage contains an *am3* stop mutation revertible at an A:T base pair, hence permitting a reversion assay. The phage may be recovered by restriction of the mouse DNA and packaged in restriction-deficient bacteria. The mutation frequency is measured as pfu (in *sup⁻ E. coli* C) per total recovered phage (in *sup⁺ E. coli* CQ2).

The spontaneous reversion rate at *am3* was determined in these animals as $2–4 \times 10^{-7}$. Fourteen days after treatment with 200 mg kg^{-1} ENU, increases in the mutation frequency were detected in the spleen (10-fold) and liver (six-fold) but not in kidney or brain. These data were taken to illustrate the role of cell proliferation in mutation fixation (since kidney and brain are largely non-dividing tissues), and as a demonstration of alkylation-induced mutation in a non-expressing gene. This system has a number of advantages: the DNA sequence of ΦX174 has no sequence analogy with *E. coli* or mammalian DNA, hence eliminating the possibility of mutations resulting from recombination. *In vitro* experiments have demonstrated that the bacterial host used for recovery of the phage DNA is unable to convert alkylthymine adducts into mutations at the *am3* site, thus indicating that the mutations observed in this system derive from mutation fixation within the host transgenic animal.

15.4.6 The application of shuttle vector systems

A growing, but as yet limited, number of chemicals have been studied using the *lacZ* and *lacI* transgenic mutation assays (for reviews see Mirsalis *et al.*, 1994;

Table 15.2. Some compounds tested in the *lacZ* transgenic mutation assay

Test agent	Negative	Positive	Reference
Ethylnitrosourea		Liver, bm, testes	
Chlorambucil		Liver, bm, testes	
Procarbazine		bm	
Cyclophosphamide		bm	
Acrylamide		bm	
Dimethylbenz[a]anthracene		bm, skin	Hoorn *et al.* (1993)
MNNG		Skin	
Acetic acid		Skin	Myhr (1991)
1,3-Butadiene	bm, liver	Lung	Recio *et al.* (1992)
Mitomycin-C	Liver	bm	Suzuki *et al.* (1993)
Diethylnitrosamine	bm	Liver	
Ethylmethanesulphonate		bm, liver	
Benzene	Liver, bm		Suzuki *et al.* (1994)

bm, bone marrow.

Table 15.3. Some compounds tested in the *lac*I transgenic mutation assay

Test agent	Negative	Positive	Reference
2-Acetylaminofluorene		Liver	
Di-(2-ethylhexyl)-phthalate	Liver		
Heptachlor	Liver		
Phenobarbital	Liver		Gunz *et al.* (1993)
Benzo[*a*]pyrene		Spleen	
Ethylnitrosourea		Spleen	
Cyclophosphamide		Spleen	Kohler *et al.* (1991a)
Methylnitrosourea		Spleen, liver	Kohler *et al.* (1991b)
Dimethylnitrosamine		Liver	
Methylmethanesulphonate	Liver		Mirsalis *et al.* (1993)
Methylmethanesulphonate		Small intestine	Tao *et al.* (1993a)
Streptozotocin		Liver, kidney	Schmezer *et al.* (1994)

Morrison and Ashby, 1994, *Tables 15.2* and *15.3*). Some studies have also used model mutagens to compare transgene mutation induction with other genetic endpoints (Suzuki *et al.*, 1993, 1994) or with mutation induction in endogenous genes (Tao *et al.*, 1993a,b).

The findings obtained with chemical mutagens generally confirm the premises of these transgenic mouse assays. Some of the results obtained, however, do not easily fit into current ideas about the mutagenic profiles of individual mutagens and will necessitate further interpretation or study (Ashby and Tinwell, 1994). Methylmethanesulphonate, for example, although active in the *in vivo* UDS assay in mouse liver, fails to induce *lac*I mutations in BigBlue™ mouse liver (Mirsalis *et al.*, 1993). On the other hand, both the clastogen acrylamide and the deletogen chlorambucil were effective in inducing *lac*Z mutations in the Muta™Mouse assay (Hoorn *et al.*, 1993).

One point which clearly emerges from the application of these systems is the critical nature of the treatment regimen and the expression time in the determination of the quantitative results obtained. The expression period reflects the time required for a series of events to occur, including the uptake, distribution and metabolism of the test agent, the formation of a DNA adduct or lesion, and at least one cell division during which mutation fixation may occur. Since in these mutation systems, the transgene is not expressed *in vivo*, there is no delay resulting from protein turnover. The available data suggest that tissue replication rates are important in determining optimum expression period. The expression pattern will also be influenced by whether the mutant cells are derived from stem cells or daughter cells (Mirsalis *et al.*, 1994; Tao and Heddle, 1994).

It is, in any case, established that the optimum expression period for mutation induction can vary between tissues. Thus, mutant frequencies in the liver following treatment with dimethylnitrosamine increased over 7, 14 and 21 day harvest times (Mirsalis *et al.*, 1993) while mutant frequencies in the small intestine induced by ENU were stable over 8 weeks (Tao *et al.*, 1993b). Male germ cells provide a critical case; the entire period of the spermatogenic cycle must pass for the formation of spermatozoa from spermatogonial cells.

The induced mutation rate in germ cells of *lac*I mice following treatment with ENU was markedly increased when the expression period was extended from 3 days to 10 weeks (Provost and Short, 1994). This finding appears to go some way to answering previous concerns about the sensitivity of the assay relative to specific locus test measures of germ cell mutation (Burkhart and Malling, 1994; Malling and Burkhart, 1992). Some concern nevertheless remains that available transgenic mutation assays may not match the sensitivity achieved with previously studied assays. Since optimal testing protocols for transgenic assays are not yet established, the comparison may not be entirely fair. This consideration is, however, central in the evaluation of transgenic assays for the routine mutagenicity testing of chemicals, and must remain under review.

There is at present no general agreement on the optimum treatment regimens for transgenic mutation assays. Much early work was conducted using single-dose or 3–5 day repeat-dose protocols. Since the multiples of increase (over spontaneous) that were achieved were often modest while the dose levels used were high, this may have contributed to an impression that the transgenic assays were relatively insensitive. There is more recent evidence that the sensitivity of the assays is substantially increased by chronic dosing protocols. Under the described conditions it compares favourably with the sensitivity of other *in vivo* assays such as the micronucleus test (Shephard *et al.*, 1994). In this context, Tao and Heddle have presented data showing the additivity of mutations at neutral loci (based on results with three model mutagens at the *dlb-1* locus). They propose that repeated dose regimens, as used in routine toxicology studies, may provide the most appropriate and sensitive testing protocols (Tao and Heddle, 1994; see also Chapter 8). This suggestion appears to be supported by published data using repeated dosing protocols in transgenic mutation assays with dimethylnitrosamine (Mirsalis *et al.*, 1993) and 2-acetylaminofluorene (Shephard *et al.*, 1993). Concern has been expressed that chronic dosing protocols may offer scope for apparent increases in mutation rates deriving from increased cell replication in target tissues and hence non-genotoxic mechanisms (Ashby and Liegibel, 1992, 1993; Mirsalis, 1993). The available evidence does not appear to support this idea; no increase in *lac*I mutations was observed following multiple dosing with carbon tetrachloride for 5 days (Mirsalis, 1993) or with heptachlor or phenobarbital for 4 months (Gunz *et al.*, 1993). As far as it is compatible with practical and economic factors, there would therefore seem to be a strong case for defined chronic dosing protocols.

It is clear that criteria for the appropriate selection of expression times and treatment regimen will be critical for the development of standardized testing protocols and routine use of these assays. At present no such standardized approach exists, and to arrive at a routine assay method other questions will also have to be addressed. These will include the criteria for the selection of dose-levels, the choice of tissues for routine examination, and evaluative criteria for a positive result (Suzuki *et al.*, 1994). It will also be necessary to standardize *in vitro* aspects of the assay, such as plaque density, agar medium composition, incubation time, etc. (Dycaico *et al.*, 1994) and to develop appropriate quality control checks for the transgenic tester strains.

The question of the statistical sensitivity of the assays and the extent of replication required has already been addressed by some investigators. A standard seems to be emerging that at least 200 000–250 000 plaques should be screened per animal, with five or six animals per treatment group, depending on the spontaneous rate for the tissue under study (Douglas *et al.*, 1994; Dycaico *et al.*, 1994). Furthermore, a working group has examined the sources of variability in these assays and made recommendations regarding the design of standard testing assays (Gorelick and Thompson, 1994; Piegorsch *et al.*, 1994).

15.4.7 Advantages and disadvantages of shuttle vector systems

The transgenic shuttle vector systems described provide new and worthwhile advances, offering *in vivo* mutation assays which can be used with almost any tissue (including somatic and germ cells: but see Lewis, 1994). Mutations of the transgene loci appear to be neutral and do not confer a selective growth advantage or disadvantage to cells *in vivo*. The shuttle systems readily permit sequencing of the mutational events. The *lac*-based systems currently developed are forward mutation systems, able to detect a wide range of mutational events, and presenting credible spontaneous mutation frequencies.

A primary problem for these assays is that of relevance. There are numerous *a priori* considerations that may qualify the relevance of the findings from these assays, including the following: integration site, transgene structure and gene expression.

(i) *Integration site.* It is well established that the site of chromosomal integration can have marked effects on the expression of transgenes. It has further been demonstrated that mutational response can be influenced by integration site: Gossen *et al.* (1991) reported that strain 35.5 *lacZ* transgenic mice (see earlier), in which the transgene was located on the X chromosome, showed a spontaneous mutation rate 25- to 100-fold greater than two other *lacZ* transgenic lines (Gossen *et al.*, 1991). This was attributed to integration in an unstable chromosomal region. It is also worth noting that the consequences for the host chromosome (and its mutational responses) of integrating more than a megabase of repetitive foreign DNA are essentially unknown.

(ii) *Transgene structure.* Where transgene targets are composed of prokaryotic DNA, and are present in multiple tandem repeat copies, they may offer an unusual substrate to the eukaryotic DNA repair apparatus. Accordingly, the transgene may be subject to mutational mechanisms (particularly recombinational) which are not relevant to endogenous genes with normal eukaryotic structure. The age-related increase in small *lacI* duplications (less than 20 bp) in BigBlue™ mice found by Lee *et al.* (1994) is a possible example.

(iii) *Gene expression.* The *lacZ* and *lacI* transgenes are extensively methylated and both constructs lack promoter sequences for expression. It is therefore assumed that they are not expressed *in vivo.* Caution may be needed in extrapolating mutation data obtained in these transgenic mutation assays to actively transcribing genes. Could factors like this (and the integration site) contribute to the apparent lack of sensitivity of transgenic mutation assays discussed earlier?

This leaves the question of how to evaluate the relevance of mutation data derived from a multicopy insert of a methylated and unexpressed prokaryotic gene, integrated randomly into a chromosomal site of unknown significance. One approach to this problem is provided by examination of the mutational sequence spectra. These can be used to demonstrate that the spontaneous mutations observed in transgenic assays are representative of spontaneous mutations in endogenous genes, and that induced mutations reflect anticipated mutational responses (Kohler *et al.*, 1991a; Provost and Short, 1994) (see also Chapter 17). Furthermore, such studies may demonstrate differences between the spectra observed and those expected from bacterial mutational mechanisms, providing additional evidence that mutation fixation occurs in the host animal and not in the bacterial host for phage recovery (Kohler *et al.*, 1991a).

The interpretation of sequence spectrum data has been complicated by recognition that mutational spectra vary between tissues (Douglas *et al.*, 1994; Knöll *et al.*, 1994), and may be markedly influenced by the nature of the genetic mutation and selection system (Dycaico *et al.*, 1994; Gordon and Holliday, 1994). This is illustrated by the study of Douglas *et al.* (1994). The sequence spectrum of 66 spontaneous *lacZ* mutants derived from 10 male animals from strain 40.6 was studied. There was no significant difference in the spontaneous mutation rate in liver, bone marrow or male germinal tissue, which was estimated at $2.23/10^5$ pfu. The predominant mutation observed was GC → AT transition: however, there were significant differences in the base-pair substitution spectra observed for liver and for bone marrow. In addition, the proportions of transitions to transversions varied between tissues. Few deletions were observed. These findings differed from *lacZ* sequence data obtained from the analysis of 35 mutant plaques isolated from strain 35.5 (Gossen *et al.*, 1993c). Mutational spectra thus show differences between tissues, between transgenic strains (i.e. according to integration site) and, on the basis of *in vitro* data, between cDNA and genomic DNA constructs (Douglas *et al.*, 1994).

A more complex view is thus emerging in which endogenous mutational spectra are modulated by a series of factors which may include the genetic background, the nature of the target gene, the transgene integration site, the tissue and the genetic selection system. Recent publications have emphasized that only limited conclusions may be drawn from mutation sequence spectra data, and that extensive data will be needed to address the concerns raised earlier (Douglas *et al.*, 1994; Knöll *et al.*, 1994).

To return to the question of the relevance of transgenic mutation assays, the problem was addressed in a different way in two studies by Heddle and collaborators. The induction of point mutations in the small intestine by methyl methane sulphonate (Tao *et al.*, 1993a) and ENU (Tao *et al.*, 1993b) was compared for the endogenous *dlb-1* locus and the *lac*I transgene (in the same animals). Broadly similar results were obtained at the two loci. In contrast, X-ray treatment was effective in inducing *dlb-1* mutations but not *lac*I mutations, suggesting that deletions are less readily detected by the *lac*I transgenic mutation system (Tao *et al.*, 1993b). The comparable findings at the two loci are encouraging, but nevertheless the authors emphasized the need for "a large body of comparative data ... to establish the validity of transgenic assays".

15.4.8 Related studies

The transgenic models described earlier for the study of point mutations all depend on recoverable shuttle vectors and none of the models discussed permits *in situ* detection of mutations. An approach to *in situ* mutation studies that could be readily applied using transgenic constructs has been illustrated by Winton *et al.* (1988). In animals that are heterozygous for the DBA lectin-binding gene *dlb-1*, mutant cells lacking binding activity can be detected by histochemical methods since they do not stain and remain clear. This technique can be applied to the intestinal villus. Since intestinal crypt formation begins at birth, if mutations at the *dlb*-1 locus occur in stem cells before birth, whole crypts or clusters of mutant crypts can be detected. If, however, mutation occurs after birth, then a ribbon of mutant cells can be visualized in the villus. This approach permitted Winton *et al.* to measure spontaneous mutation rates and chemically induced mutation produced by direct-acting alkylating agents and compounds requiring metabolic activation (Brooks *et al.*, 1994; Winton *et al.*, 1988, 1990). Mutation at this endogenous gene has been compared with transgene mutation (Tao *et al.*, 1993a,b). The principle of the assay, analogous to the specific locus test, can be extended to any heterozygous marker that causes a staining difference, especially if it can be visualized as stained mutant cells against a clear wild-type background.

15.5 Genetic background

The success of the Ames test as a mutagenicity assay was determined not only by the carefully selected revertible *his* mutations, but also by the genetic background of the tester strains. Features of the genetic background conferred greater mutability (pKM101), disabled excision repair (*uvr*A) and greater permeability to bulky molecules (*rfa*) on the tester strains. In the same way, transgenic methods in animals permit not only the development of mutation assay systems, but also the introduction of other useful genetic changes.

The manipulation of xenobiotic metabolism is an obvious area for intervention, which is currently being explored. The principle has been illustrated in the fruit fly *Drosophila*. Transgenic *Drosophila* were generated which expressed rat CYP2B1 during the larval stages (*Figure 15.1f*). This *Drosophila* strain was hypersensitive to the genetic effects of cyclophosphamide (activated by CYP2B1), as measured by the SMART assay (Jowett *et al.*, 1991). A further transgenic *Drosophila* strain expressed dog *CYP1A1* under the control of an inducible promoter (the hsp 70 promoter). Following induction, this strain showed greater sensitivity to 7,12-dimethylbenz[*a*]anthracene (activated by the CYP1A1 isozyme) in a DNA repair differential survival assay (Komori *et al.*, 1993). The rat *CYP2B2* gene has been expressed in transgenic mice both constitutively and inducibly (Ramsden *et al.*, 1993).

Such results raise the attractive notion of specific transgenic rodent strains expressing (or lacking) different cytochrome P450 subtypes or receptors (for induction). These are factors which should play a central role in explaining some significant species differences in chemical toxicity. The presence of endogenous host (rodent) cytochrome P450 and specific tissue distribution of these enzymes renders this problem more complex but not insoluble, particularly using recent gene knock-out and replacement strategies.

Other gene products playing a role in the body's defences against chemicals may be subject to manipulation. In mammals a number of antioxidant agents and enzymes are involved in defence against oxidative damage, amongst which are the superoxide dismutases (SOD). It has been shown in transgenic animals that overexpression of SOD can protect against damage caused by reactive oxygen species: in transgenic mice expressing human Cu/Zn SOD, protection has been demonstrated against a series of insults involving reactive oxygen species, including *N*-methyl-4-phenyl-1,2,3,6-tetrahydropyridine-induced neurotoxicity (Przedborski *et al.*, 1992), cerebral ischaemic damage (Kinouchi *et al.*, 1991) and brain cold-trauma injury (Chan *et al.*, 1991). Fibroblasts derived from these animals are more resistant to the toxicity of paraquat, and fewer thymine glycols can be measured in the fibroblast DNA following paraquat treatment than in non-transgenic control fibroblasts (Huang *et al.*, 1992). Similarly, transgenic mice expressing human Mn SOD were highly protected against lung damage induced by 95% oxygen treatment (Wispe *et al.*, 1992), and *Drosophila* overexpressing bovine Cu/Zn SOD are more resistant to the toxic effects of paraquat (Fleming *et al.*, 1992). These various mouse and *Drosophila* strains would offer an interesting possibility to study the effects of the modulation of SOD levels on genetic damage induced by reactive oxygen species. We may expect studies of this kind to be extended to other defensive proteins (as in knock-out mice, deficient in metallothionein; Michalska and Choo, 1993).

An elegant series of studies has been performed with alkyltransferase DNA repair enzymes (*O*⁶-alkylguanine-DNA-alkyltransferase), which remove alkyl groups from alkylated guanine residues (reviewed in Gerson *et al.*, 1994). Transgenic mice generated with the bacterial *ada* gene under the control of the metallothionein promoter were shown to express the *ada* transgene in the liver and had increased enzyme activity (Maksukama *et al.*, 1989). This transgenic strain was subsequently shown to be resistant to *N*-nitroso-dimethylamine induced hepatic tumours as compared with non-transgenics (Nakatsura *et al.*, 1993). Transgenic mice generated with the bacterial *ada* gene under the control of the rat phosphoenolpyruvate carboxykinase (PEPCK) promoter expressed *ada* in the liver and kidney (Lim *et al.*, 1990). Three-fold increased repair of methylnitrosourea-induced O^6-methylguanine adducts was demonstrated in these animals compared with non-transgenic controls (Dumenco *et al.*, 1991). Transgenic mice bearing the human methylguanine DNA methyltransferase (MGMT) alkyltransferase gene under the control of the mouse metallothionein promoter showed increased enzyme levels in a range of tissues (Fan *et al.*, 1990). Finally, transgenics bearing the *MGMT* gene under the control of the β-actin promoter, and linked with the lcr (locus control region) of the *CD2* gene (*Figure 15.1g*) showed very high expression levels (50–80 times the endogenous levels) in the thymus and spleen; these animals were strongly protected against the induction of thymic lymphomas provoked by the intraperitoneal administration of 50 mg kg^{-1} *N*-methyl-*N*-nitrosourea (Dumenco *et al.*, 1993). The incidence of these tumours was 4% in transgenics, compared with 58% in non-transgenic mice.

Limited studies have been performed with the manipulation of other DNA repair enzymes. In *Drosophila*, overexpression of rat DNA polymerase-β was studied in three different transgenic strains: overproduction of this DNA polymerase did not influence survival of transgenic *Drosophila* treated with ultraviolet radiation, methyl methanesulphonate (MMS) or mitomycin-C compared with isogenic non-transgenic strains. DNA polymerase-β apparently did not play a role in DNA repair of damage caused by these agents (Yoo *et al.*, 1994). In mice, homozygosity for targeted deletion of part of the DNA polymerase-β gene resulted in embryonic lethality (Gu *et al.*, 1994).

The various studies described in this section serve to illustrate that the use of specific transgenic lines to dissect the mechanisms of *in vivo* genetic damage is a current reality and not a future possibility. Where it is of interest, tests for induced genetic effects may therefore be performed directly on such modified strains. Furthermore, by appropriate crosses, animal lines can be prepared with multiple transgenes, permitting the use of transgene mutation assays in strains which are 'diagnostic' for particular toxicological mechanisms.

15.6 General considerations regarding transgenic models

Some general comments should be made that are of relevance to transgenic models.

15.6.1 Replication

Individual transgenic lines are each to be considered as unique. This is determined by factors relating to the (random) integration site, which may strongly influence the level of expression of the transgene, as well as its spatial and temporal pattern. Furthermore, transgenes may be subject to rearrangements on integration, or may inactivate a host gene resulting in an insertion mutation (either silent or resulting in phenotypic change). Copy number and the construct 'configuration' may also play a role in determining expression levels and expression patterns. A more detailed discussion of these factors may be found elsewhere (Grosveld and Kollias, 1992). In view of these considerations, it has become customary to confirm that experimental results obtained with transgenic constructs are representative by presenting results obtained with several independent transgenic lines. The need for replication is borne out by the variability that may be seen between transgenic lines.

This problem is equally applicable to transgenic mutation models. Preliminary data with three *lac*I transgenic lines suggests that similar spontaneous mutation rates were obtained in each (Moores *et al.*, 1992). Gossen *et al.* (1989), on the other hand, found that one transgenic line (of four *lac*Z lines generated) produced a spontaneous mutation rate which was elevated by 25- to 100-fold (Gossen *et al.*, 1991). There is little further published information to confirm the 'representative' nature of the Muta™Mouse and BigBlue™ transgenic lines. In the gene conversion and aneuploidy induction models described earlier, two or more transgenic lines were used (Murti *et al.*, 1994; Natarajan *et al.*, 1990).

15.6.2 Stability

In the general literature on transgenics information regarding the long-term stability of transgenic lines is limited, other than investigations of the host methylation of transgenes. Various mechanisms have been described which act to regulate transgenes, including transgene mutation (Trudel *et al.*, 1994), transgene amplification (Beeri *et al.*, 1994), DNA rearrangement and recombination-mediated deletions (Sandgren *et al.*, 1991); the proviso should be made that in all the cases cited, a selective advantage or disadvantage of the transgene can be hypothesized. Nevertheless, mechanisms of this kind might be expected to have an impact on the mutational responses elicited in transgenes. Detailed information on the stability and continuing integrity of transgenic constructs would therefore be desirable.

15.6.3 Pleiotropic effects

Transgene expression frequently results in pleiotropic effects, often unanticipated. Such effects may be direct results of the action of a gene product. Alternatively they may result from subtler interactions, such as competition for transcription factors, substrates or cellular mechanisms (Allison *et al.*, 1991). An unexpected consequence of oncogene expression (v-Ha-*ras*, c-*myc* and c-*neu*) in transgenic mice was a reduction in survival following treatment with reserpine compared with non-transgenic littermates (Tennant *et al.*, 1993). Unexpected interference of this kind with mechanisms involved in the response to toxic insults raises further questions for the extrapolation of safety data from studies with transgenic animals.

15.6.4 Reporter genes

In transgenic models where *lacZ* is expressed *in vivo*, there are several unexplained features of its behaviour. These include unpredictable expression patterns attributed to *cis*-acting interference with regulatory sequences (Paldi *et al.*, 1993), patchy clonal expression in tissues, and generally poor levels of expression in adult animals (Cui *et al.*, 1994). While this may not be immediately relevant to present transgenic mutation assays in which *lacZ* is not expressed *in vivo*, these unexplained findings nevertheless indicate caution in the use of *lacZ* in mutation assays where *in vivo* expression is required.

15.7 Future possibilities and conclusions

One feature of this work which distinguishes it from most other areas of genetic toxicology is a much greater emphasis on intellectual property rights and commercial exploitation. Both the Muta™Mouse and BigBlue™ transgenic mutation assays are made available on a commercial basis. This fact, together with the relatively high cost of materials and reagents for the assays, will undoubtedly influence the diffusion and development of transgenic mutation assays.

The potential of transgenic systems is clearly enormous, and we may look forward to the development of new and ingenious assay systems. Furthermore, we may expect to see animal models that can answer specific questions regarding toxicological mechanisms, or that through 'humanization' of particular genes, can provide experimental findings of greater relevance for extrapolation to man. I hope that this review has served to illustrate the rich store of future possibilities which transgenic approaches can offer for the measurement of genetic events in laboratory animals, and for enhancing the value and relevance of genetic toxicology.

References

Abeliovich A, Gerber D, Tanaka O, Katsuki M, Graybiel AM, Tonegawa S. (1992) Somatic recombination in the central nervous system of transgenic mice (Letter). *Science* 257: 404–408.

Allison J, Malcolm L, Culvenor J, Bartholomeusz RK, Holmberg K, Miller JFAP. (1991) Overexpression of beta-2-microglobulin in transgenic mouse islet beta cells results in defective insulin secretion. *Proc. Natl Acad. Sci. USA* 88: 2070–2074.

Ashby J, Liegibel U. (1992) Transgenic mouse mutation assays: potential for confusion of genotoxic and non-genotoxic carcinogenesis: a proposed solution. *Environ. Mol. Mutagen.* 20: 145–147.

Ashby J, Liegibel U. (1993) Dosing regimes for transgenic animal mutagenesis assays (Correspondence). *Environ. Mol. Mutagen.* 21: 120–121.

Ashby J, Tinwell H. (1994) Use of transgenic mouse lacI/Z mutation assays in genetic toxicology. *Mutagenesis* 9: 179–181.

Beeri R, Gnatt A, Lapidot-Lifson Y, Ginzberg D, Shani M, Soreq H, Soreq H. (1994) Testicular amplification and impaired transmission of human butyrylcholinesterase cDNA in transgenic mice. *Hum. Reprod.* 9: 284–292.

Bradley A, Hasty P, Davis A, Ramirez-Solis R. (1992) Modifying the mouse: design and desire. *Bio/Technology* **10**: 534–539.

Breitman ML, Clapoff S, Rossant J, Tsui L-C, Glode LM, Maxwell IH, Bernstein A. (1987) Genetic ablation: targeted expression of a toxin gene causes microphthalmia in transgenic mice. *Science* **238**: 1563–1565.

Brooks RA, Gooderham NJ, Zhao K, Edwards RJ, Howard LA, Boobis AR, Winton DJ. (1994) 2-Amino-1-methyl-6-phenylimidazo[4,5-b]pyridine is a potent mutagen in the mouse small intestine. *Cancer Res.* **54**: 1665–1671.

Burkhart JG, Burkhart BA, Sampson KS, Malling HV. (1993) ENU-induced mutagenesis at a single A:T base pair in transgenic mice containing ΦX174. *Mutat. Res.* **292**: 69–81.

Burkhart JG, Malling HV. (1994) Mutations among the living and the undead. *Mutat. Res.* **304**: 315–320.

Chan PH, Yang GY, Chen SF, Carlson E, Epstein CJ. (1991) Cold-induced brain edema and infarction are reduced in transgenic mice overexpressing CuZn superoxide dismutase. *Ann. Neurol.* **29**: 482–486.

COM: Committee on Mutagenicity. (1989) Guidelines for the testing of chemicals for mutagenicity. *Department of Health Report on Health and Social Subjects*, no. 35, HMSO, London.

Cui C, Wani MA, Wight D, Kopchick J, Stambrook PJ. (1994) Reporter genes in transgenic mice. *Transgen. Res.* **3**: 182–194.

Dean SW, Myhr B. (1994) Measurement of gene mutation in *vivo* using Muta™Mouse and positive selection for *lacZ*⁻ phage. *Mutagenesis* **9**: 183–185.

Doi S, Campbell C, Kuchterlapi R. (1992) Directed modification of genes by homologous recombination in mammalian cells. In: *Transgenic Animals* (eds F Grosveld, G. Kollias). Academic Press, New York.

Douglas GR, Gingerich JD, Gossen JA, Bartlett SA. (1994) Sequence spectra of spontaneous *lacZ* gene mutations in transgenic mouse somatic and germline tissues. *Mutagenesis* **9**: 451–458.

Dumenco LL, Arce C, Norton K, Yun J, Wagner T, Gerson SL. (1991) Enhanced repair of O⁶-methylguanine DNA adducts in the liver of transgenic mice expressing the *ada* gene. *Cancer Res.* **51**: 3391–3398.

Dumenco LL, Allay E, Norton K, Gerson SL. (1993) The prevention of thymic lymphomas in transgenic mice by human O⁶-alkylguanine–DNA alkyltransferase. *Science* **259**: 219–222.

Dycaico MJ, Provost GS, Kretz PL, Ransom SL, Moores JC, Short JM. (1994) The use of shuttle vectors for mutation analysis in transgenic mice and rats. *Mutat. Res.* **307**: 461–478.

Fan CY, Potter PM, Rafferty J, Watson AJ, Cawkwell L, Searle PF, O'Connor PJ, Margison GP. (1990) Expression of a human O⁶-alkylguanine–DNA-alkyltransferase cDNA in human cells and transgenic mice. *Nucleic Acids Res.* **18**: 5723–5727.

First NL, Haseltine FP. (1991) (eds) *Transgenic Animals*. Butterworth-Heinemann, Stoneham, MA.

Fleming JE, Reveillaud I, Niedzwiecki A. (1992) Role of oxidative stress in *Drosophila* aging. *Mutat. Res.* **275**: 267–279.

Gerson SL, Zaidi NH, Dumenco LL, Allay E, Fan CY, Liu L, O'Connor PJ. (1994) Alkyltransferase transgenic mice: probes of chemical carcinogenesis. *Mutat. Res.* **307**: 541–555.

Gordon A, Halliday J. (1994) Transgenic systems for *in vivo* mutational analysis (Letter to the Editor). *Mutat. Res.* **306**: 103–105.

Gorelick NJ. Thompson ED. (1994) Overview of the workshops on statistical analysis of mutation data from transgenic mice. *Environ. Mol. Mutagen.* **23**: 12–16.

Gossen JA, Vijg J. (1988) *E. coli C:* a convenient host strain for cloning highly methylated DNA. *Nucleic Acids Res.* **16**: 9343.

Gossen JA, de Leeuw WJF, Tan CHT, Zwarthof EC, Berends F, Lohman PHM, Knook DL, Vijg J. (1989) Efficient rescue of integrated shuttle vectors from transgenic mice: a model for studying mutations *in vivo*. *Proc. Natl Acad. Sci. USA* **86**: 7971–7975.

Gossen JA, de Leeuw WJF, Verwest A, Lohman PHM, Vijg J. (1991) High somatic mutation frequencies in a *lacZ* transgene integrated on the mouse X-chromosome. *Mutat. Res.* **250:** 423–429.

Gossen JA, de Leeuw WJF, Molijn AC, Vijg J. (1993a) A selective system for lacZ⁻ phage using a galactose sensitive *E. coli* host. *BioTechniques* **14:** 326–330.

Gossen JA, de Leeuw WJF, Molijn AC,Vijg J. (1993b) Plasmid rescue from transgenic mouse DNA using lacI repressor protein conjugated to magnetic beads. *BioTechniques* **14:** 624–629.

Gossen JA, de Leeuw WJF, Bakker AQ, Vijg J. (1993c) DNA sequence analysis of spontaneous mutations at a *LacZ* transgene integrated on the mouse X chromosome. *Mutagenesis* **8:** 243–247.

Gossen JA, de Leeuw WJF, Vijg J. (1994) LacZ transgenic mouse models: their application in genetic toxicology. *Mutat. Res.* **307:** 451–459.

Grosveld F, Kollias G. (1992) (eds) *Transgenic Animals.* Academic Press, New York.

Gu H, Marth JD, Orban PC, Mossmann H, Rajewsky K. (1994) Deletion of a DNA polymerase beta gene segment in T cells using cell type-specific gene targeting. *Science* **265:** 103–106.

Gunz D, Shephard SE, Lutz WK. (1993) Can non-genotoxic carcinogens be detected with the lacI transgenic mouse mutation system? *Environ. Mol. Mutagen.* **21:** 209–211.

Hoorn AJW, Custer LL, Myhr BC, Brusick D, Gossen J, Vijg J. (1993) Detection of chemical mutagens using Muta™Mouse: a transgenic mouse model. *Mutagenesis* **8:** 7–10.

Huang T-T, Carlson EJ, Leadon SA, Epstein CJ. (1992) Relationship of resistance to oxygen free radicals to CuZn-superoxide dismutase activity in transgenic, transfected and trisomic cells. *FASEB J.* **6:** 903–910.

Jones SN, Jones PG, Ibarguen H, Caskey CT, Craigen WJ. (1991) Induction of the CYP1A1 dioxin responsive enhancer in transgenic mice. *Nucleic Acids Res.* **19:** 6547–6551.

Jowett T, Wajidi MFF, Oxtoby E, Wolf CR. (1991) Mammalian genes expressed in *Drosophila:* a transgenic model for the study of mechanisms of chemical mutagenesis and metabolism. *EMBO J.* **10:** 1075–1081.

Kinouchi H, Epstein CJ, Mizui T, Carlson E, Chen SF, Chan PH. (1991) Attenuation of focal cerebral ischaemic injury in transgenic mice overexpressing CuZn superoxide dismutase. *Proc. Natl Acad. Sci. USA* **88:** 11158–11162.

Knöll A, Jacobson DP, Kretz PL, Lundberg KS, Short JM, Sommer SS. (1994) Spontaneous mutations in lacI containing lambda lysogens derived from transgenic mice: the observed patterns differ in liver and spleen. *Mutat. Res.* **311:** 57–67.

Kohler SW, Provost GS, Kretz PL, Dycaico MJ, Sorge JA, Short JM. (1990) Development of a short-term *in vivo* mutagenesis assay: the effects of methylation on the recovery of a lambda phage shuttle vector from transgenic mice. *Nucleic Acids Res.* **18:** 3007–3013.

Kohler SW, Provost GS, Fieck A, Kretz PL, Bullock WO, Sorge JA, Putnam DL, Short JM. (1991a) Spectra of spontaneous and induced mutations in the *lacI* gene in transgenic mice. *Proc. Natl Acad. Sci. USA* **88:** 7958–7962.

Kohler SW, Provost GS, Fieck A, Kretz PL, Bullock WO, Putnam DL, Sorge JA, Short JM. (1991b) Analysis of spontaneous and induced mutations in transgenic mice using a lambda ZAP/lacI shuttle vector. *Environ. Mol. Mutagen.* **18:** 316–321.

Komori M, Kitamura R, Fukuta H, Inoue H, Baba H, Yoshikawa K, Kamataki T. (1993) Transgenic *Drosophila* carrying mammalian cytochrome P-4501A1: an application to toxicology testing. *Carcinogenesis* 14: 1683–1688.

Kothary R, Clapoff S, Darling S, Perry MD, Moran LA, Rossant J. (1989) Inducible expression of an *hsp68-lacZ* hybrid gene in transgenic mice. *Development* **105:** 707–714.

Lathe R, Mullins JJ. (1993) Transgenic animals as models for human disease – report of an EC study group. *Transgen. Res.* **2:** 286–299.

Lee AT, DeSimone C, Cerami A, Bucala R. (1994) Comparative analysis of DNA mutations in *lacI* transgenic mice with age. *FASEB J.* **8:** 545–550.

Lewis SE. (1994) A consideration of the advantages and potential difficulties of the use of transgenic mice for the study of germinal mutations. *Mutat. Res.* **307:** 509–515.

Lim IK, Dumenco LL, Yun J, Donovan C, Warman B, Gorodetzkaya N, Wagner TE, Clapp DW, Hanson RW, Gerson SL. (1990) High level regulated expression of the chimeric *P*-enolpyruvate carboxykinase (GTP)-bacterial O^6-alkylguanine DNA alkyltransferase *(ada)* gene in transgenic mice. *Cancer Res.* **50:** 1701–1708.

Lundberg KS, Kretz PL, Provost GS, Short JM. (1993) The use of selection in the recovery of transgenic targets for mutation analysis. *Mutat. Res.* **301:** 99–105.

Malling HV, Burkhart JG. (1989) Use of ΦX174 as a shuttle vector for the study of *in vivo* mammalian mutagenesis. *Mutat. Res.* **212:** 11–21.

Malling HV, Burkhart JG. (1992) Comparison of mutation frequencies obtained using transgenes and specific-locus mutation systems in male mouse germ cells. *Mutat. Res.* **279:** 149–151.

Matsukama S, Nakatsuru Y, Nakagawa K, Utakoji T, Sugano H, Katoaka H, Sekiguchi M, Ishikawa T. (1989) Enhanced O^6-methylguanine–DNA-methyltransferase activity in transgenic mice containing an integrated *E. coli ada* repair gene. *Mutat. Res.* **218:** 197–206.

Matsuoka M, Nagawa F, Okazaki K, Kingsbury L, Yoshida K, Muller U, Larue DT, Winer JA, Sakano H. (1991) Detection of somatic recombination in the transgenic mouse brain. *Science* **254:** 81–86.

Michalska AE, Choo KHA. (1993) Targeting and germ-line transmission of a null mutation at the metallothionein I and II loci in mice. *Proc. Natl Acad. Sci. USA* **90:** 8088–8092.

Mientjes EJ, van Delft JHM, op't Hof BM, Gossen JA, Vijg J, Lohman PHM, Baan RA. (1994) An improved selection system for lambda-lacZ⁻ phages based on galactose sensitivity. *Transgen. Res.* **3:** 67–69.

Mirsalis JC. (1993). Dosing regimes for transgenic animal mutagenesis assays (Letter to the Editor). *Environ. Mol. Mutagen.* **21:** 118–119.

Mirsalis JC, Monforte JA, Winegar RA. (1994) Transgenic animal models for measuring mutations *in vivo. Crit. Rev. Toxicol.* **24(3):** 255-280.

Mirsalis JC, Provost GS, Matthews CD, Hamner RT, Schindler JE, O'Loughlin KG, MacGregor JT, Short JM. (1993) Induction of hepatic mutations in lacI transgenic mice. *Mutagenesis* **8:** 265–271.

Moores JC, Provost GS, Ransom SL, Hamner RT, Short JM. (1992) An analysis of spontaneous mutant frequencies for lambda/lacI transgenic mouse lineages varying in copy number and chromosomal postition. *Environ. Mol. Mutagen.* **19** (suppl.20)**:** 44 (abstract).

Morrison V, Ashby J. (1994) A preliminary evaluation of the performance of the Muta™Mouse (lacZ) and BigBlue™ (lacI) transgenic mouse mutation assays. *Mutagenesis* **9:** 367–394.

Murti JR, Bumbulis M, Schimenti JC. (1992) High frequency germ-line gene conversion in transgenic mice. *Mol. Cell. Biol.* **12:** 2545–2552.

Murti JR, Schimenti KJ, Schimenti JC. (1994) A recombination based transgenic mouse system for genotoxicity testing. *Mutat. Res.* **307:** 583–595.

Myhr BC. (1991) Validation studies with Muta™Mouse: a transgenic mouse model for detecting mutations *in vivo. Environ. Mol. Mutagen.* **18:** 308–315.

Nakatsuru Y, Matsukama S, Nemoto N, Sugano H, Sekiguchi M, Ishikawa T. (1993) O^6-Methylguanine–DNA methyltransferase protects against nitrosamine-induced hepatocarcinogenesis. *Proc. Natl Acad. Sci. USA* **90:** 6468–6472.

Natarajan AT, Vlasblom SE, Manca A, Lohman PHM, Gossen JA, Vijg J, Beermann F, Hummler E, Hansmann I. (1990) Transgenic Mouse – an *in vivo* system for detection of aneugens. In: *Mutation and the Environment: Progress in Clinical Biological Research,* **340:** Vol.B, pp. 295–299. Wiley-Liss, New York.

Paldi A, Deltour L, Jami J. (1993) Cis effect of lacZ sequences in transgenic mice. *Transgen. Res.* **2:** 325–329.

Palmiter RD, Brinster RL, Hammer RE, Trumbauer ME, Rosenfeld MG, Birnberg NC, Evans RM. (1982) Dramatic growth of mice that develop from eggs microinjected with metallothionein–growth hormone fusion genes. *Nature* **300:** 611–615.

Piegorsch WW, Lockhart A-MC, Margolin BH, Tindall KR, Gorelick NJ, Short JM, Carr GJ, Thompson ED, Shelby MD. (1994) Sources of variability in data from a *lacI* transgenic mouse mutation assay. *Environ. Mol. Mutagen.* **23:** 17–31.

Pinkert CA. (ed.). (1994) *Transgenic Animal Technology: a Laboratory Handbook.* Academic Press, New York.

Provost GS, Kretz PL, Hammer RT, Matthews CD, Rogers BJ, Lundberg KS, Dycaico MJ, Short JM. (1993) Transgenic systems for *in vivo* mutation analysis. *Mutat. Res.* **288:** 133–149.

Provost GS, Short JM. (1994) Characterisation of mutations induced by ethylnitrosourea in seminiferous tubule germ cells of transgenic B6C3F1 mice. *Proc. Natl Acad. Sci. USA* **91:** 6564–6568.

Przedborski S, Kostic V, Jackson-Lewis V, Naini AB, Simonetti S, Fahn S, Carson E, Epstein CJ, Cadet JL. (1992) Transgenic mice with increased Cu/Zn-superoxide dismutase activity are resistant to N-methyl-4-phenyl-1,2,3,6-tetrahydropyridine-induced neurotoxicity. *J. Neurosci.* **12:** 1658–1667.

Ramsden R, Sommer KM, Omiecinski CJ. (1993) Phenobarbital induction and tissue specific expression of the rat *CYP2B2* gene in transgenic mice. *J. Biol. Chem.* **268:** 21722–21726.

Recio L, Psterman-Golkar S, Csanady GA, Turner MJ, Myhr B, Moss O, Bond JA. (1992) Determination of mutagenicity in tissues of transgenic mice following exposure to 1,3-butadiene and N-ethyl-N-nitrosourea. *Toxicol. Appl. Pharmacol.* **117:** 58–64.

Sandgren EP, Palmiter RD, Heckel JL, Daughtery CC, Brinster RL, Degen JL. (1991) Complete hepatic regeneration after somatic deletion of an albumin-plasminogen activator transgene. *Cell* **66:** 245–256.

Schedl A, Montoliu L, Kelsey G, Schutz G. (1992) A yeast artificial chromosome covering the tyrosinase gene confers copy number dependent expression in transgenic mice. *Nature* **362:** 258–261.

Schmezer P, Eckert C, Liegibel UM. (1994) Tissue-specific induction of mutations by streptozotocin *in vivo. Mutat. Res.* **307:** 495–499.

Shephard SE, Sengstag C, Lutz WK, Schlatter C. (1993) Mutations in liver DNA of lacI transgenic mice (Big blue) following subchronic exposure to 2-acetylaminofluorene. *Mutat. Res.* **302:** 91–96.

Shephard SE, Lutz WK, Schlatter C. (1994) The lacI transgenic mouse mutagenicity assay: quantitative evaluation in comparison to tests for carcinogenicity and cytogenic damage *in vivo. Mutat. Res.* **306:** 119–128.

Short JM. (1994) Transgenic systems in mutagenesis and carcinogenesis. *Mutat. Res.* **307:** 427–595.

Stacey A, Bateman J, Choi T, Mascara T, Cole W, Jaenisch R. (1988) Perinatal lethal osteogenesis imperfecta in transgenic mice bearing an engineered mutant pro-α1(I) collagen gene. *Nature* **332:** 131–136.

Strouboulis J, Dillon N, Grosveld F. (1992) Efficient rejoining of large DNA fragments for transgenesis. *Nucleic Acids Res.* **20:** 6109–6110.

Suzuki T, Hayashi M, Sonufi T, Myhr BC. (1993) The concomitant detection of gene mutation and micronucleus induction by Mitomycin-C *in vivo* using *lacZ* transgenic mice. *Mutat. Res.* **285:** 219–224.

Suzuki T, Hayashi M, Sonufi T. (1994) Initial experiences and future directions for transgenic mouse mutation assays. *Mutat. Res.* **307:** 489–494.

Tao KS, Heddle JA. (1994) The accumulation and persistence of somatic mutations *in vivo. Mutagenesis* **9:** 187–191.

Tao KS, Urlando C, Heddle JA. (1993a) Mutagenicity of methyl methanesulphonate *in vivo* at the *Dlb-1* native locus and a *lacI* transgene. *Environ. Mol. Mutagen.* **22:** 293–296.

Tao KS, Urlando C, Heddle JA. (1993b) Comparison of somatic mutation in a transgenic versus host locus. *Proc. Natl Acad. Sci. USA* **90:** 10681–10685.

Tennant RW, Rao GN, Russfield A, Seilkop S, Braun AG. (1993) Chemical effects in transgenic mice bearing oncogenes expressed in mammary tissue. *Carcinogenesis* **14:** 29–35.

Tinwell H, Lefevre PA, Ashby J. (1994) Response of the Muta™Mouse lacZ/galE⁻ transgenic mutation assay to DMN: comparisons with the corresponding Big Blue™(*lacI*) responses. *Mutat. Res.* **307:** 169–173.

Trudel M, Chretien N, D'Agati V. (1994) Disappearance of polycystic kidney disease in revertant c-*myc* transgenic mice. *Mamm. Genome* **5:** 149–152.

Winegar RA, Lutze LH, Hamer JD, O'Loughlin KG, Mirsalis JC. (1994) Radiation induced point mutations, deletions and micronuclei in *lacI* transgenic mice. *Mutat. Res.* **307:** 479–487.

Winton DJ, Blount MA, Ponder BAJ. (1988) A clonal marker induced by mutation in mouse intestinal epithelium. *Nature* **333:** 463–466.

Winton DJ, Gooderham NJ, Boobis AR, Davies DS, Ponder BAJ. (1990) Mutagenesis of mouse intestine *in vivo* using the *Dlb-1* specific locus test: studies with 1,2-dimethyl-hydrazine, dimethyl-nitrosamine, and the dietary mutagen 2-amino-3,8-dimethyl-imidazo [4,5-f]quinoxaline. *Cancer Res.* **50:** 7992–7996.

Wispe JR, Warner BB, Clark JC, Dey CR, Neuman J, Glasser SW, Crapo JD, Chang L-Y, Whitsett JA. (1992) Human Mn-superoxide dismutase in pulmonary epithelial cells of transgenic mice confers protection from oxygen injury. *J. Biol. Chem.* **267:** 23937–23941.

Yoo MA, Lee WH, Ha HY, Ryu JR, Yamaguchi M, Fujikawa K, Matsukage A, Kondo S, Nishida Y. (1994) Effects of DNA polymerase beta gene overexpressed in transgenic *Drosophila* on DNA repair and recombination. *Jpn J. Genet.* **69:** 21–33.

The single cell gel/ Comet assay:

a microgel electrophoretic technique for the detection of DNA damage and repair in individual cells

Raymond R. Tice

16.1 Introduction

Techniques which permit the sensitive detection of DNA damage and repair are critically important to fields of toxicology ranging from ageing and clinical investigations to genetic toxicology and molecular epidemiology. Since the DNA damage induced by toxic agents is often tissue- and cell-type specific, an optimal technique would be one which can detect DNA damage and repair in individual cells obtained under a variety of experimental conditions. The three methods most commonly used for ascertaining DNA damage/repair involve the scoring of chromosomal aberrations, micronuclei, and/or sister chromatid exchanges in proliferating cell populations, the detection of DNA repair synthesis (so-called unscheduled DNA synthesis, UDS) in individual cells, and the detection of single-strand DNA breaks and alkali-labile sites in pooled cell populations (see Chapters 7, 8 and 12). While they provide information about damage in individual cells, cytogenetic techniques are of limited value because of the need for proliferating cell populations and because the DNA damage must be processed into microscopically visible lesions. The UDS technique is based on the excision repair of DNA lesions, as demonstrated by the incorporation of tritiated thymidine into DNA repair sites. While providing information at the level of the individual cell, the technique is technically cumbersome, requires the exposure of damaged cells to radioactivity, and is limited in sensitivity because not all DNA lesions are repaired with equal facility (see Tice and Setlow, 1985). Techniques such as alkaline elution or alkaline gel electrophoresis, which directly evaluate levels of DNA damage in pooled cell populations, circumvent many of the problems associated with the other two techniques. However, the use of pooled cells ignores the critical importance of intercellular differences in response and requires relatively large populations of cells.

Rydberg and Johanson (1978) were the first to quantitate directly DNA damage in individual cells by lysing the cells embedded in agarose on slides under mild alkaline conditions to allow the partial unwinding of DNA. After neutralization, the cells were stained with acridine orange and the extent of DNA damage quantitated by measuring the ratio of green (indicating double-stranded DNA) to red (indicating single-stranded DNA) fluorescence using a photometer. To improve the sensitivity for detecting DNA damage in individual cells, Ostling and Johanson (1984) developed a microgel electrophoresis technique. In this technique, commonly known as the Comet assay, cells embedded in an agarose gel were placed on a microscope slide, the cells lysed by detergents and high salt, and the liberated DNA electrophoresed under neutral conditions. Cells with increased damage display increased migration of DNA towards the anode. The migrating DNA was quantitated by staining with ethidium bromide and by measuring the intensity of fluorescence at two fixed positions within the migration pattern using a microscope photometer. However, as this assay can only detect the presence of DNA double-strand breaks, its utility has been limited to studies involving radiation and radiomimetic chemicals.

In 1988, Singh and coworkers introduced a microgel technique involving electrophoresis under alkaline conditions (pH >13) which was capable of detecting single-strand breaks and alkali-labile lesions in the DNA of individual cells. Since almost all genotoxic agents induce orders of magnitude more single-strand breaks and alkali-labile lesions than double-strand breaks, this alkaline assay offers greatly increased sensitivity for detecting induced DNA damage. This technique has been called the single cell gel (SCG) assay by Singh and his colleagues, although for historical reasons many investigators also refer to this method as the Comet assay. Subsequently, Olive and coworkers developed versions of the neutral technique of Ostling and Johanson (1984) which involved lysis in alkali followed by electrophoresis at either neutral (Olive *et al.*, 1990a) or mild alkaline (pH 12.3) conditions (Olive *et al.*, 1990b) to detect single-strand breaks. However, based on the respective ability of these three techniques to detect DNA damage induced by ionizing radiation, the technique of Singh *et al.* (1988) is at least one or two orders of magnitude more sensitive. Here, applications of the SCG/Comet assay for detecting DNA damage and repair in individual cells are reviewed and discussed. For an earlier review, see McKelvey-Martin *et al.* (1993)

16.2 The SCG/Comet assay methodology

While the underlying principles are identical, different microgel methodologies have been developed based either on the neutral assay of Ostling and Johanson

(1984) or the alkaline assay of Singh *et al.* (1988). In general, eukaryotic cells are mixed with molten low-melting agarose (approximately 37°C), the mixture is placed on a fully frosted microscope slide where it is allowed to harden under a coverslip, the cells are lysed using detergents and high concentrations of salt, and the liberated DNA is electrophoresed for a short time under neutral or alkaline conditions. After staining with a DNA-specific fluorochrome, individual cells are evaluated for DNA damage by fluorescence microscopy. Cells with increased DNA damage display increased migration from the nuclear region towards the anode (*Figure 16.1*).

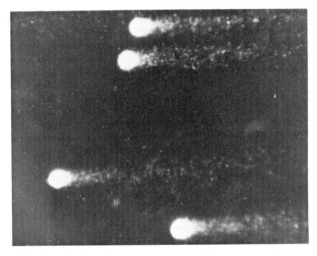

Figure 16.1. Black and white photomicrograph of human lymphocytes exhibiting DNA migration induced by ionizing radiation (×175 magnification).

The major technical variables affecting the sensitivity and resolving power of the assay include: (i) the concentration of low-melting agarose; (ii) the composition of the lysing solution; (iii) the composition and pH of the electrophoresis buffer; and (iv) electrophoretic conditions such as voltage, amperage and duration. Table 16.1 provides a summary of the major experimental conditions used when applying the neutral, pH 12.3, and pH >13 assays. As any combination of conditions (gel concentration, lysing solution, etc.) is possible and the number of variant assays is very large, only the main conditions for each assay are presented, for the sake of simplicity. Selection of which method to use should depend on the purpose of the study and the intent of the investigator. The length of unwinding (for alkaline studies) and the duration for electrophoresis are variables which depend on the cell type being investigated and the type of damage being assessed.

Other experimental variables include gel size, the DNA-specific dye used for visualization, the magnification used to examine the migrating DNA, and the method(s) used for data collection and analysis. These variables largely depend on specific needs of the investigator and presumably have little effect

Table 16.1. Major technical factors associated with the SCG/Comet assays

Technical factors	Neutral[a]	pH 12.3 alkaline[b]	pH >13 alkaline[c]
Concentration of low melting agarose	0.75%	0.75%	0.5%
Lysing solution	EDTA (0.025–0.03 M) SDS (0.5–2.5%)	NaCl (1.0 M) NaOH (0.03 M)	EDTA (100 mM) NaCl (2.5 M)
	Duration: generally >60 min	Duration: >60 min	1% N-lauroylsarcosine (10 mM) Tris (10 mM), pH 10.0 Triton X-100 (1%)
	Note: prokinase K (PK) has also been used to remove any residual protein	Note: N-lauroylsarcosine (0.5%) and EDTA (2 mM) have been added (Olive et al., 1992)	Duration: >60 min
			Note: Tice et al. (1990) added 10% DMSO to prevent DNA damage induced by radicals associated with iron when lysing samples containing erythrocytes. Under these conditions, slides can be stored in lysis solution for months. Some investigators have concluded that the N-lauroylsarcosine is unnecessary. PK has been added in some studies to remove any residual protein
Pre-electrophoresis DNA wash or unwinding	Boric acid (~90 mM) EDTA (2–5 mM) Tris (40–117 mM)	EDTA (1–2 mM) NaOH (0.03 M) pH 12.3	EDTA (1 mM) NaOH (300 mM) pH >13.0
	Duration: 2–16 h	Duration: 60 min	Duration: 20–60 min
	Note: acetic acid instead of boric acid has also been used (Olive et al., 1990b)		Note: Singh et al. (1994) changed this to 300 mM NaOH, 10 mM EDTA, 0.1% 8-hydroxyquinoline, 2% DMSO, pH >13.0
Electrophoresis	0.5–0.67 V cm^{-1}	0.5–0.67 V cm^{-1}	25 V (~0.8–1.5 V cm^{-1}, depending on gel box size), 300 mA
	Note: duration depends on voltage and amperage, but generally for 25 min	Note: duration depends on voltage, amperage, cell type, and damage, but generally for approx. 25 min	Note: duration depends on voltage, ampage, cell type, and extent of damage, but generally for from 10 to 40 min

EDTA, ethylenediaminetetraacetic acid; DMSO, dimethylsulphoxide; NaCl, sodium chloride; NaOH, sodium hydroxide; SDS, sodium dodecyl sulphate. Owing to the number of variations used by different investigators, only the conditions in general use are presented, as based on [a] Ostling and Johanson (1984); [b] Olive et al. (1990b); [c] Singh et al. (1988).

on assay sensitivity and resolving power. For example, magnification has varied from 160x to 400x, with 200–250x being used most commonly. The magnification that is most appropriate depends on the type of cell being evaluated, the range of migration responses to measure, and the constraints of the microscope and/or imaging system. Similarly, selection of a fluorescent dye depends to a large extent on limitations of the equipment and the manner in which data are collected. The dyes used most frequently are ethidium bromide (e.g. Ostling and Johanson, 1984; Singh et al., 1988), propidium iodide (e.g. Olive, 1989) and DAPI (e.g. Gedik et al., 1992). Recently, Singh et al. (1994) have reported that the use of YOYO-1 (benzoxazolium-4-quinolinum oxazole yellow homodimer) increases the sensitivity of the assay.

There are almost as many methods for quantifying DNA damage by this assay as there are scientists using the technique (Table 16.2). The most flexible approach for collecting SCG/Comet assay data involves the application of image analysis techniques to individual cells, and several commercially available software programs have been developed specifically for collecting such data. However, methods that do not rely on image analysis can be just as useful. The simplest method for collecting SCG/Comet data is based on determining the proportion of cells with damage (i.e. those exhibiting migration versus those without). However, this approach is generally limited to cell populations or electrophoretic conditions where the majority of control cells exhibit no DNA migration and fails to provide information about the extent of damage among damaged cells. A variation of this approach which addresses this latter limitation subclassifies damaged cells into ones having various degrees (e.g. none, short, medium, long) of DNA migration (e.g. Anderson et al., 1994; Gedik et al., 1992).

Table 16.2. Methods for quantifying DNA damage using the SCG/Comet assays

Endpoint	Description	Comments
% Damaged cells	Based on number of cells with tails vs. those without	Simplest method for collecting data. Does not depend on image analysis. Limited to large effects and provide no information on extent of damage
% Cells with no, short, medium, and long migration	Based on the number of cells with different size tails (short to long)	Migration length divided into arbitrary categories. Semiquantitative, probably not useful for detecting small effects
Migration length or ratio of length to width	Relates to smallest detectable fragments of DNA. Generally presented in μm	Image length and tail length should be distinguished. Criteria for identify the trailing and leading edge of the migrating DNA need to be formalized
% migrated DNA or ratio of tail DNA to head DNA	Relates to proportion of migrated DNA vs. non-migrated DNA; a measure of total damage	Assumes linearity for quantifying the amount of DNA in the tail and the head
Tail moment	Based on tail length x tail intensity or % migrated DNA. Proposed as a better parameter of DNA damage, based on data collected in neutral and pH 12.3 assays	Multiple methods for calculating makes comparisons among laboratories difficult. Use eliminates useful information on the relationship between the length of migration and the percentage of migrated DNA

The parameter most commonly used is the length of DNA migration, usually presented in micrometres. Migration length is related directly to fragment size, and would be expected to be proportional to the extent of DNA damage. This parameter has been measured using a variety of approaches, including the use of a micrometer in the microscope eyepiece, a ruler on photographic negatives/positives of cell images or a camera monitor, and the use of image analysis. Currently, the criteria used to identify the trailing and leading edge of the migrating DNA seem to depend on the investigator and/or software program. Furthermore, some investigators use the term 'DNA migration' to describe image length while others apply the term to migrated DNA only. A variant of this parameter is to present the ratio of the length to width (Jostes et al., 1993) or width to length (Fairbairn et al., 1993), with cells exhibiting no damage having a ratio of approximately 1. Olive et al. (1992) discounted the utility of DNA migration as a parameter for DNA damage in the neutral or pH 12.3 alkaline assays based on the observation that the length of DNA migration reached a plateau while the percentage of migrated DNA continued to increase. However, this limitation in migration length is not a characteristic of the pH >13 alkaline assay, where length has been reported to be the best parameter for this version of the assay (Vijayalaxmi et al., 1992).

As more investigators have begun using computerized image analysis systems to collect SCG/Comet data, a parameter based on the relative amount of migrated DNA, presented either as the percentage of migrated DNA (Olive et al., 1990a) or as the ratio of DNA in the tail to DNA in the head (Muller et al., 1994), has been used more frequently. This parameter assumes signal linearity in quantifying the amount of DNA ranging over multiple orders of magnitude and that the staining efficiency of the fluorescent dye is identical for migrated and non-migrated DNA. Both assumptions may be problematic (e.g. Olive et al., 1992).

The concept of tail moment (tail length × tail intensity or percentage migrated DNA) as a parameter for DNA migration was introduced by Olive et al. (1990a). However, a consensus among investigators as to the most appropriate manner in which to calculate tail moment has not been obtained, and the use of this parameter eliminates useful information on the relationship between the length of migration and the percentage of migrated DNA. Some agents (e.g. ionizing radiation) induce long, thin tails while others (e.g. cyclophosphamide) induce short, thick tails. Such information may provide insight into agent-specific differences in the intragenomic distribution of DNA damage within a cell.

Additional research is needed to clarify the appropriateness and limitations of various parameter under different experimental and SCG/Comet assay conditions. Until a general consensus is reached, publications using a derived parameter should also provide the original data on which it is based.

The statistical analysis of SCG/Comet data is another area of concern. A variety of parametric and non-parametric statistical methods has been used to analyse the resulting data, often without any formal evaluation of statistical propriety. The majority of analyses have been based on changes in group mean response with little attention being paid to the distribution of damage among cells. However, as these data are obtained at the level of the individual cell, ignoring such information greatly limits the sensitivity of the assay and interpretability of the results. This is especially important in situations (e.g. human biomonitoring) where a minority population of damaged cells among a majority of undamaged cells may be present. In addition, critical information on the various sources of assay variability (cell to cell, gel to gel, culture to culture, animal to animal, experiment to experiment, human to human, etc.) needs to be obtained in multiple laboratories to provide a basis for the scientific community to understand the value and limitations of the SCG/Comet assay.

To demonstrate the importance of being able to evaluate DNA damage on a cell-by-cell basis, the induction of DNA damage by γ radiation and bleomycin, a radiomimetic chemical, in human blood lymphocytes as measured by the pH >13 alkaline technique has been compared (Tice and Strauss, 1995). The experimental protocol for irradiation was as described in Vijayalaxmi *et al.* (1992), while that for treatment with bleomycin was as described by Kligerman *et al.* (1992). For both agents (*Figure 16.2*), the lowest dose tested (5 cGy for γ rays, 5 µg ml^{-1} for bleomycin) induced a significant increase in DNA migration. However, while the dose response for γ rays was linear, the dose–response for bleomycin reached a plateau at very low doses. Furthermore, while the distribution of damage among irradiated cells was relatively homogeneous, that for cells exposed to bleomycin was highly heterogeneous with only about 10–20% of the cells exhibiting an increase in DNA migration (*Figure 16.3*). These data indicate the importance of recording the distribution of damage as well as the mean response (see also Jostes *et al.*, 1993).

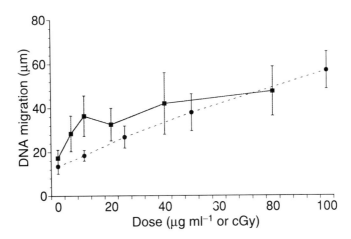

Figure 16.2. Mean DNA migration in human blood lymphocytes irradiated with γ-rays (solid squares) or exposed to bleomycin (solid circles). Data based on 50 cells scored per dose group. Error bars indicate standard error of the mean among cells. Reprinted from Tice and Strauss (1994) with permission from AlphaMed Press.

321

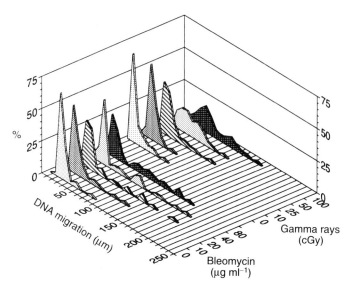

Figure 16.3. Distribution of DNA migration among human blood lymphocytes irradiated with γ-rays or exposed to bleomycin. Data based on 50 cells per dose group. Reprinted from Tice and Strauss (1994) with permission from AlphaMed Press.

16.3 Detection of DNA damage

Because DNA is maintained in a double-stranded form during electrophoresis under neutral conditions, the type of DNA lesions thus detected is limited to double-strand breaks. Double-strand breaks are reported to be the primary lesion responsible for cell lethality associated with exposure to ionizing radiation and a number of radiation studies have been conducted to evaluate the induction of such events in various types of cells (*Table 16.3*). The sensitivity of the neutral assay to detect double-strand breaks induced by sparsely ionizing radiation appears to be of the order of ≈2–3 Gy (Olive *et al.*, 1991) and depends on the stage of the cell cycle (e.g. Olive and Banath, 1993b), although Muller *et al.* (1994) have reported significantly increased DNA migration in human tumour cells at X-ray doses as low as 0.1 Gy.

In contrast, in the alkaline-based assays, the migrating DNA is single stranded which allows for the detection of single-strand breaks and any lesion capable of being transformed into a single-strand break at the alkaline pH used (i.e. alkali-labile sites). In principle, the ability to detect alkali-labile lesions depends on the pH of the electrophoretic buffer and the length of time allowed for expression before electrophoresis. For example, Singh *et al.* (1988), using 20 min of pH >13 alkali unwinding followed by 20 min of electrophoresis (also at pH >13), reported a significant increase in DNA migration in human lymphocytes induced by 25 cGy of X-rays. Subsequently, it was reported that increasing the pH >13 alkali time from 20 to 60 min, while maintaining a

Table 16.3. Agents evaluated *in vitro* for DNA damage using SCG/Comet assays

Agent	Cell type (response)	Reference
(a) *Neutral assay**		
Sparsely ionizing radiation	CHO Chinese hamster ovary cells (+)	Ostling and Johanson (1984, 1987), Olive *et al.* (1991)
	Human T lymphocytes (+)	Uzawa *et al.* (1994)
	L5178Y mouse lymphoma cells (+)	Olive *et al.* (1991)
	Mouse macrophages (+)	Olive *et al.* (1990a)
	SCCVII murine tumour cells (+)	Olive *et al.* (1990a)
	TK6 human lymphoblast cells (+)	Evans *et al.* (1993)
	Tumour cell lines (MeWo, PECA 4451, PECA 4197) (+)	Muller *et al.* (1994)
	V79 Chinese hamster lung cells (+)	Olive (1989), Olive *et al.* (1990a; 1991, 1992, 1993a), Olive and Banath (1993a)
Bleomycin	CHO Chinese hamster ovary cells (+)	Ostling and Johanson (1987)
	V79 Chinese hamster lung cells (+)	Olive and Banath (1993a)
Etoposide	V79 Chinese hamster lung cells (+)	Olive and Banath (1993a)
(b) *pH 12.3 alkaline assay*		
Sparsely ionizing radiation	V79 Chinese hamster lung cells (+)	Olive *et al.* (1992)
Ethylene oxide	Human fibroblasts	Nygren *et al.* (1994)
Etoposide	V79 Chinese hamster lung cells (+)	Olive and Banath (1992, 1993b); Olive *et al.* (1993a)
Hydrogen peroxide (also superoxide/ hydrogen peroxidegenerated by xanthine/ xanthine oxidase	Rat primary hepatocytes (+)	Higami *et al.* (1994)
	Human bladder-carcinoma cell lines (+)	Ward *et al.* (1993)
Nitrogen mustard	Chinese hamster lung cells (+)	Olive *et al.* (1992)
(c) *pH >13 alkaline assay*		
Sparsely ionizing radiation	CHO Chinese hamster ovary cells (+) A_L (+)	Jostes *et al.* (1993)
	Human fibroblasts (+)	Singh *et al.* (1991a)
	Human granulocytes (+)	Vijayalaxmi *et al.* (1993)
	Human leukocytes (+)	Vijayalaxmi *et al.* (1992), M. Green *et al.* (1994)
	Human lymphocytes (+)	Singh *et al.* (1988, 1990, 1994), Tice *et al.* (1990), Vijayalaxmi *et al.* (1992, 1993), Strauss *et al.* (1994), Tice and Strauss (1994)
Densely ionizing radiation	CHO Chinese hamster ovary cells $A_L^{(+)}$ (+)	Jostes *et al.* (1993)
UV	Human lymphocytes (+)	Tice *et al.* (1990)
UV-B	Human lymphocytes (normal, XP)(+)	Arlett *et al.* (1993)
	Human fibroblasts (+)	Arlett *et al.* (1993)
UV-C	HeLa cells (+ with aphidicolin) (+)	Gedik *et al.* (1992)
	Human lymphocytes (normal, XP) (+)	Green *et al.* (1992)
Solar radiation	Human lymphocytes (+)	Arlett *et al.* (1993)
2-Acetylamino- fluorene	Mouse hepatocytes (−) Rat hepatocytes (+)	Hirai *et al.* (1991)
4-Acetylamino- fluorene	Mouse hepatocytes (−) Rat hepatocytes (+)	Hirai *et al.* (1991)

Table 16.3 (Continued)

Agent	Cell type (response)	Reference
Bleomycin	Human lymphocytes (+)	Anderson *et al.* (1994), Tice and Strauss (1994)
Cadmium sulphate	Human leukocytes (+)	Hartmann and Speit (1994)
1-Chloromethylpyrene	Rat gastric mucosal cell (+)	Kennelly *et al.* (1993)
Cyclophosphamide	Mouse hepatocytes (+) Rat hepatocytes (+)	Hirai *et al.* (1991), Tice *et al.* (1991)
Dimethylmercury	Human lymphocytes (+) Rat lymphocytes (+) Rat gastric mucosa (+)	Betti *et al.* (1993)
Dimethylnitrosamine	Mouse hepatocytes (+) Rat hepatocytes (+)	Hirai *et al.* (1991)
Ethylmethanesulphonate	HUT-78 T-lymphocyte cell culture (+)	Shafer *et al.* (1994)
	Mouse hepatocytes (+) Rat hepatocytes (+)	Hirai *et al.* (1991)
Hydrogen peroxide	Bovine lens epithelial cells (+)	Kleiman and Spector (1993)
	HeLa (+)	Collins *et al.* (1993), O'Neill *et al.* (1993)
	HL-60 cell line (+)	Fairbairn *et al.* (1993)
	Human lymphocytes (+)	Singh *et al.* (1988, 1991a), Tice *et al.* (1990, 1991), Anderson *et al.* (1994)
	Mouse splenocytes (+)	Grigsby *et al.* (1993)
	Raji B cells (+)	Meyers *et al.* (1993)
Interleukin-1β	Rat islets of Langerhans cells (+) HIT-T15 cells (+)	Delaney *et al.* (1993), I. Green *et al.* (1994)
Lindane	Human lymphocytes (−) Rat gastric mucosa cells (+) Rat nasal mucosa cells (+)	Pool-Zobel *et al.* (1993a)
Manganese chloride	Human lymphocytes (+)	De Méo *et al.* (1991)
Methylmercury chloride	Human lymphocytes (+) Rat lymphocytes (+) Rat hepatocytes (+) Rat gastric mucosa (+)	Betti *et al.* (1993)
N-Methyl-*N*-nitro-*N*-nitrosoguanidine	Human lymphocytes (+) Rat lymphocytes (+) Rat hepatocytes (+) Rat gastric mucosa (+)	Pool-Zobel *et al.* (1993a), Betti *et al.* (1993)
	Rat gastric mucosa	Kennelly *et al.* (1993)
Morphine	HUT-78 T cell culture (+)	Shafer *et al.* (1994)
3-Morpholino-sydnonimine	Rat islets of Langerhans (−) HIT-T15 cells (−)	Delaney *et al.* (1993), I. Green *et al.* (1994)
Nitromonomethyl arginine	Rat islets of Langerhans (−) HIT-T15 cells (−)	Delaney *et al.* (1993)
S-Nitrosoglutathione	Rat islets of Langerhans (+)	Delaney *et al.* (1993)
Potassium permanganate	Human lymphocytes (+)	De Méo *et al.* (1991)
Sodium arsenite	Human leukocytes (+)	Hartmann and Speit (1994)
Vitamin C	Human lymphocytes (+)	Anderson *et al.* (1994), M. Green *et al.* (1994)
Vitamin E	Human lymphocytes (−)	Anderson *et al.* (1994)

+, significant response; −, no significant response; S-9, hepatic S-9 mix from Aroclor-treated male rats; XP, xeroderma pigmentosum.

* Based on neutral electrophoresis without prior exposure to alkali. All chemical studies were conducted in the absence of metabolic activation.

constant duration for electrophoresis, enhanced the sensitivity of the SCG assay to ionizing radiation doses as low as 5 cGy (Vijayalaxmi *et al.*, 1993). These data indicate that increased sensitivity for detecting alkali-labile damage can be obtained by increasing the length of the unwinding time before electrophoresis. These data also suggest that it may be possible to analyse multiple types of DNA damage by comparing the extent of DNA migration using different alkali-exposure times. It should be noted that Singh *et al.* (1994) recently reported several modifications to the original pH >13 alkaline technique that appear to have increased its sensitivity to the mGy range for ionizing radiation.

In addition to direct DNA damage, processes which introduce single-strand gaps in the DNA, such as incomplete excision repair events, are readily detectable (Gedik *et al.*, 1992; Green *et al.*, 1992; Tice *et al.*, 1990). The alkaline assay can also be used to detect DNA cross-linking, as demonstrated by a retardation in the extent of DNA migration (Olive *et al.*, 1992). Cross-linking, such as that induced by nitrogen mustard or *cis*-platinum, can be detected either directly by increasing the duration of electrophoresis to such an extent that the DNA of control cells exhibit significant migration (i.e. cross-linked DNA will migrate less than the DNA of control cells; Tice, unpublished data) or by using a second agent such as ionizing radiation to induce DNA damage and contrasting the extent of DNA migration in the presence and absence of this reference agent (Olive *et al.*, 1992).

An exciting approach for detecting specific classes of base damage has been introduced by Collins and coworkers. Gedik *et al.* (1992) demonstrated that the DNA of ultraviolet light (UV)-irradiated HeLa cells could be probed after the lysis step for pyrimidine dimers using a UV-damage-specific T4 endonuclease. The extent of DNA migration in irradiated cells was directly related to the number of recognized sites. From the data collected, these authors concluded that the detection limit of the pH >13 alkaline assay was as low as 0.1 DNA breaks per 10^9 Da. Subsequently, Collins *et al.* (1993) reported on the use of endonuclease III (EndoIII) to probe for oxidized pyrimidines, damage which would result from free radical attack following exposure to ionizing radiation or to other agents that release active oxygen species. The ability to detect various classes of DNA damage has important implications for DNA repair and human/environmental biomonitoring studies.

Increased DNA migration also accompanies the DNA fragmentation associated with cell death, arising through a non-DNA mediated process that may or may not involve apoptosis. For example, exposure of human lymphocytes to 60°C for up to 1 hour is accompanied by a loss in viability and an increase in DNA migration in the pH >13 alkaline assay (MacGregor *et al.*, 1994). In this study, the extent of DNA degradation increased with incubation time and was initiated before cell death as defined by a viability assay. Apoptosis (programmed cell death, e.g Carson *et al.*, 1986; Marks and Fox, 1991) results in the extensive formation of double-strand breaks and is readily detected using either neutral or alkaline electrophoretic conditions (Fairbairn *et al.*, 1994a; MacGregor *et al.*, 1994; Olive *et al.*, 1993c; Uzawa *et al.*, 1994). However, in contrast to the progressive increase in DNA migration associated with cell killing induced by heat, the extensive DNA degradation which accompanies apoptosis seems to occur quite quickly (Olive *et al.*, 1993c). The fact that apoptotic cells can be detected easily by both the neutral and the alkaline assays while the DNA damage arising from genotoxic chemicals is preferentially detected by the alkaline version makes the combined use of both assays diagnostic for apoptosis.

16.4 DNA repair studies

Because of its simplicity, the SCG/comet assay can be used to evaluate the ability of virtually any type of eukaryotic cell to repair different kinds of DNA damage, including double- and single-strand breaks, base damage and DNA cross-links.

16.4.1 Strand break repair

The neutral and alkaline assays have been used extensively to assess the repair of double-strand and single-strand breaks, respectively, in both normal and transformed cell populations. Single- and double-strand breaks induced by sparsely ionizing radiation are efficiently repaired in normal cells (e.g. G_0 human blood lymphocytes, Singh *et al.*, 1988, 1990, 1994; Tice and Strauss, 1995; human fibroblasts, Singh *et al.*, 1991a; mouse macrophages, Olive *et al.*, 1990a) and in most transformed cells (e.g. V79 and CHO cells, Olive *et al.*, 1991; SCCVII murine tumour cells, Olive *et al.*, 1990a; human tumour cell lines, Muller *et al.*, 1994). Fifty per cent of the damage is repaired within 15 min and complete repair occurs within 1–2 hours. This rate of repair has also been demonstrated for subsets of irradiated human lymphocytes, including B-, T-suppressor, and T-helper cells (Tice and Strauss, 1995).

To evaluate the effect of age of subjects on strand-break repair, Singh *et al.* (1990) employed the pH >13 method to measure DNA damage and repair in lymphocytes isolated from the peripheral blood of 31 healthy, non-smoking subjects (23 males and 8 females aged 25–91 years) and exposed *in vitro* to 2 Gy of X-irradiation. While basal (pre-irradiation) levels of damage were independent of the age of the donor, an age-dependent increase in DNA damage was observed immediately following irradiation. For all subjects, the mean level of DNA damage was restored to pre-irradiation control levels within 2 hours of incubation at 37°C. However, a distribution analysis of DNA damage among cells within each sample indicated the presence of a few highly damaged cells (4–16%) in the 2-hour sample, the occurrence of which was significantly more common among aged individuals. These data suggest an age-related decline in DNA repair competence among a small subpopulation of lymphocytes, and support the need for analysing distributions as well as means.

16.4.2 Excision repair

The alkaline assays can be used to evaluate the excision repair competence of eukaryotic cells. For most agents, however, the extent of DNA migration reflects a mixture of DNA single-strand breaks, alkali-labile lesions, and repair sites. Thus, a definitive determination of repair kinetics can be difficult. However, two methodologies can be used to evaluate specifically the kinetics of excision repair.

Firstly, the base damage induced by UV radiation is not detected as alkali-labile sites under the usual conditions employed in the alkaline assays. Thus, immediately after exposure, irradiated human lymphocytes fail to exhibit any increase in DNA migration. However, with incubation at 37°C, the irradiated lymphocytes exhibit increased migration associated with the presence of incomplete excision repair sites at the time of lysis (Gedik *et al.*, 1992; Green *et al.*, 1992; Tice *et al.*, 1990). This ability to detect excision repair sites using the alkaline assay can be enhanced by the inclusion of aphidicolin (Gedik *et al.*, 1992) or cytosine arabinoside (Andrews *et al.*, 1990) in the culture media. Aphidicolin inhibits DNA synthesis, resulting in the delayed completion of the repair sites, while cytosine arabinoside acts as a chain terminator, preventing the sites from being closed. Green *et al.* (1992) have used this response to UV to identify individuals affected with xeroderma pigmentosum, a genetic disease characterized by defects in excision repair (see Chapter 4).

Secondly, the enzymatic approach of Collins and coworkers can be used in the alkaline assays to evaluate the repair kinetics of various classes of DNA lesions in treated cells. The UV-damage-specific T4 endonuclease can be used to study the removal of UV-induced pyrimidine dimers (Gedik *et al.*, 1992), while EndoIII can be used to monitor the removal of oxidized pyrimidines (Collins *et al.*, 1993). Presumably, the repair of other classes of DNA damage can be monitored using other repair enzymes in a similar manner.

16.5 Genetic toxicology

An extremely useful application of the alkaline SCG techniques is in the area of genetic toxicology, and a number of investigators have used the pH >13 alkaline assay to evaluate the *in vitro* and/or *in vivo* genotoxicity of chemicals. As the *in vitro* SCG assays can be conducted using microculture techniques, the system is especially valuable where only limited amounts of the test chemical are available. Although a variety of normal and transformed cell types have been used for *in vitro* studies (and the number of possibilities is limited only by the number of cell types available), most genetic toxicology studies have used human lymphocytes, mouse lymphoma cells, CHO cells, and primary cultures of rodent hepatocytes (*Table 16.3*). An interesting aspect of the rodent hepatocyte system is the ability to measure DNA damage in parenchymal and non-parenchymal cells simultaneously (Hirai *et al.*, 1991). This ability permits a simple method for discriminating between direct-acting genotoxicants and those requiring metabolic activation.

An important contribution of the SCG/Comet assay to genetic toxicology is in its application to *in vivo* studies. As only a small numbers of cells are required for analysis, virtually any tissue or organ is amenable to investigation (*Table 16.4*). Investigators have used the assay to monitor age-related increases in basal levels of DNA damage in hepatocytes of ageing rats (Higami *et al.*, 1994) and the induction and persistence of DNA damage in somatic cells (e.g. Tice *et al.*, 1991; Pool-Zobel *et al.*, 1992) and germ cells (e.g. Croom *et al.*, 1991; Friend *et al.*, 1993) of chemically treated male and female rodents. The assay has proved especially valuable in situations where UDS has failed to detect the presence of DNA damage, such as that induced by 1-chloromethylpyrene in rat gastric mucosal cells (Kennelly *et al.*, 1993). As shown in *Table 16.4*, this technique has been applied to a number of rodent tissues, including blood, bone marrow, brain, gastrointestinal mucosa, kidney, liver, lung, nasal mucosa, ovaries, skin, spleen and testis. The only requirement is that a sufficient number of single cells is obtained for analysis without induction of damage or without allowing repair processes to proceed. Although collagenase or trypsin has been used to obtain suspensions of single cells from solid tissues (e.g. Betti *et al.*, 1993; Pool-Zobel *et al.*, 1992), mincing a tissue sample in a small volume of cold Hank's balanced salt solution containing 20 mM EDTA with a pair of fine scissors for a few minutes has provided sufficient cells for analysis (Tice *et al.*, 1991). To demonstrate this approach, the extent of DNA damage induced by acrylamide in various tissues of mice is presented in *Figure 16.4*. For this analysis, tissues were collected from male B6C3F1 mice (10–13 weeks of age, 25–32 g body weight, four mice per group) exposed once by gavage to 10 or 100 mg kg^{-1} acrylamide in phosphate-buffered saline and sampled 12 hours later. DNA migration data were collected separately for parenchymal and non-parenchymal cells of the liver and for round spermatids and spermatocytes of the testis. These cells can be distinguished in the gel by their image diameters. A tissue-specific difference in the magnitude of DNA damage induced by acrylamide is evident, indicating possible differences in distribution, cell sensitivity, and/or persistence of DNA damage. The distribution of damage among cells (*Figure 16.5*) provides information on the relative heterogeneity of the damage among cells in different tissues and indicates the proportion of damaged to undamaged cells for each dose and tissue.

16.6 Environmental biomonitoring

In addition to its value in laboratory studies, the SCG/Comet assay is becoming a major tool for environmental biomonitoring. Adverse exposure situations resulting from the improper disposal of hazardous wastes are generally identified by analytical techniques characterizing the levels of known pollutants. However, these techniques do not provide insight into the biological hazards associated with complex mixtures as they interact or are acted upon by various environmental pathways. One approach for assessing the possible environmental

Table 16.4. Agents evaluated *in vivo* for DNA damage using the pH >13 alkaline SCG/Comet assays

Agent (species)	Blad.	Blood leuk.	Bone marr.	Brain	GI tract	Kidney	Liver	Lung	Nasal epithel.	Ovary	Skin	Spleen	Testis	Reference
	Tissue													
Acrylamide (mouse)	+	+	+	+			+	+	+	+	+	+	+	Tice et al. (1990, 1991), Croom et al. (1991), Friend et al. (1993)
Benzene (mouse)		+	+											Plappert et al. (1994a,b)
1-Chloromethylpyrene (rat)					+		+							Kennelly et al.(1993)
Dimethylnitrosamine (mouse)		+								+				Croom et al. (1991)
Ethylmethanesulphonate (mouse)		+								+				Croom et al. (1991)
Fly ash (coal) (mouse)	−CD +CS	+CD +CS					−CD +CS	−CD −CS			−CD −CS			Andrews et al. (1994)
Lindane (rat)		−			+				+					Pool-Zobel et al. (1993a)
Methylmercury chloride (rat)					−									Betti et al. (1993)
N-Methyl-N-nitro-N-nitroso-guanidine (rat)		−			+									Betti et al. (1993), Pool-Zobel et al. (1993a, b)
N-Nitrosodimethylamine (rat)		+							+					Pool-Zobel et al. (1992)
4-(N-methyl-N-nitrosoamino)-1-(3-pyridyl)-1-butanone (rat)		−							+					Pool-Zobel et al. (1992)
Sodium arsenite (mouse)	−CD +CS	−CD −CS					−CD +CS	−CD −CS			+CD −CS			Tice et al. (1994), Yager et al. (1994)
Streptozotocin (mouse)						+	+							Schmezer et al. (1994)
Toluene (mouse)	−	−	−				−							Plappert et al. (1994b)

+, significant response; −, no significant response; blad., bladder; epithel., epithelial; GI, gastrointestinal; leuk., leukocytes; marr., marrow; CD, choline-deficient diet; CS, choline-sufficient diet.

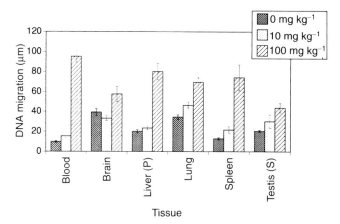

Figure 16.4. Mean DNA migration for cells sampled from various tissues of male B6C3F1 mice exposed once by gavage to acrylamide at 10 or 100 mg kg^{-1} and sampled 12 h later. Controls were treated with phosphate-buffered saline only. Each treatment group consisted of four mice. P, parenchymal; S, spermatocytes.

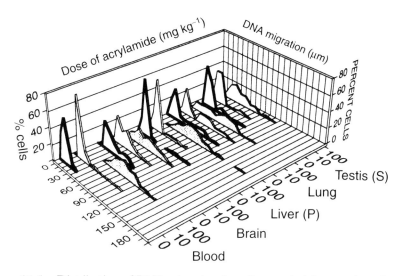

Figure 16.5. Distribution of DNA migration for cells sampled from various tissues of male B6C3F1 mice exposed once by gavage to acrylamide at 10 or 100 mg kg^{-1} and sampled 12 h later. See *Figure 16.4* for other details. Reprinted from MacGregor *et al.* (1995) with permission from Academic Press.

consequences of hazardous waste pollution involves the assessment of genotoxic damage (and other effects) in sentinel organisms. The SCG assay, again because of its simplicity, sensitivity and need for only small numbers of cells, has been suggested as an ideal technique for such studies (Tice, 1995).

To monitor for genotoxic pollutants at a hazardous waste site, Nascimbeni *et al.* (1991) conducted a pilot study using the pH >13 SCG assay to assess the extent of DNA damage in tissues (blood, bone marrow, brain and liver) of the golden mouse, *Ochrotomys nuttalli*, live-trapped at a Superfund site in North Carolina, USA, that was contaminated with a number of pollutants including trichloroethylene, chloroform, carbon tetrachloride, pesticides and laboratory solvents. Animals with similar population demographics collected at three control sites were used as the corresponding control population. The level of DNA damage, as measured by mean migration length, was increased in all four tissues of animals from the hazardous waste site, but this increase was significant only in brain cells. However, a dispersion analysis revealed that the bone marrow cells from the potentially exposed mice exhibited a significantly increased dispersion coefficient over that calculated for the control mice. This increased dispersion was due to small numbers of cells with extensive DNA damage among a majority of cells with no or little DNA migration. This result suggests the presence of low levels of genotoxic species in this organ and/or differential sensitivity among the various subpopulations of cells that comprise this tissue. The results of this small, pilot study indicate the potential usefulness of the SCG technique in evaluating DNA damage in wild or feral animals.

The pH >13 assay has also been used successfully to examine the extent of DNA damage in coelomocytes collected from earthworms (*Eisenia foetida*) maintained in different soil samples as an indicator of soil pollution (Verschaeve *et al.*, 1993). It has also been used with erythrocytes of bullheads, a species of bottom-feeding fish, collected from various regions of Lake Erie as an indicator of polluted sediment (M. Petras, personal communication), and in mammalian cells exposed to environmental water samples as an indicator of the presence of genotoxic pollutants (Fairbairn *et al.*, 1994b). In addition, techniques for the evaluation of levels of DNA damage in various organs of the medaka (*Oryzias latipes*, a small fish) (Tice, 1995) and in zebra mussels (*Dreissena polymorpha*; M. Petras, personal communication) have been developed for use in aquatic biomonitoring. Again, because of the sensitivity and simplicity of the assay and the need for only small numbers of cells, any organism is potentially suitable for investigation.

16.7 Human epidemiology

An important application of the SCG/Comet assay is in human epidemiology, either in assessing DNA damage in cells sampled from individuals exposed occupationally, clinically, or environmentally, or in evaluating differences in DNA repair competence among control and exposed individuals. Because only a few thousand cells are needed, an analysis can be conducted on the amount of blood obtained by fingerprick (i.e. a few microlitres) or in solid tissues using fine-needle biopsy techniques. For example, Ostling *et al.* (1987) used the neutral assay to evaluate levels of DNA damage in tumour cells obtained by fine-needle aspiration from patients   receiving radiotherapy for Hodgkin's disease, non-Hodgkin's lymphoma, squamous cell carcinoma or adenocarcinoma. Also using material obtained by fine-needle aspiration, Olive *et al.* (1993b) used the pH 12.3 assay to monitor numbers of hypoxic cells in irradiated human breast cancers, based on their relative insensitivity to 5–10 Gy X-rays. Other cell populations are equally amenable to analysis. For example, Klieman and Spector (1993) used the pH >13 assay to determine the DNA damage in lens epithelial cells sampled from individuals affected with cataracts and from unaffected controls. They reported that in approximately 50% of the cataractous samples, the proportion of cells containing DNA damage was significantly higher than in control lenses, and suggested an aetiological relationship. In addition, studies have been conducted or are in progress involving, for example, alveolar macrophages and lung epithelial cells collected from smokers and non-smokers (Tice, unpublished) or cells from volunteers exposed to ozone (Tice, unpublished); bladder epithelial cells obtained from the urine of bladder cancer patients (McKelvey-Martin *et al.*, 1992); and nasal and gastric mucosa cells obtained from biopsy material (Pool-Zobel *et al.*, 1994).

Most human studies have evaluated DNA damage in nucleated blood cells. Singh *et al.* (1991b) reported an age-related increase in 'spontaneous' levels of DNA damage in lymphocytes isolated from the blood of non-smoking subjects, using a procedure based on increasing the sensitivity of the pH >13 assay by increasing the duration of electrophoresis from 20 to 40 minutes.

The pH >13 technique has also been used to evaluate DNA damage in cryopreserved peripheral blood lymphocytes from 11 breast cancer patients treated with high doses of cyclophosphamide and cisplatin and given autologous bone marrow transplantation after treatment (Tice *et al.*, 1992). The pretreatment samples of several patients contained lymphocytes with increased levels of DNA damage, presumably reflecting persistent DNA damage induced by previous treatment regimens. Chemotherapy resulted in a significant but variable increase in DNA damage in cells from all patients. Among the samples collected after transplantation, increased levels of DNA damage were absent in most but not all patients. Whether the extent of DNA damage correlates with outcome of treatment remains to be established. However, the fact that SCG/Comet data can be obtained within a few hours of sampling suggests that this method may be used to monitor levels of damage in individual patients associated with a treatment regimen and that the regimen could be changed accordingly. This study also demonstrated the lymphocytes could be cryopreserved without an accompanying increase in DNA damage.

In a population study involving 100 healthy individuals, Betti *et al.* (1994) found that the extent of DNA migration measured using the pH >13 assay was significantly increased in blood lymphocytes of smokers, with a greater increase occurring in males than in females. The lack of a corresponding increase in sister chromatid exchange (SCE) frequency in mitogen-stimulated lymphocytes from the same individuals led the investigators to conclude that the SCG assay may be more sensitive. This differential sensitivity may be due, in part, to the fact that the gel assay used unstimulated lymphocytes while the SCE assay depended on mitogen-stimulated T lymphocytes.

The sensitivity and applicability of the pH >13 alkaline assay is clearly demonstrated in two recent intriguing studies. First, Tice *et al.* (1990) had suggested that vigorous exercise may result in increased levels of DNA damage in blood leukocytes. Subsequently, Hartmann *et al.* (1994) reported that physical activity above the aerobic–anaerobic threshold caused DNA damage in blood leukocytes, with the increase being detected 6 hours after cessation, reaching a maximum at 24 hours, and returning to control levels by 72 hours. This finding should not be interpreted as an indictment of exercise, but rather that very strenuous exercise can lead to increased levels of free radicals capable of damaging cells and tissues. These data are supported by studies on muscle cell damage showing that maintenance of adequate levels of radical scavengers during exercise prevents DNA damage (Hartmann, personal communication).

Secondly, M. Green *et al.* (1994) demonstrated the value of a good breakfast (combined with vitamin C) by its ability to decrease consistently the *in vitro* sensitivity of lymphocytes to ionizing radiation, as measured by a decrease in DNA migration in cells irradiated with 2 Gy. This effect peaked at 4 hours after breakfast and the magnitude of the inhibition varied among individuals. The investigators concluded that variation in normal diet may not only alter individual susceptibility to endogenous oxidative damage but may also affect individual responses to radiation.

While biomonitoring studies employing cytogenetic techniques are limited to lymphocytes, the SCG technique can be applied to any cell population. This fact needs to be appreciated when studies involving whole blood are conducted. Leukocytes are a heterogeneous mixture of cells, with some populations (i.e. B and T lymphocytes) having a lifespan which can vary from weeks to decades while other cells (i.e. granulocytes) have a short half-life ranging from 7 to 24 hours. These differences in lifespan among different cell types are critical to the design and interpretation of biomonitoring studies. For a retrospective epidemiological study to be informative, the analysis of DNA damage in blood cells (or any cell population) should focus selectively on the population(s) of cells present during the exposure period, and preferentially on those cells with the least capacity for DNA repair. For chronic exposures, comparing the levels of DNA damage in cells with short versus long lifespans (i.e. granulocytes versus lymphocytes, respectively) may provide information about concurrent versus past exposure levels. However, for such data to be interpretable, information about the relative sensitivity of different subtypes of leukocytes to genotoxic agents and their ability to repair such damage needs to be obtained.

Several investigators have started collecting such information. For example, Ficoll hypaque-isolated granulocytes and lymphocytes are equally sensitive to γ radiation, as measured by migration length using the pH >13 alkaline gel assay, and both are more sensitive than leukocytes in whole blood (Vijayalaxmi *et al.*, 1993). Greater specificity as to cell type, however, can be obtained by using cell populations purified with a cell sorter based on differences in membrane markers. Using this approach, Uzawa *et al.* (1994) demonstrated that human CD45RO$^+$ memory T cells were more radiosensitive *in vitro* than CD45RO$^-$ naive T cells. Recently, Strauss *et al.* (1994) introduced an exciting application of an immunological typing technique which circumvents the need for cell sorting by allowing the identification of subtypes of blood leukocytes in the gel matrix. In this method, specific cell types are recognized individually in the agarose gel at the time of analysis by the co-presence of immunomagnetic beads which had adhered to a selected membrane marker before processing. This approach has been used to demonstrate comparable repair kinetics for single-strand breaks induced in B, T, T-helper, and T-suppressor cells irradiated with 1.75 Gy of γ rays (Tice and Strauss, 1995).

The potential utility of the SCG/Comet assay to human studies, especially when combined with an ability to identify selected cell populations and to recognize different classes of DNA damage using specific repair enzymes, is enormous. However, care must be taken to ensure the adequacy of the study design and that the data obtained are interpreted correctly.

16.8 Future directions and conclusions

As discussed in the Introduction, an optimal technique for evaluating DNA damage and repair should be sensitive and quantitative, provide data at the level of the single cell, and be applicable to diverse cell populations. More than any other assay currently available, the SCG/Comet assay fulfils these criteria.

The advantages of the technique include: (i) data are collected at the level of the individual cell, providing information on the intercellular distribution of damage and repair; (ii) only small numbers of cells are required (i.e. a few thousand); (iii) virtually any eukaryotic cell population can be used; (iv) the assay is sensitive, simple, and cost-effective; and (v) data can be obtained within a few hours of sampling. The ability of the assay to identify 'sensitive' or 'responding' cells in an otherwise normal population of cells is of critical importance in determining low dose–response relationships, in identifying genotoxic/cytotoxic damage, and in conducting quantitative risk assessments which account for cell, tissue, organ and individual differences. An additional advantage is flexibility. The DNA damage detected by the alkaline versions can arise through various mechanisms, including DNA interstrand cross-linking, DNA single-strand breaks, alkali-labile sites, and incompletely repaired excision sites present at the time of lysis. Increased DNA migration also accompanies the DNA fragmentation associated with cell death arising through a non-DNA mediated process or apoptosis. Although the alkaline assay can be used to detect generic damage, it can also be modified in such a way that specific classes of damage in selected cell types can be easily investigated.

The potential applications of the SCG/Comet assay in such areas as radiobiology, genetic toxicology, environmental biomonitoring, and human studies is almost unlimited. However, its ease of application also ensures that the assay will be misused and the resulting data misinterpreted. To minimize such occurrences, several important issues need to be addressed. These include understanding the relationship between DNA migration and the processes which lead to cell death, the effect of toxicity being the single most important artefact associated with this assay. Other issues include the nature and limitations of the different versions of the assay (neutral, pH 12.3, pH >13). Moreover, the choice of cell, gel, culture, tissue and animal all impose variability. Finally, the most appropriate means of statistical analysis needs to be evaluated. The next few years should see these issues resolved.

Acknowledgements

This review was supported by NIEHS through SBIR Phase II grant 2-R44-ES05884-02 to Integrated Laboratory Systems. Individuals interested in obtaining a detailed protocol for the pH >13 alkaline assay may contact Dr R. Tice at Integrated Laboratory Systems, PO Box 13501, Research Triangle Park, NC 27709, USA (fax: 919-544-0380).

References

Anderson D, Yu T-W, Phillips BJ, Schmezer P. (1994) The effect of various antioxidants and other modifying agents on oxygen-radical-generated DNA damage in human lymphocytes in the Comet assay. *Mutat. Res.* **307:** 261–271.

Andrews PW, Tice RR, Nauman CH. (1990) *In vitro* DNA damage in peripheral blood leukocytes as measured by the single cell gel (SCG) assay. *Environ. Mol. Mutagen.* **15** (S17): 6.

Andrews PW, Tice RR, Schmitt MT, Yager JW. (1994) DNA damage in normal and choline deficient male B6C3f1 mice treated by oral gavage with fly ash. *Environ. Mol. Mutagen.* **23** (S23): 2.

Arlett CF, Lowe JE, Harcourt SA, Waugh APW, Cole J, Roza L, Diffey BL, Mori T, Nikaido O, Green MHL. (1993) Hypersensitivity of human lymphocytes to UV-B and solar radiation. *Cancer Res.* **53:** 609–614.

Betti C, Barale R, Pool-Zobel BL. (1993) Comparative studies on cytotoxic and genotoxic effects of two organic mercury compounds in lymphocytes and gastric mucosa cells of Sprague–Dawley rats. *Environ. Mol. Mutagen.* **22:** 172–180.

Betti C, Davini T, Giannessi L, Loprieno N, Barale R. (1994) Microgelelectrophoresis assay (Comet test) and SCE analysis in human lymphocytes from 100 normal individuals. *Mutat. Res.* **302:** 323–333.

Carson DA, Seto S, Wasson DB, Carrera CJ. (1986) DNA strand breaks, NAD metabolism, and programmed cell death. *Exp. Cell Res.* 273–281.

Collins AR, Duthie SJ, Dobson VL. (1993) Direct enzymatic detection of endogenous base damage in human lymphocyte DNA. *Carcinogenesis* **14:** 1733–1735.

Croom DK, Andrews PW, Nascimbeni B, Tice RR. (1991) Evaluation of chemically induced DNA damage in germ cells of male mice using the single cell gel (SCG) electrophoretic assay. *Environ. Mol. Mutagen.* **17**(S19): 19.

Delaney CA, Green MHL, Lowe JE, Green IC. (1993) Endogenous nitric oxide induced by interleukin-1β in rat islets of Langerhans and HIT-T15 cells causes significant DNA damage as measured by the 'comet' assay. *FEBS Lett.* **333:** 291–295.

De Méo M, Laget M, Castegnaro M, Dumenil G. (1991) Genotoxic activity of potassium permanganate in acidic solution. *Mutat. Res.* **260:** 295–306.

Evans HH, Ricanati M, Horng M, Jiang Q, Mencl J, Olive PL. (1993) DNA double-strand break repair deficiency in TK6 and other human B lymphoblast cell lines. *Radiat. Res.* **134:** 307–315.

Fairbairn DW, O'Neill KL, Standing MD. (1993) Application of confocal laser scanning microscopy to analysis of H_2O_2-induced DNA damage in human cells. *Scanning* **15:** 136–139.

Fairbairn DW, Carnahan KG, Thwaits RN, Grigsby RV, Holyoak GR, O'Neill KL. (1994a) Detection of apoptosis induced DNA cleavage in scrapie infected sheep brain. *FEMS Microbiol. Lett.* **115:** 341–346.

Fairbairn DW, Meyers D, O'Neill KL. (1994b) Detection of DNA damaging agents in environmental water samples. *Bull. Environ. Contam. Toxicol.* **52:** 687–690.

Friend JH, Carpenter TD, Tice RR. (1993) Evaluation of chemically-induced DNA damage in germinal tissue of female mice using the single cell gel (SCG) assay. *Environ. Mol. Mutagen.* **21 (S22):** 20.

Gedik CM, Ewen SWB, Collins AR. (1992) Single-cell gel electrophoresis applied to the analysis of UV-C damage and its repair in human cells. *Int. J. Radiat. Biol.* **62:** 313–320.

Green IC, Cunningham JM, Delaney CA, Elphick MR, Mabley JG, Green MHL. (1994) Effects of cytokines and nitric oxide donors on insulin secretion, cyclic GMP and DNA damage; relation to nitric oxide production. *Biochem. Soc. Trans.* **22:** 30–37.

Green MHL, Lowe JE, Harcourt SA, Akinluyi P, Rowe T, Cole J, Anstey AV, Arlett CF. (1992) UV-C sensitivity of unstimulated and stimulated human lymphocytes from normal and xeroderma pigmentosum donors in the comet assay: a potential diagnostic technique. *Mutat. Res.* **273:** 137–144.

Green MHL, Lowe JE, Waugh APW, Aldridge KE, Cole J, Arlett CF. (1994) Effect of diet and vitamin C on DNA strand breakage in freshly-isolated human white blood cells. *Mutat. Res.* **316:** 91–102.

Grigsby RV, Fairbairn D, O'Neill KL. (1993) Differential DNA damage detected in hybridomas. *Hybridoma* **12:** 755–761.

Hartmann A, Speit G. (1994) Comparative investigations of the genotoxic effects of metals in the single cell gel (SCG) assay and the sister chromatid exchange (SCE) test. *Environ. Mol. Mutagen.* **23:** 299–305.

Hartmann A, Plappert U, Raddatz K, Grunert-Fuchs M, Speit G. (1994) Does physical activity induce DNA damage? *Mutagenesis* **9: 23:** 269–272.

Higami Y, Shimokawa I, Okimoto T, Ikeda T. (1994) An age-related increase in the basal level of DNA damage and DNA vulnerability to oxygen radicals in the individual hepatocytes of male F344 rats. *Mutat. Res.* **316:** 59–67.

Hirai O, Andrews PW, Tice RR, Nauman CH. (1991) DNA damage evaluation using the rodent in vitro hepatocyte culture system and the single cell gel (SCG) electrophoretic assay. *Environ. Mol. Mutagen.* **17** (S19): 31.

Jostes RF, Hui TE, Cross FT. (1993) Single-cell gel technique supports hit probability calculations. *Hlth Phys.* **64:** 675–679.

Kennelly JC, Lane MP, Barker JA, Barber G, Tinwell H, Gallagher JT, Pool-Zobel B, Schmezer P, Ashby J. (1993) Genotoxic activity of 1-chloromethylpyrene in stomach epithelium. *Carcinogenesis* **14:** 637–643.

Kleiman NJ, Spector A. (1993) DNA single strand breaks in human lens epithelial cells from patients with cataract. *Curr. Eye Res.* **12:** 423–431.

Kligerman AD, Bryant MF, Doerr CL, Halperin EC, Kwanyuen P, Sontag MR, Erickson GL. (1992) Interspecies cytogenetic comparisons: studies with X-radiation and bleomycin sulfate. *Environ. Mol. Mutagen.* **19:** 235–243.

MacGregor JT, Farr S, Tucker JD, Heddle JA, Tice RR, Turtletaub KW. (1995) New molecular endpoints and methods for routine toxicity testing. *Fund. Appl. Toxicol,* in press.

Marks D, Fox RM. (1991) DNA damage, poly(ADP-ribosyl)ation and apoptotic cell death as a potential common pathway of cytotoxic drug action. *Biochem. Pharmacol.* **42:** 1859–1867.

McKelvey-Martin VJ, Butler M, Stewart LH. (1992) Analysis of DNA content and integrity in cells extracted from bladder washing and voided urine specimens, in bladder cancer patients, using the comet assay. *Mutat. Res.* **271**: 163.

McKelvey-Martin VJ, Green MHL, Schmezer P, Pool-Zobel BL, De Meo MP, Collins A. (1993) The single cell gel electrophoresis assay (comet assay): a European review. *Mutat. Res.* **288**: 47–63.

Meyers CD, Fairbairn DW, O'Neill KL. (1993) Measuring the repair of H_2O_2 induced DNA single strand breaks using the single cell gel assay. *Cytobios* **74**: 144–153.

Müller W-U, Bauch T, Streffer C, Niedereicholz F, Böcker W. (1994) Comet assay studies of radiation-induced DNA damage and of DNA repair in various tumor cell lines. *Int. J. Radiat. Biol.* **65**: 315–319.

Nascimbeni B, Phillips MD, Croom DK, Andrews PW, Tice RR. (1991) Evaluation of DNA damage in golden mice (*Ochrotomys nutalli*) inhabiting a hazardous waste site. *Environ. Mol. Mutagen.* **17** (S19): 55.

Nygren J, Cedervall B, Eriksson S, Dusinska M, Kolman A. (1994) Induction of DNA strand breaks by ethylene oxide in human diploid fibroblasts. *Environ. Mol. Mutagen.* **24**: 161–167.

Olive PL. (1989) Cell proliferation as a requirement for development of the contact effect in Chinese hamster V79 spheroids. *Radiat. Res.* **117**: 79–92.

Olive PL, Banath JP. (1992) Tumour growth fraction using the comet assay. *Cell Prolif.* **25**: 447–457.

Olive PL, Banath JP. (1993a) Detection of DNA double-strand breaks through the cell cycle after exposure to X-rays, bleomycin, etoposide and [125]IdUrd. *Int. J. Radiat. Biol.* **64**: 349–315.

Olive PL, Banath JP. (1993b) Induction and rejoining of radiation induced DNA single-strand breaks: 'tail moment' as a function of position in the cell cycle. *Mutat. Res.* **294**: 275–283.

Olive PL, Banath JP, Durand RE. (1990a) Heterogeneity in radiation-induced DNA damage and repair in tumor and normal cells using the 'comet' assay. *Radiat. Res.* **122**: 86–94.

Olive PL, Banath JP, Durand RE. (1990b) Detection of etoposide resistance by measuring DNA damage in individual Chinese hamster cells. *J. Natl Cancer Inst.* **82**: 779–783.

Olive PL, Wlodek D, Banath JP. (1991) DNA double-strand breaks measured in individual cells subjected to gel electrophoresis. *Cancer Res.* **51**: 4671–4676.

Olive PL, Wlodek D, Durand RE, Banath JP. (1992) Factors influencing DNA migration from individual cells subjected to gel electrophoresis. *Exp. Cell Res.* **198**: 259–267.

Olive PL, Banath JP, Evans HH. (1993a) Cell killing and DNA damage by etoposide in Chinese hamster V79 monolayers and spheroids: influence of growth kinetics, growth environment and DNA packaging. *Br. J. Cancer* **67**: 522–530.

Olive PL, Durand RE, Le Riche J, Olivotto I, Jackson SM. (1993b) Gel electrophoresis of individual cells to quantify hypoxic fraction in human breast cancers. *Cancer Res.* **53**: 733–736.

Olive PL, Frazer G, Banath JP. (1993c) Radiation-induced apoptosis measured in TK6 human B lymphoblast cells using the comet assay. *Radiat. Res.* **136**: 130–136.

O'Neill KL, Fairbairn DW, Standing MD. (1993) Analysis of single-cell gel electrophoresis using laser-scanning microscopy. *Mutat. Res.* **319**: 129–134.

Östling O, Johanson KJ. (1984) Microelectrophoretic study of radiation–induced DNA damages in individual mammalian cells. *Biochem. Biophys. Res. Commun.* **123**: 291–298.

Östling O, Johanson KJ. (1987) Bleomycin in contrast to gamma irradiation induces extreme variation of DNA strand breakage from cell to cell. *Int. J. Radiat. Biol.* **52**: 683–691.

Östling O, Johanson KJ, Blomquist E, Hagelqvist E. (1987) DNA damage in clinical radiation therapy studied by microelectrophoresis in single tumour cells. *Acta Oncol.* **26**: 45–48.

Plappert U, Barthel E, Raddatz K, Seidel HJ. (1994a) Early effects of benzene exposure in mice. Hematological versus genotoxic effects. *Arch. Toxicol.* **68**: 284–290.

Plappert U, Barthel E, Seidel HJ. (1994b) Reduction of benzene toxicity by toluene. *Environ. Mol. Mutagen.*, in press.

Pool-Zobel BL, Klein RG, Liegibel UM, Kuchenmeister F, Weber S, Schmezer P. (1992) Systemic genotoxic effects of tobacco-related nitrosamines following oral and inhalational administration to Sprague–Dawley rats. *Clin. Invest.* **70:** 299–306.

Pool-Zobel BL, Guigas C, Klein R, Neudecker Ch, Renner HW, Schmezer P. (1993a) Assessment of genotoxicity effects by lindane. *Fd. Chem. Toxic.* **4:** 271–283.

Pool-Zobel BL, Bertram B, Knoll M, Lambertz R, Neudecker C, Schillinger U, Schmezer P, Holzapfel WH. (1993b) Antigenotoxic properties of lactic acid bacteria *in vivo* in the gastrointestinal tract of rats. *Nutrition Cancer* **20:** 271–281.

Pool-Zobel BL, Lotzmann N, Knoll M, Kuchenmeister F, Lambertz R, Leucht U, Schroder H-G, Schmezer P. (1994) Detection of genotoxic effects in human gastric and nasal mucosa cells isolated from biopsy samples. *Environ. Mol. Mutagen.* **24:** 271–281.

Rydberg B, Johanson KJ. (1978) Estimation of single strand breaks in single mammalian cells. In: *DNA Repair Mechanisms* (eds PC Hanawalt, EC Friedberg, CF Fox). Academic Press, New York, pp. 465–468.

Schmezer P, Eckert C, Liegibel UM. (1994) Tissue-specific induction of mutations by streptozotocin *in vivo. Mutat. Res.* **307:** 495–499.

Shafer DA, Xie Y, Falek A. (1994) Detection of opiate enhanced increases in DNA damage, HPRT mutants and the mutation frequency in human HUT-78 cells. *Environ. Mol. Mutagen.* **23:** 37–44.

Singh NP, McCoy MT, Tice RR, Schneider EL. (1988) A simple technique for quantitation of low levels of DNA damage in individual cells. *Exp. Cell Res.* **175:** 184–191.

Singh NP, Danner DB, Tice RR, Pearson JB, Brant LJ, Schneider EL. (1990) DNA damage and repair with age in individual human lymphocytes. *Mutat. Res.* **237:** 123–130.

Singh NP, Tice RR, Schneider EL. (1991a) A microgel electrophoresis technique for the direct quantitation of DNA damage and repair in individual fibroblasts cultured on microscope slides. *Mutat. Res.* **252:** 289–296.

Singh NP, Danner DB, Tice RR, Pearson JD, Brant LJ, Morrel CH, Schneider EL. (1991b) Basal DNA damage in individual human lymphocytes with age. *Mutat. Res.* **248:** 285–289.

Singh NP, Stephens RE, Schneider EL. (1994) Modifications of alkaline microgel electrophoresis for sensitive detection of DNA damage. *Int. J. Radiat. Biol.* **66:** 23–28.

Strauss GHS, Peters WP, Everson RB. (1994) Measuring DNA damage in individual cells of heterogeneous mixtures: a novel application of an immunological typing technique. *Mutat. Res.* **304:** 211–216.

Tice RR. (1995) Applications of the single cell gel assay to environmental biomonitoring for genotoxic pollutants. In: *Biomonitors and Biomarkers as Indicators of Environmental Change* (ed. FM Butterworth). Plenum Press, New York, in press.

Tice RR, Setlow RB. (1985) DNA repair and replication in aging organisms and cells. In: *Handbook of the Biology of Aging*, 2nd edn (eds CE Finch, EL Schneider). Von Nostrand Reinhold, New York, pp. 173–224.

Tice RR, Strauss GHS. (1995) The single cell gel electrophoresis/comet assay: a potential tool for detecting radiation-induced DNA damage in humans. In: *Assessment of Radiation Effects by Molecular and Cellular Approaches* (eds TM Fliedner, EP Cronkite, VP Bond). Alphamed Press, Dayton, OH, in press.

Tice RR, Andrews PW, Singh NP. (1990) The single cell gel assay: a sensitive technique for evaluating intercellular differences in DNA damage and repair. In: *Methods for the Detection of DNA Damage in Human Cells* (eds B Sutherland, A Woodhead). Plenum Press, New York, pp. 291–301.

Tice RR, Andrews PW, Hirai O, Singh NP. (1991) The single cell gel (SCG) assay: an electrophoretic technique for the detection of DNA damage in individual cells. In: *Biological Reactive Intermediates IV, Molecular and Cellular Effects and Their Impact on Human Health* (eds CR Witmer, RR Snyder, DJ Jollow, GF Kalf, JJ Kocsis, IG Sipes). Plenum Press, New York, pp. 157-164.

Tice RR, Strauss GHS, Peters WP. (1992) High-dose combination alkylating agents with autologous bone marrow support in patients with breast cancer: preliminary assessment of DNA damage in individual peripheral blood lymphocytes using the single cell gel electrophoresis assay. *Mutat. Res.* **271:** 101–113.

Tice RR, Schmitt MT, Andrews PW, Yager JW. (1994) DNA damage in normal and choline deficient male B6C3f1 mice treated by oral gavage with sodium arsenite. *Environ. Mol. Mutagen.* **23** (S23): 66.

Uzawa A, Suzuki G, Nakata Y, Akashi M, Ohyama H, Akanuma A. (1994) Radiosensitivity of CD45RO+ memory and CD45RO- naive T cells in culture. *Radiat. Res.* **137:** 25–33.

Verschaeve L, Gilles J, Schoctors J, Van Cleuvenbergen R, De Fre R. (1993) The single cell gel electrophoresis technique or comet test for monitoring dioxin pollution and effects. In: *Organohalogen Compounds 11* (eds H Fiedler, H Frank, O Hutzinger, W Parzefall, A Riss, S Safe). Federal Environmental Agency, Austria, pp. 213–216.

Vijayalaxmi, Tice RR, Strauss GHS. (1992) Assessment of radiation-induced DNA damage in human blood lymphocytes using the single cell gel electrophoresis technique. *Mutat. Res.* **271:** 243–252.

Vijayalaxmi, Strauss GHS, Tice RR. (1993) An analysis of gamma-ray-induced DNA damage in human blood leukocytes, lymphocytes and granulocytes. *Mutat. Res.* **292:** 123–128.

Ward AJ, Rosin MP, Olive PL, Burr AH. (1993) A sensitivity to oxidative stress is linked to chromosome 11 but is not due to a difference in single-strand DNA damage or repair. *Mutat. Res.* **294:** 299–308.

Yager JW, Schmitt MT, Andrews PW, Tice RR, Crecelius E. (1994) Urinary excretion kinetics and DNA damage in normal and choline deficient male B6C3F1 mice treated once by oral gavage with sodium arsenite. *Environ. Mol. Mutagen.* **23** (S23): 75.

Monitoring for somatic mutation in humans

Richard J. Albertini and J. Patrick O'Neill

17.1 Introduction

Gene mutations occur *in vivo* in human somatic cells. Methods for their recognition and quantitation, if reliable, are of considerable value for assessing effects and consequences of exposure to environmental mutagens or carcinogens. When, in addition, the methods allow recovery of mutant cells in sufficient numbers for cellular and molecular analyses, they become invaluable tools for research into human mutation. There are now several methods for measuring these genetic events; some permit molecular investigations.

It was recognized as early as 1957 that methods for assessing *in vivo* somatic mutations must meet two challenges (Russell and Major, 1957). First, the methods must be capable of demonstrating that the measured variation truly has a genetic (i.e. a mutational) basis. Over the last four decades, at least three examples have shown that non-genetic phenotypic somatic cell variations *in vivo* can be very difficult to distinguish from true genetic changes, even when the variations respond to mutagen exposures as would be expected for mutations (Atwood and Scheinberg, 1958; Papayannopoulou *et al.*, 1977a, b; Sutton, 1972). Early assays were abandoned for this reason. Fortunately, current methods almost certainly measure true genetic events; some even allow verification at the DNA level.

A slightly more subtle challenge for *in vivo* mutation studies is to define the relationship between the mutant cells, which are the endpoints measured, and the underlying mutations, which are the events of usual interest. Defining this relationship underlies all mutation research and is even more complex *in vivo* where some cells rest while others divide. Although the issue of 'bursts' must be addressed in mutation research, historically its recognition has provided basic insights into the mutation process (Nicklas *et al.*, 1986). Some of the current methods for measuring *in vivo* somatic mutations can distinguish between events and their results, and are providing insights into mechanisms.

There is now a third challenge for those who measure *in vivo* somatic mutations, at least those studied by most genetic toxicologists, that is mutations in reporter genes. Genetic toxicologists use mutations in reporters as surrogates for mutations in disease genes, usually those of relevance to cancer. Although unrelated to disease, mutations in reporter genes have several advantages for studying the mutation process itself. They are usually selectable by physical, chemical or immunological methods and do not themselves alter cell proliferation characteristics. Also, some reporter gene mutations have finite lifespans *in vivo* so that mutational events are usually related temporally to mutant cell production. By contrast, mutations in cancer genes are usually not selectable at the rare-cell stage, often do alter cell proliferation characteristics, and have very long lifespans, making primary frequencies of *in vivo* mutations difficult to measure. However, for mutations in reporter genes to be useful in preventive medicine as more than *in vivo* dosimeters, they must be valid surrogates for cancer and other diseases associated with mutations. Actually, monitoring for environmental genotoxicity involves two levels of surrogacy – one concerning the gene and the other the tissue in which mutations are measured. The third challenge, therefore, is to establish how well surrogate mutations in surrogate tissues mimic mutations in genes that predispose to disease in target tissues.

This chapter considers the several methods currently used for investigating *in vivo* somatic mutations in humans. Quantitative results of studies in human populations are summarized, as are molecular analyses of the underlying mutational events. Evidence is presented that mutations in reporter genes may serve as valid surrogates for at least some pathogenic mutational mechanisms. Ways in which this last claim can be verified, making these methods truly valuable tools in public health, are discussed.

17.2 The assays for *in vivo* somatic mutation

We shall focus on the use of somatic cell mutation assays for monitoring of environmental exposures of humans to potentially genotoxic agents. Studies of somatic cell mutation *per se* in humans are clearly valuable for understanding mechanisms of mutation events occurring *in vivo*, especially in comparison with germ cell mutations (Cooper and Krawczak, 1990; Sculley *et al.*, 1992; Sommer, 1992) and mutations involved in carcinogenesis (Hollstein *et al.*, 1991). For human monitoring, a mutation assay should fulfil the following criteria in order to allow extrapolation to risk of disease, which is generally considered to be the goal of mutagenicity monitoring: the biomarker must provide an adequate measure of the actual *in vivo* mutation frequency; the assay must show the required sensitivity to measure an increase in frequency induced by an environmental exposure; and the endpoint must somehow provide a signature of the specific exposure. These last two characteristics, sensitivity and specificity, are elusive and provide the ultimate challenge for genetic toxicology. Finally, mechanisms of mutagenicity that underlie disease processes such as cancer should be captured in the reporter genes used. The five assays currently used for studies of *in vivo* human somatic cell gene mutations measure changes in cells obtained from peripheral blood samples. Erythrocytes can be used to measure the frequency of haemoglobin variants or glycophorin A variants, the latter only in individuals heterozygous for the M and N blood group antigens. The mononuclear cell fraction can be used to measure the frequency of mutants in T lymphocytes. At present, mutations can be studied at two autosomal loci, HLA and the CD3/T-cell receptor complex, and at one X-chromosomal locus, *hprt*. Aspects of these five assays have been reviewed in recent years (Albertini *et al.*, 1990; Cole and Skopek, 1994; Grant and Bigbee, 1993).

17.2.1 The erythrocyte assays

Erythrocytes are abundant in human blood and are easily analysed for the presence of rare variants containing altered protein antigens. However, since erythrocytes have no nucleus and thus no DNA, any gene mutation must have arisen in a precursor cell, and this places an unknown number of cell divisions (estimated to be maximally 20–25) between the mutation event and the measured variant frequency. Because the mutant phenotype of an erythrocyte cannot actually be verified as a defined mutant genotype, it is customary to refer to a 'variant frequency' (Vf) with these erythrocyte assays. Both haemoglobin (β gene on chromosome 11 contains three exons and spans 2 kb) and glycophorin A (*GPA* gene on chromosome 4 contains seven exons and spans 44 kb) variants have been detected by immunochemical methods.

The assay for haemoglobin variants is best exemplified by the mutation of the β gene from normal (Hb) to HbS (an A → T transversion in codon 6). Such a mutation in a person with normal haemoglobin can be detected by antibodies specific for HbS and the rare, fluorescent variant cell recognized by either flow cytometry (Bigbee *et al.*, 1981) or automated image analysis (Verwoerd *et al.*, 1987). Limited studies have shown these variants to be present at a frequency of $1 \times 10^{-8} - 10 \times 10^{-8}$ in unexposed adults (aged 25–58). This assay has the advantage of measuring a specific altered protein thus making it almost certainly a measure of genetic mutation. However, only a single specific mutation can be detected.

The other erythrocyte assay is based on the erythrocyte GPA surface molecules which constitute the M and N blood group antigens. The M and N proteins differ by two non-adjacent amino acids and highly specific anti-M and anti-N antibodies have been developed. In MN heterozygous individuals, these antibodies can be used in conjunction with flow cytometry to enumerate rare variant erythrocytes which lack expression of one of the antigens with either single or double expression of the other. The former are termed hemizygous variants, due to the simple loss of one allele and the latter, homozygous variants, due to both a loss of one allele and double expression of the other. In practice, the widest application has concentrated on the detection of hemizygous

N0 and homozygous NN variants in heterozygous MN individuals (Jensen *et al.*, 1986, 1987; Langlois *et al.*, 1987a). The hemizygous variants are thought to result from the loss of antigenicity through deletion of the M allele or from a point mutation. The homozygous variants are thought to result from somatic recombination, gene conversion or chromosome missegregation leading to the loss of heterozygosity. The frequencies of these two types of variants were similar in an unexposed population of 377 humans with mean Vf values of 7.3 and 11.7 x 10^{-6} for the N0 and NN class, respectively (Grant and Bigbee, 1993).

17.2.2 The T-lymphocyte assays

T lymphocytes are nucleated cells that are also readily available in peripheral blood samples, constituting about 70% of the mononuclear cell fraction. This fraction is obtained by density sedimentation and the cells can be cryopreserved for future assays. Peripheral blood T lymphocytes are primarily arrested in the G_0 phase of the cell cycle *in vivo* and are stimulated to enter the cell cycle and proliferate in response to a specific antigen. The antigen specificity is determined by T-cell receptor (TCR) molecules on the cell surface. This binding receptor is composed of a heterodimer protein with approximately 90% of the T cells bearing the α/β chains and the remainder the γ/δ chains. The heterogeneity of the TCR proteins is due to the rearrangement of the *TCR* genes during intrathymic differentiation. (These rearrangement patterns are readily defined on Southern blots after appropriate restriction of genomic DNA. It is estimated that there are more than 10^7 different *TCR* gene rearrangement patterns possible per individual.)

T lymphocytes can be stimulated to enter the cell cycle and proliferate *in vitro* by incubation with the mitogen phytohaemagglutinin (PHA). The ability to stimulate cell division allows the assay of mutant phenotypes as well as the ability to analyse the gene mutations. Three gene loci are being studied for mutation analysis in T lymphocytes at present.

Mutations in the T cell receptor genes (α gene on chromosome 14 and β gene on chromosome 7) are determined by flow cytometry to detect the rare cells which express the CD4 antigen (approximately 60% of T lymphocytes), but not the CD3 antigen normally present on all T lymphocytes. The CD3/TCR α/TCR β complex is normally expressed on the cell surface. However, if one of the TCR molecules is defective, the CD3/TCR complex is not present on the cell surface. The mean frequency in 127 unexposed humans (aged 20–80 years) is approximately 250×10^{-6} (Kyoizumi *et al.*, 1990, 1992). The reason for this high mutant frequency value is currently unknown, and it remains to be demonstrated that all CD3 loss results from genetic mutation. However, because these genes are rearranged to create the diversity of the immune system, a high mutation rate may be intrinsic due to mechanisms specific to T cells.

Mutations at the human leukocyte antigen (HLA) histocompatibility gene family on chromosome 6 have been measured at the HLA-A locus (seven exons spanning 5 kb of DNA) in individuals heterozygous for the A2, A3 or A24 alleles. The basis of the assay is an immunocytotoxicity step to kill selectively the normal, non-mutant cells. This entails the use of antibodies specific for the HLA antigen plus complement and then either cell cloning (described later) or flow cytometry analysis with fluorescent antibodies to quantify the mutant frequency (Grist *et al.*, 1992; Janatipour *et al.*, 1988; Kushiro *et al.*, 1992; Morley, 1991; Morley *et al.*, 1990; Turner and Morley, 1990). Mean mutant frequencies for A2 and A3 loss of $25 \times 10^{-6} - 30 \times 10^{-6}$ have been reported for adults aged 18–50 (Turner and Morley, 1990).

Mutations at the *hprt* locus (located on the X chromosome, consisting of nine exons and spanning 44 kb) can be measured by taking advantage of the cytotoxicity of purine analogues such as 6-thioguanine (2-amino-6-mercaptopurine, TG). Cells which lack HPRT enzyme activity are not killed by this analogue because cytotoxicity requires metabolism of the base to the nucleotide. In addition, the existence of a human disease which is due to the lack of this enzyme, namely Lesch–Nyhan syndrome, has greatly facilitated research concerning mutations in this gene (Sculley *et al.*, 1992).

There are three different assays for *hprt* mutations, using three different approaches based on the ability of mutant cells to replicate DNA or proliferate in the presence of TG. Two short-term assays determine the frequency of variant cells which replicate DNA during 40 h after *in vitro* stimulation with PHA. DNA synthesis is measured either by the incorporation of tritium-labelled thymidine and autoradiography, or by the incorporation of 5-bromodeoxyuridine and differential staining (Albertini *et al.*, 1988; Ostrosky-Wegman *et al.*, 1988; Strauss and Albertini, 1979). The ratio of labelled cells in the presence of 2×10^{-4} M TG to that in the absence of TG defines the Vf. The third approach uses cell cloning in 96-well microtitre dishes, with cells cultured in the presence and absence of 1×10^{-5} M TG (Albertini *et al.*, 1982; Morley *et al.*, 1983). Several laboratories have developed slightly different versions of this assay (Hakoda *et al.*, 1988c; Henderson *et al.*, 1986; Morley *et al.*, 1985; O'Neill *et al.*, 1987; Tates *et al.*, 1991b). In general, T lymphocytes grow from single cells into colonies in the presence of interleukin-2 growth factor and irradiated lymphoblastoid accessory cells. Most procedures also include PHA. Cells are usually plated in the absence of TG at 1–10 cells/well and in the presence of TG at $1 \times 10^4 - 2 \times 10^4$ cells/well. Cloning efficiencies are determined from the Poisson relationship, $P_0 = e^{-x}$, where P_0 is the proportion of wells negative for colony growth and x is the calculated average number of cloning cells per well (cloning efficiency, CE). The ratio of CE in the presence and absence of TG defines the Mf. In unexposed adults aged 20–40 years, most investigators report Vf or Mf in the range of $2 \times 10^{-6} - 15 \times 10^{-6}$. One study determined Vf by autoradiography and Mf by the cloning assay in 27 adults and found a good correlation between the two assays (Albertini *et al.*, 1988).

In summary, there are five assays available at present for measuring *in vivo* somatic cell mutations in humans. The two erythrocyte assays provide a measure of the Vf, but because erythrocytes do not have nuclei, there is no possibility of defining the actual mutation frequency or of analysing the mutational events. The T-lymphocyte assays allow these last studies.

17.3 Quantitative results with *in vivo* somatic mutation assays

Five assays for human somatic mutations have been described. If an assay is to be useful for monitoring the genotoxicity of environmental exposures of humans and for assessing the genetic risk of these exposures, it should provide quantitative measurements of mutant frequency and reasonable estimates of mutation frequency, as well as qualitative information on the actual mutational events. This last requirement is essential to identify the nature of the genotoxic agent. The ability of an assay to measure increases in mutant frequency defines its sensitivity, while the ability to provide a unique mutational spectrum for a given exposure defines its specificity (Section 17.4). A perfect somatic mutation assay would provide a measure of the frequency of mutations induced as a result of an exposure to any genotoxic agent and a signature mutation event specific for that agent. This information would make risk assessment feasible by allowing extrapolation of specific mutational events in these 'indicator' genes to specific mutational events in genes involved in diseases, such as cancer or birth defects, that are caused by genotoxic agents.

17.3.1 Mutant frequencies in non-exposed individuals

The five assays have been used to measure mutant frequencies in humans not known to be exposed to specific mutagenic agents. The values obtained are usually considered to represent the background frequency resulting from spontaneous mutation. It seems reasonable that the probability of a mutation occurring spontaneously would be dependent on the age of the individual, based on the idea that errors in DNA replication or repair would increase with increasing rounds of cell division or increasing levels of DNA damage respectively. This age-related increase in mutant frequency is, in fact, observed in the four assays which addressed this question. (The Hb assay has not addressed the age relationship.) Clearly, the early tendency to refer to mean mutant frequencies in a given population was an oversimplification. Each assay is evolving as the number of individuals studied increases and its sensitivity characteristics are being defined.

The erythrocyte assays provide a frequency of cells with a variant phenotype. The target for mutation in the Hb assays is a single base pair and the measured Vf is accordingly low, in the range of 1×10^{-8}–10×10^{-8} in humans aged 25–58 years (Bernini *et al.*, 1990; Tates *et al.*, 1989). Studies with this assay have been limited (Cole and Skopek, 1994). The GPA assay has been used primarily by one group (Jensen *et al.*, 1986; Langlois *et al.*, 1987a). There is a small age-related increase in the Vf for both homozygous (NN) and hemizygous (N0) variants; the mean Vf in umbilical cord blood is $3 \times 10^{-6} - 4 \times 10^{-6}$ and in adults $6 \times 10^{-6} - 10 \times 10^{-6}$ (Cole and Skopek, 1994).

The T-lymphocyte assays give characteristic frequencies for each of the three loci. The CD3/TCR assay yields the highest Vf ranging from 50×10^{-6} to 650×10^{-6} in adults aged 20–80 years (Kyoizumi *et al.*, 1992). An age-related increase of approximately 30×10^{-6} per decade has been estimated (Cole and Skopek, 1994). Sorting of the variant cells and growth of clones has had limited success and so the mutant nature of the variants is difficult to confirm.

The HLA assay relies on an immunocytotoxicity selection step to allow recovery of the cells which have lost the A2, A3 or A24 antigen, followed by cell cloning in 96-well microtitre dishes (Janatipour *et al.*, 1988). The efficiency of the immunoselection is usually not 100% and so many of the variant colonies are not mutants, but phenocopies. However, since the surviving variant cells are cloned, the mutant phenotype can be confirmed by flow cytometry analysis with the appropriate HLA antibodies. The need to confirm the mutant phenotype because of incomplete selection may limit the usefulness of this assay for monitoring studies. A clear age-related effect has been found, with the mean frequency increasing from 7.1×10^{-6} in newborn cord bloods to 65.3×10^{-6} in adults older than 60 years (Grist *et al.*, 1992).

The *hprt* assay relies on the ability of mutant T lymphocytes to replicate DNA in the presence of TG. The short-term assay measures replication during the first S phase *in vitro* (Vf), while the cloning assay measures colony formation (Mf). The short-term assay provides a measure of the Vf similar to the erythrocyte assays. The cloning assay provides a mutant frequency measured as colony formation in the presence of TG and necessitates a heritable phenotype. Both assays yield similar frequencies and both show an age-related increase (summarized in Cole and Skopek, 1994).

A large database for *hprt* Mf with the cloning assay is available. One example of the relationship between Mf and age is shown in *Figure 17.1*. The increase in Mf with age appears to be biphasic, with a higher slope from 0 to 18 years. These relationships are described in detail in Branda *et al.* (1993) and Finette *et al.* (1994). Similar relationships between Mf and age are found in three other datasets and the four show good agreement in a recent comparative analysis (Robinson *et al.*, 1994). It is comforting to find that four research groups report similar age-related increases in Mf in unexposed populations. Reproducible results in unexposed populations are an essential requirement for any assay which hopes to supply evidence for exposure-related increases in Mf.

17.3.2 Relationship between the measured in vivo Mf and the underlying mutation frequency

These somatic mutation assays provide a measure of the frequency of variant or mutant cells in an individual. However, the goal of human monitoring studies is to determine the frequency of *in vivo* mutations (i.e. the events that produced the mutant cells). The relationship between mutant frequency and

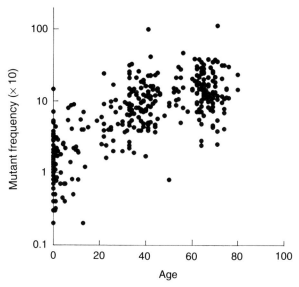

Figure 17.1. The relationship between *hprt* mutant frequency and age. The results with 378 assays are shown for individuals aged 0 (newborn cord bloods) to 80 years.

mutation frequency may not be one-to-one in cell populations that are capable of division. If the mutation does not inhibit cell division, one mutation event could yield two or more mutant cells by proliferation *in vivo*. In fact, mutation fixation requires at least one round of DNA replication and cell division and so cell proliferation becomes an integral part of the mutation process. Since both erythrocytes and T lymphocytes proliferate *in vivo*, the potential that a measured mutant frequency is higher than the actual mutation frequency must be addressed.

The use of non-nucleated erythrocytes for *in vivo* mutation studies suffers from at least two deficiencies. As discussed earlier, it is not possible to confirm that the variant cells are actually mutant cells. The second limitation is that there is also no way to address the mutant/mutation paradox in erythrocytes.

The T lymphocyte assays possess the advantage of using nucleated cells, which can be propagated *in vitro*. This allows three types of studies: confirmation of the mutant phenotype by cell growth, analysis of the mutant genotype (i.e. the mutation) by DNA analysis (discussed in Section 17.4) and analysis of an independent marker of clonality to determine the magnitude of *in vivo* proliferation. The basis for lymphocyte proliferation *in vivo* is activation by specific antigens and antigen specificity is conferred through the T-cell receptor protein. The diversity of the T-cell receptor proteins is due to genomic rearrangement of the TCR genes. The α, β, γ, and δ TCR genes contain constant (C), variable (V), diversity (D) and joining (J) segments which are rearranged during thymic differentiation through V(D)J recombinase-mediated recombinations (Van Dongen and Wolvers-Tettero, 1991; Van Dongen et al.,

1990). These rearrangements of genomic DNA provide a marker of uniqueness of T lymphocytes. If a mature T lymphocyte proliferates *in vivo*, all progeny cells will have identical TCR gene rearrangement patterns. These patterns can be differentiated by several methods of TCR gene analysis (de Boer *et al.*, 1993; Caggana *et al.*, 1991; Curry *et al.*, 1993; Nicklas *et al.*, 1986; Trainor *et al.*, 1991). Determining the TCR gene rearrangements of a group of mutant clones from a single individual defines the degree of T-cell clonality in that sample and allows correction of the mutant frequency into a mutation frequency.

The most extreme case of clonality reported to date consisted of 61 of 66 *hprt* mutants with identical TCR patterns (Nicklas *et al.*, 1988). An analysis of 413 wild-type and 1736 *hprt* mutants from 58 unexposed individuals has been reported (O'Neill *et al.*, 1994b). This study found no clonality of wild-type colonies, but clonality of mutant isolates in 35 of the 58 individuals (60.3%). In general, the magnitude of the clonality was usually small, with only nine individuals showing greater than 30% clonality (i.e. 70% of the mutants had unique TCR patterns, consistent with independent mutations). *Figure 17.2* shows the magnitude of the clonality which was seen in 418 of the 1736 mutant colonies (24.1%). In 51 of the 93 observations, the clonality was a 2mer (i.e. two colonies with the identical TCR gene pattern). Clonality of more than six was rarely seen (10/58 = 17%). In only seven of the 58 individuals was the calculated mutation frequency significantly lower than the measured mutant frequency (O'Neill *et al.*, 1994b). In these seven individuals, the measured Mf

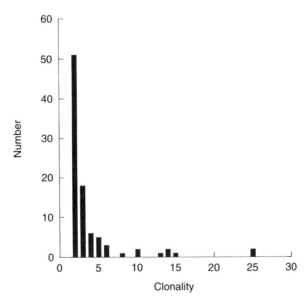

Figure 17.2. Clonality of *hprt* mutants from 57 unexposed individuals. The number of colonies with identical TCR patterns (clonality) and the number of observations of this clonality are shown. Clonality was observed 93 times involving a total of 418 colonies. (The clonality of 61 of 66 colonies is not included in the figure.)

was greater than 40×10^{-6}. These results demonstrate the usefulness of a T-lymphocyte-based assay to allow analysis of actual mutation frequency.

In general, intra-individual clonality is frequently observed in T-lymphocyte mutants from unexposed individuals but it is low in magnitude and infrequently alters the mutant/mutation frequency relationship. However, it clearly can be a significant influence with elevated Mf. In monitoring studies, it is important to confirm that an elevated *mutation frequency* has been observed. At present, only a T-lymphocyte cloning assay can provide this information.

17.3.3 Mutant frequencies in exposed populations

In order to develop a mutation assay for human monitoring, researchers eventually have to demonstrate that the assay can show expected increases in mutation in a human population. This is not a trivial undertaking. The results with the five assays in unexposed populations give baseline frequencies and define some of the variables inherent in each assay. For two of the lymphocyte assays (*hprt* and HLA), these results show that Mf is being measured and that it approximates to a mutation frequency. The similarity of the results with the two *hprt* assays suggest that the short-term assay, which could be automated, is an appropriate alternative to the more costly cloning assay (Albertini *et al.*, 1988). But can these assays measure increases in mutations in exposed populations?

Two approaches for 'validation' have been used: analysing the frequencies in humans with cancer-prone, genetic disorders and in cancer patients receiving genotoxic chemotherapy or radiotherapy. Cancer-prone genetic disorders such as ataxia telangiectasia (AT), Bloom's syndrome (BS), Fanconi's anaemia (FA) and xeroderma pigmentosum (XP) exhibit altered DNA repair/processing characteristics, and are thought to be cancer-prone because of somatic mutation hypersensitivity, either spontaneous or induced (e.g. by ultraviolet radiation in XP individuals; see Chapter 4). It is thus reasonable to expect to detect increased somatic mutation in individuals with these syndromes. The results of biomarker assays in cancer-prone individuals are reviewed in Cole and Skopek (1994). The Hb assay did not detect increases in Vf in individuals with either AT or XP, while the GPA assay detected increases with AT or BS, but not XP. The *hprt* assay detected elevated Vf or Mf in all four syndromes. The CD3/TCR assay has found elevated frequencies in FA and AT individuals, and the HLA assay in BS individuals. Overall, these results are consistent with the proposal that these assays are sensitive measures of genetic alterations occurring in humans. Patients receiving cancer therapy agents with known mutagenicity *in vitro* can act as positive controls for mutation induction *in vivo*. Measurements of mutation induction in these patients may also have predictive value for the risk of subsequent treatment-induced cancers (Branda *et al.*, 1991). A variety of chemotherapy and radiotherapy regimens have been shown to induce somatic mutation at the

GPA locus (Bigbee *et al.*, 1990) and at the *in vitro* locus (Albertini, 1985; Ammenheuser *hprt*, 1988, 1991; Branda *et al.*, 1991; Nicklas *et al.*, 1990, 1991a). Such studies have not been performed using the other three loci.

Chemotherapy and radiotherapy usually involve exposure to acute doses of a genotoxic agent. Another acute exposure which has been studied is radiation exposure received by survivors of the Hiroshima atomic bomb. Linear increases in Vf with estimated radiation exposure have been found in the GPA assay; similar but much lower increases in Mf were observed with the *hprt* assay (Akiyama *et al.*, 1990; Hakoda *et al.*, 1988a, b; Langlois *et al.*, 1987b). No effect was seen with the HLA or CD3/TCR assay (Kyoizumi *et al.*, 1992). These *in vivo* mutation assays were performed 45-50 years after the radiation exposure of the study population and the low-to-absent mutant responses in the T-cell assays probably reflect the shorter temporal relationship between mutation induction and persistence of mutants in peripheral lymphocytes than in bone marrow erythropoietic stem cells.

The above results do suggest that these assays for somatic mutation are sensitive to acute exposures to genotoxic agents. (Clearly, factors such as age must be considered and are usually controlled for in monitoring studies.) The crucial question is whether these assays are sensitive to changes induced by more chronic exposures, typical of environmental effects. One environmental exposure which has been investigated both indirectly (i.e. evaluation of existing data) and directly (i.e. through a planned study for specific effects) is the effect of tobacco smoke on somatic mutation. Both types of red blood cell (RBC) assay have detected an elevated (although not significantly increased) Vf in smokers (Bernini *et al.*, 1990; Jensen *et al.*, 1990). The *hprt* assay detected a significantly elevated Vf in adult smokers compared with non-smokers (Ammenheuser *et al.*, 1988, 1991). Recently, elevated *hprt* Vf has been reported in cord-blood samples from newborns born of smoking mothers compared with those born of non-smokers (Ammenheuser *et al.*, 1994). A significant increase in Mf due to smoking has also been reported with the cloning assay (Cole and Skopek, 1994; Vrieling *et al.*, 1992). An analysis of the *hprt* Mf database from four laboratories shows a variable but significant contribution of smoking to the age-related increase in Mf (Robinson *et al.*, 1994).

Several recent studies of occupational exposures have used the *hprt* assay. Exposure to ethylene oxide (EO) was monitored in hospital workers and factory workers engaged in sterilization procedures using EO. Average exposure was estimated by air sampling and haemoglobin adducts. A significant increase in Mf was found in the factory worker population (Tates *et al.*, 1991a). A study of workers exposed to 1,3-butadiene (BD) showed an increase in *hprt* Vf in the exposed workers compared with age-matched controls (Ward *et al.*, 1995). The magnitude of the increase was proportional to the amount of a BD-specific metabolite in the urine. Evidence for a specific mutational spectrum has been reported in studies with mice exposed to BD (Cochrane and Skopek, 1994a, b). A study of foundry workers exposed to polycyclic hydrocarbons (PAH) showed a strong positive correlation between *hprt* Mf and PAH DNA adducts in these workers (Perera *et al.*, 1993).

These are the first three reports of apparent inductions of somatic mutations resulting from environmental chemicals, which in these studies resulted from occupational exposures. These results, which demonstrate the sensitivity of the *hprt* T-lymphocyte assays, offer the possibility that exposure to these agents will result in specific mutations. Biological monitoring for genotoxic effects could then be accomplished by screening for signature mutations in the *hprt* gene as evidence for agent-specific induction of mutation.

17.4 Mutational spectra in humans

The molecular nature of mutations occurring at the HLA and *hprt* loci in human T lymphocytes have been analysed by several different methods. With both the *hprt* and the HLA mutations, Southern blot analysis of genomic DNA digested with restriction enzymes has used probes for both the selected gene and for flanking genes. This analysis detects primarily large deletions and gross structural alterations. With the *hprt* mutations, polymerase chain reaction (PCR) amplification of the nine exons (multiplex PCR) was also used to define large deletions, as well as small deletions (Gibbs *et al.*, 1990). Reverse transcriptase (RT)-mediated synthesis of *hprt* cDNA from *hprt* mRNA followed by PCR amplification and cDNA sequencing has been used to define both base substitutions and small additions and deletions (i.e. point mutations; Gibbs *et al.*, 1989). The combination of cDNA sequencing and multiplex PCR can differentiate splice-site mutations. These can then be defined by exon-specific genomic PCR amplification and sequencing (Andersson *et al.*, 1992; Rossi *et al.*, 1992; Steingrimsdottir *et al.*, 1992). This ability to define specific mutations is the major advantage of the T-lymphocyte assays.

17.4.1 Mutation analysis in unexposed humans

The *HLA* loci on chromosome 6 are autosomal and mutations are selected as the loss of function of one of the heterozygous alleles, usually the *HLA-A2* or *A3* allele (McCarron *et al.*, 1989; Morley *et al.*, 1990). Southern blot analysis uses an *HLA-A* gene probe as well as a number of linked polymorphic chromosome-6 probes determined to be heterozygous in the individual studied. By this approach, mutations can be classified into one of seven categories as listed in *Table 17.1*. The class 1 mutants represent 64.7% of the total (281/434) and probably are single base substitutions or small deletions/additions (small relative to the 5 kb size of the gene). The class 2 gene deletions comprise 2.8% of the total (12/434) or 4.1% (12/293) of the non-recombination type mutations (classes 4 and 5). Large deletions involving both the selected *HLA* allele and flanking markers (class 3) have not been observed. The recombination/gene conversion mutants represent 32.4% (141/434) of the

Table 17.1. Molecular characterization of HLA loss mutations in human T-lymphocyte colonies

Class designation	Number (n = 434)	Description	Southern blot results
1	281	Point mutation	No change in selected HLA-A band
2	12	Small deletion	Absence of selected HLA-A band only
3	0	Large deletion	Like 2 (above) with LOH of some flanking markers
4	2	Gene conversion	Absence of selected HLA-A band and homozygosity for other HLA-A allele (double dosage)
5	139	Mitotic recombination	Like 4 (above) with LOH of some flanking markers
6	0	Chromosome 6 loss	LOH for all chromosome 6 genes and loss of selected HLA-A band
7	0	Chromosome 6 duplication	Like 6 (above) but with double dosage for all chromosome 6 genes

Adapted from Cole and Skopek (1994).

total. This type of event is formally similar to the GPA NN phenotype, in which the heterozygosity for the HLA alleles is lost (loss of the selected allele) and the non-selected allele is present in two copies (double dosage). Most often, flanking markers are also no longer heterozygous (loss of heterozygosity, LOH) and this LOH of flanking markers ranges in extent (Grist *et al.*, 1992; Turner *et al.*, 1988).

One limitation of the HLA assay is the lack of information on putative 'point mutations' which comprise 64.7% of the total. While LOH can clearly be measured with this autosomal gene assay, and not with an X-chromosomal gene such as *hprt*, it is not known if environmental exposures can induce these events which appear to result from mitotic recombination. Of note, the relatively low frequency of HLA gene deletions as well as the absence of deletions involving flanking genes seems contrary to the conventional theory that an autosomal locus is more efficient for detection of deletion than an X-chromosome locus because the loss of essential genes would be lethal events with hemizygous genes (Cole and Skopek, 1994).

Information on mutations at the *hprt* locus is being collected in a computer database (Cariello and Skopek, 1993; Cariello *et al.*, 1992). The advantage of such a repository of information is that a spectrum of mutations occurring in unexposed humans evolves as new data are added.

Initial studies of *hprt* mutants by Southern blot techniques showed that 16.9% (125/739) of the *hprt* mutations in unexposed adults were large deletions or rearrangements (Cole and Skopek, 1994). The suggestion that large deletions cannot be recovered at the hemizygous *hprt* locus because of the potential problem of deleting linked essential genes led to an analysis of deletion size by defining the co-deletion of X-linked anonymous sequences known by genetic studies to map in the Xq26 region near the *hprt* gene (Nicklas *et al.*, 1991b). These studies mapped the co-deletion spectrum and showed that a range of deletions involving the *hprt* gene occurs in humans. Pulsed field gel electrophoresis studies have demonstrated that deletions as large as 3.5 Mb can occur (Lippert *et al.*, 1995a, b). Thus, *hprt* is a sensitive target for deletion events.

The analysis of the mutations which are not due to large structural alterations is proceeding through DNA sequencing studies. A detailed analysis of the database is presented in Cole and Skopek (1994). To date, base-substitution mutations have been reported in 208 of the 657 coding bases with a total of 291 different base substitutions. (In theory, an amino acid change could be produced by a base change at 570 of the 657 bases.) The background spectrum for *in vivo* base-substitution mutations is clearly not yet complete since 156 different mutations have been observed only once. An additional 41 different mutations have been observed in 2–10 different individuals. At present, there are no obvious hotspots other than a disproportionate predominance of the base-substitution mutations at GC base pairs (358/548 = 65%). One-quarter of the total *in vivo* somatic mutations cause aberrant splicing of *hprt* mRNA primarily due to single base changes in either the donor or acceptor splice site sequences. Loss of exons 4 or 8 occurs most frequently.

A unique spectrum of *hprt* mutations has been found in newborn cord blood T lymphocytes (considered in more detail in Section 17.4.2). In these cells, 83% of the mutants have large structural alterations, with deletion of exons 2 + 3 predominating (McGinniss *et al.*, 1989, 1990). These mutations are specific deletions which contain all the hallmarks of a V(D)J recombinase mechanism (Fuscoe *et al.*, 1991, 1992). This is the enzyme that mediates the recombination of TCR gene sequences during intrathymic maturation of T lymphocytes and it appears that these cord-blood mutants are the result of illegitimate V(D)J activity. Recently, a mutation resulting from the inclusion of TCR α gene sequences into the same region of the *hprt* gene (intron 1) has been reported (Hou, 1994).

In summary, the *hprt* mutation spectrum shows a predominance of large structural alterations in newborn cord-blood samples and samples from children (McGinniss *et al.*, 1989; B. Finette, personal communication). By contrast, samples from individuals aged 25 and older show a predominance of 'point mutations'. The limited spectrum in cord-blood samples may allow more sensitive detection of induced 'point mutations' since the latter are relatively infrequent in unexposed newborns. The spectrum seen in adults so far is random with no obvious predominant mutations. This suggests that *hprt* is a sensitive target for mutation which will allow detection of all classes of mutation events in individuals.

17.4.2 Mutation analysis in exposed individuals

Monitoring a human population for mutation induction by an environmental exposure would be greatly facilitated if the agent of exposure induced a specific mutational event. Mutagen specificity would allow screening individuals for that mutation, defined as a specific biomarker of exposure. Such signature mutations in a reporter gene could then be tested for relevance to mutations in a disease-related gene.

There are now three examples of mutation specificity in human population studies. One is the finding of a difference in mutation class in two populations of newborns, where the 'exposed' population exhibits an elevated Mf, accounted for almost entirely by an increase in 'point' mutations (Manchester *et al.*, personal communication). The second is the finding of an increase in Mf and in *hprt* total gene deletion mutations in patients receiving radioimmunotherapy (RIT; Nicklas *et al.*, 1990). The third is the finding of an increase in both Mf and in a specific base-substitution mutation in workers exposed to ethylene oxide (Tates *et al.*, 1991a). These are considered in more detail below.

A study of Mf in newborn cord-blood samples was undertaken to determine whether maternal environmental/lifestyle factors contribute to transplacental mutagenesis. This study used samples from newborns born of women drawn from a population of low socioeconomic status and with high exposure to tobacco smoke [designated CO(UHD) in *Table 17.2*]. The Mfs were compared with those found in babies born of two groups of women of higher socioeconomic status [designated CO(PHD) and VT in *Table 17.2*]. The CO(UHD) samples showed a significantly elevated Mf compared with the other two populations (*Table 17.2a*). In addition, molecular analysis revealed a significant decrease in the relative contribution of gross structural alterations detected by Southern blot studies (25/38 = 66% vs. 34 of 41 = 83%) suggesting that the induced mutations differ from the background mutations. The primary difference found in the calculated frequency of the three classes of mutations (*Table 17.2b*) was a five-fold increase in the CO(UHD) population in mutations showing no Southern blot changes (i.e. point mutations). The frequency of

Table 17.2. *Hprt* mutations in newborn cord-blood samples

(a) Mutation frequency and *hprt* mutation analysis

Sample	Number	Mean Mf (± SD) x 10^{-6}	No alteration (%)	V(D)J (%)	Other alteration (%)
CO(UHD)	66	1.4 (± 1.1)	13/38 (34)	9/38 (24)	16/38 (42)
CO(PHD)	10	0.7 (± 0.5)	–	–	–
VT	45	0.6 (± 0.4)	7/41 (17)	20/41 (49)	14/41 (34)

(b) Mutation frequency for the three classes of mutations

Sample	Mean Mf x 10^{-6}	No alteration* Mf x 10^{-6}	V(D)J Mf x 10^{-6}	Other alteration Mf x 10^{-6}
CO(UHD)	1.40	0.5	0.3	0.6
VT	0.64	0.1	0.3	0.2

* Indicative of point mutation.

V(D)J mutations was identical in the two populations, with a three-fold increase in the CO(VHD) population in the other types of deletion events. These results demonstrate the unique advantages of mutation monitoring in cord-blood samples, including a low Mf and a relatively limited spectrum of background mutations.

The genotoxic effects of exposure to ionizing radiation have been studied in patients receiving ^{131}I-radiolabelled antihuman ferritin antibody therapy for treatment of non-resectable hepatoma or cholangiocarcinoma (Nicklas *et al.*, 1990, 1991a). This internal exposure to ionizing irradiation clearly resulted in increased Mf. *Table 17.3* lists the results for a pretreatment group and three post-treatment groups (i.e. a group that received a single treatment and was sampled 2 months later and groups who received multiple treatments and were sampled at 2 or 4–27 months after the last treatment). These two groups are defined by their last treatment and not the total treatment because linear regression of each group showed a better correlation of Mf with the single treatment or the last of multiple treatments (O'Neill *et al.*, 1994a). These results also show that Mf remains elevated for up to 2 years after the exposure.

Table 17.3. *Hprt* in patients receiving ^{131}I-radiolabelled antihuman ferritin antibody therapy (RIT)

Group (*n*)	Last treatment (mCi)		Sampling time after last treatment	Mean Mf ($\pm$ SD) x 10^{-6}
	Mean $\pm$ SD	Range		
Pretreatment (13)				11.5 $\pm$ 5.1
Post-treatment				
Single treatment (11)	35.1 $\pm$ 8.4	29–51	2 months	82.7 $\pm$ 80.1
Multiple treatment (14)	32.9 $\pm$ 7.9	28–51	2 months	74.9 $\pm$ 61.6
Multiple treatment (7)	35.9 $\pm$ 9.3	29–51	4–27 months	32.3 $\pm$ 18.9

Adapted from O'Neill *et al.* (1994b).

An analysis of the TCR gene rearrangement patterns showed that the elevated Mf reflected an elevated mutation frequency, as 86–89% of the mutants had unique TCR patterns (*Table 17.4*, line 3). Molecular analysis by Southern blot studies showed an increase in the fraction of mutations with *hprt* structural alterations, especially total gene deletions (*Table 17.4*, lines 6–8). These results demonstrate a signature mutation for exposure of humans to ionizing irradiation.

The third report showing evidence for mutagen specificity is a study of workers exposed to EO. A group of factory workers showed a significant increase in Mf compared with a control population not exposed to EO (Tates *et al.*, 1991a). Recent preliminary studies of the mutational spectrum have revealed specific mutations in these EO workers compared with the cumulative background spectrum in unexposed individuals (Tates, personal communication). Overall (unexposed plus exposed) a total of 41 different *hprt*

Table 17.4. *Hprt* mutations in patients receiving ^{131}-radiolabelled antihuman ferritin antibody therapy (RIT)

	'Normal'	Pretreatment	Post-treatment	
			2 months	4–27 months
No. of mutants	326	203	–	308
'Clonality'	17–2mers	2–2mers	–	10–2mers
	2–3mers	6–3mers	–	3–3mers
	3–4 mers	1–6 mers	–	2–4mers
	1–9mers	1–10mers	–	1–6 mers
			1–8mers	
% Unique mutants	88%	86%	–	89%
No. of mutations	288	175	–	274
No. of mutations analysed	288	118	89	136
No. of complex alterations	12	7	9	16
(%)	(4%)	(6%)	(10%)	(12%)
No. of alterations	41	20	32	57
(%)	(14%)	(17%)	(36%)	(42%)
No. of partial deletions	20	10	11	16
(%)	(7%)	(8%)	(12%)	(12%)
No. with total deletions	9	3	12	25
(%)	(3%)	(3%)	(13%)	(18%)

Adapted from O'Neill *et al.* (1994b), Nicklas *et al.* (1990, 1991a).

mutations have been shown to occur in more than one individual. An additional 156 mutations have been recorded as occurring in only one individual (Cole and Skopek, 1994). In particular, the first G in two GTGT dinucleotide repeats (G_{197} and G_{617}) appears to be a hotspot for G → A transitions. At G_{197}, four G → A transitions have been found in unexposed humans and six in exposed humans, three of whom were exposed to EO. At G_{617}, one G → A transition has been found in unexposed humans and five in exposed humans, three of whom were exposed to EO. (The other five G → A transition 'exposure' mutations were found in humans who received platinum-based chemotherapy.) In these EO workers, six of 22 mutations were found in these two GTGT repeats, all G → A transitions. This appears to be growing evidence for a mutation specificity for EO-induced mutations in humans.

Signature mutations of environmental exposures are being discovered within the *hprt* gene. Such specificity will facilitate population studies and provide direct evidence for a genotoxic effect in potentially exposed individuals. These signature mutations can then be used in studies to understand the molecular basis of the disease linked to the environmental exposure.

17.5 Reporter gene mutations as surrogates for cancer mutations

Among somatic mutations with relevance to disease, of most concern are those associated with cancer. Reporter genes that capture mutagenic mechanisms known to be responsible for carcinogenic mutations have an additional advantage for human monitoring in that they may be predictive of disease risk in individuals. It is worth examining, therefore, how well reporter gene mutations mimic those underlying cancer, given the current state of knowledge.

There is now abundant evidence that somatic recombination, with resultant loss of heterozygosity, is an important mechanism for loss of function of tumour suppressor genes, especially when one inactive mutant copy of the gene is inherited via the germ line. As noted earlier, this mechanism is reflected in two kinds of reporter gene mutations – GPA and HLA. This mechanism represents approximately 50% of the former and 30% of the latter mutations in unexposed, normal adults. The frequencies of these kinds of mutations rise in an age-related manner in normal subjects and, at least for the GPA mutations, predominate in the BS repair-deficiency disease, which is characterized by leukaemias with associated chromosome translocations (Langlois *et al.*, 1989). This mechanism is therefore of relevance to cancer and important to record in human monitoring studies. There are, however, no reports to date that somatic recombinations detected in these two reporter genes have been induced by environmental exposures.

Another general mechanism of mutation relevant to cancer is partial or complete gene deletion. Although such events are relatively infrequent among HLA mutations, they undoubtedly constitute some of the 'hemizygous' GPA mutations. As noted earlier, large deletions constitute approximately 17% of the background *hprt* mutations in normal adults and increase in a dose-dependent manner following exposures to ionizing radiation (Nicklas *et al.*, 1990, 1991a; O'Neill *et al.*, 1994a). Thus, a general mechanism by which such exposures may cause leukaemias and other malignancies is reflected in the damage they induce in the *hprt* gene.

Specific as well as general mutagenic mechanisms may be reflected in reporter gene mutations. This is best illustrated by the illegitimate V(D)J recombinase-mediated deletions of *hprt* noted earlier. The V(D)J recombinase mechanism is critical for the rearrangement of immunoglobulin genes in B lymphocytes and TCR genes in T lymphocytes. Normally, these processes occur only during the maturation stages of B and T cells. However, several lymphoid malignancies are characterized by non-random chromosome rearrangements with one of the breakpoints near the immunoglobulin or TCR genes (Finger *et al.*, 1986). The precise site of this breakpoint is often the heptamer–nonamer consensus sequence that directs canonical rearrangements of these genes. The other breakpoint required to produce the translocation, which is usually in the region of an oncogene, is often in a cryptic consensus heptamer. Junctional regions of the new, translocated chromosome also frequently bear the hallmarks of a V(D)J-mediated event. Therefore, this mechanism of immunoglobulin maturation and TCR gene rearrangement may proceed illegitimately, with carcinogenic consequences.

There are many examples of an aberrant V(D)J recombinase mechanism in lymphoid malignancies. Of particular relevance here is a characteristic translocation t(1;14)(p32;q11) seen in approximately 3% of childhood T-cell acute lymphoblastic leukaemias (T-ALL). In the leukaemic cells, most of the breakpoints on chromosome 14 cluster in the D_δ–J_δ region of the TCR δ locus while those on chromosome 1 are in a 1 kb 5' region of the so-called *tal*-1 gene (Aplan *et al.*, 1990; Bernard *et al.*, 1991; Breit *et al.*, 1993). The *tal*-1 gene codes for a protein thought to be important in early haematopoietic cells. Its dysregulation by the t(1;14) translocation probably produces an oncogene function in the genesis of T-ALL (Chen *et al.*, 1990).

It has recently been found that in an additional 20–30% of childhood T-ALL a submicroscopic deletion of approximately 90 kb of the 5' region of *tal*-1 juxtaposes its coding sequences to the first non-coding exon of an upstream *sil* gene (Breit *et al.*, 1993). This puts the fused *sil*–*tal*-1 gene (now called *tal*^{d}) under the transcriptional control of the *sil* gene promoter. Again, dysregulation of *tal*-1 occurs, leading to leukaemia. Importantly, the 5' breakpoint of this deletion occurs in only one V(D)J recombinase heptamer sequence while the 3' breakpoints cluster in one of four locations, each however in a V(D)J recombinase heptamer–nonamer consensus sequence. The junctional sequences show the hallmarks of V(D)J recombinase activity (Breit *et al.*, 1993).

As noted earlier, approximately 40%–50% of *hprt* T-cell mutations that arise during normal fetal development are due to illegitimate V(D)J recombinase-mediated deletions (Fuscoe *et al.*, 1991). This frequency persists during early childhood and falls off during later life, although such *hprt* mutations constitute a small percentage of the total even into adult life (Fuscoe *et al.*, 1992). In these mutations, the 5' breakpoint (in *hprt* intron 1) always occurs in a V(D)J recombinase heptamer sequence while the 3' breakpoints (in *hprt* intron 3) arise in one of three clustered regions, each showing a V(D)J recombinase heptamer–nonamer consensus sequence. The junctional sequences show the hallmarks of V(D)J recombinase activity. Thus, the specific mutational and carcinogenic mechanism seen in several lymphocyte leukaemias-lymphomas, and studied in detail in the *tal* ^{d} mutation, is captured precisely in the *hprt* reporter gene. Although this mutational mechanism is usually of spontaneous origin, there is also suggestive *in vivo* and *in vitro* evidence that it may be induced by environmental agents (Fuscoe *et al.*, 1992; McGregor *et al.*, 1994).

Even at this early stage of knowledge, there is evidence that mutational mechanisms relevant to disease can be detected in the reporter genes currently available. Whether or not such mutations are more predictive of subsequent disease attributable to genotoxic agents remains to be determined by studies at the population level. Recent reports that non-specific chromosome aberrations may be predictive in individuals of increased risk for subsequent development of cancer does support the hypothesis that mutations in reporter genes, at least of some kinds, may some day provide an estimate of risk that can be applied to individuals (Hagmar *et al.*, 1994; Kleinerman *et al.*, 1994). While this clearly is not possible at present, it should be an area of intensive research.

17.6 Summary and conclusions

Humans can be monitored for the occurrence of *in vivo* somatic cell gene mutations. Somatic mutations are 'effect biomarkers' which can be used to determine if a given exposure did induce genetic damage. Heterogeneity of human susceptibility should be reflected in differences in the magnitude of the effect between individuals.

It is evident that the endpoints used in the T-lymphocyte assays have a genetic basis and that the use of the TCR gene rearrangement patterns can provide a mutation frequency. The third challenge remains: how well do surrogate mutations in a surrogate tissue (peripheral blood T lymphocytes) mimic disease gene mutations in target tissues? If they are valid surrogates, these assays become predictors of diseases caused by genotoxins and may even identify individuals for medical surveillance and suggest possible preventative intervention. If they are not valid surrogates for mutations in genes that predispose to disease, these assays remain dosimeters of exposure, supplying information only on the genotoxic effect of an exposure. However, the inability to predict diseases in an individual would be quite disappointing to genetic toxicologists involved in biomonitoring.

The challenge of validation of these mutation assays is primarily a jump from populations down to individuals. Populations are exposed to genotoxicants in the environment. At present, biomonitoring for somatic mutations defines the genetic effect and may serve as an impetus to reduce or abolish these exposures. If not, individuals will subsequently get genetic diseases such as cancer. In order to determine if these assays are valid surrogates for mutations in cancer genes there are at least three approaches to validation.

One approach takes advantage of the recent advances in detecting mutations in cancer. Mutations in reporter genes and cancer genes can now be measured in the same individuals, for example, in cancer patients receiving mutagenic chemotherapy agents. Positive correlations between the two endpoints would be evidence that reporter gene mutations are valid surrogates.

A second approach is to add mutagenicity studies to human cancer chemoprevention trials. If such prospective studies showed a correlation between decreased cancer incidence and a suppression of reporter gene mutations, one more piece of evidence would be obtained in support of the latter as valid surrogates for events leading to cancer.

The third approach may be the most relevant, but it will require the collaboration of genetic toxicologists with epidemiologists, physicians and statisticians and a 5–10 year duration. It will entail the study of cancer outcomes in individuals, for example, cancer patients receiving chemotherapy or radiotherapy or people exposed to chemicals or radiation through environmental disasters. Blood samples would be collected, and all components and cells should be cryopreserved and stored in a repository. Once the incidence of disease has been determined, the clinical outcome will be used to design nested 'case-control' studies to validate the biomarkers used. The study will ask if the assays accurately predicted the occurrence of cancer in exposed individuals.

If these studies show positive correlations, these assays can be used meaningfully for human risk assessments and in designing strategies for prevention. If not, they remain valuable as effect dosimeters and as biomarkers of population exposure. Of course, these assays will still be valid for basic research into human mutational mechanisms. Whatever the outcome, the results will define the future use of human mutation monitoring assays.

Acknowledgements

The research of the authors has been supported by NCI RO1 CA30688, NIEHS PO1 E505249 and DOE FG028760502. This support does not constitute an endorsement of the views expressed. The authors are indebted to Inge Gobel for her patient preparation of this chapter, and to Dr Janice A. Nicklas for her critical reading and thoughtful suggestions.

References

Akiyama M, Kyoizumi S, Hirai Y, Hakoda M, Nakamura N, Awa AA. (1990) Studies on chromosome aberrations and HPRT mutations in lymphocytes and GPA mutation in erythrocytes of atomic bomb survivors. In: *Mutation and the Environment Part C: Somatic and Heritable Mutation, Adduction, and Epidemiology* (eds ML Mendelson, RJ Albertini). Wiley-Liss, New York, pp. 69–80.

Albertini RJ. (1985) Somatic gene mutations *in vivo* as indicated by the 6-thioguanine-resistant T-lymphocytes in human blood. *Mutat. Res.* **150:** 411–422.

Albertini RJ, Castle KL, Borcherding WR. (1982) T-cell cloning to detect the mutant 6-thioguanine-resistant lymphocytes present in human peripheral blood. *Proc Natl Acad. Sci. USA* **79:** 6617–6621.

Albertini RJ, Sullivan LS, Berman JK, Greene CJ, Stewart JA, Silveira JM, O'Neill JP. (1988) Mutagenicity monitoring in humans by autoradiographic assay for mutant T-lymphocytes. *Mutat. Res.* **204:** 481–492.

Albertini RJ, Nicklas JA, O'Neill JP, Robison SH. (1990) *In vivo* somatic mutations in humans: measurement and analysis. *Annu. Rev. Genet.* **24:** 305–326.

Ammenheuser MM, Ward JB Jr, Whorton EB Jr, Killian JM, Legator MS. (1988) Elevated frequencies of 6-thioguanine resistant lymphocytes in multiple sclerosis patients treated with cyclophosphamide: a prospective study. *Mutat. Res.* **204:** 509–520.

Ammenheuser MM, Au WW, Whorton EB Jr, Belli JA, Ward JB Jr. (1991) Comparison of *hprt* variant frequencies and chromosome aberration frequencies in lymphocytes from radiotherapy and chemotherapy patients: a prospective study. *Environ. Mol. Mutagen.* **18:** 126–135.

Ammenheuser MM, Berenson AB, Stiglich NJ, Whorton EB Jr, Ward JB Jr. (1994) Elevated frequencies of *hprt* mutant lymphocytes in cigarette-smoking mothers and their newborns. *Mutat. Res.* **304:** 285–294.

Andersson B, Hou S-M, Lambert B. (1992) Mutations causing defective splicing in the human *hprt* gene. *Environ. Mol. Mutagen.* **20:** 89–95.

Aplan PD, Lombardi DP, Ginsberg AM, Cossman J, Bertness VL, Kirsch IR. (1990) Disruption of the human SCL locus by "illegitimate" V-(D)-J recombinase activity. *Science* **250:** 1426–1429.

Atwood KC, Scheinberg SL. (1958) Somatic variation in human erythrocyte antigens. *J. Cell. Comp. Physiol.* **52:** 97–123.

Bernard O, Lecointe N, Jonveaux P, Souyri M, Mauchauffe M, Berger R, Larsen CJ, Mathieu-Mahul D. (1991) Two site-specific deletions and t(1:14) translocation restricted to human T-cell acute leukemias disrupt the 5′ part of the *tal*-1 gene. *Oncogene* **6:** 1477–1488.

Bernini LF, Natarajan AT, Schreuder-Rotteveel AHM, Giordano PC, Ploem JS, Tates A. (1990) Assay for somatic mutation of human hemoglobins. In: *Mutation and the Environment Part C: Somatic and Heritable Mutation, Adduction, and Epidemiology* (eds ML Mendelson, RJ Albertini). Wiley-Liss, New York, pp. 57–68.

Bigbee WL, Branscombe EW, Weintraub HB, Papayannapoulou T, Stamatoyannopoulos G. (1981) Cell sorter immunofluorescence detection of human erythrocytes labeled in suspension with antibodies specific for hemoglobin S and C. *J. Immunol. Methods* **45:** 117–127.

Bigbee WL, Wyrobek RG, Langlois RG, Jensen RH, Everson RB. (1990) The effect of chemotherapy on the *in vivo* frequency of glycophorin A 'null' variant erythrocytes. *Mutat. Res.* **240:** 165–175.

Branda RF, O'Neill JP, Sullivan LM, Albertini RJ. (1991) Factors influencing mutation at the *hprt* locus in T-lymphocytes: women treated for breast cancer. *Cancer Res.* **51:** 6603–6607.

Branda RF, Sullivan LM, O'Neill JP, Falta MT, Nicklas JA, Hirsch B, Vacek PM, Albertini RJ. (1993) Measurement of HPRT mutant frequencies in T-lymphocytes from healthy human populations. *Mutat. Res.* **285:** 267–279.

Breit TM, Mol EJ, Wolvers-Tettero ILM, Ludwig W-D, van Wering ER, Van Dongen JJM. (1993) Site-specified deletions involving the *tal*-1 and *sil* genes are restricted to cells of the T-cell αβ receptor lineage: T-cell receptor δ gene deletion mechanism affect multiple genes. *J. Exp. Med.* **177:** 965–977.

Caggana M, Benjamin MB, Little JB, Liber HL, Kelsey KT. (1991) Single-strand conformation polymorphisms can be used to detect T cell receptor gene rearrangements: an application to the *in vivo hprt* mutation assay. *Mutagenesis* **6:** 375–379.

Cariello NF, Skopek TR. (1993) Analysis of mutation occurring at the human *hprt* locus. *J. Mol. Biol.* **231:** 41–57.

Cariello NF, Craft TR, Vrieling H, van Zeeland AA, Adams T, Skopek TR. (1992) Human HPRT mutant database: software for data entry and retrieval. *Environ. Mol. Mutagen.* **20:** 81–83.

Chen Q, Ying-Chuan Yang C, Tsou Tsan J, Xia Y, Ragab AH, Peiper SC, Carroll A, Baer R. (1990) Coding sequences of the *tal*-1 gene are disrupted by chromosome translocation in human T-cell leukemia. *J. Exp. Med.* **172:** 1403–1408.

Cochrane JE, Skopek TR. (1994a) Mutagenicity of butadiene and its epoxide metabolites: I. Mutageneic potential of 1,2-epoxybutene, 1,2,3,4-diepoxybutane, and 3,4-epoxy-1,2-butanediol in cultured human lymphoblasts. *Carcinogenesis* **15:** 713–717.

Cochrane JE, Skopek TR. (1994b) Mutagenicity of butadiene and its epoxide metabolites: II. Mutational spectra of butadiene, 1,2-epoxybutene, and diepoxybutane at the *hprt* locus in splenic T-cells from exposed B6C3F1 mice. *Carcinogenesis* **15:** 719–723.

Cole J, Skopek TR. (1994) ICPEMC Committee on spontaneous mutation: working paper 3: somatic mutant frequency, mutation rates and mutational spectra in the human population *in vivo*. *Mutat. Res.* **304:** 33–106.

Cooper DN, Krawczak M. (1990) The mutational spectrum of single base pair substitutions causing human genetic disease: patterns and predictions. *Hum. Genet.* **85:** 55–74.

Curry J, Skandalis A, Hollcraft J, de Boer JG, Glickman BW. (1993) Coamplification of *hprt* cDNA and *gamma* T-cell receptor sequences from 6-thioguanine-resistant human T-lymphocytes. *Mutat. Res.* **288:** 269–275.

de Boer JG, Curry JD, Glickman BW. (1993) A fast and simple method to determine the clonal relationship among human T-cell lymphocytes. *Mutat. Res.* **288:** 173–180.

Finette BA, Sullivan LM, O'Neill JP, Nicklas JA, Vacek PM, Albertini RJ. (1994) Determination of HPRT mutant frequencies in T-lymphocytes from a healthy pediatric population: statistical comparison between newborn, children and adult mutant frequencies, cloning efficiency and age. *Mutat. Res.* **308:** 223–231.

Finger LR, Harvey RC, Moore RCA, Showe LC, Croce CM. (1986) A common mechanism of chromosomal translocation in T- and B-cell neoplasia. *Science* **234:** 982–985.

Fuscoe JC, Zimmerman LJ, Lippert ML, Nicklas JA, O'Neill JP, Albertini RJ. (1991) V(D)J recombinase-like activity mediates *hprt* gene deletion in human fetal T-lymphocytes. *Cancer Res.* **51:** 6001–6005.

Fuscoe JC, Zimmerman LJ, Harrington-Brock K, Burnette L, Moore MM, Nicklas JA, O'Neill JP, Albertini RJ. (1992) V(D)J recombinase-mediated deletion of the *hprt* gene in T-lymphocytes from adult humans. *Mutat. Res.* **283:** 13–20.

Gibbs RA, Nguyen PN, McBride LJ, Koepf SM, Caskey CT. (1989) Identification of mutations leading to the Lesch–Nyhan syndrome by automated direct DNA sequencing of *in vitro* amplified cDNA. *Proc. Natl Acad. Sci. USA* **86:** 1919–1923.

Gibbs RA, Nguyen PN, Edwards A, Civitello AB, Caskey CT. (1990) Multiplex DNA deletion detection and exon sequencing of the hypoxanthine phosphoribosyltransferase gene in Lesch–Nyhan families. *Genomics* **7:** 235–244.

Grant S, Bigbee WL. (1993) *In vivo* somatic mutation and segregation at the human glycophorin A (GPA) locus. *Mutat. Res.* **288:** 163–172.

Grist SA, McCarron M, Kutlaca A, Turner DR, Morley AA. (1992) *In vivo* human somatic mutation: frequency and spectrum with age. *Mutat. Res.* **266:** 189–196.

Hagmar L, Brøgger A, Hansteen I-L et al. (1994) Cancer risk in humans predicted by increased levels of chromosomal aberrations in lymphocytes: Nordic study group on the health risk of chromosome damage. *Cancer Res.* **54:** 2919–2922.

Hakoda M, Akiyama M, Hirai Y, Kyoizumi S, Awa AA. (1988a) *In vivo* mutant T cell frequency in atomic bomb survivors carrying outlying values of chromosome aberration frequencies. *Mutat. Res.* 202: 203–208.

Hakoda M, Akiyama M, Kyoizumi S, Awa AA, Yamakido M, Otake M. (1988b) Increased somatic cell frequency in atomic bomb survivors. *Mutat. Res.* **201:** 39–48.

Hakoda M, Akiyama M, Kyoizumi S, Kobuke K, Awa AA, Yamakido M. (1988c) Measurement of *in vivo* HGPRT-deficient mutant cell frequency using a modified method for cloning human peripheral blood T-lymphocytes. *Mutat. Res.* **197:** 161–169.

Henderson L, Cole H, Cole J, James SE, Green M. (1986) Detection of somatic mutations in man: evaluation of the microtitre cloning assay for T-lymphocytes. *Mutagenesis* **1:** 195–200.

Hollstein MC, Sidransky D, Vogelstein B, Harris CC. (1991) *p53* mutations in human cancer. *Science* **253:** 49–53.

Hou S-M. (1994) Novel types of mutation identified at the *hprt* locus of human T-lymphocytes. *Mutat. Res.* **308:** 23–31.

Janatipour M, Trainor KJ, Kutlaca R, Bennett G, Hay J, Turner DR, Morley AA. (1988) Mutations in human lymphocytes studied by an HLA selection system. *Mutat. Res.* **198:** 221–226.

Jensen RH, Langlois RG, Bigbee WL. (1986) Determination of somatic mutations in human erythrocytes by flow cytometry. *Prog. Clin. Biol. Res.* **209B:** 177–184.

Jensen RH, Bigbee WL, Langlois RG. (1987) *In vivo* somatic mutations in the glycophorin A locus of human erythroid cells. In: *Mammalian Cell Mutagenesis* (eds MM Moore, DM DeMarini, KR Tindall). Cold Spring Harbor Laboratory Press, Cold Spring Harbor, NY, pp. 149–159.

Jensen RH, Bigbee WL, Langlois RG. (1990) Multiple endpoints for somatic mutations in humans provide complementary views for biodosimetry, genotoxicity and health risks. In: *Mutation and the Environment, Part C* (eds ML Mendelsohn, RJ Albertini). Wiley-Liss, New York, pp. 81–92.

Kleinerman RA, Littlefield LG, Tarone RE, Sayer AM. (1994) Chromosome aberrations in lymphocytes from women irradiated for benign and malignant gynecological disease. *Radiat. Res.* **139:** 40-46.

Kushiro J-i, Hirai Y, Kusunoki Y, Kyoizumi S, Kodama Y, Wakisaka A, Jeffreys A, Cologne JB, Dohi K, Nakamura N, Akiyama M. (1992) Development of a flow-cytometric HLA-A locus mutation assay for human peripheral blood lymphocytes. *Mutat. Res.* **272:** 17–29.

Kyoizumi S, Akiyama M, Hirai Y, Kusunoki Y, Tanabe K, Umeki S. (1990) Spontaneous loss and alteration of antigen receptor expression in mature CD4+ T cells. *J. Exp. Med.* **171:** 1981–1999.

Kyoizumi S, Umeki S, Akiyama M, Hirai Y, Kusunoki Y, Nakamura N, Endoh K, Konishi J, Sasaki MS, Mori T, Fujita S, Cologne JB. (1992) Frequency of mutant T lymphocytes defective in the expression of the T-cell antigen receptor gene among radiation-exposed people. *Mutat. Res.* **265:** 173–180.

Langlois RG, Bigbee WL, Jensen RH. (1987a) Measurements of the frequency of human erythrocytes with gene expression loss phenotypes in the glycophorin A locus. *Hum. Genet.* **74:** 353–362.

Langlois RG, Bigbee WL, Kyoizumi S, Nakamura N, Bean MA, Akiyama M, Jensen RH. (1987b) Evidence for increased somatic cell mutations at the glycophorin A locus in atomic bomb survivors. *Science* **236:** 445–448.

Langlois RG, Bigbee WL, Jensen RH, German J. (1989) Evidence for increased *in vivo* mutation and somatic recombination in Bloom's Syndrome. *Proc. Natl Acad Sci USA* **86:** 670–674.

Lippert MJ, Albertini RJ, Nicklas JA. (1995a) Physical mapping of the *hprt* chromosomal region (Xq26). *Mutat. Res.* **326:** 39–49.

Lippert MJ, Nicklas JA, Hunter TC, Albertini RJ. (1995b) Pulsed field analysis of *hprt* T-cell large deletions: telomeric region breakpoint spectrum. *Mutat. Res.* **326:** 51–64.

McCarron MA, Kutlaca A, Morley AA. (1989) The HLA-A mutation assay: improved technique and normal results. *Mutat. Res.* **225:** 189–193.

McGinniss MJ, Nicklas JA, Albertini RJ. (1989) Molecular analyses of *in vivo hprt* mutations in human T-lymphocytes IV. Studies in newborns. *Environ. Mol. Mutagen.* **14:** 229–237.

McGinniss MJ, Falta MT, Sullivan LM, Albertini RJ. (1990) *In vivo hprt* mutant frequencies in T-cells of normal human newborns. *Mutat. Res.* **240:** 117–126.

McGregor WG, Maher VM, McCormick JJ. (1994) Kinds and locations of mutations induced in the hypoxanthine–guanine phosphoribosyltransferase gene of human T-lymphocytes by 1-nitrosopyrene, including those caused by V(D)J recombinase. *Cancer Res.* **54:** 4207–4213.

Morley AA, (1991) Mitotic recombination in mammalian cells *in vivo. Mutat. Res.* **250:**345–349.

Morley AA, Grist SA, Turner DR, Kutlaca A, Bennett G. (1990) Molecular nature of *in vivo* mutations in human cells at the autosomal HLA-A locus. *Cancer Res.* **50:** 4584–4587.

Morley AA, Trainor KJ, Seshadri R, Ryall RG. (1983) Measurement of *in vivo* mutations in human lymphocytes. *Nature* **302:** 155–156.

Morley AA, Trainor KJ, Dempseh JL, Seshadri RS. (1985) Methods for study of mutations and mutagenesis in human lymphocytes. *Mutat. Res.* **147:** 363–367.

Nicklas JA, O'Neill JP, Albertini RJ. (1986) Use of T-cell receptor gene probes to quantify the *in vivo hprt* mutations in human T-lymphocytes. *Mutat. Res.* **173:** 65–72.

Nicklas JA, O'Neill JP, Sullivan LM, Hunter TC, Allegretta M, Chastenay B, Libbus BL, Albertini RJ. (1988) Molecular analyses of *in vivo* hypoxanthine–guanine phosphoribosyltransferase mutations in human T-lymphocytes: II. Demonstration of a clonal amplification of *hprt* mutant T-lymphocytes *in vivo. Environ. Mol. Mutagen.* **12:** 271–284.

Nicklas JA, Falta MT, Hunter TC, O'Neill JP, Jacobson-Kram D, Williams J, Albertini RJ. (1990) Molecular analysis of *in vivo hprt* mutations in human lymphocytes – V. Effects of total body irradiation secondary to radioimmunoglobulin therapy (RIT). *Mutagenesis* **5:** 461–468.

Nicklas JA, O'Neill JP, Hunter TC, Falta MT, Lippert MJ, Jacobson-Kram D, Williams JR, Albertini RJ. (1991a) *In vivo* ionizing irradiations produce deletions in the *hprt* gene of human T-lymphocytes. *Mutat. Res.* **250:** 383–396.

Nicklas JA, Hunter TC, O'Neill JP, Albertini RJ. (1991b) Fine structure mapping of the *hprt* region of the human X chromosome (Xq26). *Am. J. Hum. Genet.* **49**: 267–278.

O'Neill JP, McGinniss MJ, Berman JK, Sullivan LM, Nicklas JA, Albertini RJ. (1987) Refinement of a T-lymphocyte cloning assay to quantify the *in vivo* thioguanine-resistant mutant frequency in humans. *Mutagenesis* **2**: 87–94.

O'Neill JP, Albertini RJ, Nicklas JA. (1994a) Molecular analysis of mutations induced *in vivo* in humans. In: *Molecular Environmental Biology* (ed. SJ Garte). Lewis, Boca Raton, FL, pp. 207–223.

O'Neill JP, Nicklas JA, Hunter TC, Batson OB, Allegretta M, Falta MT, Branda RF, Albertini RJ. (1994b) The effect of T-lymphocyte 'clonality' on the calculated *hprt* mutation frequency occurring *in vivo* in humans. *Mutat. Res.* **313**: 215–225.

Ostrosky-Wegman P, Montero RM, Cortinas de Nova C, Tice RR, Albertini RJ. (1988) The use of bromodeoxyuridine labeling in the human lymphocyte HGPRT somatic mutation assay. *Mutat. Res.* **191**: 211–214.

Papayannopoulou TH, Brice M, Stamatoyannopoulos G. (1977a) Hemoglobin F synthesis *in vitro:* evidence for control at the level of primitive erythroid stem cells. *Proc. Natl Acad. Sci. USA* **74**: 2923–2927.

Papayannopoulou TH, Nute PE, Stamatoyannopoulos G, McGuire TG. (1977b) Hemoglobin ontogenesis: test of the gene exclusion hypothesis. *Science* **197**: 1215–1216.

Perera FP, Tang DL, O'Neill P, Bigbee WL, Albertini R, Santella R, Ottman R, Tsai WY, Dickey C, Mooney LA, Savela K, Hemminki K. (1993) HPRT and glycophorin A mutations in foundry workers: relationship to PAH exposure and to PAH-DNA adducts. *Carcinogenesis* **14**: 969–973.

Robinson DR, Goodall K, Albertini RJ, O'Neill JP, Sala-Trepat M, Tates AD, Beare D, Green MHL, Cole J. (1994) An analysis of *in vivo hprt* mutant frequency in circulating T-lymphocytes in the normal human population: a comparison of four databases. *Mutat. Res.* **313**: 227–247.

Rossi AM, Tates AD, van Zeeland AA, Vrieling H. (1992) Molecular analysis of mutations affecting *hprt* mRNA splicing in human T-lymphocytes *in vivo*. *Environ. Mol. Mutagen.* **19**: 7–13.

Russell LB, Major MH. (1957) Radiation induced presumed somatic mutants in the house mouse. *Genetics* **42**:161-175.

Sculley DG, Dawson PA, Emmerson BT, Gordon RB. (1992) A review of the molecular basis of hypoxanthine-guanine phosphoribosyltransferase (HPRT) deficiency. *Hum. Genet.* **90**: 195–207.

Sommer SS. (1992) Assessing the underlying pattern of human germline mutations: lessons from the factor IX gene. *FASEB J.* **6**: 2767–2774.

Steingrimsdottir H, Rowley G, Dorado G, Cole J, Lehmann AR. (1992) Mutations which alter splicing in the human hypoxanthine-guanine phosphoribosyltransferase gene. *Nucleic Acids Res.* **20**: 1201–1208.

Strauss GH, Albertini RJ. (1979) Enumeration of 6-thioguanine resistant peripheral blood lymphocytes in man as a potential test for somatic cell mutations arising *in vivo*. *Mutat. Res.* **61**: 353–379.

Sutton HE. (1972) In: *Mutagenic Effects of Environmental Contaminants* (eds HE Sutton, MI Harris). Academic Press, New York, pp. 121-128.

Tates AD, Grummt T, Tornqvist M, Farmer PB, van Dam FJ, van Mossel H, Schoemaker HM, Osterman-Golkar S, Uebel C, Tang YS, Zwinderman AH, Natarajan AT, Ehrenberg L. (1991a) Biological and chemical monitoring of occupational exposure to ethylene oxide. *Mutat. Res.* **250**: 483–497.

Tates AD, van Dam FJ, van Mossel H, Schoemaker H, Thijssen JCP, Woldring VM, Zwinderman AM, Natarajan AT. (1991b) Use of the clonal assay for the measurement of frequencies of HPRT mutants in T-lymphocytes from five control populations. *Mutat. Res.* **253**: 199–213.

Trainor KJ, Brisco MJ, Wan JH, Neoh S, Morley AA. (1991) Gene rearrangement in B and T lymphoproliferative disease detected by the polymerase chain reaction. *Blood* **78:** 192–197.

Turner DR, Morley AA. (1990) Human somatic mutation at the autosomal HLA-A locus. In: *Mutation and the Environment Part C: Somatic and Heritable Mutation, Adduction, and Epidemiology* (eds ML Mendelson, RJ Albertini). Wiley-Liss, New York, pp. 37–46.

Turner DR, Grist SA, Janatipour M, Morley AA. (1988) Mutations in human lymphocytes commonly involve gene duplication and resemble those seen in cancer cells. *Proc. Natl Acad. Sci. USA* **85:** 3189–3193.

Van Dongen JJM, Wolvers-Tettero ILM. (1991) Analysis of immunoglobulin and T cell receptor genes. Part I: basic and technical aspects. *Clin. Chim. Acta* **198:** 1–92.

Van Dongen JJM, Comans-Bitter WM, Wolvers-Tettero ILM, Borst J. (1990) Development of human T lymphocytes and their thymus-dependency. *Thymus* **16:** 207–234.

Verwoerd NP, Bernini LF, Bonnet J, Tanke HJ, Natarajan AT, Tates AD, Sobels FH, Ploem JS. (1987) Somatic cell mutations in humans detected by image analysis of immunofluorescently stained erythrocytes. In: *Clinical Cytometry and Histometry* (eds G Burger, JS Ploem, K Gorttler). Academic Press, San Diego, pp. 465–469.

Vrieling H, Thijssen JCP, Rossi AM, van Dam FJ, Natarajan AT, Tates AD, van Zeeland AA. (1992) Enhanced *hprt* mutant frequency but no significant difference in mutation spectrum between a smoking and a non-smoking human population. *Carcinogenesis* **13:** 1625–1631.

Ward JB, Ammenheuser MM, Bechtold WE, Whorton EB Jr, Legator MS. (1995) *hprt* mutant lymphocyte frequencies in workers at a 1,3-butadiene production plant. *Environ. Hlth Perspect.*, in press.

Protein and DNA adducts as biomarkers of exposure to environmental mutagens

David H. Phillips and Peter B. Farmer

18.1 Introduction

Mutagens and many human carcinogens exert their biological activity by damaging DNA. In many cases this is through the formation of electrophilic species that can react covalently with nucleophilic sites on the purine or pyrimidine bases. Monitoring DNA isolated from human or animal tissues is thus a means of determining the levels of prior exposure to environmental mutagens. The same electrophiles also react with nucleophilic sites in proteins, and the abundance of haemoglobin and albumin in readily accessible human red blood cells and plasma makes these attractive alternatives to DNA for monitoring carcinogen exposure.

The study of protein and DNA adducts can be used in determining the pathways of activation of carcinogens and mutagens. It can also be used to assess the genotoxic potential of environmental mixtures and to identify their biologically significant components. It can be used to test new chemicals for potential carcinogenic and mutagenic activity and to investigate whether an established carcinogen acts through a genotoxic or non-genotoxic mechanism. Also, it can be seminal in monitoring human exposure to carcinogens and mutagens and in the identification of the causative agents of human cancer.

This chapter describes the methods available for detecting and characterizing protein and DNA adducts. The relationship between adduct formation and carcinogenic activity, determined in experimental studies, is described, together with examples of the formation of these lesions in humans and other species as a result of environmental exposure to genotoxic agents.

18.2 Detection methods

18.2.1 Protein adducts

For reasons of availability and long lifetime, two proteins have been extensively used for carcinogen adduct detection. These are haemoglobin, present at 140 mg ml^{-1} in blood with a lifetime of around 120 days in humans, and albumin, present at 45 mg ml^{-1} in blood with $t_{1/2}$ of about 20 days. The analytical methods that are currently most used to detect adducts in these proteins are based on immunoassay, mass spectrometry (MS), fluorescence or (in the case of experimental systems only) radiochemical measurements. Quantitation of protein–carcinogen adducts is, in theory, most accurately performed using mass spectrometry coupled with a stable isotope-labelled internal standard, although poor choice of standard or inaccurate estimation of standard adduct concentrations may lead to quantitative errors. Quantitation using immunoassay is based on standard inhibition curves made using solutions of known amounts of adduct. A possible source of inaccuracy in such determinations arises if the antibody used is not 100% specific in its binding to the adduct.

The absolute analytical detection limits of adducts for these methods are normally in the low picogram range, although some gas chromatography (GC)–MS approaches can exceed these sensitivities. This corresponds, for example, to a detection limit of about 10 pmol g^{-1} protein for immunoassay of the aflatoxin B$_1$–albumin adduct (uncorrected for recovery 23%; Wild et al., 1990a). This is similar to the limit of detection for a similar-sized sample by HPLC–fluorescence assay (although the recovery is lower, 5.5%; Wild et al., 1990a). GC–MS analysis of adducts of the N-terminal valine of haemoglobin with low-molecular-weight alkylating agents also reaches limits of detection of 10 pmol g^{-1} protein although, in this case, these are corrected for recovery (Bailey et al., 1988; Bergmark et al., 1993). A much greater sensitivity of detection has, however, been reported for the analysis of tobacco-specific nitrosamine–haemoglobin adducts, which may be measured by negative ion chemical ionization MS at concentrations down to 5 fmol g^{-1} haemoglobin (Carmella et al., 1990).

Mass spectrometric detection of protein adducts. As the technology of MS has developed, there has been a continual increase in the possible approaches for the detection of protein adducts. Initially, the MS analytical methods depended upon procedures involving complete acidic hydrolysis of the protein to amino acids, followed by purification and derivatization of the latter, and capillary GC–MS using electron impact (EI) or chemical ionization (CI). The advent of softer ionization conditions (e.g. fast atom bombardment, FAB) allowed adducts to be detected at the peptide level following enzymatic degradation of the protein. Tandem MS (in which ions separated by one MS may be fragmented in a collision cell and then reanalysed in a second MS), enabled sequences to be obtained of adducted peptides. Finally, the more recent development of electrospray ionization (ESI) and matrix-assisted laser desorption ionization (MALDI) gives the possibility of detecting adducts on intact proteins. The highest sensitivity of detection (and ability to quantitate) is associated with the GC–MS methods, although one gains no information regarding the location of the adduct within the sequence of the protein chain. For example, negative ion CI GC–MS may detect adducts derived from tobacco-specific nitrosamines with a detection limit for the standard compound of about 100 amol (Carmella *et al.*, 1990; Hecht *et al.*, 1994). In contrast, detection of adducts on intact haemoglobin by ES–MS (Springer *et al.*, 1993), or on peptides by FAB MS–MS requires a much higher level of modification of the protein (Kaur *et al.*, 1989), and these methods are, therefore, unlikely to find application for routine monitoring of human exposure to carcinogens. They are, however, of great value for the structural elucidation of adducts produced in *in vitro* experiments.

The amino acids where carcinogen adducts are produced in proteins are those containing nucleophilic centres (e.g. cysteine, histidine, aspartic acid, glutamic acid, N-terminal amino acid and lysine; Farmer, 1994). The mass spectral procedures for analysing these are summarized in *Table 18.1*. Stable adducts (e.g. *S*-alkylcysteines, *N*-alkylhistidines) may be determined by GC–MS following acidic hydrolysis of the protein chain, chromatographic separation of the adduct and derivatization. Less stable adducts such as the esters formed by alkylating agents on aspartic and glutamic acids, and the sulphinamides formed by aromatic amines on cysteine sulphydryl groups may be analysed following mild hydrolysis of the adducted protein, which cleaves carcinogen-derived residues from these adducts that may then be extracted and subjected to GC–MS. Other chemical degradation methods which have been used for adduct detection are those for cysteine thioethers, which may be reductively cleaved with Raney nickel, and for N-terminal amino acid adducts, which may be cleaved from the protein chain by a modified Edman degradation, using pentafluorophenyl isothiocyanate.

Little has been done, to date, to check the accuracy and reproducibility in different laboratories of these analytical approaches, the exception being the analysis of the adduct of ethylene oxide with the N-terminal valine in haemoglobin. In an interlaboratory study using the same adducted proteins, variations of eight-fold were seen in adduct determinations (Törnqvist *et al.*, 1992). This is believed to be due to the use of different internal standards in the participating laboratories.

Immunochemical analysis of protein adducts. Although many antibodies have been produced against DNA-carcinogen adducts (see later), the use of immunoassay for determining protein adducts has been relatively limited. The leading use of this technique for human biomonitoring has been for the major adduct of aflatoxin B_1 with albumin, which is on lysine (Wild *et al.*, 1993). Following precipitation of the protein from plasma, enzymatic hydrolysis is

Table 18.1. GC–MS procedures for determining protein adducts

Amino acid	Method	Adduct	Material analysed	Example	Reference
Cysteine	Strong acid hydrolysis	S-alkylcysteine	S-alkylcysteine	S-(2-carboxyethyl)-cysteine	Bailey et al. (1986)
	Mild hydrolysis	S-arylcysteine Sulphinamides with aromatic amines	S-arylcysteine Aromatic amines	S-phenylcysteine 4-Aminobiphenyl 4,4'-methylene dianiline	Bechtold et al. (1991 Bryant et al. (1987) Bailey et al. (1990)
	Raney nickel	S-alkylcysteine (e.g. from interaction with stryene 7,8-oxide benzoquinone conjugates)	Alcohol	Phenyl ethanol	Ting et al. (1990) MacDonald et al. (1993)
Carboxylic acids (aspartic, glutamic, C-terminal)	Mild hydrolysis	Esters (e.g. from interaction with PAH epoxides or tobacco-specific nitrosamine metabolites)	Alcohol	Benzo[a]pyrene 7,8,9,10-tetrahydrotetrol 4-Hydroxy-1-(3-pyridyl)-1-butanone	Weston et al. (1989), Day et al. (1990) Carmella et al. (1990)
Histidine	Strong acid hydrolysis	N-alkylhistidine	N-alkylhistidine	N-(2-hydroxyethyl) histidine	Osterman-Golkar et al. (1983)
N-terminal valine in haemoglobin	Modified Edman degradation	N-alkylvaline	Pentafluorophenyl thiohydantoin of alkyl valine	N-(2-hydroxyethyl) valine pentafluorophenyl thiohydantoin	Törnqvist et al. (1986)

used to liberate the adduct which is purified by cartridge chromatography and quantitated by ELISA. (Analysis of such an adduct would not be feasible by GC–MS for reasons of its non-volatility and polarity.) Other antibodies which have been used to detect protein adducts include those against the adduct of ethylene oxide with the N-terminal valine in haemoglobin (Wraith *et al.*, 1988), and against benzo[*a*]pyrene adducts with albumin (Santella *et al.*, 1986; Sherson *et al.*, 1990). In the latter case, the antibody recognizes benzo[*a*]pyrene 7,8,9,10-tetrahydrotetrol, which is released from the protein by acidic hydrolysis. Adducts with proteins are also formed by acetaldehyde, the major product of alcohol metabolism, and antibodies are available to monitor such damage (Lin *et al.*, 1993; Niemala and Israel, 1992).

Other analytical approaches for protein adducts. HPLC with fluorescence detection has been used to detect both the aflatoxin B_1–lysine adduct in albumin hydrolysates (Wild *et al.*, 1990a) and, using normal or synchronous fluorescence spectrometry, benzo[*a*]pyrene 7,8,9,10-tetrahydrotetrol released by acidic hydrolysis of benzo[*a*]pyrene–haemoglobin adducts (Bechtold *et al.*, 1991; Weston *et al.*, 1989). Extensive purification of the fluorophore is necessary, for example by immunoaffinity chromatography, in order to give satisfactory fluorescence spectra. Owing to the lower sensitivity of other HPLC detection systems compared with fluorescence, they are not generally suitable for human biomonitoring of protein adducts. Electron capture detection coupled with gas chromatography has, however, found application for monitoring adducts of ethylene oxide and propylene oxide with the N-terminal valine of haemoglobin (sensitivity limit 100 pmol g^{-1} globin; Kautiainen and Törnqvist, 1991).

18.2.2 DNA adducts

Current methods for the detection of DNA adducts, which have been reviewed more comprehensively elsewhere (Bartsch *et al.*, 1988; Groopman and Skipper, 1991; Phillips, 1990; Strickland *et al.*, 1993) fall into three broad categories, each with its own strengths and weaknesses. All have received a wide range of application and in many cases the different methods can be used to complement each other in order to provide additional information on the nature of environmentally induced adducts. Furthermore, combinations of different procedures, for example by using immunoaffinity chromatography (IAC) to concentrate adducts before their analysis by other methods, can both improve the sensitivity of the assays and also increase their selectivity, if such is desired.

Immunochemical methods. There has been widespread use of antisera elicited against carcinogen-modified DNA, or against the carcinogen–DNA adducts themselves, to detect and quantify adducts in human or environmental samples.

A critical requirement is for chemically synthesized adducts, or highly adducted DNA, in order to make antibodies, which limits the method to those carcinogens whose interactions with DNA have been thoroughly characterized. Nevertheless, immunoassays have been used to detect human exposure to many carcinogens, including aflatoxins, aromatic amines, polycyclic aromatic hydrocarbons (PAHs), nitrosamines, UV light and several chemotherapeutic agents that damage DNA (Poirier, 1994). Such antibodies have been used in competitive assays, whereby the antiserum competes with a standard radioactive adduct (RIA, radioimmunoassay) or with standard adduct bound to the bottom of microtitre plate wells (ELISA, enzyme-linked immunosorbent assay). In the ELISA, a second antibody is used that is linked to an enzyme whose substrate is converted to a fluorescent or coloured product. A recent, more sensitive, version of this assay is DELFIA (dissociation-enhanced lanthanide fluoroimmunoassay), in which an avidin–biotin amplification system is made highly fluorescent by the inclusion of bound europium (Schoket et al., 1993a).

A relatively large effort needs to be put into the preparation of antisera, and the amount of DNA required (around 50 µg for a single determination) restricts its use in some human biomonitoring studies. Cross-reactivity of antisera with other carcinogen–DNA adducts complicates interpretation of the analysis of human or environmental samples, but this is confined mostly to adducts formed by carcinogens of the same class; for example, antibodies to DNA modified by benzo[a]pyrene diol-epoxide (BPDE) recognize adducts formed by other PAH diol-epoxides (Poirier, 1991). There are a number of uncertainties inherent in the quantification of adducts by immunoassay, due to variable cross-reactivity and the dependency of the efficiency of antibody binding on the level of DNA modification (Poirier, 1991).

Antibody cross-reactivity can be exploited in biomonitoring studies by using IAC, in which the antibodies are bound to a matrix, to concentrate adducts of a particular class of carcinogens from DNA digests (Poirier, 1993, 1994). Further analytical procedures can then be applied, such as fluorescence spectroscopy or GC–MS (Shuker and Bartsch, 1994).

Antibodies to carcinogen–DNA adducts can be used to locate adducts in tissue samples by immunohistochemistry (Poirier, 1994). Although less sensitive than immunoassay, it has been used, for example, to detect UV-induced thymine dimers (Roza et al., 1991) and adducts formed by 8-methoxypsoralen (Yang et al., 1989) in human skin, and cisplatin–DNA adducts in buccal cells of cisplatin-treated cancer patients (Terheggen et al., 1988).

Postlabelling methods. The ^{32}P-postlabelling assay (Beach and Gupta, 1992) for carcinogen–DNA adducts is the mostly widely used and involves the following steps:

(i) Enzymatic digestion of DNA to 3'-mononucleotides;
(ii) T4 polynucleotide kinase-mediated 5'-phosphorylation using [γ-^{32}P]ATP as the donor molecule;
(iii) Chromatographic separation of the labelled adducts;
(iv) Their detection and quantification by monitoring for decay of the radiolabel.

In most studies, adduct resolution has been by means of anion-exchange thin-layer chromatography (TLC; Beach and Gupta, 1992), which affords great sensitivity when combined with autoradiography and which allows the adducts to be visualized as two-dimensional 'maps'. Increasingly, however, the higher resolving power of high-performance liquid chromatography (HPLC) is finding favour (Gorelick, 1993).

A key stage in the ^{32}P-postlabelling process is the enhancement of the sensitivity of the technique by preselecting the adducted nucleotides for labelling. This can be achieved in a number of ways, the applicability of the method being determined by the nature of the adducts under investigation. One straightforward method for both large and small adducts is to resolve them on HPLC before labelling (Dunn and San, 1988; Kato et al., 1993). Another method, applicable to a broad range of aromatic adducts, is to extract them into an organic phase, such as 1-butanol, leaving the normal nucleotides in the aqueous phase (Gupta, 1985). A third widely used method is prior digestion with nuclease P$_1$ (Reddy and Randerath, 1986). This dephosphorylates normal nucleotides, leaving only those adducts that are resistant to be substrates for the kinase-mediated labelling reaction.

Where DNA samples containing unknown adducts are being analysed, some information can be gained on their nature by comparing adduct recovery after different enhancement procedures. For example, PAH–DNA adducts are detectable after nuclease P$_1$ and after butanol extraction enhancement methods, whereas only the latter method is suitable for the recovery of aromatic amine adducts (Gallagher et al., 1989; Gupta and Earley, 1988).

Alternative digestion and labelling procedures can generate labelled 5'-mononucleotides or labelled dinucleotides (Randerath et al., 1989a), which can aid adduct characterization. Some investigators have explored labelling with ^{35}S-phosphorothioate as a safer alternative to ^{32}P-orthophosphate (Lau and Baird, 1991), and the increasing commercial availability of ^{33}P-labelled ATP offers a more direct alternative than ^{35}S to ^{32}P (Baird et al., 1993), but it is significantly more expensive.

The great strengths of ^{32}P-postlabelling are its sensitivity and versatility. It has the capability to detect adduct frequencies as low as 1 in 10^9 nucleotides using less than 10 μg of DNA, and has been used to detect a wide range of lesions, including bulky/aromatic adducts, cyclic adducts, oxidative DNA damage, apurinic sites, radiation-induced damage and adducts containing small aliphatic moieties (Phillips et al., 1993).

A newer radioactive postlabelling method with the potential for widespread application is acylation with [^{35}S]methionine (Sheabar et al., 1994). Other, non-radioactive, postlabelling methods include derivatizing adducts with fluorescent labels such as dansyl chloride (Jain and Sharma, 1993) or phenylmalondialdehyde (Shuker *et al.*, 1993).

Physical methods. Physical methods offer a more chemical-specific approach to adduct detection, identification and quantification (Weston, 1993), but they are generally less sensitive than immunochemical and postlabelling methods.

Fluorescence spectroscopy has been used to detect DNA adducts formed by several aromatic carcinogens that are highly fluorescent, namely PAHs and aflatoxins. In addition, some methylated nucleosides, such as O^6-methyldeoxyguanosine and N-7-methyldeoxyguanosine, can be detected by fluorescence, but with limited sensitivity. A general prerequisite is knowledge of the structure and/or spectral properties of the adduct of interest. The challenge in human biomonitoring is to identify a particular fluorescent adduct in what is frequently a complex mixture of adducts. One approach is to use synchronous fluorescence spectroscopy (SFS), combined with HPLC, to resolve and detect adducts. By scanning the sample with a fixed wavelength difference between excitation and emission, different adducts can be selectively detected. For example, for the analysis of benzo[*a*]pyrene diol-epoxide–DNA adducts in human DNA, the sample is hydrolysed with acid and the liberated benzo[*a*]pyrene tetrols are resolved on HPLC and detected by SFS using a $\Delta\lambda$ of 34 nm (Weston *et al.*, 1989).

Fluorescence line-narrowing spectroscopy (FLNS; Jankowiak and Small, 1991) combines low temperature and laser excitation in order to produce spectra that can give diagnostic evidence, when compared with a standard, of the presence of a particular adduct in DNA (or protein). It should, in theory, be possible to observe selectively the presence of different adducts in a complex mixture, but to date the analysis by FLNS of environmental or human DNA samples has been limited.

Mass spectrometry has the greatest potential for structural identification of adducts, but the low levels of adducts in human DNA has limited its application in human biomonitoring to circumstances where large amounts of DNA are available. In theory, similar methods to those used to analyse protein adducts are applicable, including selected ion monitoring and negative ion chemical ionization (Talaska *et al.*, 1992). Mass spectrometry is often linked to a chromatography system, including gas or liquid chromatography, capillary electrophoresis or capillary chromatography, and may require adducts to be derivatized in order to render them volatile. Rapidly improving instrumentation and procedures for sample preparation (Giese and Vouros, 1993) indicate that these methods will find widespread future application to DNA adduct monitoring, but at present their sensitivity commonly falls short of that required for routine applications.

Among the most successful procedures that have been attempted for monitoring DNA adducts by mass spectrometry are those that involve hydrolysis of the adducted DNA to yield smaller molecules that are then subjected to high-sensitivity GC–MS procedures. These products are either derived from the carcinogen residue itself or are carcinogen-modified bases. An example of the former is the release by hydrolysis of benzo[a]pyrene tetrol from benzo[a]pyrene-adducted DNA, which can be analysed by GC–MS following conversion to the trimethylsilyl derivative (Weston et al., 1989). A novel procedure using derivatization by permethylation appears to have considerable scope for monitoring mixtures of PAH diols after their hydrolysis from DNA (Melikian et al., 1995). Aromatic amine adducts are also unstable to hydrolysis (which yields the free amines) and may be similarly analysed after derivatization by GC–MS. For example, Friesen et al. (1994) have detected the heterocyclic food mutagen 2-amino-1-methyl-6-phenylimidazo[4,5-b]pyridine (PhIP) in hydrolysates of human colon DNA.

Alternatively, MS analysis may be carried out on adducted nucleic acid bases, following their hydrolytic cleavage from DNA. Thus, the cyclic adduct formed by malondialdehyde with guanine has been quantitated in human liver DNA using GC-electron capture negative chemical ionization MS (Chaudhary et al., 1994). The adduct was obtained following mild acidic hydrolysis of the deoxyribonucleoside mixture resulting from digestion of DNA. DNA bases oxidized by reactive oxygen radicals have also been successfully analysed by GC–MS techniques (Halliwell and Dizdaroglu, 1992).

DNA repair systems are effective in cleavage of some N-7-alkylguanines and N-3-alkyladenines from DNA, and the alkylated bases are then excreted in urine (see Section 18.4.4). These may be purified by chromatographic or immunoaffinity procedures and then derivatized and quantitated by GC–MS (reviewed in Shuker and Farmer, 1992). This procedure has demonstrated, for example, the presence of low-molecular-weight N-3-alkyladenine adducts in cigarette smokers' urine.

Electrochemical detection of 8-oxo-guanine, an abundant lesion in DNA (accounting for around 1 in 10^5 guanine residues) formed through oxidative processes, offers a 1000-fold more sensitive assay than absorbance methods (Weston, 1993). Although 8-oxo-guanine is naturally present in DNA, its levels increase with ageing and as a result of exposure to some genotoxic agents. Thus its presence in DNA isolated from tissues, or its concentration in urine, gives a measure of oxidative damage to the tissue or individual. However, care must be taken to avoid further oxidation of DNA during its isolation (Harris et al., 1994). Electrochemical detection combined with immunoaffinity purification has also been proposed as a method of detecting N-7-methylguanine in human DNA (Bianchini et al., 1993).

Atomic absorbance spectrometry is a means of measuring concentrations of individual elements in a sample. Its main use in DNA adduct dosimetry has been the detection and quantification of platinum bound to DNA in cancer patients treated with platinum drugs, but the method gives no structural information on the adducts present (Weston, 1993).

18.3 Experimental studies

18.3.1 Dose response for protein adducts

The available data for the dose–response relationship for carcinogen interactions with proteins relates exclusively to haemoglobin binding (reviewed by Farmer, 1993). For most electrophilic carcinogens, the relationship between dose and binding is linear, although deviations from linearity are sometimes observed at high doses. Thus, for ethylene oxide administered to rats or mice by inhalation (6 h/day, 4 weeks, 5 days/week), the dose–response relationship for haemoglobin adduct measurement is linear up to 33 ppm, but above this dose, the slope of the dose–response curve increases (Walker et al., 1992a). Effects of this type could be caused by saturation of detoxifying pathways. Other compounds which show upward deviations from linearity in their dose–response curves include dimethylnitrosamine, acrylonitrile and acrylamide. Downward deviations from linearity in the curves (which could be caused, for example, by saturation of activating metabolic pathways), seem to be less common, but have been demonstrated for benzene and benzo[a]pyrene.

18.3.2 Persistence of protein adducts: acute exposure

Unlike the situation for DNA–carcinogen adducts (see later), there do not appear to be repair enzymes for carcinogen-modified amino acids in proteins. However, chemical factors, such as the lability of the adduct towards hydrolysis, could affect protein adduct lifetimes. For adducts which are stable in vivo, their persistence would be predicted to be related to the life span of the protein, and this does appear to be the case. Thus, following a single administration of ethylene oxide to mice, the degree of alkylation of haemoglobin decreased to zero over 40 days, the lifetime of mouse erythrocytes (Osterman-Golkar et al., 1976). Similar remarks apply to the alkylation of cysteine and histidine in mouse haemoglobin by methyl methanesulphonate (eliminated in 40 days; Segerbäck et al., 1978), and the alkylation of N-terminal valine in rat haemoglobin by ethylene oxide, which was lost at a steady rate of 1.4%/day (Walker et al., 1992a), equivalent to total loss of 71 days, which is close to the average rat erythrocyte lifetime, reported as 40–100 days.

In contrast to these eliminations with apparently zero-order kinetics, there have been several reports of first-order reaction rates for adduct loss [e.g. adducts of benzidine ($t_{1/2}$ 11.5 days; Neumann, 1984), and tobacco-specific nitrosamine ($t_{1/2}$ 9.1 days; Carmella and Hecht, 1987)] with rat haemoglobin. This could be due to lack of stability of the adducts (which are believed to be sulphinamides and carboxylic esters for these carcinogens, respectively). Neumann et al. (1993) have, however, pointed out that rat erythrocytes are randomly destroyed, independent of their age, and that the elimination rate is satisfactorily described by an exponential function up to 40 days, concluding that benzidine adduction does not alter erythrocyte elimination rates. Work with 4,4'-methylenedianiline (MDA) adducts ($t_{1/2}$ 9.8 days with rat haemoglobin) supports this hypothesis (E. Bailey *et al.*, unpublished data).

18.3.3 Persistence of protein adducts: chronic exposure

Chronic exposure to electrophilic carcinogens that form stable adducts would be predicted to result in an accumulation of alkylations of the protein up to a plateau level. This has been shown to occur for exposures of experimental animals, for example, to methyl methanesulphonate (Segerbäck *et al.*, 1978), and ethylene oxide (Walker *et al.*, 1992a). In contrast to the linear loss of haemoglobin adduct levels observed following a single dose of ethylene oxide to rats (Walker *et al.*, 1992a), the adducts, following repeated doses of ethylene oxide, show an initial rapid loss, followed by a slower decline. This was suggested to be due to a combination of a first-order elimination of adducts with an increased rate of loss of older erythrocytes. The loss of 4-aminobiphenyl–haemoglobin adducts in humans following the cessation of cigarette smoking has also been observed to be faster than predicted on the basis of an erythrocyte life-span of 120 days (Bryant *et al.*, 1987; Maclure *et al.*, 1990).

18.3.4 Dose response for DNA adducts

Extensive studies on the DNA binding of aflatoxin B_1 in rodent liver and the induction of hepatocellular carcinomas in rats indicate that there is a linear dose–response relationship for both (Choy, 1993), and there was no evidence of a threshold. Another example of the linear proportionality to dose of both adduct and tumour formation is 2-acetylaminofluorene in female mouse liver (Poirier and Beland, 1992). These findings are consistent with the view that for humans the risk of cancer from exposure to a genotoxic carcinogen is proportional to the dose and support the case that DNA adduct measurement is a valid means of measuring the internal dose. In other cases, however, the relationship is not so simple. With 4-aminobiphenyl in female mouse liver, diethylnitrosamine in male rat liver and 4-(*N*-methyl-*N*-nitrosamino)-1-(3-pyridyl)-1-butanone (NNK), in rat lung, both DNA adduct and tumour incidence increase as a linear function of dose at lower doses but approach a plateau at higher doses (Poirier and Beland, 1992). In two other cases, 2-acetylaminofluorene in female mouse bladder and 4-aminobiphenyl in male mouse bladder, adduct formation was proportional to dose at low doses, but tumour formation was not (Poirier and Beland, 1992).

Thus, in all these cases, at low chronic doses, adduct formation is linearly related to dose. If these findings can be extrapolated to human exposures, they suggest that chronic human exposure to carcinogens will also cause adducts linearly related to dose. However, DNA adduct formation alone is not an accurate indicator of tumorigenic risk, which will depend on additional factors such as the degree of cell proliferation in the target tissue, interindividual variations in the efficiency of DNA repair mechanisms and concomitant factors imposed by various aspects of lifestyle, such as diet.

18.3.5 Accumulation and persistence of DNA adducts

Adducts, once formed, can be lost from DNA by depurination (for example, many adducts formed at the N-7 position of guanine weaken the N-9 glycosidic bond) or through DNA repair processes (although not all adducts are removed with equal efficiency by repair enzymes). Also, in tissues undergoing DNA replication and cell division, adduct levels will become lower due to 'dilution' of the modified DNA.

Many studies have demonstrated that a single administration of a carcinogen to experimental animals results in maximum levels of DNA adducts within 48 h. This is followed by a rapid loss of up to 80% of the adducts in the following 7 days, after which loss of adducts is very much slower. While the first phase of adduct loss is indicative of DNA repair, it is not clear why a fraction of the damage persists (Culp et al., 1993), even in proliferating tissues such as skin (Hughes and Phillips, 1990). Nevertheless, DNA damage can be detected in experimental animals many months after a single dose of a carcinogen (Randerath et al., 1985), or after chronic exposure has ceased (Culp et al., 1993). Where exposures are continuous (akin to probable human environmental exposures) there is evidence that the level of DNA adducts reaches a plateau, with the formation of new adducts being roughly balanced by the rate of removal of existing adducts (Culp et al., 1993).

Adduct persistence may vary between tissues. Studies of the kinetics of loss of 7-(2-hydroxyethyl)guanine from DNA of target and non-target tissues of mice and rats administered ethylene oxide showed evidence of adduct loss by depurination in mouse kidney and rat brain and lung, but the more rapid loss from other tissues was consistent with removal by depurination and by DNA repair processes (Walker et al., 1992b).

Within a class of carcinogens, such as polycyclic aromatic hydrocarbons or aromatic amines, the extent of DNA adduct formation in the target tissue or tissues is proportional to the carcinogenic potency of the compounds (Phillips et al., 1979; Segerbäck et al., 1993). This implies that where exposure to a mixture of different agents of the same class occurs, those that account for most of the DNA damage will be responsible for most of the biological effects.

18.3.6 The quantitative relationship between protein and DNA adducts, and between different tissues

As indicated above, dose–response relationships for DNA adducts are similar to those for haemoglobin adducts, in that for most compounds they are linear at low doses, but may deviate from linearity at high doses. Consequently, the amount of haemoglobin adduct, especially at low doses of carcinogen, may be used as a proportional marker of DNA adduct formation. The proportionality factor may, in theory, be determined in experimental systems; however, it will vary according to a large number of parameters, including the species, tissue, type and length of exposure, time since exposure and efficiency of DNA repair systems. Consequently, in a situation where the exposure conditions are unknown (e.g. human exposure to environmental pollution), it may not be possible to predict accurately the extent of DNA adduct formation in a particular human tissue based on the haemoglobin adduct level.

With DNA adducts, there is the additional consideration of the relationship between adduct levels in target tissues, which may not be readily obtainable from human subjects, and adduct levels in accessible non-target (surrogate) tissues used as the source of DNA for biomonitoring studies. Nesnow et al. (1993) found that when administered intraperitoneally to rats, benzo[a]pyrene formed adducts in white blood cells at levels that correlated with the adduct levels in lung and liver, but another compound, benzo[b]fluoranthene, gave rise to adduct levels in the blood that were less predictive of adduct levels in the lung than in the liver. Thus, in human studies, there will be uncertainties in the relationship between adducts in the surrogate tissue and those in the target tissue (when the latter is not available or accessible) particularly where the routes of exposure (ingestion, inhalation or absorption) are variable.

18.4 Protein and DNA adducts in humans

18.4.1 Tobacco smoking

The prevalence of tobacco smoking among the human population (despite clear medical evidence of its harm) has given researchers a valuable opportunity to study adduct dosimetry in humans. Adducts of aromatic amines, tobacco-specific nitrosamines and ethylene oxide have been detected in the haemoglobin of smokers. For example, Bryant et al. (1987) quantitated the sulphinamide adduct formed by 4-aminobiphenyl with cysteine in haemoglobin by hydrolysis of the adduct, followed by negative ion chemical ionization GC–MS. Smokers contained 911 ± 278 fmol 4-aminobiphenyl g^{-1} haemoglobin, compared to the non-smoker value of 166 ± 13 fmol g^{-1} haemoglobin in non-smokers, and there was no overlap between the groups.

Acetylator phenotype appears to be an important parameter in governing the extent of adduct formation. Vineis et al. (1990) and Yu et al. (1994) have shown that, for smokers, slow acetylators have higher mean levels of the 4-aminobiphenyl adduct compared to fast acetylators, independent of race. Exposure to environmental tobacco smoke results in an elevated level of the aminobiphenyl adduct in the haemoglobin of non-smokers (Hammond et al., 1993).

The tobacco-specific nitrosamines NNK and N'-nitrosonornicotine (NNN) are both metabolized to an electrophilic species 4-(3-pyridyl)-4-oxobutanediazohydroxide, which forms adducts with carboxylic acid groups in globin. Hydrolysis of these yields 4-hydroxyl-1-(3-pyridyl)-1-butanone (HPB), which may be quantitated by negative ion chemical ionization GC–MS, and can thus be used as a marker for formation of the active metabolites of NNK and NNN. Smokers and snuff dippers yielded more HPB from their globin [mean levels 163 fmol g^{-1} ($n=100$) and 329 fmol g^{-1} ($n = 35$) respectively] than did non-smokers [68 fmol g^{-1} ($n=68$); Hecht et al., 1993]. However, the range of values observed in these populations was wide and there was overlap between the groups.

Cigarette smoke also contains ethylene, which is metabolized to the electrophilic species ethylene oxide (which is also present at much lower levels than ethylene in the smoke). Ethylene oxide forms an adduct with the N-terminal valine of haemoglobin, N-(2-hydroxyethyl) valine, which may be determined by GC–MS following a modified Edman degradation procedure on the adducted protein (Törnqvist et al., 1986). Much higher levels of this adduct are seen both in non-smokers and in smokers, compared with the 4-aminobiphenyl and the tobacco-specific nitrosamine adducts. Thus, non-smokers contain $46.4 \pm SD$ 26.1 pmol N-(2-hydroxyethyl)valine g^{-1} globin ($n = 47$), and cigarette smoking increases this by about 71 pmol g^{-1} globin per 10 cigarettes/day (Bailey et al., 1988).

Haemoglobin adduct measurements have also been used to demonstrate transplacental transfer of tobacco-related carcinogens. Umbilical venous blood from smoking mothers contained elevated levels of the 4-aminobiphenyl adduct, compared to the corresponding blood from non-smoking mothers (Coghlin et al., 1991). Newborn babies of smoking mothers also contained more N-(2-hydroxyethyl)valine in their haemoglobin compared with babies of non-smokers (Tavares et al., 1994).

DNA adducts levels are higher in many tissues of tobacco smokers compared with non-smokers. This has been demonstrated for peripheral lung (Phillips et al., 1988b; Randerath et al., 1989b), bronchus (Dunn et al., 1991; Phillips et al., 1990), larynx (Degawa et al., 1994; Szyfter et al., 1994), bladder (Talaska et al., 1991a), cervix (Phillips and Ni Shé, 1994; Simons et al., 1993) and placenta (Everson et al., 1986; Manchester et al., 1992). Postlabelling analysis indicates the formation of a complex mixture of aromatic/bulky adducts, and immunoassays have indicated that at least some of these are recognized by antibodies to PAH–DNA adducts. The levels of adducts typically lie in the range of one adduct per 10^6–10^8 nucleotides. Aromatic amines are also likely to contribute to the overall DNA binding seen in bladder (Phillips and Hewer, 1993; Talaska et al., 1991a,b). Higher levels of N-7-methylguanine residues are found in smokers' bronchial DNA (Mustonen et al., 1993). Adducts formed by the tobacco-specific nitrosamines NNN and NNK are also present at higher levels in smokers' lung DNA than in that of non-smokers (Foiles et al., 1992). Analyses of DNA from peripheral blood cells have shown, with some exceptions, higher levels of bulky adducts in smokers' lymphocytes and monocytes (Jahnke et al., 1990; Mustonen et al., 1993; Santella et al., 1992; Savela and Hemminki, 1991).

The presence of benzo[a]pyrene–DNA adducts in the tissues of smokers has been demonstrated by several techniques that involve detecting the fluorescence, phosphorescence and/or mass spectrum of benzo[a]pyrene tetrols released from DNA by hydrolysis. Among the tissues thus analysed are lung (Alexandrov et al., 1992; Corley et al., 1995; Weston and Bowman, 1991), peripheral white blood cells (Rojas et al., 1994) and placenta (Manchester et al., 1988). A major adduct in bladder biopsies from smokers has been tentatively identified, on the evidence of co-chromatography with the synthetic standard in ^{32}P-postlabelling experiments, as the C-8-deoxyguanosinyl adduct of 4-aminobiphenyl (Talaska et al., 1991a).

18.4.2 Occupational exposure

There have been many studies in recent years involving monitoring the blood protein and DNA of workers occupationally exposed to genotoxic agents. As in the case of smoking-related exposure to genotoxic agents, the extent of adduct formation in occupationally exposed individuals is highly dependent upon the compound considered. Amongst the highest protein adduct levels that have been found are those in ethylene oxide-exposed workers, from a factory in the former German Democratic Republic, with individual levels of N-(2-hydroxyethyl)valine as high as 16.1 nmol adduct g^{-1} globin (Tates et al., 1991). In this study, there was a good correlation between adduct levels and sister chromatid exchange (SCE) and, to a lesser extent, chromosome aberrations and micronuclei. In a similar study of Mayer et al. (1991) on sterilization workers in a hospital, exposure to ethylene oxide caused a roughly three-fold increase in N-(2-hydroxyethyl)valine level over control values. (NB. As indicated above, the numerical values supplied by this laboratory are not consistent with other laboratories, being about eight-fold higher.)

In contrast to the situation with ethylene oxide, adducts formed by styrene oxide resulting from exposure to styrene have been detected at very much lower levels. Thus a study of workers in the reinforced plastic industry revealed no levels greater than 52 pmol styrene oxide–valine adducts g^{-1} globin (Christakopoulos et al., 1993). However, one individual with a level of adduct about 5000 times the control value was observed by Brenner et al. (1991) in a population of boat builders. Styrene oxide–cysteine adduct levels in albumin (0.19–3.8 nmol g^{-1} protein) in another group of boat builders have been shown by Rappaport et al., (personal communication) to correlate with exposure to styrene.

Occupational exposure to acrylamide and acrylonitrile results in increased levels of the adducts of these compounds with N-terminal valine in haemoglobin, the levels being 0.3–34 nmol adduct g^{-1} haemoglobin for acrylamide and 0.18–39.9 nmol g^{-1} haemoglobin for acrylonitrile (Calleman et al., 1994). Indicators of peripheral neuropathy correlated with adduct levels.

Some attempts have also been made to monitor occupational exposure to aromatic amines by measurement of their cysteine adducts in haemoglobin (e.g. for aniline and MDA). In the latter case, adduct levels in a group of workers (n = 12) were in the range 5.9–55.3 pmol g^{-1} haemoglobin for MDA and 0–91.8 pmol g^{-1} haemoglobin for the adduct of N-acetyl-MDA (Bailey et al., 1990; and unpublished data).

Occupational exposure to benzo[a]pyrene has been measured through immunoassay of its albumin adduct. Adduct levels were significantly higher in foundry workers (both smokers and non-smokers) than in corresponding control individuals (Sherson et al., 1990). Other similar studies have shown results varying from an increased level of adducts in foundry workers through to similar adduct levels in the workers and controls (Omland et al., 1994). The exposure of taxi drivers to PAHs in urban air has been demonstrated by their increased levels of PAH–protein adducts (Hemminki et al., 1994). Benzo[a]pyrene–haemoglobin adducts have been measured, using HPLC with fluorescence detection of a haemoglobin hydrolysate, by Ferreira et al. (1994) in workers from steel foundries and a graphite electrode-producing plant, and a correlation with atmospheric PAH levels was observed.

A number of biomonitoring studies have examined DNA adduct formation in white blood cells or mononuclear blood cells of groups occupationally exposed to PAHs. Higher levels of adduct have been found in iron foundry workers (Phillips et al., 1988a; Santella et al., 1993), coke oven workers (Hemminki et al., 1990; van Schooten et al., 1990), aluminium plant workers (Schoket et al., 1991, 1993a), roofers (Herbert et al., 1990) and bus drivers exposed to diesel exhausts (Hemminki et al., 1994), compared with controls. In most studies, adduct levels of around 1 adduct per 10^8 nucleotides are detected in controls, with levels in the exposed population showing a wide range of values from as low as the control values to as much as 100 times higher. Some studies, however, have indicated the possible confounding influence of dietary sources of PAHs. Thus, a study of firefighters showed that adduct levels in white blood cells correlated not with firefighting activity but with consumption of barbecued food (Rothman et al., 1993), and the results of a study of military personnel extinguishing oil-field fires in Kuwait showed that their adduct levels were actually lower when on duty in Kuwait than when stationed at bases in Europe (Poirier et al., 1994).

Studies conducted on people living in the Upper Silesia region of Poland also show evidence of adducts arising from environmental (that is, non-occupational) exposure to genotoxic agents. It was found that groups of coke oven workers and control residents from the area both had significantly higher levels of white blood cell DNA adducts, detected by ^{32}P-postlabelling, than residents of rural eastern Poland (Hemminki et al., 1990; Perera et al., 1992). Furthermore, adduct levels are subject to seasonal variation, the higher levels of adducts in the winter sampling being attributed to the widespread use throughout the region of coal for domestic heating and cooking (Grzybowska et al., 1993).

18.4.3 Medicinal exposures

Attempts to monitor exposure to cancer chemotherapeutic methylating drugs have been hindered by the presence of large amounts of 'background' methylated amino acids in human haemoglobin (e.g. 16.4 nmol of S-methylcysteine g^{-1} globin; Bailey *et al.*, 1981). The background level of the N-terminal hydroxyethylated valine adduct is much lower, which permits the use of measurement of this adduct to monitor human therapy with hydroxyethylating agents. A variety of anticancer 1-(2-chloroethyl)-1-nitrosoureas has been shown to produce N-(2-hydroxyethyl)valine in haemoglobin, although large interindividual variation was observed (Bailey *et al.*, 1991).

Lidocaine, which is used for anaesthesia and as an antiarrhythmic agent, is metabolized to 2,6-dimethylaniline. Using a GC–MS method, the production of the cysteine sulphinamide adduct of 2,6-dimethylaniline has been monitored in lidocaine-treated patients (Bryant *et al.*, 1994).

Cancer patients treated with cytotoxic platinum-based drugs [*cis*-diamminedichloroplatinum II (cisplatin) and diamminecyclobutane-carboxylatoplatinum II (carboplatin)] form DNA adducts in many tissues. Although the drugs do not require metabolic activation, wide interindividual variations in the extent of DNA adduct formation have been observed. Adducts have been detected by immunoassay in patients as long as 15 months after the last treatment, indicating the high degree of adduct persistence (Poirier *et al.*, 1992). Moreover, several studies have shown a significant correlation between the extent of adduct formation in white blood cells and the response to treatment, those patients with higher adduct levels having the better prognosis (Blommaert *et al.*, 1993; Gupta-Burt *et al.*, 1993; Reed *et al.*, 1987, 1988). These findings indicate that DNA adduct determination may have some use in predicting patient response to treatment, or even in tailoring the dosage to improve an individual's prognosis.

Antibodies raised against the adducts formed by 8-methoxypsoralen (8-MOP), which is used in combination with UVA irradiation to treat psoriasis, have been used to detect by immunofluorescence staining the presence of 8-MOP–DNA adducts in skin biopsies of patients undergoing treatment. Adduct levels were at about 1 adduct/10^6 nucleotides (Yang *et al.*, 1989), but adducts were not detectable in white blood cell DNA with a limit of sensitivity for the assay of 1 adduct/10^8 nucleotides.

Other studies that illustrate the ability to detect DNA adducts in patients receiving chemotherapy include the measurement, by ^{32}P-postlabelling, of 7-methylguanine residues in white blood cell DNA following treatment with dacarbazine or procarbazine (Mustonen *et al.*, 1991); the detection, also by ^{32}P-postlabelling, of mitomycin C adducts in patients treated with the drug (Kato *et al.*, 1988); and the detection by ELISA of acrolein-modified DNA in patients treated with cyclophosphamide (McDiarmid *et al.*, 1991), and of melphalan–DNA adducts in leukaemia patients undergoing high-dose melphalan therapy (Tilby *et al.*, 1993).

18.4.4 Analysis of urine for adducts

Many DNA adducts can be lost from DNA through depurination and excreted in the urine. Thus, urine can be a suitable source of human material with which to monitor human exposure to carcinogens and mutagens (Shuker and Farmer, 1992).

Urine analysis, employing immunoaffinity columns to concentrate the N-7-guanine adduct and HPLC with fluorescence detection for quantification, has been used to monitor human exposure to aflatoxins. Such studies have demonstrated a linear correlation between daily aflatoxin ingestion and adduct excretion in areas of Africa and China where aflatoxin contamination of food is high (Groopman et al., 1992a, b). Furthermore, prospective studies of Chinese men from Shanghai have indicated that levels of urinary aflatoxin adducts correlate with risk of developing hepatocellular carcinoma (Qian et al., 1994; Ross et al., 1992), demonstrating the potential for adduct measurements as biomarkers of cancer risk in humans.

A problem with measuring carcinogen–purine adducts in urine is that they could have arisen from DNA or RNA. Not all adducts in urine are the result of exposure to genotoxic agents. For example, 3-methyladenine in urine is largely the result of its presence in the diet, and 7-methylguanine derives partly from degradation of tRNA (Shuker and Farmer, 1992). However, 3-ethyladenine may be a useful urinary biomarker of exposure to ethylating agents (Prevost et al., 1993).

The special case of a worker who received an acute exposure to 4,4'-methylene-bis(2-chloroaniline) (MOCA) afforded the opportunity to investigate DNA adduct formation by this compound (Kaderlik et al., 1993). Exfoliated urothelial cells were recovered from urine samples collected from the worker at several time-points in the days following the exposure. From these, sufficient DNA was isolated to enable the demonstration, by ^{32}P-postlabelling, of the formation and removal of MOCA–DNA adducts in the bladder.

A further demonstration of the use of exfoliated urothelial cells for human biomonitoring is provided by a study of smoking-related adducts in their DNA (Talaska et al., 1991a). An adduct with the chromatographic characteristics of N-(deoxyguanosin-8-yl)-4-aminobiphenyl was detected by ^{32}P-postlabelling. As mentioned in Section 18.4.1, this adduct is also found in bladder biopsy samples from smokers (Talaska et al., 1991a).

18.5 Adducts due to environmental exposures in non-human species

In principle, any wild, domesticated or feral species can be used as a source of DNA or protein with which to monitor environmental exposure to genotoxic agents. To date, the marine environment has been investigated most extensively in this way, with most of the studies involving the measurement of DNA adducts in liver. The presence of DNA adducts, detected by ^{32}P-postlabelling, in fish from polluted rivers and coastal waters has been reported (Dunn *et al.*, 1987; Varanasi *et al.*, 1989). The sediments from these areas had high concentrations of PAHs, and the adduct profiles have the characteristics of a complex mixture of bulky, aromatic DNA adducts.

More recently, high levels of oxidative damage in liver DNA from fish in polluted waters have been reported (Malins and Gunselman, 1994). GC–MS with selective ion monitoring (SIM) together with Fourier-transform infrared spectroscopy revealed hydroxyl radical-induced 2,6-diamino-4-hydroxy-5-formamidopyridine (Fapy-G) and 8-hydroxyguanine in DNA, and the equivalent adenine lesions. These findings indicate that oxidative damage to DNA may be a suitable biomarker for monitoring environmental exposure to toxic/genotoxic agents.

Pilot studies have been carried out to investigate the feasibility of measuring DNA adducts in earthworms as a means of monitoring soil pollution. Large sums of money have been allocated to the task of decontaminating disused industrial sites in order to make them habitable and usable for other purposes, and there is a clear need for an objective means of assessing the effectiveness of the clean-up procedures. Initial results indicate that measuring adducts in earthworms by ^{32}P-postlabelling will be useful in this regard (Walsh *et al.*, 1995).

18.6 Identifying the environmental origin of adducts in humans

Much work has been undertaken to assess the contribution that carcinogens in the environment or the diet make to the spectrum of adducts that are found in normal human populations. For example, exposure to the active metabolite of dietary aflatoxin B_1 may be monitored by immunoassay or HPLC with fluorescence detection of the albumin lysine adduct (Wild *et al.*, 1993) or the purine adduct (see earlier). In some countries (e.g. The Gambia), levels greater than 0.4 nmol g^{-1} albumin of the lysine adduct have been observed and their seasonal variation monitored (Wild *et al.*, 1990b). The effect of the environment on protein adduct levels is illustrated by a recent demonstration that exposure to diesel exhaust in a group of individuals in a bus garage increased significantly the levels of *N*-(2-hydroxyethyl)valine in haemoglobin (P.B. Farmer *et al.*, unpublished data).

Despite attempts to identify possible sources for the 'background' levels of adducts that are found in haemoglobin or albumin, the causes for many of these are unknown. Typical 'background' levels of haemoglobin adducts in supposedly unexposed populations are (in nmol adduct g^{-1} globin) *S*-methylcysteine (16.4), *N*-(2-hydroxyethyl)histidine (1.6), *N*-methylvaline (0.5), *N*-(2-hydroxyethyl)valine (0.05), *N*-(2-cyanoethyl)valine (0–0.04) and 4-aminobiphenyl-cysteine sulphinamide (0.00017) (Farmer *et al.*, 1993). For some of these adducts, endogenous processes might be postulated, but for others, environmental exposure to the carcinogen seems the only option.

There are still only a few examples of the definitive identification of DNA adducts in humans resulting from environmental exposure (see *Table 18.2*). The problem is that, at present, the sensitivity of those methods that can *characterize* DNA adducts in human tissues falls short of the sensitivity of those methods that can *detect* them; this has tended to limit the successful application of the former in human biomonitoring studies.

Table 18.2. Examples of characterized DNA adducts in human tissues

Agent	Tissue or excretion	Method of detection and identification	Reference
Benzo[*a*]pyrene	Lung Placenta White blood cells	HPLC/fluorescence/ phosphorescence and mass spectrometry of hydrolysis product, benzo[*a*]pyrene tetrol	Alexandrov *et al.* (1992), Corley *et al.* (1995), Manchester *et al.* (1988), Rojas *et al.* (1994), Weston *et al.* (1989)
Aflatoxin B_1	Urine	HPLC/fluorescence	Groopman *et al.* (1992a,b)
2-Amino-1-methyl-6-phenylimidazo [4,5-*b*] pyridine (PhIP)	Colon	^{32}P-Postlabelling/ mass spectrometry	Friesen *et al.* (1994)
Malondialdehyde	Liver	Mass spectrometry	Chaudhary *et al.* (1994)
4-(*N*-methyl-*N*-nitrosamino)-1-(3-pyridyl)-1-butanone (NNK)	Lung	Mass spectrometry of hydrolysis product, 4-hydroxyl-1-(3-pyridyl)-1-butanone (HPB)	Foiles *et al.* (1991, 1992)

As is the case with protein adducts, human tissues clearly contain 'background' adducts of unknown source and identity. It is a plausible hypothesis that they result, in part, from environmental exposure to carcinogens and mutagens, but some of these adducts are apparently of endogenous origin (Chaudhary *et al.*, 1994; Marnett and Burcham, 1993) and their biological significance is presently unknown.

Some types of adducts that have been shown to be induced by exogenous carcinogen exposure are also thought to arise through endogenous processes; for example etheno adducts in DNA are formed by compounds like vinyl chloride (Nath *et al.*, 1994), but it is possible that they may also result from endogenous lipid peroxidation (Nair *et al.*, 1995). Furthermore, there is evidence that carcinogens such as aflatoxin B_1 (Shen *et al.*, 1995) and benzo[*a*]pyrene (Mauthe *et al.*, 1995) can induce oxidative damage as well as form covalent DNA adducts.

Clearly, while DNA and protein adducts are useful markers of human exposure to genotoxic agents and, conversely, the demonstration that a chemical forms adducts in experimental systems indicates its genotoxic potential, there is still much to be learned about the origin of adducts in human tissues and, more importantly, the relative biological significance of those that are 'endogenous' and those that are 'environmental'.

References

Alexandrov K, Rojas M, Geneste O, Castegnaro M, Camus A-M, Petruzzelli S, Giuntini C, Bartsch H. (1992) An improved fluorimetric assay for dosimetry of benzo[a]pyrene diol-epoxide–DNA adducts in smokers' lung: comparisons with total bulky adducts and aryl hydrocarbon hydroxylase activity. *Cancer Res.* **52:** 6248–6253.

Bailey E, Brooks AG, Bird I, Farmer PB, Street B. (1990) Monitoring exposure to 4,4'-methylenedianiline by the gas chromatography–mass spectrometry determination of adducts to hemoglobin. *Anal. Biochem.* **190:** 175–181.

Bailey E, Brooks AGF, Dollery CT, Farmer PB, Passingham BJ, Sleightholm MA, Yates DW. (1988) Hydroxyethylvaline adduct formation in haemoglobin as a biological marker of cigarette smoke intake. *Arch. Toxicol.* **62:** 247–253.

Bailey E, Connors TA, Farmer PB, Gorf SM, Rickard J. (1981) Methylation of cysteine in hemoglobin following exposure to methylating agents. *Cancer Res.* **41:** 2514–2517.

Bailey E, Farmer PB, Bird I, Lamb JH, Peal JA. (1986) Monitoring exposure to acrylamide by the determination of S-(2-carboxyethyl)cysteine in hydrolyzed hemoglobin by gas chromatography-mass spectrometry. *Anal. Biochem.* **157:** 241–248.

Bailey E, Farmer PB, Tang Y-S et al. (1991) Hydroxyethylation of hemoglobin by 1-(2-chloroethyl)-1-nitrosoureas. *Chem. Res. Toxicol.* **4:** 462–466.

Baird WM, Lau HHS, Schmerold I, Coffing SL, Brozich SL, Lee H, Harvey RG. (1993) Analysis of polycyclic aromatic hydrocarbon–DNA adducts by postlabelling with the weak β-emitters ^{35}S-phosphorothioate and ^{33}P-phosphate, immobilized boronate chromatography and high-performance liquid chromatography. In: *Postlabelling Methods for Detection of DNA Adducts* (eds DH Phillips, M Castegnaro and H Bartsch). IARC, Lyon pp. 217–227.

Bartsch H, Hemminki K, O'Neill IK. (eds) (1988) *Methods for Detecting DNA Damaging Agents in Humans: Applications in Cancer Epidemiology and Prevention.* IARC Scientific Publications no. 89. IARC, Lyon.

Beach AC, Gupta RC. (1992) Human biomonitoring and the ^{32}P-postlabelling assay. *Carcinogenesis* **13:** 1053–1074.

Bechtold WE, Sun JD, Wolff RK, Griffith WS, Kilmer JW, Bond JA. (1991) Globin adducts of benzo[a]pyrene: markers of inhalation exposure as measured in F344/N rats. *J. Appl. Toxicol.* **11:** 115–118.

Bergmark E, Calleman CJ, He F, Costa LG. (1993) Determination of hemoglobin adducts in humans occupationally exposed to acrylamide. *Toxicol. Appl. Pharmacol.* **120:** 45-54.

Bianchini F, Montesano R, Shuker DEG, Cuzick J, Wild CP. (1993) Quantitation of 7-methyldeoxyguanosine using immunoaffinity purification and HPLC with electrochemical detection. *Carcinogenesis* **14:** 1677–1682.

Blommaert FA, Michael C, Terheggen MAB, Muggia FM, Kortes V, Schornagel JH, Hart AAM, den Engelse L. (1993) Drug-induced DNA modification in buccal cells of cancer patients receiving carboplatin and cisplatin combination chemotherapy, as determined by an immunocytochemical method: interindividual variation and correlation with disease response. *Cancer Res.* **53:** 5669–5675.

Brenner DD, Jeffrey AM, Latriano L et al. (1991) Biomarkers in styrene-exposed boatbuilders. *Mutat. Res.* **261:** 225–236.

Bryant MS, Simmons HF, Harrell RE, Hinson JA. (1994) 2,6-Dimethylaniline–hemoglobin adducts from lidocaine in humans. *Carcinogenesis* **15:** 2287–2290.

Bryant MS, Skipper PL, Tannenbaum SR, Maclure M. (1987) Hemoglobin adducts of 4-aminobiphenyl in smokers and non-smokers. *Cancer Res.* **47:** 602–608.

Calleman CJ, Wu Y, He F, Tian G, Bergmark E, Zhang S, Deng H, Wang Y, Crofton KM, Fennell T, Costa LG. (1994) Relationships between biomarkers of exposure and neurological effects in a group of workers exposed to acrylamide. *Toxicol. Appl. Pharmacol.* **126:** 361–371.

Carmella SG, Hecht SS. (1987) Formation of hemoglobin adducts upon treatment of F344 rats with the tobacco specific nitrosamines 4-(methylnitrosamino)-1-(2-pyridyl)-1-butanone and N-nitrosonornicotine. *Cancer Res.* **47:** 2626–2630.

Carmella SG, Kagan SS, Kagan M, Foiles PG, Palladino PG, Quart AM, Quart E, Hecht SS. (1990) Mass spectrometric analysis of tobacco-specific nitrosamine hemoglobin adducts in snuff-dippers, smokers, and non-smokers. *Cancer Res.* **50:** 5438–5445.

Chaudhary AK, Nokubo M, Reddy GR, Yeola SN, Morrow JD, Blair IA, Marnett LJ. (1994) Detection of endogenous malondialdehyde–deoxyguanosine adducts in human liver. *Science* **265:** 1580–1582.

Choy WN. (1993) A review of the dose–response induction of DNA adducts by aflatoxin B$_1$ and its implications to quantitative cancer-risk assessment. *Mutat. Res.* **296:** 181–198.

Christakopoulos A, Bergmark E, Zorcec V, Norppa H, Maki-Paakkanen J, Osterman-Golkar S. (1993) Monitoring occupational exposure to styrene from hemoglobin adducts and metabolites in blood. *Scand. J. Work Environ. Hlth* **19:** 255–263.

Coghlin J, Gann PH, Hammond K, Skipper PL, Taghizadeh K, Paul M, Tannenbaum SR. (1991) 4-Aminobiphenyl hemoglobin adducts in fetuses exposed to the tobacco smoke carcinogen *in utero. J. Natl Cancer Inst.* **83:** 274–280.

Corley J, Hurtubise RJ, Bowman ED, Weston A. (1995) Solid matrix, room temperature phosphorescence identification and quantitation of the tetrahydrotetrols derived from the acid hydrolysis of benzo[*a*]pyrene–DNA adducts from human lung. *Carcinogenesis* **16:** 423–426.

Culp SJ, Poirier MC, Beland FA. (1993) Biphasic removal of DNA adducts in a repetitive DNA sequence after dietary administration of 2-acetylaminofluorene. *Environ. Hlth Perspect.* **99:** 273–275.

Day BW, Naylor S, Gan L-S, Sahali Y, Nguyen TT, Skipper PL, Wishnok JS, Tannenbaum SR. (1990) Molecular dosimetry of polycyclic aromatic hydrocarbon epoxides and diol epoxides via hemoglobin adducts. *Cancer Res.* **50:** 4611–4618.

Degawa M, Stern SJ, Martin MV, Guengerich FP, Fu PP, Ilett KF, Kaderlik RK, Kadlubar FF. (1994) Metabolic activation and carcinogen–DNA adduct detection in human larynx. *Cancer Res.* **54:** 4915–4919.

Dunn BP, San RHC. (1988) HPLC enrichment of hydrophobic DNA adducts for enhanced sensitivity of ^{32}P-postlabeling analysis. *Carcinogenesis* **9:** 1055–1060.

Dunn BP, Black JJ, Maccubbin A. (1987) ^{32}P-Postlabeling analysis of aromatic DNA adducts in fish from polluted areas. *Cancer Res.* **47:** 6543–6548.

Dunn BP, Vedal S, San RHC, Kwan W-F, Nelems B, Enarson DA, Stich HF. (1991) DNA adducts in bronchial biopsies. *Int. J. Cancer* **48:** 485–492.

Everson RB, Randerath E, Santella RM, Cefalo RC, Avitts TA, Randerath K. (1986) Detection of smoking-related covalent DNA adducts in human placenta. *Science* **231:** 54–57.

Farmer PB. (1993) Biomarkers as molecular dosimeters of genotoxic substances. In: *Use of Biomarkers in Assessing Health and Environmental Impacts of Chemicals* (ed. CC Travis). Plenum Press, New York, pp. 53-62.

Farmer PB. (1994) Carcinogen adducts: use in diagnosis and risk assessment. *Clin. Chem.* **40:** 1438–1443.

Farmer PB, Bailey E, Naylor S, Anderson D, Brooks A, Cushnir J, Lamb JH, Sepai O, Tang Y-S. (1993) Identification of endogenous electrophiles by means of mass spectrometric determination of protein and DNA adducts. *Environ. Health Perspect.* **99:** 19–24.

Ferreira MJ, Tas S, Dell'Omo L, Goormams G, Buchet JP, Lauwerys R. (1994) Determinants of benzo[*a*]pyrene diol epoxide adducts to haemoglobin in workers exposed to polycyclic aromatic hydrocarbons. *Occup. Environ. Med.* **51:** 451–455.

Foiles PG, Akerkar SA, Carmella SG, Kagan M, Stoner GD, Resau JH, Hecht SS. (1991) Mass spectrometric analysis of tobacco-specific nitrosamine–DNA adducts in smokers and nonsmokers. *Chem. Res. Toxicol.* **4:** 364–368.

Foiles PG, Murphy SE, Peterson LA, Carmella SG, Hecht SS. (1992) DNA and hemoglobin adducts as markers of metabolic activation of tobacco-specific carcinogens. *Cancer Res.* **52:** 2698s–2701s.

Friesen MD, Kaderlik K, Lin D, Garren L, Bartsch H, Lang NP, Kadlubar FF. (1994) Analysis of DNA adducts of 2-amino-1-methyl-6-phenylimidazo[4,5-b]pyridine in rat and human tissues by alkaline hydrolysis and gas chromatography/electron capture mass spectrometry: validation by comparison with ^{32}P-postlabelling. *Chem. Res. Toxicol.* **7:** 733–739.

Gallagher JE, Jackson MA, George MH, Lewtas J, Robertson IGC. (1989) Differences in detection of DNA adducts in the ^{32}P-postlabelling assay after either 1-butanol extraction or nuclease P_1 treatment. *Cancer Lett.* **45:** 7–12.

Giese RW, Vouros P. (1993) Methods development toward the measurement of polyaromatic hydrocarbon–DNA adducts by mass spectrometry. *Res. Rep. Hlth Effects Inst.* **61:** 1–25.

Gorelick NJ. (1993) Application of HPLC in the ^{32}P-postlabeling assay. *Mutat. Res.* **288:** 5–18.

Groopman JD, Skipper PL. (eds) (1991) *Molecular Dosimetry and Human Cancer.* CRC Press, Boca Raton, FL.

Groopman JD, Hall AJ, Whittle H, Hudson GJ, Wogan GN, Montesano R, Wild CP. (1992a) Molecular dosimetry of aflatoxin-N7-guanine in human urine obtained in The Gambia, West Africa. *Cancer Epidemiol., Biomarkers Preven* **1:** 221–227.

Groopman JD, Jiaqi Z, Donahue PR, Pikul A, Lisheng Z, Jun-shi C, Wogan GN. (1992b) Molecular dosimetry of urinary aflatoxin–DNA adducts in people living in Guangxi Autonomous Region, People's Republic of China. *Cancer Res.* **52:** 45–52.

Grzybowska E, Hemminki K, Szeliga J, Chorazy M. (1993) Seasonal variation of aromatic DNA adducts in human lymphocytes and granulocytes. *Carcinogenesis* **14:** 2523–2526.

Gupta RC. (1985) Enhanced sensitivity of ^{32}P-postlabeling analysis of aromatic carcinogen:DNA adducts. *Cancer Res.* **45:** 5656–5662.

Gupta RC, Earley K. (1988) ^{32}P-adduct assay: comparative recoveries of structurally diverse DNA adducts in the various enhancement procedures. *Carcinogenesis* **9:** 1687–1693.

Gupta-Burt S, Shamkhani H, Reed E, Tarone RE, Allegra CJ, Pai LH, Poirier MC. (1993) Relationship between patient response in ovarian and breast cancer and platinum drug–DNA adduct formation. *Cancer Epidemiol. Biomarkers Preven.* **2:** 229–234.

Halliwell B, Dizdaroglu M. (1992) The measurement of oxidative damage to DNA by HPLC and GC/MS techniques. *Free Rad. Res. Commun.* **16:** 75–87.

Hammond SK, Coghlin J, Gann PH, Paul M, Taghizadeh K, Skipper PL, Tannenbaum SR. (1993) Relationship between environmental tobacco smoke exposure and carcinogen–hemoglobin adduct levels in nonsmokers. *J. Natl Cancer Inst.* **85:** 474–478.

Harris G, Bashir S, Winyard PG. (1994) 7,8-Dihydro-8-oxo-2'-deoxyguanosine present in DNA is not simply an artefact of isolation. *Carcinogenesis* **15:** 411–413.

Hecht SS, Carmella SG, Foiles PG, Murphy SE, Peterson LA. (1993) Tobacco-specific nitrosamine adducts: studies in laboratory animals and humans. *Environ. Hlth Perspect.* **99:** 57–63.

Hecht SS, Carmella SG, Foiles PG, Murphy SE. (1994) Biomarkers for human uptake and metabolic activation of tobacco-specific nitrosamines. *Cancer Res.* **54:** 1912–1917.

Hemminki K, Grzybowska E, Chorazy M et al. (1990) DNA adducts in humans environmentally exposed to aromatic compounds in an industrial area of Poland. *Carcinogenesis* **11:** 1229–1231.

Hemminki K, Zhang LF, Krüger J, Autrup H, Törnqvist M, Norbeck H-E. (1994) Exposure of bus and taxi drivers to urban air pollutants as measured by DNA and protein adducts. *Toxicol. Lett.* **72:** 171–174.

Herbert R, Marcus M, Wolff MS et al. (1990) Detection of adducts of deoxyribonucleic acid in white blood cells of roofers by ^{32}P-postlabeling. *Scand. J. Work Environ. Hlth* **16:** 135–143.

Hughes NC, Phillips DH. (1990) Covalent binding of dibenzpyrenes and benzo[a]pyrene to DNA: evidence for synergistic and inhibitory interactions when applied in combination to mouse skin. *Carcinogenesis* **11:** 1611–1619.

Jahnke GD, Thompson CL, Walker MP, Gallagher JE, Lucier GW, DiAugustine RP. (1990) Multiple DNA adducts in lymphocytes of smokers and nonsmokers determined by ^{32}P-postlabeling analysis. *Carcinogenesis* **11:** 205–211.

Jain R, Sharma M. (1993) Fluorescence postlabeling assay of DNA damage induced by *N*-methyl-*N*-nitrosourea. *Cancer Res.* **53:** 2771–2774.

Jankowiak R, Small GJ. (1991) Fluorescence line narrowing: a high-resolution window on DNA and protein damage from chemical carcinogens. *Chem. Res. Toxicol.* **4:** 256–269.

Kaderlik KR, Talaska G, DeBord DG, Osorio AM, Kadlubar FF. (1993) 4,4′-Methylene-bis(2-chloroaniline)–DNA adduct analysis in human exfoliated urothelial cells by ^{32}P-postlabeling. *Cancer Epidemiol. Biomarkers Preven.* **2:** 63–69.

Kato S, Yamashita K, Kim T, Tajiri T, Onda M, Sato S. (1988) Modification of DNA by mitomycin C in cancer patients detected by ^{32}P-postlabeling analysis. *Mutat. Res.* **202:** 85–91.

Kato S, Petruzzelli S, Bowman ED, Turtletaub KW, Blomeke B, Weston A, Shields PG. (1993) 7-Alkyldeoxyguanosine adduct detection by two-step HPLC and the ^{32}P-postlabeling assay. *Carcinogenesis* **14:** 545–550.

Kaur S, Hollander D, Haas R, Burlingame AL. (1989) Characterization of structural xenobiotic modifications in proteins by high sensitivity tandem mass spectrometry. *J. Biol. Chem.* **264:** 16981–16984.

Kautiainen A, Törnqvist M. (1991) Monitoring exposure to simple epoxides and alkenes through gas chromatographic determination of hemoglobin adducts. *Int. Arch. Occup. Environ. Health* **63:** 27–31.

Lau HHS, Baird WM. (1991) Detection and identification of benzo[*a*]pyrene–DNA adducts by [^{35}S]phosphothioate labeling and HPLC. *Carcinogenesis* **12:** 885–893.

Lin RC, Shahidi S, Kelly TJ, Lumeng C, Lumeng L. (1993) Measurement of hemoglobin–acetaldehyde adduct in alcoholic patients. *Alcoholism: Clin. Exp. Res.* **17:** 669–674.

Maclure M, Bryant MS, Skipper PL, Tannenbaum SR. (1990) Decline of the hemoglobin adduct of 4-aminobiphenyl during withdrawal from smoking. *Cancer Res.* **50:** 181–184.

Malins DC, Gunselman SJ. (1994) Fourier-transform infrared spectroscopy and gas chromatography–mass spectrometry reveal a remarkable degree of structural damage in the DNA of wild fish exposed to toxic chemicals. *Proc. Natl Acad. Sci. USA* **91:** 13038–13041.

Manchester DK, Weston A, Choi JS, Trivers GE, Fennessey P, Quintana E, Farmer PB, Mann DL, Harris CC. (1988) Detection of benzo[*a*]pyrene diol epoxide–DNA adducts in human placenta. *Proc. Natl Acad. Sci. USA* **85:** 9243–9247.

Manchester DK, Bowman ED, Parker NB, Caporaso NE, Weston A. (1992) Determinants of polycyclic aromatic hydrocarbon–DNA adducts in human placenta. *Cancer Res.* **52:** 1499–1503.

Marnett LJ, Burcham PC. (1993) Endogenous DNA adducts: potential and paradox. *Chem. Res. Toxicol.* **6:** 771–785.

Mauthe RJ, Cook VM, Coffing SL, Baird WM. (1995) Exposure of mammalian cell cultures to benzo[*a*]pyrene and light results in oxidative DNA damage as measured by 8-hydroxydeoxyguanosine formation. *Carcinogenesis* **16:** 133–137.

Mayer J, Warburton D, Jeffrey AM et al. (1991) Biologic markers in ethylene oxide-exposed workers and controls. *Mutat. Res.* **248:** 163–176.

McDiarmid MA, Iype PT, Kolodner K, Jacobson-Kram D, Strickland PT. (1991) Evidence for acrolein-modified DNA in peripheral blood leukocytes of cancer patients treated with cyclophosphamide. *Mutat. Res.* **248:** 93–99.

Melikian AA, Sun P, Amin S, Hecht SS. (1995) Gas chromatography–mass spectrometry (GC–MS) characterization of polycyclic aromatic hydrocarbon (PAH)-derived globin adducts in smokers. *Proc. Am. Assoc. Cancer Res.* **36:** 112.

Mustonen R, Försti A, Hiatenen P, Hemminki K. (1991) Measurement by ^{32}P-postlabelling of 7-methylguanine levels in white blood cell DNA of healthy individuals and cancer patients treated with dacarbazine and procarbazine. Human data and method development for 7-alkylguanines. *Carcinogenesis* **12:** 1423-1431.

Mustonen R, Schoket B, Hemminki K. (1993) Smoking-related DNA adducts: ^{32}P-postlabeling analysis of 7-methylguanine in human bronchial and lymphocyte DNA. *Carcinogenesis* **14:** 151–154.

Nair J, Barbin A, Guichard Y, Bartsch H. (1995) 1,N^6-Ethenodeoxyadenosine and 3,N^4-ethenodeoxycytidine in liver DNA from humans and untreated rodents detected by immunoaffinity/^{32}P-postlabelling. *Carcinogenesis* **16**: 613–617.

Nath RG, Chen H-JC, Nishikawa A, Young-Sciame R, Chung FL. (1994) A ^{32}P-postlabeling method for simultaneous detection and quantification of exocyclic etheno and propano adducts in DNA. *Carcinogenesis* **15**: 979–984.

Nesnow S, Ross J, Nelson G, Holden K, Erexson G, Kligerman A, Gupta RC. (1993) Quantitative and temporal relationships between DNA adduct formation in target and surrogate tissues: implications for biomonitoring. *Environ. Hlth Perspect.* **101** (Suppl. 3): 37–42.

Neumann H-G. (1984) Dosimetry and dose–response relationships. In: *Monitoring Human Exposure to Carcinogenic and Mutagenic Agents* (eds H Berlin, M Draper, K Hemminki, H Vainio). International Agency for Research on Cancer, Lyon, pp. 115–126.

Neumann H-G, Birner G, Kowallik P, Schutze D, Zwirner-Baier I. (1993) Hemoglobin adducts of *N*-substituted aryl compounds in exposure control and risk assessment. *Environ. Hlth Perspect.* **99**: 65–69.

Niemala O, Israel Y. (1992) Hemoglobin–acetaldehyde adducts in human alcohol abusers. *Lab. Invest.* **67**: 246–252.

Omland O, Sherson D, Hansen WM, Sigsgaard T, Autrup H, Overgaard E. (1994) Exposure of iron foundry workers to polycyclic aromatic hydrocarbons: benzo(*a*)pyrene–albumin adducts and 1-hydroxypyrene as biomarkers for exposure. *Occup. Environ. Med.* **51**: 513–518.

Osterman-Golkar S, Ehrenberg L, Segerbäck D, Hällström I. (1976) Evaluation of genetic risks of alkylating agents. II. Haemoglobin as a dose monitor. *Mutat. Res.* **34**: 1–10.

Osterman-Golkar S, Farmer PB, Segerback D, Bailey E, Calleman CJ, Svensson K, Ehrenberg L. (1983) Dosimetry of ethylene oxide in the rat by quantitation of alkylated histidine in hemoglobin. *Teratogen. Carcinogen, Mutagen.* **3**: 395–405.

Perera FP, Hemminki K, Gryzbowska E et al. (1992) Molecular and genetic damage in humans from environmental pollution in Poland. *Nature* **360**: 256–258.

Phillips DH. (1990) Modern methods of DNA adduct determination. In: *Chemical Carcinogenesis and Mutagenesis I, Handbook of Experimental Pharmacology,* Vol. 94/I (eds CS Cooper, PL Grover). Springer-Verlag, Berlin, pp. 503–546.

Phillips DH, Hewer A. (1993) DNA adducts in human urinary bladder and other tissues. *Environ. Hlth Perspect.* **99**: 45–49.

Phillips DH, Ni Shé M. (1994) DNA adducts in cervical tissue of smokers and non-smokers. *Mutat. Res.* **313**: 277–284.

Phillips DH, Grover PL, Sims P. (1979) A quantitative determination of the covalent binding of a series of polycyclic hydrocarbons to DNA in mouse skin. *Int. J. Cancer* **23**: 201–208.

Phillips DH, Hemminki K, Alhonen A, Hewer A, Grover PL. (1988a) Monitoring occupational exposure to carcinogens: detection by ^{32}P-postlabelling of aromatic DNA adducts in white blood cells from iron foundry workers. *Mutat. Res.* **204**: 531–541.

Phillips DH, Hewer A, Martin CN, Garner RC, King MM. (1988b) Correlation of DNA adduct levels in human lung with cigarette smoking. *Nature* **336**: 790–792.

Phillips DH, Schoket B, Hewer A, Bailey E, Kostic S, Vincze I. (1990) Influence of cigarette smoking on the levels of DNA adducts in human bronchial epithelium and white blood cells. *Int. J. Cancer* **46**: 569–575.

Phillips DH, Castegnaro M, Bartsch H. (eds) (1993) *Postlabelling Methods for Detection of DNA Adducts.* IARC Scientific Publications no. 124, Lyon.

Poirier MC. (1991) Development of immunoassays for the detection of carcinogen–DNA adducts. In: *Molecular Dosimetry and Human Cancer* (eds JD Groopman, PL Skipper). CRC Press, Boca Raton, FL, pp. 211–229.

Poirier MC. (1993) Antisera specific for carcinogen–DNA adducts and carcinogen-modified DNA: applications for detection of xenobiotics in biological samples. *Mutat. Res.* **288**: 31–38.

Poirier MC. (1994) Human exposure monitoring, dosimetry, and cancer risk assessment: the use of antisera specific for carcinogen–DNA adducts and carcinogen-modified DNA. *Drug Metab. Rev.* **26**: 87–109.

Poirier MC, Beland FA. (1992) DNA adduct measurements and tumor incidence during chronic carcinogen exposure in animal models: implications for DNA adduct-based human cancer risk assessment. *Chem. Res. Toxicol.* **5**: 749–755.

Poirier MC, Reed E, Litterst CL, Katz D, Gupta-Burt S. (1992) Persistence of platinum-ammine–DNA adducts in gonads and kidneys of rats and multiple tissues from cancer patients. *Cancer Res.* **52**: 149–153.

Poirier MC, Schoket B, Weston A, Rothman N, Scott B, Deeter DP. (1994) Blood cell polycyclic aromatic hydrocarbon (PAH)–DNA adducts and PAH urinary metabolites in soldiers exposed to Kuwaiti oil well fires. *Proc. Am. Assoc. Cancer Res.* **35**: 95.

Prevost V, Shuker DEG, Friesen MD, Eberle G, Rajewsky MF, Bartsch H. (1993) Immunoaffinity purification and gas chromatography–mass spectrometric quantification of 3-alkyladenines in urine: metabolism studies and basal excretion levels in man. *Carcinogenesis* **14**: 199–204.

Qian G-S, Ross RK, Yu MC, Yuan J-M, Gao Y-T, Henderson BE, Wogan GN, Groopman JD. (1994) A follow-up study of urinary markers of aflatoxin exposure and liver cancer risk in Shanghai, People's Republic of China. *Cancer Epidemiol. Biomarkers Preven.* **3**: 3–10.

Randerath E, Agrawal HP, Weaver JA, Bordelon CB, Randerath K. (1985) ^{32}P-Postlabeling analysis of DNA adducts persisting for up to 42 weeks in the skin, epidermis and dermis of mice treated topically with 7,12-dimethylbenz[a]anthracene. *Carcinogenesis* **6**: 1117–1126.

Randerath K, Randerath E, Danna TF, van Golen KL, Putman KL. (1989a) A new sensitive ^{32}P-postlabeling assay based on the specific enzymatic conversion of bulky DNA lesions to radiolabeled dinucleotides and nucleoside 5'-monophosphates. *Carcinogenesis* **10**: 1231–1239.

Randerath E, Miller RH, Mittal D, Avitts TA, Dunsford HA, Randerath K. (1989b) Covalent DNA damage in tissues of cigarette smokers as determined by ^{32}P-postlabelling assay. *J. Natl Cancer Inst.* **81**: 341–347.

Reddy MV, Randerath K. (1986) Nuclease P1-mediated enhancement of sensitivity of ^{32}P-postlabeling test for structurally diverse DNA adducts. *Carcinogenesis* **7**: 1543–1551.

Reed E, Ozols RF, Tarone R, Yuspa SH, Poirier MC. (1987) Platinum–DNA adducts in leukocyte DNA correlate with disease response in ovarian patients receiving platinum-based chemotherapy. *Proc. Natl Acad. Sci. USA* **84**: 5024–5028.

Reed E, Ozols RF, Tarone R, Yuspa SH, Poirier MC. (1988) The measurement of cisplatin–DNA adduct levels in testicular cancer patients. *Carcinogenesis* **9**: 1909–1911.

Rojas M, Alexandrov K, van Schooten F-J, Hillebrand M, Kriek E, Bartsch H. (1994) Validation of a new fluorometric assay for benzo[a]pyrene diolepoxide–DNA adducts in human white blood cells: comparisons with ^{32}P-postlabeling and ELISA. *Carcinogenesis* **15**: 557–560.

Ross RK, Yuan J-M, Yu MC, Wogan GN, Qian G-S, Tu J-T, Groopman JD, Gao Y-T, Henderson BE. (1992) Urinary aflatoxin biomarkers and risk of hepatocellular carcinoma. *Lancet* **339**: 943–946.

Rothman N, Correa-Villaseñor A, Ford DP, Poirier MC, Haas R, Hansen JA, O'Toole T, Strickland PT. (1993) Contribution of occupation and diet to white blood cell polycyclic aromatic hydrocarbon-DNA adducts in wildland firefighters. *Cancer Epidemiol. Biomarkers Preven.* **2**: 341–347.

Roza L, de Gruijl FR, Bergen Henegouwen JB, Guikers K, van Weelden H, van der Schans GP, Baan RP. (1991) Detection of photorepair of UV-induced thymine dimers in human epidermis by immunofluorescence microscopy. *J. Invest. Dermatol.* **96**: 903–907.

Santella RM, Lin CD, Dharmaraja N. (1986) Monoclonal antibodies to a benzo[a]pyrene diolepoxide modified protein. *Carcinogenesis* **7**: 441–444.

Santella RM, Grinberg-Funes RA, Young TL, Dickey C, Singh VN, Wang LW, Perera FP. (1992) Cigarette smoking related polycyclic aromatic hydrocarbon–DNA adducts in peripheral mononuclear cells. *Carcinogenesis* **13:** 2041–2045.

Santella RM, Hemminki K, Tang D-L et al. (1993) Polycyclic aromatic hydrocarbon–DNA adducts in white blood cells and urinary 1-hydroxypyrene in foundry workers. *Cancer Epidemiol. Biomarkers Preven.* **2:** 59–62.

Savela K, Hemminki K. (1991) DNA adducts in lymphocytes and granulocytes of smokers and nonsmokers detected by the ^{32}P-postlabelling assay. *Carcinogenesis* **12:** 503–508.

Schoket B, Phillips DH, Hewer A, Vincze I. (1991) ^{32}P-Postlabelling detection of aromatic DNA adducts in peripheral blood lymphocytes from aluminium production plant workers. *Mutat. Res.* **260:** 89–98.

Schoket B, Doty WA, Vincze I, Strickland PT, Ferri GM, Assennato G, Poirier MC. (1993a) Increased sensitivity for determination of polycyclic aromatic–DNA adducts in human DNA samples by dissociation-enhanced lanthanide fluoroimmunoassay (DELFIA). *Cancer Epidemiol. Biomarkers Preven.* **2:** 349–353.

Schoket B, Phillips DH, Poirier MC, Vincze I. (1993b) DNA adducts in peripheral blood lymphocytes from aluminium production plant workers determined by ^{32}P-postlabelling and by enzyme-linked immunosorbent assay (ELISA). *Environ. Hlth Perspect.* **99:** 307–309.

Segerbäck D, Calleman CJ, Ehrenberg L, Löfroth G, Osterman-Golkar S. (1978) Evaluation of genetic risks of alkylating agents. IV. Quantitative determination of alkylated amino acids in haemoglobin as a measure of the dose after treatment of mice with methyl methanesulfonate. *Mutat. Res.* **49:** 71–82.

Segerbäck D, Kaderlik KR, Talaska G, Dooley KL, Kadlubar FF. (1993) ^{32}P-Postlabeling analysis of DNA adducts of 4,4′-methylenebis(2-chloroaniline) in target and nontarget tissues in the dog and their implications for human risk assessment. *Carcinogenesis* **14:** 2143–2147.

Sheabar FZ, Morningstar ML, Wogan GN. (1994) Adduct detection by acylation with [^{35}S]methionine: analysis of DNA adducts of 4-aminobiphenyl. *Proc. Natl Acad. Sci. USA* **91:** 1696–1700.

Shen H-M, Ong C-N, Lee B-L, Shi CY. (1995) Aflatoxin B$_1$-induced 8-hydroxydeoxy-guanosine formation in rat hepatic DNA. *Carcinogenesis* **16:** 419–422.

Sherson D, Sabro P, Sigsgaard T, Johansen F, Autrup H. (1990) Biological monitoring of foundry workers exposed to polycyclic aromatic hydrocarbons. *Br. J. Ind. Med.* **47:** 448–453.

Shuker DEG, Bartsch H. (1994) Detection of human exposure to carcinogens by measurement of alkyl–DNA adducts using immunoaffinity clean-up in combination with gas chromatography–mass spectrometry and other methods of quantitation. *Mutat. Res.* **313:** 263–268.

Shuker DEG, Farmer PB. (1992) Relevance of urinary DNA adducts as markers of carcinogen exposure. *Chem. Res. Toxicol.* **5:** 450–460.

Shuker DEG, Durand M-J, Molko D. (1993) Fluorescent postlabelling of modified DNA bases. In: *Postlabelling Methods for Detection of DNA Adducts* (eds DH Phillips, M Castegnaro and H Bartsch). IARC, Lyon, pp. 227–232.

Simons AM, Phillips DH, Coleman DV. (1993) Smoking-related DNA damage in cervical epithelium; molecular evidence consistent with smoking as a cause of cervical cancer. *Br. Med. J.* **306:** 1444–1448.

Springer DJ, Bull RJ, Goheen SC, Sylvester DM, Edmonds CG. (1993) Electrospray ionization mass spectrometric characterization of acrylamide adducts to hemoglobin. *J. Toxicol. Environ. Hlth* **40:** 161–176.

Strickland PT, Routledge MN, Dipple A. (1993) Methodologies for measuring carcinogen adducts in humans. *Cancer Epidemiol. Biomarkers Preven.* **2:** 607–619.

Szyfter K, Hemminki K, Szyfter W, Szmeja Z, Banaszewski J, Yang K. (1994) Aromatic DNA adducts in larynx biopsies and leukocytes. *Carcinogenesis* **15:** 2195–2199.

Talaska G, Al-Juburi AZSS, Kadlubar FF. (1991a) Smoking related carcinogen–DNA adducts in biopsy samples of human urinary bladder: identification of N-(deoxyguanosin-8-yl)-4-aminobiphenyl as a major adduct. *Proc. Natl Acad. Sci. USA* **88**: 5350–5354.

Talaska G, Schamer M, Skipper P et al. (1991b) Detection of carcinogen–DNA adducts in exfoliated urothelial cells of cigarette smokers: association with smoking, hemoglobin adducts, and urinary mutagenicity. *Cancer Epidemiol. Biomarkers Preven.* **1**: 61–66.

Talaska G, Roh JH, Getek T. (1992) ^{32}P-Postlabelling and mass spectrometric methods for analysis of bulky, polyaromatic carcinogen–DNA adducts in humans. *J. Chromatog.* **580**: 293-323.

Tates AD, Grummt T, Törnqvist M et al. (1991) Biological and chemical monitoring of occupational exposure to ethylene oxide. *Mutat. Res.* **250**: 483–497.

Tavares R, Ramos P, Palminha J, Bispo MA, Paz I, Bras A, Rueff J, Farmer PB, Bailey E. (1994) Transplacental exposure to genotoxins. Evaluation in haemoglobin of hydroxyethylvaline adduct levels in smoking and non-smoking mothers and their newborns. *Carcinogenesis* **15**: 1271–1274.

Terheggen PMAB, Dijkman R, Begg AC, Dubbelman R, Floot BGJ, Hart AAM, den Engelse L. (1988) Monitoring of interaction products of *cis*-diamminedichloroplatinum(II) and *cis*-diammine(1,1-cyclobutanedicarboxylato)platinum(II) with DNA in cells from platinum-treated cancer patients. *Cancer Res.* **48**: 5597–5603.

Tilby MJ, Newell DR, Viner C, Selby PJ, Dean CJ. (1993) Application of a sensitive immunoassay to the study of DNA adducts formed in peripheral blood mononuclear cells of patients undergoing high-dose melphalan therapy. *Eur. J. Cancer* **29A**: 681–686.

Ting D, Smith MT, Doane-Setzer P, Rappaport SM. (1990) Analysis of styrene oxide-globin adducts based upon reaction with Raney nickel. *Carcinogenesis* **11**: 755–760.

Törnqvist M, Mowrer J, Jensen S, Ehrenberg L. (1986) Monitoring of environmental cancer initiators through hemoglobin adducts by a modified Edman degradation method. *Anal. Biochem.* **154**: 255–266.

Törnqvist M, Magnusson A-L, Farmer PB, Tang Y-S, Jeffrey AM, Wazneh L, Beulink GDT, van der Waal H, van Sittert NJ. (1992) Ring test for low levels of N-(2-hydroxyethyl)valine in human hemoglobin. *Anal. Biochem.* **203**: 357–360.

van Schooten FJ, van Leeuwen FE, Hillebrand MJX, deRijke ME, Hart AAM, van Veen HG, Oosterink S, Kriek E. (1990) Determination of benzo[*a*]pyrene diol epoxide–DNA adducts in white blood cell DNA from coke-oven workers: the impact of smoking. *J. Natl Cancer Inst.* **82**: 927–933.

Varanasi U, Reichert WL, Stein JE. (1989) ^{32}P-Postlabeling analysis of DNA adducts in liver of wild English sole (*Parophrys vetulus*) and winter flounder (*Pseudopleuronectes americanus*). *Cancer Res.* **49**: 1171–1177.

Vineis P, Caporaso N, Tannenbaum S, Skipper P, Glogowski J, Bartsch H, Coda M, Talaska G, Kadlubar FF. (1990) The acetylation phenotype, carcinogen–hemoglobin adducts, and cigarette-smoking. *Cancer Res.* **50**: 3002–3004.

Walker VE, MacNeela P, Swenberg JA, Turner MJJ, Fennell TR. (1992a) Molecular dosimetry of ethylene oxide: formation and persistence of N-(2-hydroxyethyl)valine in hemoglobin following repeated exposures of rats and mice. *Cancer Res.* **52**: 4320–4327.

Walker VE, Fennell TR, Upton PB, Skopek TR, Prevost V, Shuker DEG, Swenberg JA. (1992b) Molecular dosimetry of ethylene oxide: formation and persistence of 7-(2-hydroxyethyl)guanine in DNA following repeated exposures of rats and mice. *Cancer Res.* **52**: 4328–4334.

Walsh P, El Adlouni C, Mukhopadhyay MJ, Viel G, Nadeau D, Poirier GG. (1995) ^{32}P-Postlabelling determination of DNA adducts in the earthworm *Lumbricus terrestris* exposed to PAH-contaminated soils. *Bull. Environ. Contamination Toxicol.*, in press.

Weston A. (1993) Physical methods for the detection of carcinogen–DNA adducts in humans. *Mutat. Res.* **288**: 19–29.

Weston A, Bowman ED. (1991) Fluorescence detection of benzo[a]pyrene–DNA adducts in human lung. *Carcinogenesis* **12**: 1445–1449.

Weston A, Rowe ML, Manchester DK, Farmer PB, Mann DL, Harris CC. (1989) Fluorescence and mass spectral evidence for the formation of benzo[*a*]pyrene anti-diol-expoxide–DNA and –hemoglobin adducts in humans. *Carcinogenesis* **10:** 251–257.

Wild CP, Jiang Y-Z, Sabbioni G, Chapot B, Montesano R. (1990a) Evaluation of methodologies for quantitation of aflatoxin–albumin adducts and their application to human exposure assessment. *Cancer Res.* **50:** 245–251.

Wild CP, Jiang Y-Z, Allen SJ, Jansen LAM, Montesano R. (1990b) Aflatoxin–albumin adducts in human sera from different regions of the world. *Carcinogenesis* **11:** 2271–2274.

Wild CP, Jansen LAM, Cova L, Montesano R. (1993) Molecular dosimetry of aflatoxin exposure: contribution to understanding the multifactorial etiopathogenesis of primary hepatocellular carcinoma with particular reference to hepatitis B virus. *Environ. Hlth Perspect.* **99:** 115–122.

Wraith MJ, Watson WP, Eadsforth CV, van Sittert NJ, Törnqvist M, Wright AS. (1988) An immunoassay for monitoring human exposure to ethylene oxide. In: *Methods for Detecting DNA-Damaging Agents in Humans: Applications in Cancer Epidemiology and Prevention* (eds H Bartsch, K Hemminki, IK O'Neill). IARC, Lyon, pp. 271–274.

Yang XY, Gasparro FP, DeLeo VA, Santella RM. (1989) 8-Methoxypsoralen–DNA adducts in patients treated with 8-methoxypsoralen and ultraviolet A light. *J. Invest. Dermatol.* **92:** 59–63.

Yu MC, Skipper PL, Taghizadeh K, Tannenbaum SR, Chan KK, Henderson BE, Ross RK. (1994) Acetylator phenotype, aminobiphenyl–hemoglobin adduct levels, and bladder cancer risk in white, black, and Asian men in Los Angeles, California. *J. Natl Cancer Inst.* **86:** 712–716.

Index

Human Gene Mutation

D.N. Cooper & M. Krawczak
respectively Charter Molecular Genetics Laboratory, Thrombosis Research Institute, London; and Institut für Humangenetik, Medizinsiche Hochschule, Hannover, Germany

An informative, readable and highly-praised reference work. Reprinted in paperback with a new section on cancer mutations of direct relevance to clinicians.

"... hardly a wasted word ... a remarkable achievement for both the authors and publisher ... obligatory reading to anyone involved in mammalian mutation." *J. Medical Genetics*

"This is an informative, readable reference work, and I recommend it to anyone interested in the nature and source of variation in the human genome." *Science*

Contents

An historical view of research into the nature of mutation; An introduction to the structure, function and expression of human genes; Human genetic disease and its analysis - an overview; The methodology of mutation detection; Indirect analysis of human genetic disease; Single base-pair substitutions; Gene deletions; Gene insertions, duplications and inversions; Single base-pair substitutions in human gene mRNA splice junctions and their phenotypic consequences; Regulatory mutations; Mutations affecting RNA processing and translation; The genotype-phenotype relationship; Mutation rates in humans. *Appendices:* Direct and indirect analysis of human genetic disease; Single base-pair substitution database; Deletion database (< 20 bp); Examples of splice site mutations; Amino acid symbols and the genetic code; Cancer-associated somatic mutations in the human TP53 gene.

Human Molecular Genetics; Published pages;   1859960553; 1995

Functional Analysis of the Human Genome

F. Farzaneh & D.N. Cooper (Eds)
respectively King's College School of Medicine and Dentistry, London; and Charter Molecular Genetics Laboratory, Thrombosis Research Institute, London, UK

Analysis of the structure and organization of the human genome is proceeding apace, bringing with it new insights into its function. This book describes some of the most important methodological approaches to the study of the relationship between structure and function in the human genome. The introductory chapters provide an up-to-date account of progress in the mapping of the human genome and an analysis of the role of specific DNA sequences in mediating particular cellular functions. Later chapters provide detailed accounts of some of the most important strategies aimed at the cloning and functional analysis of the genome. Topics covered include cDNA synthesis, gene transfer analysis of DNA methylation and imprinting, homologous recombination strategies for targeted gene replacement and inactivation, somatic cell hybridization and complementation studies, transgenic animals including their use in the development of models for the study of human disease, and antisense oligonucleotides and ribozymes as powerful tools for experimental and therapeutic inhibition of expression of specific gene products. Together, these strategies promise to allow the functional analysis of the human genome to keep pace with, and complement, its structural analysis.

Contents

Structure and function in the human genome, *D.N.Cooper*; Mapping the human genome, *D.N.Cooper*; Cloning the transcribed portion of the genome, *P.Towner*; Retroviral insertional mutagenesis, *F.Farzaneh et al*; Gene entrapment, *H.von Melchner & H.E.Ruley*; Gene transfer studies, *D.Darling & M.Kuiper*; Foreign DNA integration and DNA methylation patterns, *W.Doerfler*; Transgenic animals in human gene analysis, *F.Theuring*; Homologous recombination, *A.Mansouri*; Complementation analysis, *A.Patel*; Antisense oligonucleotides: a survey of recent literature, possible mechanisms of action and therapeutic progress, *D.Pollock & J.Gaken*.

Of interest to:

Advanced undergraduates; medical students; postgraduates; researchers.

Human Molecular Genetics; Published pages;   1872748465; 1995

From Genotype to Phenotype

S.E. Humphries & S. Malcolm (Eds)
respectively University College Medical School, London, UK; and Institute of Child Health, University of London, UK

The study of how the effects of different mutations - the 'genotype' of the individual - are modified by other genetic factors and by the environment to produce variable clinical symptoms - the 'phenotype' - is one of the fastest growing areas of human molecular genetics. *From Genotype to Phenotype* provides a unique review of the mechanisms of interaction between genotype and phenotype, for both common and rare genetic disorders. A detailed understanding of common human phenotypes will improve disease diagnosis and help determine specific therapeutic measures for the future.

"The excitement that comes from making new, often unexpected, observations in the area of clinical genetics is conveyed superbly by the editors and individual authors throughout this book. It is a pleasure to read a book that describes these recent advances so well and which is also so full of novelty." Prof. Kare Berg, Institute of Medical Genetics, Oslo. - "...the book will be of primary interest to those concerned with genetic etiology of clinical conditions ...the references are laudably current ...a worthwhile resource for those interested in current information about human genetics." *Choice*

Contents

Mutations and human disease, *S.Malcolm*; Cystic fibrosis, *P.F.Pignatti*; Mutations in type I and type III collagen genes, *R.Dalgleish*; Genotype-phenotype correlation in Gaucher disease, *M.Horowitz & A.Zimran*; Familial hypercholesterolaemia, *A.K.Soutar*; The molecular basis of Charcot-Marie-Tooth disease, *F.Baas et al*; The genetics of Wilms' tumour, *J.K.Cowell*; How a dynamic mutation manifests in myotonic dystrophy, *C.L.Winchester & K.J.Johnson*; Length variation in fragile X, *M.C.Hirst*; Somatic mosaicism, chimerism and X inactivation, *A.O.M.Wilkie*; Mitochondrial DNA-associated disease, *S.R.Hammans*; Diabetes - from phenotype to genotype and back to phenotype, *G.A.Hitman et al*; Coronary artery disease and the variability gene concept - the effect of smoking on plasma levels of high density lipoprotein and fibrinogen, *S.E.Humphries*; Genetic predisposition to dyslipidaemia and accelerated atherosclerosis - environmental interactions and modification by gene therapy, *E.Boerwinkle & L.Chan*.

Of interest to:

Clinicians; researchers in molecular genetics; genetic counsellors.

Human Molecular Genetics; Published pages;   1872748627; 1994

From Genetics to Gene Therapy
The molecular pathology of human disease

D.S. Latchman (Ed.)
University College and Middlesex School of Medicine, London, UK

In this book a team of distinguished scientists provide an overview of the molecular pathology of human disease. Each chapter provides an analysis of the molecular biological approaches to individual diseases, such as leukaemia, cardiovascular disease and cancer, and includes a discussion on the likely impact of gene therapy.

Contents

What is molecular pathology? *D.S.Latchman*; Apolipoprotein B and coronary heart disease, *J.Scott*; Prospects for gene therapy of X-linked immunodeficiency diseases, *C.Kinnon*; Duchenne muscular dystrophy, *S.C.Brown & G.Dickson*; Molecular genetics of leukaemia, *M.F.Greaves*; The molecular pathology of neuroendocrine tumours, *A.E.Bishop & J.M.Polak*; Genetic predisposition to breast cancer, *M.R.Stratton*; Gene therapy for cancer, *M.K.L.Collins*; Retrovirus receptors on human cells, *R.A.Weiss*; Viral vectors for gene therapy, *G.W.G.Wilkinson et al*; Direct gene transfer for the treatment of human disease, *G.J.Nabel & E.G.Nabel*; Processing of membrane proteins in neurodegenerative diseases, *R.J.Mayer et al*; Herpes simplex - once bitten, forever smitten? *D.S.Latchman*.

Of interest to:

Medical researchers and clinicians.

UCL Molecular Pathology; Published pages;   1872748368; 1994

Molecular Genetics of Cancer

J.K. Cowell (Ed.)
ICRF Oncology Group, Institute of Child Health, London, UK

An up-to-date review of cancer genetics. Leading experts from North America and Europe provide current information on those genes which are known to be critical in the development or progression of human cancer.

Contents

Molecular genetics of retinoblastoma, *B.Gallie;* The genetics of Wilm's tumour, *D.Haber;* Neurofibromatosis, *M.R.Wallace*; Multiple endocrine neuplasia, *M.Nordenskjold*; Genetics of breast cancer, *T.Bishop*; Genetics of colon cancer, *M.Dunlop*; p53, *T.Soussi*; Genetics of small cell lung carcinoma, *F.Kaye*; Rhabdomyosarcoma, *D.Shapiro*; Transgenic models for cancer, *V.Pachnis*

Of interest to:

Researchers in genetics and cancer; also of interest to clinical geneticists and those working in paediatric oncology.

Human Molecular Genetics; Published pages;   1872748090; 1995

ORDERING DETAILS

Main address for orders

BIOS Scientific Publishers Ltd
9 Newtec Place, Magdalen Road,
Oxford OX4 1RE, UK
Tel: (0)1865 726286
Fax: (0)1865 246823

Australia and New Zealand
DA Information Services
648 Whitehorse Road, Mitcham, Victoria 3132, Australia
Tel: (03) 873 4411
Fax: (03) 873 5679

India
Viva Books Private Ltd
4325/3 Ansari Road, Daryaganj, New Delhi 110 002, India
Tel: 11 3283121
Fax: 11 3267224

Singapore and South East Asia
(Brunei, Hong Kong, Indonesia, Korea, Malaysia, the Philippines,
Singapore, Taiwan, and Thailand)
Toppan Company (S) PTE Ltd
38 Liu Fang Road, Jurong, Singapore 2262
Tel: (265) 6666
Fax: (261) 7875

USA and Canada
Books International Inc
PO Box 605, Herndon, VA 22070, USA
Tel: (703) 435 7064
Fax: (703) 689 0660

Payment can be made by cheque or credit card (Visa/Mastercard, quoting number and expiry date). Alternatively, a *pro forma* invoice can be sent.

Prepaid orders must include £2.50/US$5.00 to cover postage and packing for one item. Prepaid orders for two or more books are delivered postage free.